Basics of data Communications

Edited by Harry R. Karp
Editor-in-chief, Data Communications

ELECTRONICS BOOK SERIES

Also published by *Electronics*
Microprocessors
Large scale integration

Library of Congress Catalog Card No. 76-16475

Copyright © 1976 by McGraw-Hill, Inc. All rights reserved. Printed in the United States of America. No part of this publication may be reproduced, stored in a retrieval system, or transmitted, in any form or by any means, electronic, mechanical, photocopying, recording, or otherwise, without the prior written permission of the publisher.

 McGraw-Hill Publications Co.
1221 Avenue of the Americas
New York, New York 10020

BASICS OF data COMMUNICATIONS

data
data
data
data
data
data
data
data

Edited by Harry R. Karp
Editor-in-chief, Data Communications

Electronics Book Series

1 Part I—Introduction

2 Perspective on present practices in data communications
4 Management mates systems objectives to business goals
7 Basics of network design
15 What the future has in store for data communications

31 Part II—Terminals

32 The broad array of data terminals for communications
38 Facsimile terminals send the documents with the data
41 A new read-only terminal: your TV set

46 Part III—Acoustic couplers and modems

47 Acoustic couplers lend portability to data terminals
49 Modems marry data terminals to telephone lines
54 Transmitting data pulses over short distances

59 Part IV—Communications processors

60 Multiplexing cuts cost of communications lines
66 Programable front-end processors
72 Programable front ending beats computer upgrading
80 Programable remote data concentrators
83 Estimating buffer size and queuing delay in a remote concentrator
89 Message switcher links diverse data services, speeds
96 Making the most of simultaneous voice-plus-teleprinter technique
102 Fundamentals of software for communications processors

110 Part V—Networking

111 Computer networking—the giant step in data communications
118 New pricing structures upset data network strategies
125 How transaction cost declines as data networks get larger
131 Breaking the logjam in dynamic data base systems
138 Packet switching with assorted computers in a private network
146 The user's role in connecting to a value-added network
152 Where to spend money to improve system availability
157 How to calculate network reliability
164 Design workshop: configuring an actual data system

table of contents

168 Part VI—Channel performance

169 Error control: keeping data messages clean
174 Hit analysis leads to reduced errors from a microwave link
179 Getting peak performance on a data channel
188 Taking a fresh look at voice-grade line conditioning

193 Part VII—Data link controls

194 Taking a fresh look at data link controls
201 Synchronous data link control
211 Advanced link control runs full and half duplex
220 Burroughs data link control

227 Part VIII—Network diagnostics

228 Network-management centers help keep the system going
232 Line monitors open a window on data channels
240 Centralized diagnostics cuts line downtime to minutes

245 Part IX—Interfaces

246 Using standard interfaces doesn't always assure trouble-free performance
254 Don't overlook electrical compatibility when mating equipment

258 Part X—Regulations and policy

259 Regulations and tariffs
262 Dilemmas in the nation's telecommunications policy
266 AT&T's new attitude might have forestalled Justice's suit
268 Wanted! A computer/communications resale industry

275 Part XI—Miscellaneous subjects

276 Standards for data communications
286 Microcomputers unlock the next generation
294 What's needed today in a high-level data communications language

302 Key Subject Locator

Data communications as it's practiced today is fairly sophisticated and requires knowledge of subjects ranging from common-carrier offerings, to communications processors, to regulatory matters. But the field is really in its infancy compared with the kinds of networks and applications for data communications that are predicted for the future. Thus, a new breed of professional is becoming more important in industry, commerce, and Government: the data communications specialist.

Aimed at that specialist and the manager, who both need to know about the present state of the art of today's networks, this book contains the information that will help them contribute to and participate in this growth field. It presents the important material that will help individuals off to a fast start. The material in it is based on selected articles from DATA COMMUNICATIONS, a McGraw-Hill magazine, since it started regular publication with the May/June issue in 1974, as well as from two preceding pilot issues. The articles have been arranged into cohesive units with a logical progression of information and depth.

Many of these articles contain discussions of costs, which are so important in the design and operation of any data communications network. Costs, particularly those related to tariffed services, are in a constant state of change. The cost values included, therefore, should be construed as informational and comparative, rather than as a true reflection of prices for today's equipment and services.

—Harry Karp

FOREWORD

part 1
data
data
data
data
data
data
data
data

introduction

Perspective on present practices in data communications

Tailoring a data communications system to the needs of a specific business operation requires a familiarity with available methods and equipment that this publication is designed to provide

Harry R. Karp
Data Communications

The payoff for a company's investment in a data communications system is a more responsive, more competitive business operation. To obtain this payoff takes much hard work by a dedicated team. For one thing, management must set realistic objectives based on actual and projected business needs. For another, data communications and data processing specialists must establish detailed plans for developing, installing, and testing the system. Finally, vendors must be found who will supply reliable and cost-effective hardware, software, and communications links.

How will management know the system performs properly? A single, all-inclusive measure of performance is how transparent the data communications system is to the people in the business operation—that is, how free the interaction is of frustrations and delays. Ideally, the people who use the system should not even know it's there, but just enjoy its benefits.

Despite the specific problems inherent in developing a complex computer-based data communications system that is adequately responsive and highly reliable, many such systems have been installed for a variety of businesses. Large and small companies throughout the country presently rely on their systems to provide private, accurate, rapid, and low-cost transmission of messages and to permit the processing of remote-source data at a central site. And many more companies are slated to enjoy the benefits of data communications systems in the future.

Universality spurs growth

Accelerating this growth is the fact that data communications is a functional technology, in that the same equipment and methodology serve all users, no matter what kinds of businesses they are in. In other words, the equipment in a system for a food-distribution company is substantially the same as that for an oil company. By the same token, because the same kinds of equipment can be used by many different types of businesses, a large market results that has induced makers to invest in designing improved, more universal equipment while simultaneously reducing costs.

This synergistic situation has existed for several years. It has not only been of economic benefit to both systems users and the manufacturers of a wide variety of equipment from terminals to minicomputers and computers—it has also improved the state of the art of data communications systems, which today is well developed, acceptable to users, and constantly advancing.

All the major aspects of current practice are covered

in this book on data communications.

The many and varied considerations that have to be weighed in setting up a data communications system—from management's role to the intricacies and impact of error control—are put in perspective and then explained in detail. Every major type of equipment, from modems to programable front-end processors, is discussed, and its place in the over-all system described.

Future developments notwithstanding, only discussions of current equipment and methodology are included, because companies who need data communications systems to carry out their businesses cannot wait for projected improvements. The time from its initial planning to the date a data communications system becomes fully operational may be as long as three or four years. Thus, to get such a system up and running in the near future, companies will have to design their systems substantially around present equipment, communications links, and methodology. Later, when feasible, they can modify their plans—or their present installations—to embrace a new development.

Helping with answers

The articles in this book contain many solutions to major problems in modern data communications systems. To understand the basic nature of these problems, though, it is instructive to go back to the mid-1960s. At that time many companies had big investments in underutilized central batch-data processing computers, and they wanted to put the valuable unused computing resources to work with such remote, outlying locations as warehouses and branch offices.

However, the only quick way to accomplish computer-sharing required the integration of two separate, massive, entrenched, and—to a great extent—technically incompatible resources: the nationwide telephone network, and the array of data processing computers dispersed throughout the country. The integration called for a set of accommodations implemented by equipment and software, and by and large the need for accommodation will continue for many more years.

These accommodations are basically the answers to problems arising from a variety of sources. Computers are designed to handle digital data—but digital data has to be converted to analog signals to meet telephone-line requirements. Control signals are needed to establish and maintain connections, and this requires a line protocol acceptable to terminals, telephone equipment, and computers—but executing the protocol wastes time and reduces line utilization. Electrical disturbances on telephone lines create errors in messages—so resource-consuming techniques for error control must be incorporated in the system. And there is a drastic mismatch in speed, since digital computers can manipulate data characters about 1,000 times faster than voice-grade telephone lines can transmit them.

Though highly workable solutions exist to all these problems, neither users nor suppliers of data communications equipment appear content with the status quo. Spurred by a projected rapid growth in the use of data communications by business and industry, significant efforts are being mounted to lower technical barriers still further and thus improve over-all cost-effectiveness in computer communications.

But because of the multi-billion dollar investment in the presently installed data processing computers and data communications equipment, not including the massive investment in the telephone network, most future developments will have to mesh with, and gradually modify, current equipment, links, and procedures.

For example, minicomputers have already started to play an important role in improving over-all system efficiency. They increase the versatility of a terminal, enabling it to match a variety of business applications, and they reduce the communications load on the transmission lines. Minicomputers can serve at a distance from the host computer, as remote data concentrators; then they act like multiplexers and perform such functions as traffic smoothing to handle peak loads with fewer lines. Or a minicomputer can function very close to the host computer, as a programable front-end processor; then it relieves the host computer of expensive communications overhead, as for message switching and error control, and frees the host computer for its primary task of rapidly processing large batches of data.

More transmission links are now available that carry digital data, and they will certainly have an economic and technical impact on the design and performance of data communications systems.

Management mates system objectives to business goals

Before a data communications system can be properly designed, managers must specify such operational demands as traffic routes, message volume, urgency, accuracy, and language

Rolf Vickner
*American Telephone & Telegraph Co.
New York, N.Y.*

In business and industry, data-communications systems are proposed and paid for by corporate and operating management. But the systems are planned, designed, specified, installed, and tested by a host of systems planners, engineers, programmers, and technicians working for the owner-company and vendors. Whether these specialists deliver, on time, the kind of system that is actually needed for a given application depends, initially, on whether user-management adequately communicates company objectives both to the in-house staff and to outside vendors and consultants.

In other words, management's obligation in system development is to prepare a clear and succinct statement of what it wants to accomplish. Management need not, in the planning stage, be too concerned about technological matters, about available equipment, or, for that matter, about costs. Objectives that are stated too tightly may hamper planning and implementation, while objectives stated too loosely may force planners and designers to guess at the final goals and thus perhaps introduce unneeded and expensive capabilities.

But management cannot remain ignorant about how data-communications equipment operates, about the kinds of equipment—such as terminals and minicomputers—that make up a system, and about the intricacies of such other factors as tariffs, regulations, and vendor relationships. Management will need this knowledge when the planning staff returns with several alternative configurations that, hopefully, meet the company's objectives. Often, these alternatives will then give management a chance to rethink the objectives and—with a deeper understanding—perhaps change them if the cost increments can be justified.

In most businesses, the need for a data-communications system may be indicated by such insidious situations as deterioration in customer services, increased inventory because changes are not reported fast enough, or management reports that are too late for timely action.

While it is the job of the systems and technical staff to design and implement the system, management (or a team reporting to management) must set the objectives from the facts gathered and analyzed from the present and planned operations. Among the major areas in fact-gathering as a prelude to specifying system objectives are distribution, volume, urgency, language, and accuracy of messages.

Determining the most efficient pattern for distribution of company information includes the traffic at headquarters, and the data flow to and from all branch offices, warehouses, distributors, and other remote facilities. Some type of graphic representation should record present and planned locations that receive or send such essential operating information as sales orders, production reports, administrative messages, and payroll data. One simple representation is a matrix chart listing all locations along both the vertical and horizontal edges of the matrix. A dot placed at the intersection of two locations means that traffic flows between them. Another clear way of showing traffic patterns is to actually draw lines between sending-receiving locations on a geographical map.

Traffic volume

With the traffic pattern established, the next step is to calculate the information volume to be handled at each location. The most common measure of volume in the planning stage is the total number of characters transmitted and received at each location each day. Later, when the terminal and transmission code have been selected, a more precise measure of traffic volume—bits per day—is obtained by multiplying the total characters by the number of bits per character associated with the particular code to be used in data transmission.

Actually, traffic volume has four different aspects, each of which must be determined to provide insight into communications requirements: the average daily volume of messages, the number of characters in an average message, the average daily total transmission time, and the peak volume.

The peak-volume calculation plays a vital role in planning any large system in which reduced cost can be balanced against increased and acceptable message delays. On the one hand, a data-communications system that can handle instantaneous peak volume on demand and without delay would require so many telephone lines and so much equipment to transmit and process data from each terminal that the high cost would be completely unacceptable. On the other hand, a system configured to service only the average traffic could lose so many transactions during peak periods as to seriously impact business activity.

Considerable money can be saved by designing a system to handle traffic volume somewhere between the average and peak values. But doing so means that at certain times during the day, people will have to tolerate some delay in sending or receiving data. In general terms, the longer the tolerable delay, the lower the system cost.

Arriving at traffic volume, in characters per day, for each location first requires a count of the traffic as determined from sampling the activity for a prescribed number of days, the number depending on an estimate of the monthly traffic. In general, the more messages sent each day, the fewer the number of days that need be sampled to arrive at an accurate monthly volume. The table relates estimated monthly message volume to a suggested number of days to be studied. These sampling days should be selected at random.

At each location, a separate survey should be made for transmitted messages and for received messages. Such studies can reveal a disparity between incoming and outgoing message volume that can justify using a low-speed, low-cost, sending terminal and a high-speed, high-cost receiving terminal. Besides a savings on overall terminal costs, the volume disparity could also lead to reduced cost of communications links.

The message survey should include such details as point of origin, destination, message length in characters, filing time at which the originator wanted to send the message, and actual sending time, as well as charges, if billed separately.

While revealing actual data-communications activity, the count is subject to adjustment under certain circumstances. For example, a count taken during a seasonal high or low in business activity will not be representative. Also, the actual count may reflect present, but undesirable, constraints. For example, the count may show a large volume between New York City and Chicago, and New York City and New Orleans, but no traffic between Chicago and New Orleans—simply because no data-communications link between Chicago and New Orleans exists in the system under study. Finally, the traffic count may reflect bad practice; perhaps messages that should go air mail are, in fact, loading the data-communications system.

One of the more cost-sensitive aspects of data communications is the urgency of responses, the allowable delay between the time an input or inquiry is sent and when the required information must be returned. A request by a business manager for a printout of an inventory, where inventory status is stored in a remote data base, may not be urgent and he can be satisfied with an overnight response. In an airline reservation system, though, a response that takes more than, for example, 10 seconds may be unacceptable because it will waste the seller's time and reduce customer service.

The specified urgency for each type of transaction does influence the ultimate decisions about the required type and number of lines, the speed of transmission, and the system delay at the host computer. If a given number of lines into a location results in excessive de-

| SAMPLING DAYS FOR SURVEYING TRAFFIC ||
Estimated monthly message volume	Number of days to be studied
Less than 1,000	20
1,000 to 2,000	10
2,000 to 5,000	5
5,000 to 10,000	3
10,000 and over	2

lay, because too many busy signals or inputs are being ignored completely, then the number of lines will have to be increased to meet urgency specifications.

Urgency is another way of specifying tolerable delay. Since a host computer will not respond to an inquiry or other type of input until the input message is complete, then the elapsed transmission time for the message is the minimum actual delay. That is, a message that requires a response in 10 seconds cannot be sent at such low speed as to use up 30 seconds for transmission. For a given message, the transmission time is the quotient of the message length, in characters, and the transmission speed, in characters per second.

In general, though, the controlling factor in setting the transmission speed is the volume of information. There is little practical difference in most cases whether a single message travels from its source to its destination at low speed or high speed. For example, transmitted at 10 characters per second, a 50-character message will take 5 seconds. Transmitted at 300 characters per second, the same message will arrive in 0.167 second. If a messenger in the company takes an hour to deliver the message, then the difference in transmission times is not meaningful. Thus, in a system where short and occasional administrative messages are being transmitted, a low transmission speed will suffice.

However, when volume is large, faster transmission speed is needed. For example, using the slow transmission speed of 10 characters per second the total volume might take two hours to transmit—most likely an intolerable delay. But the transmission time would be only 4 minutes at the higher speed. Often a high transmission speed is mandated not by the urgency of the actual message, but by the requirement of keeping expensive computer memory efficiently utilized while accumulating the total message.

The language problem

The term language has two meanings in data communications: the physical form or medium on which the information is recorded, or the code used to record the information. The information medium at the sending site, say a branch office, usually will not satisfy the medium requirements at the receiving site, usually the host data processing computer. For example, an application for a loan on a life-insurance policy will be handwritten on a paper form. The computer, though, will want the information on punched cards or magnetic tape.

The codes for recording information vary both as to the number of two-state, or binary, code elements (or levels or bits) that represent a character, and the assignment of code combinations that define a particular character. Present codes in data communications use from four to 12 code elements to define a character. In general, the more elements (or bits) in a code, the more unique characters in the code's repertoire. A four-level code can define 2^4 or 16 different characters. A seven-level code can define 2^7 or 128 different characters, which can therefore include all numerals from 0 to 9, the complete alphabet, and many special characters.

As a rule, information on one medium employs a code that is different from the code used on another medium. For example, the code used on punched-paper tape differs from the code used with cathode-ray-tube (CRT) displays. Some codes originate from communications applications, others from computer requirements, and some satisfy both computer and communications standards. As a result, a computer-based data-communications system will employ many different codes. Therefore, the data-communications system must be able to accept all the information in the form and code in which it occurs, and deliver it in the required forms and codes. Performing code conversion is one of the system's essential, but resource-consuming, tasks.

Accuracy of message content, a paramount factor in data communications, is usually rated in terms of error, or the loss of accuracy. Errors in messages spring from three sources: human errors from inputting, or keying, wrong data; errors developed within a malfunctioning terminal; and random errors occurring on the communications link.

Terminal equipment in good repair operates virtually without error. But if an error should occur it will probably be consistent, due to a particular malfunction, and thus readily noticeable and repairable.

Several techniques for controlling error arising from random transmission disturbances have been developed, and the more common approaches are covered in pp.169-173 on error control. Briefly, though, the tighter the error control, the higher the cost of the system. Thus, the kind and cost of error control depends on such factors as the context of the message, whether the message will be "read" by a human being or by a computer, and the probability of error occurrence.

An administrative message, addressed to a person, containing one or two letters in error usually will not prevent the reader from understanding the message content. That is, for messages read by people, a high correlation between content and context will require little if any type of error control scheme.

The amount of error introduced by humans entering data at a keyboard depends on such factors as the readability of the source document and the skill of the operator in keying messages on the terminal. Obviously, people who use terminals occasionally will introduce more errors than skilled operators, particularly when the message's content and context are somewhat unrelated, such as in keying in a 10- or 12-digit number.

Then even assuming that all data is entered correctly into the system through a terminal, errors may occur on the transmission link. Technically, it is possible to control these errors, basically by the receiver—a terminal or the host computer—rejecting data until it is received correctly. In many computer-based data-communications systems, a significant portion of transmission time and computer load is due to the transmission and retransmission of erroneous messages.

As with the other major tradeoffs in data communications, the system can be specified for tight error control at high cost, or lesser error control at lower cost. The judgment depends on the economic losses due to errors, which could range from simply the cost of retransmitting a block of data, to losing a large sale because of an error in quoting the order. Thus, establishing accuracy requirements is another of the system objectives that must be set by management. ☐

basics of network design

An efficient and reliable data-communications system involves many interdependent factors—from terminals to lines to error control

Dixon R. Doll
Consultant
Ann Arbor, Mich.

Faced with an urgent demand to increase utilization of assets and introduce practices that cut operating costs, today's business manager may be called on without warning to participate in the over-all planning of a data-communications system for his company. Such a task may at first appear to be formidable. However, a clear understanding of system objectives and price-performance tradeoffs in the system components will enable the manager to evaluate application requirements and thus contribute to the network configuration. Here a discussion of the major design criteria puts the project into perspective for easy analysis. The main objective is to provide the minimum-cost installation that satisfies the present operational requirements, yet allows for foreseeable expansion.

Regardless of the particular business application, a data-communications network's fundamental task is to economically and reliably interconnect remote terminals to a host computer, which may perform data-processing activities and simple routing of messages between terminals. Once business management has defined its operational requirements, the planner/analyst faced with system design responsibility will find that he has available numerous options in equipment, software, and operating procedures. The major choices will generally involve decisions about:

- Terminals
- Modems (or datasets)
- Communications lines
- Multiplexers and remote data concentrators
- Software and error-control procedures
- Transmission-control units and front-end processors
- Fault isolation and backup features

Before any design decisions are made, some type of analysis must clearly establish the economic and performance benefits of the planned network. Involved in the plan are the kinds of transactions to be processed and their urgency, volume, geographic disposition of company sites, expected growth rates, and the system accuracy needed. In fact, operational management requirements, cost justification, and technical factors are so interrelated that several alternative plans may be formulated and adjusted as project planning proceeds—to assure that management and the analyst do not include performance features that increase equipment, software, and communications burdens.

In simple applications that involve a single type of transaction and a relatively small number of remote terminal sites, the configuration to solve a seemingly complex design problem will often become intuitively apparent from a tabulation of system requirements. More complex data-communications systems may profitably rely on special computer-based procedures and models that assess the relative costs and performance characteristics of hundreds of alternative network layouts. Optimizing programs are available from certain vendors, common carriers, time-sharing services, and independent communications consultants.

Setting the goals

The objective of most networks is to provide a minimum-cost layout that satisfies such performance requirements as net data throughput, reliability, and availability. Response-time statements such as "95% of all input messages must receive a reply from the computer within 3 seconds" are common measures of performance in leased-line networks. In switched-link systems involving dial-up calls through the ordinary telephone network, performance measures commonly stipulate an acceptable busy-signal probability level and a satisfactory net data throughput, once a connection is established with the computer.

Reliability and availability levels are often stipulated via the traditional mean-time-to-repair or mean-time-between-failure parameters. Such critical elements in the system as heavily utilized terminals, links, and computers are often configured in redundant pairs to achieve desired standards. Of course, such duplication raises the cost.

The problems and tradeoffs in choosing the most suitable system elements can be illustrated by tracing a typical transaction through a data communications network from start to finish (Fig. 1). Data is typically entered from some type of remote data-communications terminal. Among the more widely used terminals are the familiar Teletype machine, the alphanumeric cathode-ray-tube (CRT) keyboard/display, the remote-batch terminal, and the remote data-entry station that contains a minicomputer and operates with such peripherals as card reader/punches and line printers.

In the terminal, message characters are encoded into groups of binary digits (bits) and transmitted serially, bit by bit, into the transmitting modem. As shown in Fig. 2, the transmitting modem converts the rectangular waveshape of binary 1s and 0s into analog or continuous waveshapes, usually composed of frequencies suited to the transmission properties of the particular communications link being used. Traditionally, this link has been provided by the common carriers in the form of voice-grade lines—installed to provide quality transmission of human conversations.

The receiving modem performs the opposite function of the transmitting modem by reconstructing the originally transmitted data stream and feeding it to the receiving terminal, which may be either another terminal or the host computer. The electrical interface between the terminals and modems usually conforms to an accepted industry standard such as the Electronic Industries Association (EIA) RS 232C interface standard.

Such standards define common connections for voltage levels, indicate which electrical connector pins are to be used for the transmitted and received data bits, and define the numerous control functions and "handshaking" needed to establish connections.

The data bits leaving the receiving modems enter some type of line adapter or line-termination unit on a transmission-control unit or front-end processor (see Fig. 1) where the bit stream is reassembled into the originally entered characters. When a specified number of characters (data block) has been accumulated, or when some special control character is detected in the controller, the host computer is notified. Then, the data block is transferred either directly to the computer or, if the sending terminal does not have a current task under active processing in the computer, to secondary storage or mass memory (see Fig. 1).

Most installed transmission control units are hard-

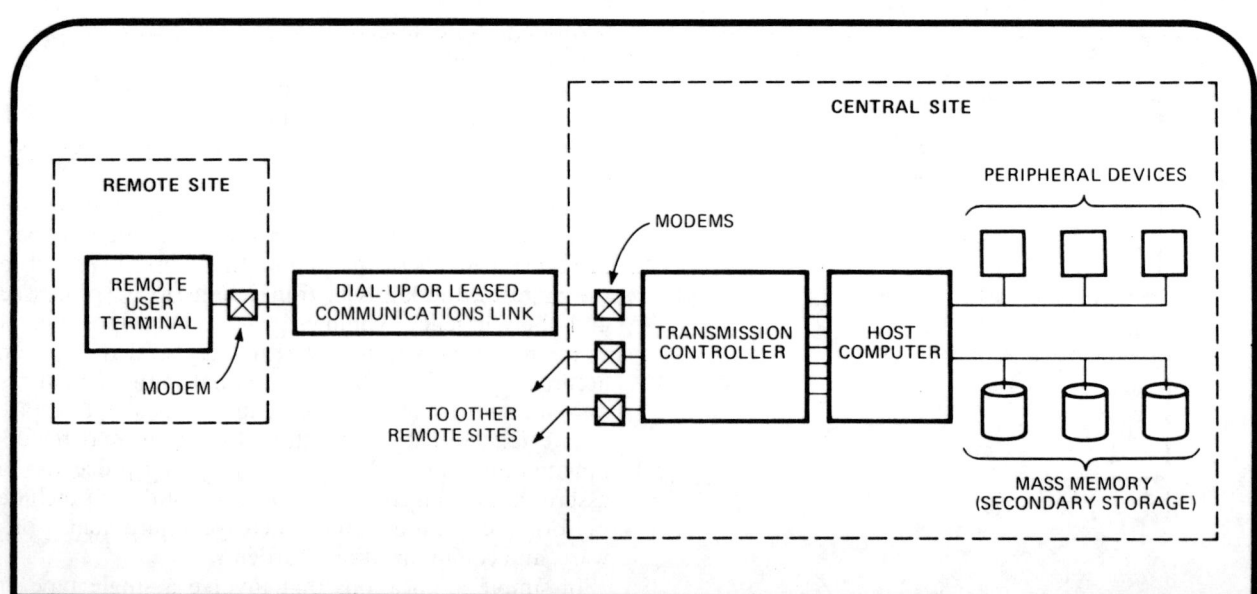

1. Basic. In typical computer-based communications system, data flows over telephone links between terminals and computer. Modems interface terminals to the lines and to transmission controller that sorts out multiple data streams for entry into the computer.

wired, and thus not easily modified if system requirements change. Further, most are IBM models 2701, 2702, or 2703. Although each has different features, generally they are classified as 270X controllers.

A recent innovation in transmission control is based on the programable minicomputer, which in data-communications systems performs transmission control, as well as other functions that can improve system efficiency. In this activity, the minicomputer is called a programable front-end processor. Many minicomputer companies offer front-end processors, and recently IBM announced a similar programable unit, the IBM 3705 communications controller.

Reply generation

Whenever enough characters have been received in the controller and a processing task is activated in the host computer, some type of output reply will typically be generated for transmission back to the remote station. This reply message is usually produced by the applications or operating system (OS) program being executed in the host computer. The controller is notified by an interrupt signal that tells the controller the length and starting location of the output message in a buffer storage area. This buffer, which is accessible to both the controller and the host computer, frees the computer for further applications processing tasks coming from other remote terminals and from local peripherals connected to the computer. The characters of the reply message are then converted into the proper serial-bit patterns for transmission back out to the remote station.

Messages and replies entering the system thus pass sequentially through numerous hardware devices and are subjected to a variety of software-controlled processing steps in the round trip from remote terminal into the central computer and back out again. Consequently, the art of successfully planning, designing, and implementing a system involves the evaluation of many tradeoffs for each of these interdependent hardware and software functions that contribute to response time, reliability, and availability of the entire on-line system.

Selection of a terminal allows for many alternative options that, besides many other factors, permit a tradeoff of higher terminal cost for lower communications cost. Some terminals contain local buffers—essentially small memories that accumulate and assemble a character, a line, or a message—that add to the cost of the terminal, but, on the other hand, permit faster transmission speeds and improve line utilization.

Most terminal applications involving transmission at speeds up to 1,200 b/s (or about 120 characters a second) use the asynchronous—or start-stop—method of encoding and sending characters. In asynchronous transmission (Fig. 3), the actual data bits of each character are preceded by a start bit and followed by a longer stop bit to separate the characters. These start-stop bits can carry no useful data and, compared with synchronous transmission, they reduce net data throughput.

Synchronous transmission, on the other hand, is usually used for applications involving speeds of 2,000 b/s or faster. In the synchronous mode, a constant-rate digital pulse generator, called a clock and usually located in the modem, determines the exact instant at which each bit is sent and received, eliminating the need for the special start and stop bit sequences necessary in the asynchronous mode.

Selecting lines

Either dial-up-connection (switched) lines (Fig. 4) or private lines (Fig. 5) may be selected for communications links, and often a network contains both kinds of links. Dial-up connections, by using the national telephone system, enable a particular terminal site to communicate selectively with well over 100,000,000 other locations. Dial-line facilities, generally limited to line speeds of 3,600 to 4,800 b/s, usually are two-wire cir-

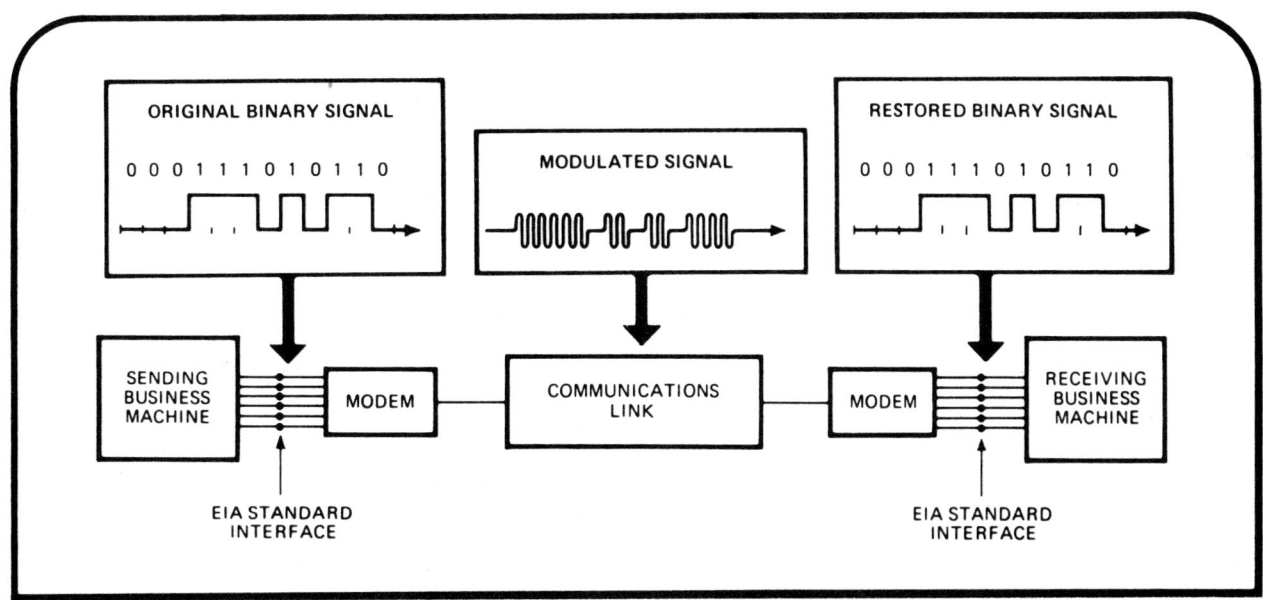

2. Conversion. Modem at sending business machine converts original data stream into modulated signals that can be transmitted by telephone lines. Then modem at receiving terminal demodulates analog signal back into binary-pulse stream for input to the computer.

cuits which are unable to sustain bona fide full-duplex transmission (simultaneous data transfer in both directions) at synchronous speeds.

Further, special properties of the ordinary telephone network—for example, the echo suppressors needed for voice communications, but which interfere with data communications—prevent dial network lines from being established or turned around in less than a few hundred milliseconds. Dial-up lines may also be subject to larger error rates than private lines, but suitable error-control techniques, either in hardware or software, may offset the need for higher-cost leased lines.

Private, or leased, lines are available in either point-to-point or multipoint configurations, as shown in Fig. 5. In point-to-point, a separate private line connects a terminal directly to the computer. Since most private lines are four-wire circuits, control signals can be sent without line turnarounds. Also the usable bandwidth, which relates to data transmission speed, can be extended. However, the cost is higher, since charges on each line are based on relevant common-carrier tariffs. Multipoint lines, often called multidropped lines, interconnect three or more data stations (including the computer) and are mainly used to reduce line and modem costs by enabling two or more remote stations to share a link.

The costs of separate point-to-point lines from a data-processing center to each remote location are thus avoided, as are the costs of a separate modem and termination equipment for each remote terminal at the processing center. One disadvantage of multidropped lines is that only one of the remote terminals on that line may communicate with the central site at any moment. Another disadvantage is the requirement for line-control software for polling or contention to determine which terminal seizes the line at any time.

Duplexing

The terms half-duplex, full-duplex, two-wire circuit, and four-wire circuit—widely used in data communications—need to be clearly understood. Half-duplex and full-duplex apply to a terminal's mode of operations (see Fig. 6). The terms two-wire and four-wire apply to the links provided by the common carrier. Regardless of the communications link, terminals are said to be operating in the half-duplex mode when the terminal can transmit or receive, but not do both at the same time. Full-duplex terminals, on the other hand, do transmit and receive simultaneously.

Most data terminals operate in the half-duplex mode. Even though not widely used yet, full-duplex remote-batch terminals are becoming more popular because of their ability to process more data in a given time than half-duplex stations.

As for communications channels, two-wire links may operate in either direction, but not both at the same time. The four-wire circuits, basically a pair of two-wire circuits, can operate in both directions simultaneously. Thus a half-duplex terminal may use either two-wire or four-wire links, but a full-duplex terminal requires a four-wire connection. The four-wire link can be obtained in a leased line, but the ordinary switched telephone network provides only two-wire links. Thus, if

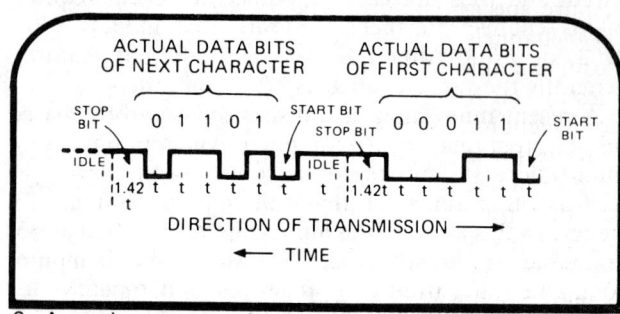

3. Asynchronous code. Low-speed terminals use asynchronous transmission, in which each character, shown here as a five-level code, is preceded by a start bit and followed by a stop bit that advises the receiver when a character arrives.

the switched network is to service a full-duplex terminal, two separate dial-up connections are required, with each connection transmitting in the opposite direction. Although most leased lines are four-wire circuits, they often connect half-duplex terminals when frequent changes in transmission direction and consequent turn-around delays cannot be tolerated.

The modems (called datasets by the common carriers), which must be used on carrier-provided links longer than several hundred feet, are available from many of the carriers and independent suppliers. Since the end-to-end performance of a data-communications channel will be governed by the link and the attached modems, special requirements and features of modems for particular applications should be analyzed thoroughly.

At asynchronous speeds up to 1,200 b/s, ordinary voice-grade lines are usually more than adequate for even the simplest modems. However, at synchronous speeds, some type of special conditioning in the lines and/or equalizing circuits in the modem are often required for satisfactory operation. Modems also directly affect the network's operational efficiency, since they must take the time to establish connections and change the direction of transmission—both of these factors contribute to overhead losses in synchronous dial network applications and in multidropped configurations.

Line-transmission speeds from 7,200–9,600 b/s may be achieved with highly conditioned private voice-grade lines using automatically equalized modems, and transmission speeds from 3,600–4,800 b/s may be achieved by using appropriate modems on most dialed connections.

Channel selection

In selecting the transmission speed for a particular application, the planner will generally choose from three main classes of channels offered by such common carriers as the Bell System telephone companies, General Telephone & Electronics, and Western Union. As examples, Series 1000 channels from the Bell System carry asynchronous transmission at speeds up to 150 b/s. Bell System Series 3000, known as voice-grade channels, are generally required for asynchronous applications faster than 150 b/s and for synchronous speeds from 2,000 to 9,600 b/s.

Wideband channels operating faster than 9,600 b/s

require special facilities, such as Bell's Series 5000 or 8000 channels, often known as Telpak offerings. Telpak C, for example, corresponds to the transmission capabilities of 60 voice-grade channels, and Telpak D is equivalent to 240 channels.

Rates for leasing different types of channels increase with distance, but successive incremental distances cost less than preceding increments. For example, for an interstate link, the incremental price for the 400th mile of a private-line voice channel is less than half the incremental rate for the 20th mile. Intrastate rates, of course, apply to any circuit with all drops or access points in the same state. These rates differ from state to state; generally, however, an intrastate leased link will cost a user at least twice as much as a comparable-distance interstate link.

The entire subject of monthly lease rates for interstate private lines is being reevaluated by the existing established carriers in response to the nationwide service offerings promised by such specialized carriers as MCI and Datran. Rates have already been reduced by some carriers on a regional basis—for example between St. Louis and Chicago—by MCI and Western Union. It is safe now to conclude that rates along other high-density routes will come down, while rates for local distribution links off the major routes will increase.

Line-sharing equipment, such as multiplexers and remote data concentrators, connect clusters of remote terminals to data-processing centers by employing voice-grade or wideband lines rather than a multiplicity of individual low-speed circuits.

The primary reason for multiplexing and concentrating, in many ways similar techniques, is to exploit the economies of scale prevalent in today's telephone-line tariffs. Voice-grade line (Series 3000) for example, can carry the equivalent (in number of bits per second) of 20 to 50 low-speed (Series 1000) lines, yet they cost only about twice as much. Regardless of how tariffs may be modified in the future, multiplexing and concentrating will always be important cost-reduction techniques, as long as there are economies of scale in the relative cost of bandwidth.

Other important considerations in planning the network include whether remote terminals will gain access to the system by polling or by contention, which transmission-error procedure will be implemented, what code formats and types of terminals are suitable, and what block lengths (total number of bits in a record or group of records) will be transmitted.

In most private-line systems, polling procedures govern terminal access to the network. With polling, the line-control program contains an internal table of terminal addresses to be queried in prescribed sequence. The program continuously cycles through this list, soliciting messages from those terminals that are ready to send. Polling procedures are inherently flexible because the order of the polling list, the number of times a terminal address appears in the list, and the priority assignments may be easily changed via software modification. Polling makes possible relatively tight control of systems hardware and software resources, since access is monitored by the communications-control software and not by the system's users.

A different type of polling is used in such dial-up applications as collection of sales and inventory data. Here, the central facility does not poll continuously, but instead selectively polls the remote stations, which then transmit their messages or data. A combination of hardware and software initiates each call without human intervention, so that polling can take place even when the terminal and computer sites are unattended. On completion of a data transmission, the call is disconnected, a new call is placed, and data is received from the next remote station on the polling list.

Contention procedures provide for access to the computer in many dial-up time-sharing systems, whether commercial utilities or user-owned installations. Here, the remote users call in at random and have to contend for service. Thus, terminals may gain access on a demand (first-come, first-served) basis, depending on contractual arrangements. Contention is also employed in

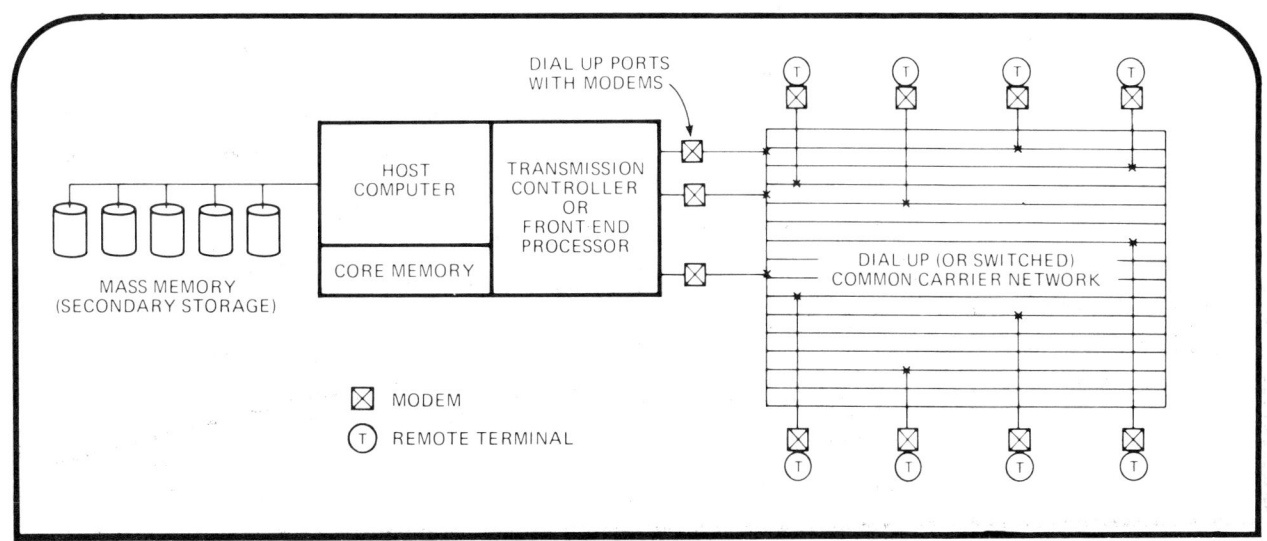

4. Dial-up. Using the national switched-telephone network, any phone connection can reach the host computer from anywhere in the country, provided the terminal user knows the numbers assigned to modems at the computer's interface transmission controller.

such unique communications installations as voice-response systems and facsimile networks.

Error correction

In data communications, transmission errors are inevitable. Errors result from such natural causes as lightning, relay switching noise, rain that may cause moisture to accumulate in telephone cable, and wind that may cause microwave antennas temporarily to get out of alignment. A bit error occurs when a bit becomes inverted so that a 1 changes to a 0, or vice versa. Such an inversion changes the meaning of a character so that the receiving terminal might type a Q, even though the sending terminal actually sent an F. Barring a drastic line failure, errors usually occur randomly and last long enough to affect a sequence of bits.

Most on-line systems require integrity of data and need at least some effort to correct these errors. The most popular practices involve some type of error detection and retransmission scheme: manual, echo-back, or automatic.

Manual correction methods require the critical information fields in a message to be entered several times by the operator at the sending station. The receiving operator or the computer checks the content of these redundant fields. If they match, it is assumed that no errors occurred. If the fields don't match, the sending station is requested to retransmit the message. These manual procedures are widely used in lower-speed applications involving teletypewriters and similar unbuffered terminals where automatic correction cannot be readily implemented.

A semimanual approach to error control is echo-back, or echoplexing: each character entered from a low-speed unbuffered terminal is promptly returned to the sending operator by the distant computer. Characters are verified one at a time, perhaps on a CRT display, and the user may make corrections whenever the locally keyed characters do not correspond to the echoed ones. This approach is not flawless, since an error may occur during echo transmission. However, echo-back has been used with considerable success in highly interactive applications where a human operator is engaged in a real-time conversation with the computer—say during the on-line solution of a computer-aided engineering design problem.

In synchronous applications, a terminal usually contains a buffer—a memory that assembles and temporarily stores a character, word, or record, along with some added bits for parity check. These parity bits permit an automatic approach for control of transmission errors. Here, data is transmitted in blocks, with the redundant parity bits being sent along with the actual data.

The receiving terminal performs parity calculations identical to those performed in the transmitting terminal and compares its computed redundant bits with those actually received from the link. If they match, it is assumed no error took place, and a positive acknowledgement (ACK) is sent back to the transmitting terminal, thereby initiating transmission of the next data block.

If an error is detected, a negative acknowledgement (NAK) is returned to the transmitter, requesting a retransmission of the erroneously received block. Powerful codes have been developed so that with modest amounts of redundant bits—perhaps 5-10% of those in a transmitted block—virtually no errors will go undetected. Figure 7 shows the one-way time profile for line usage of a half-duplex error-detection and retransmission connection in which each block is separated by a time interval, T. This interval is the amount of time required for the line and modem turnaround, plus the time for an ACK or NAK control signal to go back to the sending terminal. When an error has occurred, a block is retransmitted. The total lost time, called overhead, is shown at the bottom of Fig. 7.

Error detection and retransmission is implemented in IBM's widely-used binary synchronous communications (BCS) protocol—or line-control discipline—and in software offered by other vendors of computers and terminals. Other codes detect and correct transmission errors without the slowdown for retransmission, but generally they are more costly in hardware utilization and less powerful.

In addition to transmission slowdown created by the line protocol, many other factors directly affect the net data throughput (NDT), the rate at which usable data bits arrive at the receiving end. NDT is obtained by deducting from the rated operating speed of a modem-line combination the overhead losses that arise from call-connection time, retransmission times for erroneously received data blocks, and all types of control,

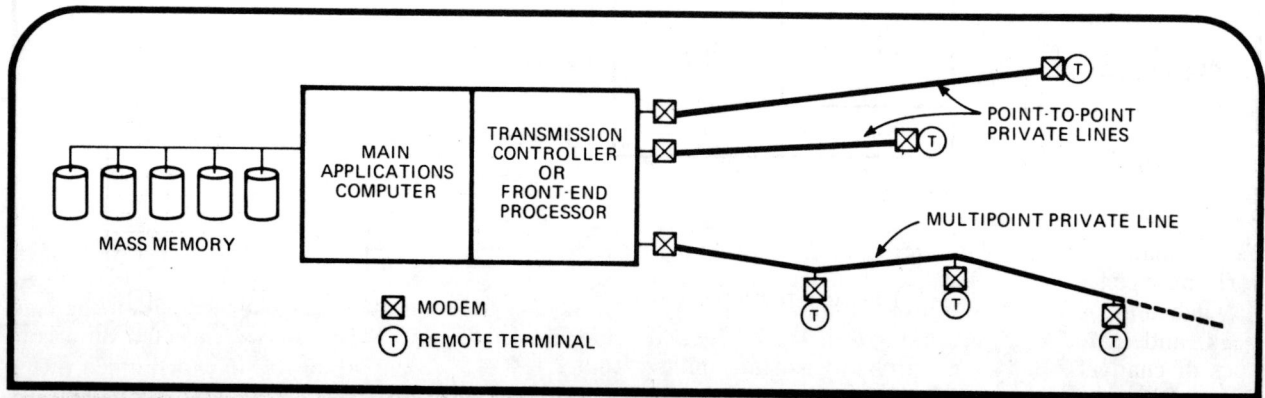

5. Private lines. Common carriers will lease dedicated or private lines for exclusive operation by users. Private lines come in two configurations: point-to-point, connecting end-city locations, and multipoint, to drop data at terminals located between end-cities.

signaling, and redundant bits. For example, on a channel normally rated for 4,800-b/s transmission, the overhead for call connection might be 5%, for retransmission 5%, and for control signaling 30%, to give a total overhead of 40%. Thus, this line will actually have a net data throughput of 60%, or $0.6 \times 4,800 = 2,880$ bits per second. These factors, noted in the equation below, can be balanced against each other to obtain some optimum value. For an error-detection and block-retransmission system, as is usually the case, the NDT in bits per second is:

$$\text{NDT} = M(1-K)(1-P)/(M/R+T)$$

where R is the modem/channel nominal operating speed in b/s; P is the block-retransmission probability, statistically related to the channel's error rate; M is the transmitted block length of the message, including redundant bits, in bits; T is the time between blocks, in seconds; and K is the fraction of noninformation bits (including redundant and synchronizing bits) in each block.

Optimum block length

An optimum block length exists for any particular application. Or, stated another way, there is no universally best block length. Moreover, the optimum block length can be determined only when all factors affecting net data throughput are known quantitatively. Figure 8 shows a typical change in data throughput as a function of block length, based on the preceding equation for a nominally rated 4,800-b/s modem/line combination.

Here, some given value of K and error rate are prescribed. Note that at short block lengths, any given error rate implies a very small value of P; that is, many short blocks can be transmitted before one of them is in error and must be retransmitted. Thus, for short blocks, the lower NDT is primarily due to the total control signaling time between blocks. For the same error rate, blocks longer than optimum have a higher probability of containing error bits, thereby increasing retransmissions and again reducing the net data throughput.

The control programs in most data-communications systems determine actual block lengths. This length will usually be specified as an operating system (OS) or communications-control program input parameter. Should the selected block length later prove to be less than optimum, the software control can readily be changed.

Given these many hardware, software, and system-design options, the planner should consider essentially two kinds of network organizations—centralized or distributed—to determine the most suitable configuration for his application (Fig. 9). In centralized networks, all applications processing and information files are kept in a single center, and the network traffic will consist exclusively of communications between remote terminals and the main processor/data base site. Centralized networks may generally be arranged through combination of switched links, leased multidropped links of the same speed, and leased-line networks consisting of several types of channels that would probably include multiplexer and remote data concentrators.

Distributed networks are generally more complicated, but they offer the possibilities of reducing communi-

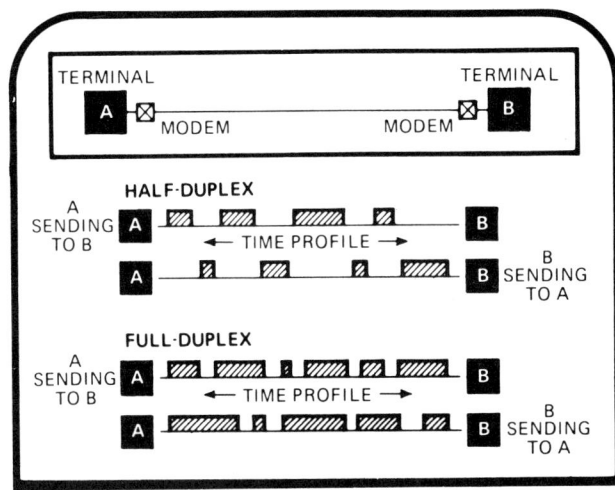

6. Duplexing. In half-duplex mode using two wires, only one terminal can send at any time, but in full-duplex mode using four wires, both terminals can send data simultaneously.

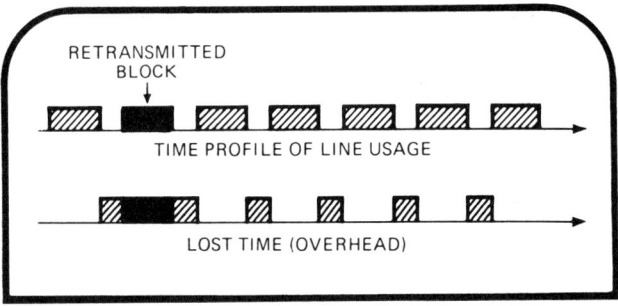

7. Error detection. In the half-duplex mode's detection-and-retransmission protocol, lost time results from retransmitted blocks and ACK/NAK gaps between blocks.

cations costs by locating large portions of the files and data-processing capabilities in regional centers for such applications as banking and insurance. The key cost tradeoff here is simply whether or not the total cost of a properly designed multiple-center, distributed network is less than the cost of a single large center served by a larger communications network.

A major obstacle to distributed networks has been the lack of reasonably priced switched or leased wideband communications links for interconnecting the regional centers. Other deterrents are the hidden costs for operating and maintaining multiple processing centers, the costs and complexities of the operating systems in distributed nets, the politics of large organizations, and unresolved questions about the security features of such decentralized networks.

Computerized evaluation

In the design phase, it is desirable to evaluate the alternative link combinations available for centralized networks against one or more possible distributed-network configurations. Various network-optimizing techniques, rapidly and inexpensively executed on a computer, can play an important role in expediting a variety of design-related tasks. One of the simplest, most useful computer-aided routines determines costs of alternative network layouts. Other programs design least-costly lay-

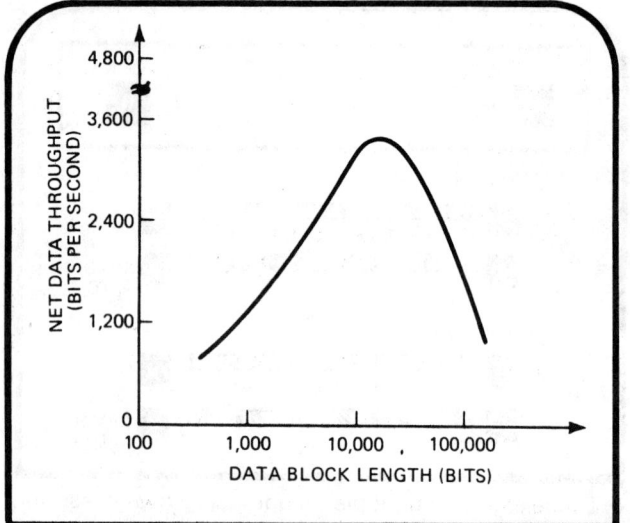

8. Data efficiency. Net data throughput, the useful number of bits transmitted each second, depends on block length, and such other factors as error probability and redundancy.

outs of multidrop lines, determine the best locations of remote multiplexers and concentrators, and estimate the response-time characteristics of particular configurations.

In centralized applications involving the dial network, such bulk tariffs as the Bell System's WATS (Wide Area Telecommunications Service) offerings should be considered as a way of reducing cost. Here, calls may be placed either into (Inward WATS) or out of (Outward WATS) a central location.

Computer programs can evaluate WATS networks and perform these design tasks for all types of leased-line networks. Required input data for these optimizing programs generally includes:
- Terminal and data-processing locations
- Terminal traffic volume, average and peak
- Maximum acceptable line loading
- Maximum acceptable response time, including computer processing
- Costs of lines, modems, multiplexers, and other equipment and services
- Possible locations of multiplexers and concentrators
- Types of line layouts to be evaluated

Outputs—in the form of lists and even routing maps—will generally contain cost itemization for the best layout, including specific line routings, line loading, and the best remote locations for the multiplexers and concentrators. Multiplexers and concentrators will usually be used only when they produce lower net costs than an optimally configured switched network layout or an optimally designed network based on multidrop lines.

The final burden of data-communications systems design and implementation falls on the planner/analyst working in behalf of his company. He should use computer-aided methods when indicated, add his experience and intuition, and listen to all advice from suppliers of equipment and communications lines. But he should refrain from carte blanche acceptance of system-layout recommendations obtained from parties having vested interests, particularly when these recommendations are intimately related to negotiations for equipment and services. □

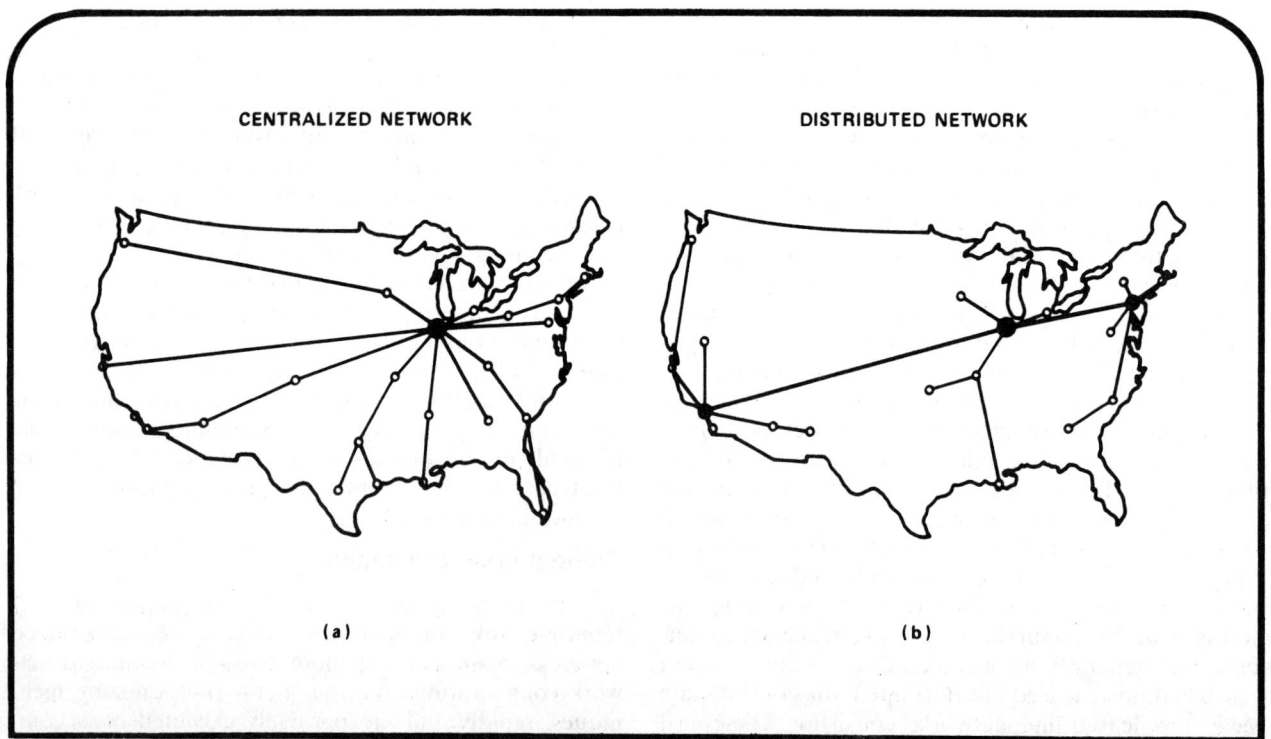

9. Centralized or decentralized. In centralized network (a), all terminals connect to a host computer, in this case located in Chicago. In decentralized network (b) data processing is done at regional centers in Los Angeles, New York, and Chicago; Chicago also serves as the main data center that connects with regional centers for specific data-base storage and retrieval.

What the future has in store for data communications

Momentous changes lie ahead for almost every aspect of data communications

Harry R. Karp and
Gerald Lapidus
Data Communications

The next five years certainly will be the most exciting period so far in the history of data communications. It almost goes without saying that data communications will continue its rapid expansion as a key element in the infrastructure of modern business and industry. And the importance of the data communications manager likewise will grow.

More to the point, however, the technology of data communications is in for some major changes. Together, these changes will have profound effects on the practice and the marketplace of data communications as we know it today. Longstanding predictions and promises will be manifesting themselves in a very practical fashion.

Satellites will routinely handle data traffic. All-digital terrestrial transmission links offering a wide range of services will form a web from one end of the country to the other. The value-added network will be a viable entity. The already-ubiquitous microprocessor will improve almost every hardware device employed in data communications networks. And, eventually, optical fibers will carry ultra-high-speed data streams over short distances.

Making predictions is an easy enough pastime. But data communications itself is a serious business. Credible predictions are becoming more important because large networks can take two to five years from initial concept until the last terminal is on line. So a degree of confidence in technology trends, costs, new services, new network-architecture concepts, the direction of competitive products and the like, lends weight to the decision processes of network application and design.

In researching the information for this report, we identified 10 major forces that will be the key factors in shaping the growth of data communications over the next five years. These influences are technology, research, networks, software, satellites, regulation, terminals, communications processors, multiplexers, and modems.

Some areas directly interact with others, while others are relatively isolated from the rest. Knowing how these areas are likely to be affected reduces the risk of making poor decisions. Here, each area is presented individually, to form an independent scene of what and, more importantly, why certain activities are likely to take place before the end of this decade. In all, then, the individual scenes form a tapestry of technology, equipment capabilities, carrier offerings, and system architecture, depicting data communications practices.

TECHNOLOGY

Technology being paced by microprocessors

Ten years ago developers of integrated circuits (ICs), and later of large-scale integration (LSI), were prophesying the eventual birth of a computer-on-a-chip. Today the microprocessor-on-a-chip represents the nearest thing to a fulfillment of the prophecy, even though it falls rather short of being a complete computer. The latter stage of development comes with the addition of a read-only memory to store an instruction set; a read-write memory ("main memory") for dynamic storage; input/output drivers, and a clock. Only then does it merit the term microcomputer.

Practically speaking, however, "The average microcomputer will contain a microprocessor and 11 other parts," says Douglas Powell of Motorola's Semiconductor Products division. With a power supply, a control panel, a cabinet, and a few other items, it is up to a computer in a box, but still cheaper than a standard minicomputer.

Because of its low cost and versatility, the microprocessor has become a prime factor in the numerous changes sweeping data communications.

The microprocessor is the core of the microcomputer. It consists of the arithmetic and logic unit to process instructions, general purpose registers to temporarily hold data being transferred, an accumulator to temporarily store computed results, and circuits to control the computer's operation.

Microprocessors range in cost from less than $30 for 4-bit units to several hundred dollars for 16-bit devices. And although a completely outfitted microcomputer might cost $1,000, the addition of intelligence to a terminal which already has, say, a power supply, I/O circuits, some memory, a control panel, and a cabinet, may not entail much extra hardware expense beyond that of the microprocessor. Software development, however, can still add substantially to the cost of a microprocessor-based product.

But with its low-cost computing power, the microprocessor might function as a line interface unit to implement a protocol, including a cyclic redundancy check for error control, and it might also perform code conversion and data formating. When such a unit is built into a terminal, along with a modem-on-a-chip (also related to the microprocessor) the terminal is no longer a simple I/O device, but a complete remote station. Considering that the input and display portions of the terminal may cost several thousand dollars, the addition of this intelligence will someday involve only hundreds more. In contrast, the same functions built with present-day minis and modems would double or triple the terminal price.

Compact model. The microcomputer, however, will not replace the minicomputer except in cases where the minicomputer is underutilized. Minicomputers offer faster instruction speeds, larger instruction sets, and more sophisticated software libraries. And since minicomputers also are built with low-cost integrated circuits, some are priced at only $1,000 more than microcomputers.

Therefore, microcomputers are not really a new type of equipment, but a more miniaturized and less expensive minicomputer. And just as minis have not replaced large-scale computers, microcomputers will not replace minis. But the low cost of microprocessors has added a powerful new tool to the hands of the cost-sensitive designer.

The microprocessor will also foster radical changes in modems, which are now generally analog devices. An analog modem requires expensive filtering circuits which may not be readily adjustable for different line speeds. Microprocessors, on the other hand, perform the modulation and demodulation function digitally. Analog signals entering from the line are converted to digital form by an analog-to-digital converter. Then the microprocessor performs mathematical operations that are equivalent to demodulation in an analog modem. This prepares the signal for use by the data processing system.

In the modulation mode, signals to be transmitted from the data-processing system are modulated mathematically rather than with oscillators and filters as in analog modems, and then the signals are passed through a digital-to-analog converter for transmission by the common carrier. In addition to basic modulation and demodulation, the microprocessors also can perform line equalization, speed changing, and diagnostics.

Besides lower costs, microprocessors promise to improve the error rate for a given signal-to-noise ratio. This is possible because the accuracy of the digitized signal (the number of places in the a-to-d converter output) can be better than the accuracy of analog circuits. Modems specially built with LSI circuits will be so small, that 10 to 15 could fit on a 12-inch by 12-inch printed circuit board.

Viable alternatives. Several substitutes for voice-grade telephone service will be in widespread use in the next few years. All-digital terrestrial carriers such as AT&T's Dataphone Digital Service, Datran's Datadial and Dataline, and Bell Canada's Dataroute—all presently in limited service—provide higher data rates and promise fewer errors and lower costs than analog lines.

The cost reductions result primarily from the

use of microwave transmission on intercity links instead of cables. For AT&T and Bell Canada, these economies are great because services can be piggybacked onto existing microwave facilities.

Higher data rates than with voice lines are offered because the 3-kHz bandwidth of voice lines restrict the practical data rate to 9,600 bits per second for today's modems. However, present microwave facilities provide the potential for expansion to 1.544 megabits per second (T-1 carrier), provided that wide-band local loops can be made available by the common carrier.

Satellite is another reliable communications medium which can be cost-effective, but the long transmission delays can cause problems. (A separate section on satellite systems appears later on page 23.

In the 1980s a new transmission medium may offer much larger bandwidths than cables for data communications. Information will be carried by modulated light beams down thin pipes called optical fibers. Fiber optics, as this transmission method is called, will initially be used over short distances, such as inside or between buildings, but eventually they may see service as intercity trunk facilities.

One of the early applications of fiber optics could be to carry high-speed data from a roof-top receiving antenna to a computer in the building. Corning Glass Works is now marketing a 19-strand bundle of fiber optics in single pieces as long as 500 meters (1,640 feet). It uses a light source and photodiode detector made by Texas Instruments. A link of this type sells for $1,000 for the electronics package, plus $28.50 per meter and because it works on photons, not electrons, optical fibers are immune to the electromagnetic noise that plagues both copper and microwave links.

Bell Laboratories has an extensive investigation of fiber optics under way. It wants to develop this medium for wideband transmission between cities, for digital exchange trunk systems, and for inexpensive broadband subscriber loop systems. Bell's approach is to use a one-strand optical fiber, measuring about two thousandths of an inch in diameter, and a laser light source. Such an arrangement could carry 4,000 one-way voice circuits, each digitally modulated to carry data. Meanwhile, Western Electric is developing production methods for fiber optics. ∎

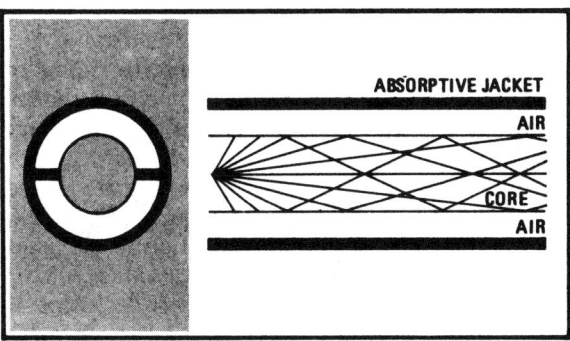

Loss problem in optical fibers is attacked with proper geometric design and material selection.

Security and radio links head up research list

As in any diverse and highly technical field, the lines of inquiry into ways of improving data communications are numerous and wide ranging. Among the more prominent areas of research these days are those that focus on systems security, alternatives to telephone lines for packet-switching networks, and academic studies of global communications problems.

Thomas N. Pyke Jr. of the Institute for Computer Science and Technology, a part of the National Bureau of Standards (NBS), notes that, "computers in the Federal Government are a valuable resource, and they must be shared efficiently and in a controlled manner over networks."

In this regard, one research project being handled by Pyke's group concerns the confidentiality of personal dossiers filed in computer data bases. Just who will have authority to access a data base is one thing, says Pyke, and then it's up to the network itself to make sure that only authorized personnel are using it. On a distributed-computer network, he emphasizes, "privacy becomes aggravated."

Fingerprints, therefore, are being seriously considered as a key to personal access, with some kind of fingerprint identifying device as part of a terminal. Other approaches being investigated include magnetic-tape cards issued to authorized personnel, or the use of passwords. "Whatever the final method for personal identification," concludes Pyke, "it will have to be a low-cost device attached to a remote terminal."

Even with personal identification, however, the data surely will also undergo encryption to make sure information cannot be intercepted or compromised. An encryption algorithm, which seems to satisfy security requirements and has found favor with the NBS, has been developed by IBM personnel. Its details were published in the Federal Register of March 17, 1975. (The Federal Register item also states the conditions under which IBM will issue a royalty-free license.) The goal of this project is to have the algorithm declared a Federal information processing standard.

The encryption algorithm can be implemented in hardware, probably as a large-scale integrated circuit. Thus it should be small enough and cheap enough to fit on every terminal and front-end processor on the network.

The IBM algorithm requires a 64-bit password for

RESEARCH

enciphering and deciphering the data. It "is so powerful," says Pyke, "that it would take several years for a large computer, completely devoted to the task, to break the code, and that's for just one password combination. Long before then the password would have been changed."

In the commercial research area, two noteworthy projects are exploring alternatives to telephone lines for packet-switching networks. One involves the use of microwave links, the other cable TV lines. Further in the future might be the sharing of commercial television channels. Data will be superimposed either on spare scan lines in a television signal or else on the picture portion of the signal, but the data will not be visible to view-

NBS'
Thomas N. Pyke, Jr.

ers of the conventional television broadcast.

The microwave study is part of a large-scale project being carried out by several companies for the Advanced Research Projects Agency of the U.S. Department of Defense. The goal is a radio network that uses digital channels for packet switching. Howard Frank, president of Network Analysis Corp., says that a radio network will someday be a reality. "Five years from now, this approach will be the thing people are talking about."

The microwave system under study is intended for localized distribution networks; its wireless, mobile terminals will have a range of up to 25 or 30 miles before needing a relay station or other type of collection point. The units will be useful in police cars and ambulances, by military troops on the move, or general business applications where there is no access to telephone lines. Miniature hardcopy printers and video readouts for receiving data are also being developed.

Another objective of the project is to make digital radio communications more efficient, says Frank, whose company is working on the network architecture. Other organizations on the project are Collins Radio (for initial hardware), Stanford Research (communications aspects), UCLA (measurement studies) and Bolt, Beranek & Newman (communications software).

The second project, concerning the use of cable TV for local distribution of data, has the benefit of cost breaks, says Frank. "The speed capacity is considerably higher than conventional telephone lines. Depending on the modems, a 6-megahertz channel can run from the 1-kilobit-per-second to 6-megabit-per-second range, or even higher with expensive hardware."

CATV capacity today is far from being filled with video broadcasts, so that "the incremental costs of providing channels for other uses is relatively modest," says Frank.

The problem at present is that CATV has not yet arrived in most big cities. When it does, within five years or so, it will have a major impact, as the bulk of data communications is conducted between urban areas. Furthermore, all new CATV systems now being installed must be capable, according to an FCC ruling, of two-way transmission. Eventually, says Frank, localized CATV systems "will be tied together by satellites to provide the connection between the local networks and the rest of the nation."

Academic work. Raymond L. Pickholtz, professor of electrical engineering and computer science at George Washington University, observes that, "Data communications networks have not been amenable to neat synthesis and analysis." Nevertheless the academic community, which has been criticized for not responding to industry needs in computer/communications, can now boast of graduate courses and research activities in data communications. Besides GWU, some of the more active institutions in this field are Massachusetts Institute of Technology, and the University of California at Los Angeles.

At George Washington for example, Pickholtz teaches a graduate course in "Data Communications Networks" to 20 students who come with either a communications or a computer background. "Even here, communications people rarely speak the same language as computer people," says Pickholtz, and to get over that hurdle each of the students must complete prerequisite courses like queuing theory, information theory, digital coding theory, and modulation and demodulation.

Pickholtz also directs three students doing research into the development of analytical models for such fundamental problems as adaptive routing, satellite networks, and line protocols.

Other areas ripe for academic inquiry are the development of better understanding in multi-user types of network, the interconnection of such networks, and the minimization of economic disturbances when interconnected networks react to each other.

Says Pickholtz, universities are getting on the "computer/communications bandwagon." More students will come out of school with more formal training in data communications and networking, and their research projects should soon provide succinct and fundamental models for the solutions to the "global" problems in networks. ∎

Networks: future is here, but designers waiting

Most attributes of tomorrow's networks are technically feasible today. Low-cost intelligence, efficient protocols, all-digital carrier facilities, satellite links, and distributed networking techniques, to name a few, are all possible now, but years may pass before their powers are harnessed in a concerted fashion, to bring about radical changes.

The reason is not without a certain irony, as James W. McNabb, vice president of teleprocessing systems and services at Sanders Associates Inc., explains it. McNabb says the designer of data communications equipment, "has fallen into a candy jar. All the things he ever dreamed are happening, but what he designed last month is almost obsolete. So," he says, "although many companies can already build some of these things, they're waiting for technology to shake out."

As examples, he cites the incomplete information on IBM's SDLC implementations and the problem of deciding which functions to build into hardware and what to do in software.

"To be perfectly safe, "McNabb says, "you tend to allocate most of it in software, so that when you obtain the necessary information you can easily retrofit it. However, with complete knowledge of SDLC, designers might decide to implement it most economically in faster hardware.

"That is what's frustrating most companies," he says. "All of us hope that next week, next month, or later this year, certain things will be made clear so that we can make better decisions."

The dust is showing some signs of settling. Designers are firming up the architecture for the next generation of computer networks. And the trend will be for computer networks to reduce reliance on large corporate data-processing centers. The growth in numbers and computing power of intelligent terminals will result in more processing locally or in regional centers rather than at central sites. As a result the data communications facilities of organizations will continue to grow because of the need for far-flung computers to converse with each other when exchanging data and programs, and the requirement to maintain orderly traffic patterns in the network.

Making do. This form of interchange suggests that communication between local sites might use wide-band private-line carrier facilities more heavily because of the speeds and low error rates needed, but just the opposite may be true in certain instances. Locations having substantial local computing power and storage capacity may not produce enough traffic to justify expensive communications facilities. Lower-cost carrier services such as dial-up lines, low-speed private lines, or packet-switching services may suffice.

Of course many tightly-knit networks will undoubtedly require high-speed carrier facilities. But the growth of networks will not necessarily engender more expensive carrier facilities, although the number of these facilities will grow with the complexity of the networks.

Decentralized computing means that the role of the corporate data processing center will become more of a centralized data base for a network rather than the company's main number-cruncher. It is this changing system architecture that is causing the data communications function to evolve from a cumbersome mainframe add-on to an integral busing element in a geographically-spread network of computers.

As envisioned by McNabb, the radical differences between advanced network architectures of today and those of tomorrow are shown in the diagrams on p 20 and 21. Of course, systems with architectures very similar to those of today will exist in the future. Conversely, some networks already are configured much like the one shown for tomorrow. What these diagrams illustrate is the general shape of present and future systems.

It is important to bear in mind that not all data communications systems will need to be at the state-of-the-art level. Some of the most popular data communications systems in use today—a few simple asynchronous, keyboard terminals connected to a computer—probably will continue their popularity in the foreseeable future because they fully satisfy many user needs. (Teletype Corp. has goldplated its 500,000th teletypewriter and they're still selling).

LSI a factor. The intelligent terminal of today performs routine jobs involving data formating and error control, and some also relieve the central computer of routine processing tasks. However, the LSI revolution is making the addition of intelligence to terminals and communications processors less costly, causing central computing functions to migrate to local sites.

As time goes by the network designer will be faced with an increasing number of critical choices as to which function to allocate to remote sites and which to central computers. In a company with scattered manufacturing and warehousing facilities, for example, each facility may perform all of the data processing functions needed to sustain its routine daily operations, while files holding business transactions, production rates, inventory information, and financial data may be continuously updated at central locations. In addition, special

NETWORKS

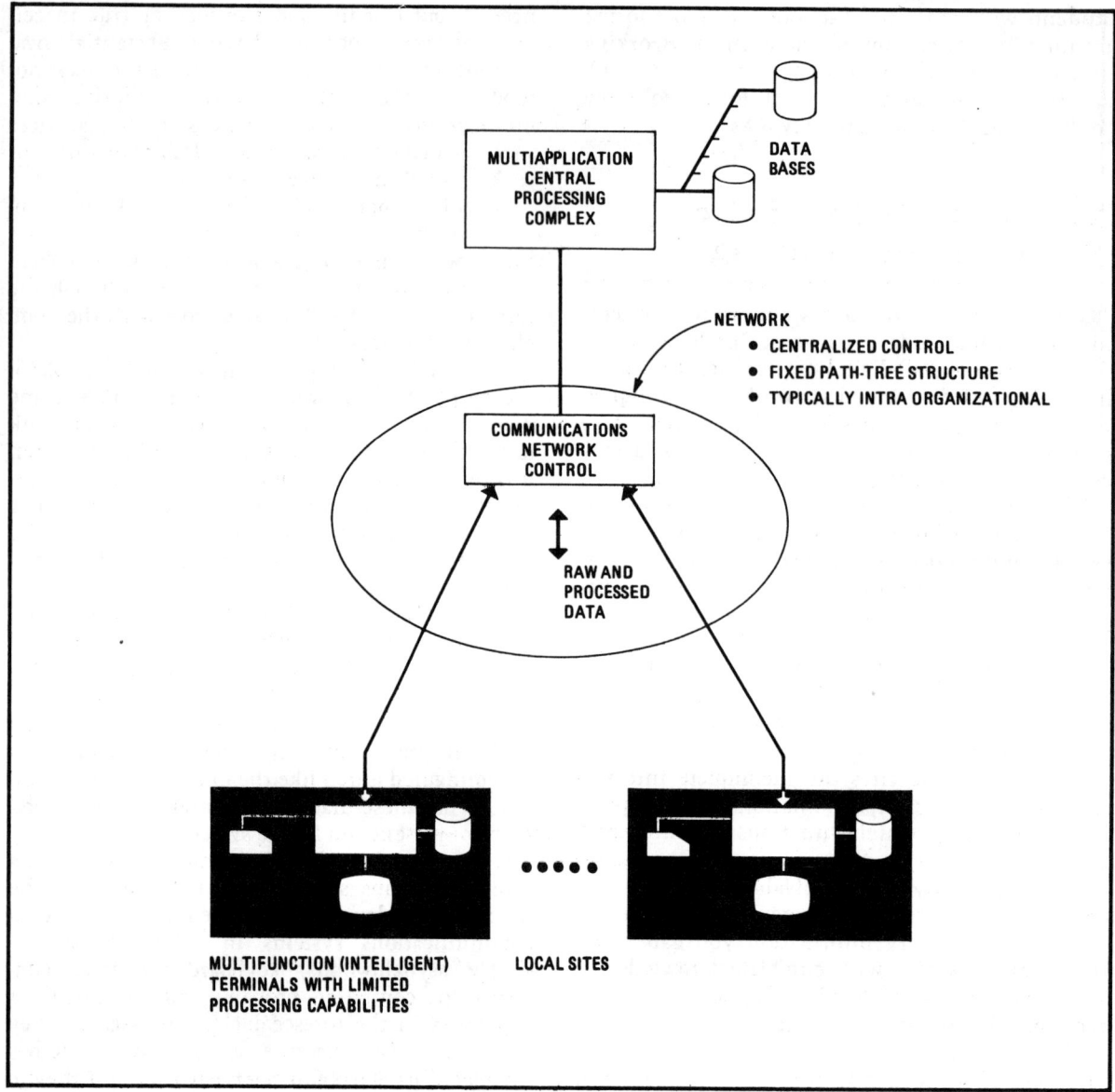

Today's hierarchical networks have limited intelligence nodes and a fixed communications structure.

processing jobs that unnecessarily load local computers will be transmitted to the central site's number cruncher. Similarly, machines with special capabilities, such as scientific computers or timesharing systems, will be accessed by other network nodes, as needed.

In the distributed network of tomorrow, data will flow freely among the nodes, rather than exclusively between remote sites and the central computer. One of these nodes might even be an external timesharing service.

The future network will have distributed data bases, of which the largest will probably continue be that of the headquarters site. The user will ask for a block of data by name without having to know where the data base is and the system will locate and access the site holding that data and then process it possibly at yet another site.

A significant distinction between the distributed network and the centralized network will lie in the hierarchies of computing power, as indicated by the primary and secondary networks in the "tomorrow" diagram. The primary network will interconnect the major computing sites, which may correspond to divisions in a company or regional offices. The secondary sites will connect to terminals located at smaller offices, much like today's remote sites.

The communications controller at a remote site, in addition to servicing its own peripherals, will also manage a secondary network. If a device at an end point requests that a specific job be done, the nodal processor will route the request along the appropriate lines after performing whatever portion

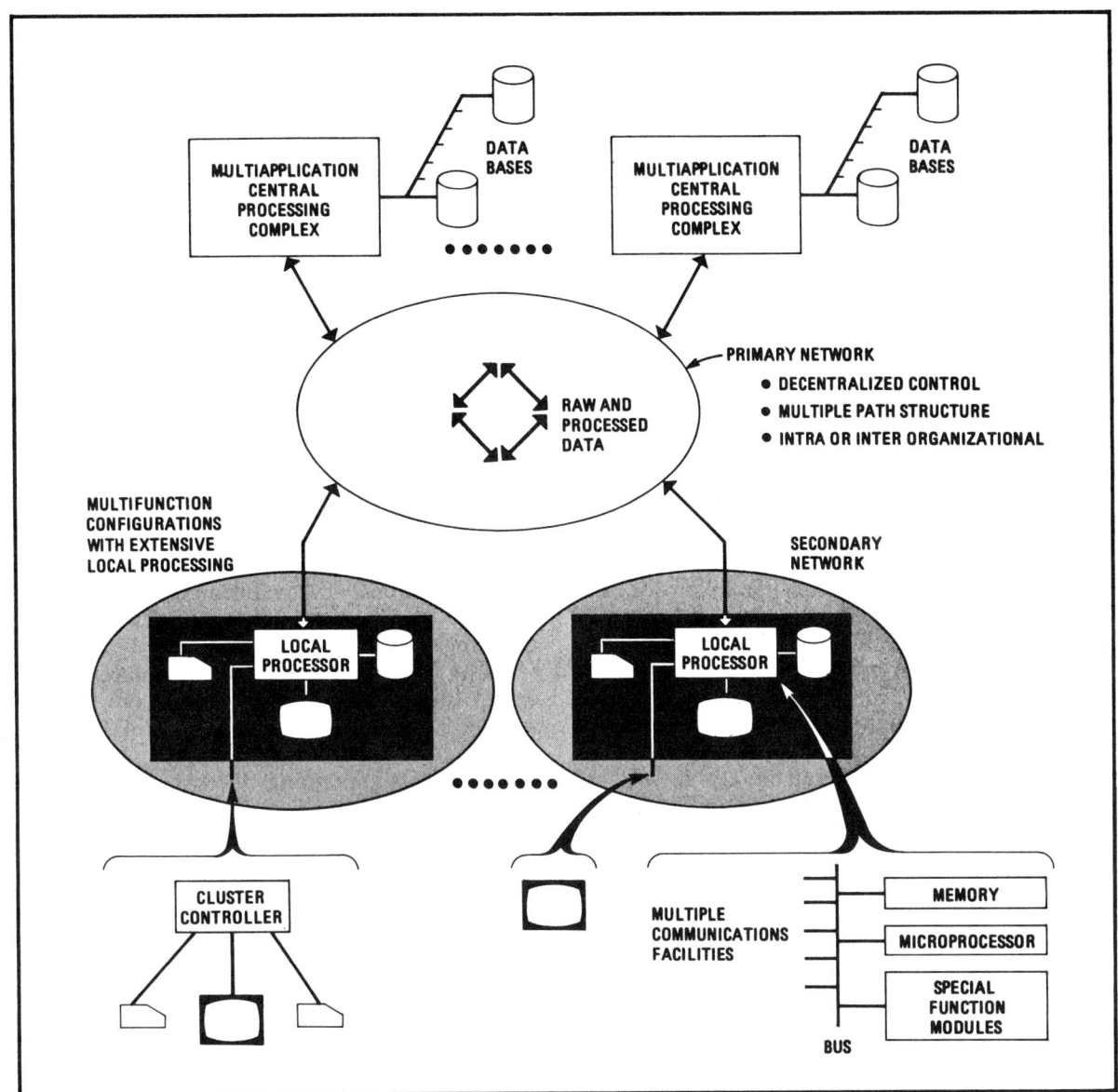

Tomorrow's distributed networks will have powerful local processors and flexible communications.

of the processing job it can handle locally. The secondary network may contain a wide variety of terminals ranging in complexity from teletypewriters to large remote job entry systems. As is true today, management of large numbers of terminals will require the use of cluster controllers or other multiplexing hardware.

Any carrier. The secondary network may communicate on private, hardwired cable as might be the case in an industrial complex, hospital, or university, or else through conventional carrier services. As mentioned earlier, the primary network may utilize any type of carrier: private network, public switched network, satellite, or a value-added network—depending on traffic requirements, reliability, error rates, and economics. Large primary networks with relatively light traffic loads may use message switching computers to minimize the use of carrier facilities.

The "data processing box" at local sites may be characterized as a hybrid of today's product categories. The terms minicomputer, microcomputer, intelligent terminal, and communications controller cannot accurately describe this box because its functions may contain all or some of those presently associated with the forms that these product categories have today.

The transition to a distributed network will not necessarily require the user to discard all of his present equipment. As is true today, upgrading packages consisting of new software and add-on hardware modules will be available to add new functions.

The road to such capabilities will be evolu-

SOFTWARE

tionary. Initially the user operating the system will have to select his computing resources and data bases. But eventually all he will have to do is request a particular function and the system will automatically allocate the required computing and communications facilities based on availability, capabilities to perform the function, and economical routing. In other words, the network and computing resources will become totally transparent to the system user.

Developer decides. Not all networks will be fully distributed; many will be hierarchical. The approach selected will depend largely on the preferences and experience of the equipment manufacturer or systems house that develops the software. Of course, the direction taken by IBM will probably be the strongest influence on the shape that most systems will take, but the computer giant does not discuss its plans.

Nathan Teichholtz, corporate program manager for computer networks of Digital Equipment Corp., says of IBM: "As far as I can tell, it is still making every effort to support networks with large, centralized 370s, although IBM is moving towards distributing application programs and data bases into smaller machines.

"We at DEC can do the same thing with our Decsystem 10," he added, "but deep down most of us feel that it would be putting all our eggs in one basket. If the central system fails, you're in trouble. So our approach probably will focus more on distributed control."

One criticism of fully distributed network control is that it may require more communications overhead costs than hierarchical operation because of the frequent need to exchange status information. Teichholtz counters this argument with the view that, as channel bandwidths increase, communications will become less expensive on a per bit basis. Also, added economies will result from communications processing making line utilization more efficient. Therefore, the network overhead will be easily tolerated.

"It's like using more memory today," Teichholtz observes. "People would rather use more memory than spend two months optimizing and coding an assembler to save 1k of core." ∎

Software: less for user to develop

Suppliers of data communications systems are aiming to reduce the user's software requirements, now and in the future, by providing standardized program modules and by converting routine software functions, such as protocols, into hardware.

The circuits are being tailored to specific applications with written addresses and other specialized information stored in read-only memories. At the same time, efficient operation of the communications networks of the future will require that software be extremely fast and totally transparent to the operator of a terminal.

Many industry observers see the difference between tomorrow's software and today's as analogous to the difference between virtual memory techniques and conventional memory management in data processing. Virtual storage has, of course, made routine memory management an automated affair, no longer the continuous responsibility of the programer.

Recently some communications software packages, like Interdata's Telecommunications Access Method, have enabled users to address any remote terminal in a system as if it were local. The only difference for the operator might be in the response time. To obtain such transparent communications capabilities, future versions of higher-level languages will have to include communications subroutines.

Distributed software. Nathan Teichholtz of DEC says that, in a full implementation of the distributed-network concept, the operating system also will be distributed and there may not be a master controller. Different computers will be controlled by different versions of the network operating system, according to their individual capabilities and their functions in the network. Each nodal computer will be responsible for its local equipment and for making connections to adjacent nodes. Messages will be passed from one node to the next according to routing algorithms developed for each site and based on traffic patterns.

Thus, as in packet switching, a message might be routed between one or more intermediate nodes as it travels from the source to the destination.

Teichholtz depicts the following version of the way such an operating system will handle user programs: "The local compiler will be really smart and

Sanders' James W. McNabb.

say that this kind of a program runs best on a Decsystem 10 or else on an IBM 370. Then the program will be compiled in that way and routed to the appropriate computer. And the user may not even be aware of whether the program is being processed here or there."

To know more. Most industry observers are satisfied with the new full-duplex protocols, but they would like to know more about IBM's synchronous data link control (SDLC). The use of full-duplex carrier facilities, the variety of languages that can be supported in the standard format, and the efficiency gained in permitting a large number of unacknowledged characters (although only up to seven unacknowledged messages), makes SDLC a viable protocol for the foreseeable future. In contrast, the older binary synchronous communications (BSC) is available in several versions and must await a positive or negative acknowledgement in response to a transmission before another block can be sent.

Jim McNabb of Sanders comments: "It isn't that people do not understand SDLC. What isn't clear is what they will do with various bits in a frame. Everybody knows there are 8-bit flags at the beginning and at the end. But IBM can choose to allocate the bits in the various fields in special ways and we must be able to accommodate them. Also, IBM can utilize these bits in one way for a 3270 terminal system and still another way for the 3600-type terminal networks."

But even the skeptics would be surprised if SDLC should be yanked from under them in the near future, says Harold Buchanan, data communications product manager of Interdata. Buchanan says, "I think IBM has blessed SDLC as their format for years to come." He points out that it took SDLC seven years to reach the marketplace after the first proposal was written. IBM certainly will not scuttle this investment, says an IBMer close to SDLC.

Meanwhile, what of the large installed base of BSC and other half-duplex systems? The answer seems to be that they will be around for many years. But Donald J. Alusic, marketing manager for DEC's Deccomm line believes that completely new systems will use the new protocols even if they need only half-duplex lines. The reason that designers of half-duplex systems may opt for underutilized, full-duplex protocols is that the new protocols will support faster terminals and computer-to-computer communications. ∎

Satellite-technology readied for full-scale data use

Today, some 10 years after launch of the world's first commercial communications satellite (Intelsat I), the prospect of data communications via satellite at commercially viable cost has arrived. All indicators point to satellite transmission of data as being quite commonplace in the next five years.

The pace at which satellite communications actually impacts on the data communications field depends on two intimately related factors. One is the speed with which the satellite common carriers invest in the electronics equipment to provide the kinds of services—"switched" networks of dedicated earth stations, for example—that they know how to build from a technological viewpoint. The other factor is the speed with which users are willing to submit certain networks to the distressing situation of long propagation delay—which, if not properly taken into account, can degrade network throughput.

Two ways. Eugene R. Cacciamani of American Satellite explains the basic alternatives to accessing a satellite. One is through interconnection via dedicated terrestrial links between the customer site and the carrier's earth station. In this approach, says Cacciamani, a satellite could be a lower-cost substitute for the all-terrestrial private, voice-grade line, thus permitting data traffic at up to 9,600 bits per second. Here, the satellite's ultrawideband transponder channel will be divided into, say, 1,200 one-way voice channels, in much the same way that wideband coaxial-cable and microwave links accommodate many voice-grade channel customers.

The alternative is through the dedication of a substantial bandwidth for just one user—thus permitting data rates as high as 1.544 megabits per second, if not higher—with the user owning small earth stations located on his own property. This direct-access approach can be used for fast transmission of voluminous data, as in facsimile-image, or computer-to-computer applications.

The user must examine particular needs to select the appropriate method of accessing the communications satellite. However, the obligations of the user and those of the carrier are somewhat different in each alternative, for the most part stemming from the problem of propagation delay and its impact on link efficiency (throughput).

The problem right now is that most terrestrial voice-grade networks use a positive-acknowledgment, stop-and-wait protocol—binary synchronous communications (BSC) being the most popular example—to assure that no blocks of data that are in error are accepted by the receiving station. Each block is transmitted, then the transmitter stops and waits for an acknowledgment from the receiver as to whether the block has been received correctly. The transmission resumes for another block. With an 800 millisecond round-trip propagation delay, and depending on the error rate on the link, the actual usable data getting through can drop to a small percentage of the nominal bit rate.

The satellite carrier has ways of minimizing the

SATELLITES

error rate of its link. One is to employ the highest practical transmitting power and the largest practical antennas at the earth station. Redundant coding of the data going over the link, through a technique known as forward error correction, is another way of reducing the error rate. (While at first it may appear that redundant coding actually reduces net data throughput, there is a positive tradeoff in power, antenna size, and available bandwidth, so that coding actually benefits the situation, says Cacciamani.)

Negotiation. When a user opts for the direct access to a satellite channel through his own "rooftop" antennas and earth stations, he will be leasing all facilities from the satellite carrier, or a channel from someone else. In either case, the user is obtaining "plant" from his vendors. What he wants to accomplish will be a matter of system specifications and negotiation with the relevant vendors. The matter of error control is an illustration.

For some applications, such as facsimile transmission of newspaper pages, no error control may be needed because an average error rate of one bit for every 100 million bits transmitted, which satellite carriers are quoting, might be of little consequence to how the finished page looks. But when error control is needed, a systems solution becomes mandatory.

However, when a user accesses a satellite via a terrestrial link, which most users will do in the near term, the problem of error control falls into his lap, because only the user can do anything about end-to-end error control. For one thing, the user, not the carrier, determines the line protocol. And, while errors originate in the links supplied by the various interconnected carriers in the link, the control of errors occurs at one or more places in the user's network.

In particular, interconnecting to a satellite means the use of short local loops to the in-town interconnection point of a terrestrial carrier, who then has a microwave link from his in-town office to the "far-out" site of the satellite carrier's earth station. In such a situation, the average error rate that customarily occurs on the short local loops will be about one bit in error for every hundred thousand bits transmitted (10^{-5}).

As mentioned, for such a situation the conventional means of error control, through a positive acknowledgment protocol, will become quite inefficient when burdened with both the 10^{-5} error environment of the local loops and the propagation delay of the satellite link.

The newer data link controls, or line protocols, are likely to resolve this problem. A line protocol does not require a halt to forward transmission while it waits for a block-acknowledgment signal from a receiving station. IBM's synchronous data link control (SDLC) and Digital Equipment Corp.'s digital data communications message protocol (DDCMP) are two such line protocols. But as far as is known, they have not yet been tried on satellite data links.

Another alternative for improving throughput efficiency on satellite links is a modification of the popular request-for-repeat protocol. In general, this approach calls for the continuous transmission of blocks and for an acknowledgment to be returned to the sending station only when a block has been received in error.

One continuous version, which has been converted to hardware by Codex under a contract from Comsat, is called the selective repeat method. Here, the transmitter sends blocks of data (with identification numbers) continuously to the receiver. As long as the receiver receives "good" blocks, the transmission continues without interruption.

Suppose the transmitter sends blocks 0, 1, 2, and 3 in good condition, but block 4 is in error. Then the receiver returns an error signal to the transmitter, and during the propagation interval of the returning error signal the transmitter continues to send blocks 5, 6, and 7. The instant the transmitter is advised that block 4 was in error, it retrieves block 4 from its buffer and retransmits it to the receiver. In the meanwhile, the receiver has retained blocks 5, 6, and 7 in its buffer. When block 4 arrives (and is good) it is placed ahead of blocks 5, 6, and 7. And continuous transmission resumes.

Bench tests at Codex on a simulated satellite link have demonstrated that transmission efficiency remains fairly high, despite high error rate, even under conditions that would have driven the customary stop-and-wait method to zero throughput.

Present plans include testing the Codex selective-repeat hardware on an actual satellite link, probably in conjunction with American Satellite Corp. The equipment, which is now being test-marketed, is fairly expensive, perhaps about $10,000 at each end of the link. What happens next depends on

Direct satellite access will require user to lease facilities directly or indirectly from carrier.

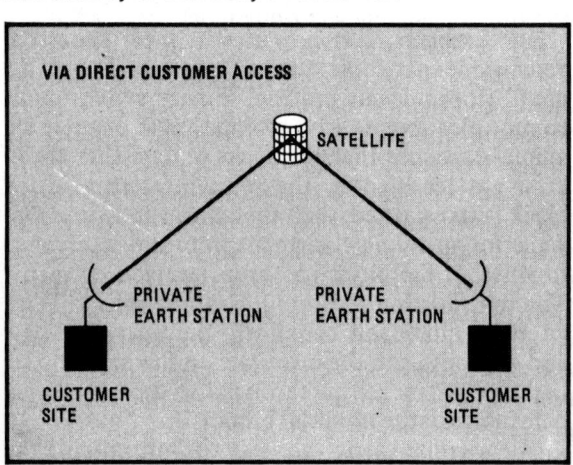

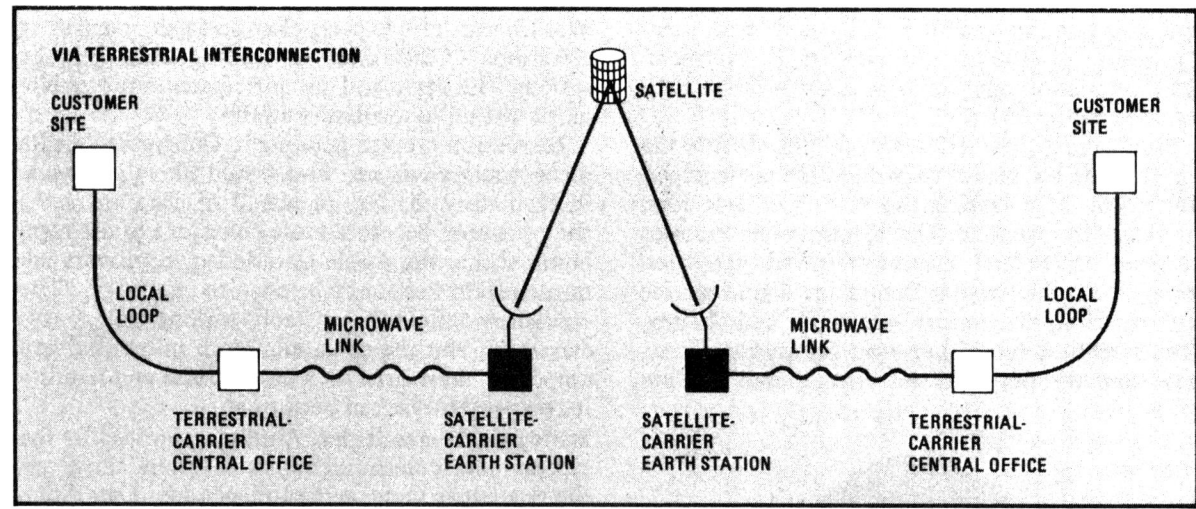

Terrestrial interconnections to satellites will make the user responsible for error control.

whether new protocols will obviate the need for such hardware, whether the user would be willing to pay the price for hardware, and whether the satellite carriers themselves would install the selective-repeat hardware as part of their plant and provide its functions as part of a private line service.

Dedicated stations. In an area of major significance, small earth stations are feasible now and will go down in cost as their usage increases. Estimates on the cost of a 15-foot-diameter-antenna earth station range from $60,000 to $150,000. The reason for the difference is the varying sophistication of the communications electronics.

A high dispersion of private earth stations holds out the prospect for "broadcast" services—for example, to simultaneously update widely scattered data bases or to down-load programs.

Jerome G. Lucas, of Comsat Laboratories, says that, "For a small number of earth stations the monthly cost of service is determined primarily by the amortized cost of the satellite and its launching expense. [Satellites are good for about seven years.] As the number of earth stations increases the monthly cost approaches that of the earth station itself." He predicts that by 1980 not only will there be integrated data, digitized voice, and facsimile services but also innovative tariffs for weekend and night high-speed data, and charges by the minute or by burst of data.

There's also a strong prospect for a "switched data network in the sky." As more and more transmit/receive earth stations are installed on customer premises, then demand will grow for a satellite carrier to provide a way for any such earth station to connect with any other station. The key word here is "connectivity." The more connectivity, the greater the access capability and the better will satellites serve the nation. One way connectivity will come about is through a technique known as time-division multiple access (TDMA), in which two stations will be connected through an assigned time window in the wideband channel. In this way, one satellite data system will be readily available to a multitude of users.

What will probably *not* happen, at least in the near future, will be the ability of a user to make a data call via direct distance dialling (DDD) and a satellite hop for the long-distance portion of the call. Robert L. Wagner, executive director of the data communications division of Bell Laboratories, says, "Private-line data communications over satellite is controllable because you know where you're at, but a DDD/satellite channel is another matter."

When AT&T's satellite becomes operational, the network itself will determine the routing of any given call—whether it will go totally over terrestrial links or partially over satellite. This arrangement will be all right for voice, but, again, the propagation delay could raise havoc with a data call that needs error control. While it is technically feasible to alert the network as to whether a particular call should be allowed on the satellite, there would be a cost for disallowing access and that extra cost might not be palatable.

Wagner says the common carrier must bear the burden on DDD service. "Obviously, then, the thing to do is keep DDD data calls on the ground. It's the most desirable thing for the user so he doesn't suffer lack of throughput." ■

Beleaguered FCC poises for progress

The present Federal Communications Commission has been accused of "being the weakest yet," of "being unwilling to do its homework" when faced with complex issues, and remiss in its own organizational management, particularly with re-

TERMINALS

gard to attracting qualified staff people and assuring orderly promotions and transfers. Furthermore, the Commission appears to be more reactive than active in facing issues.

True or not, these criticisms do not obscure the fact that the FCC stands as perhaps the most important single force shaping the destiny of data communications practice. The Commission remains involved in several as-yet-unresolved activities, among them the two-tier Dataphone digital service rate situation, the inquiry into resale and sharing, AT&T's request for higher rates for interstate services, and the proposed IBM venture into satellite

FCC's
Richard E. Wiley

communications. These and other issues will affect the kinds of services that communications users will be able to obtain, and the prices charged.

Certain defenses can be raised about the Commission's performance, most of them based on the fact that the FCC has been understaffed and underbudgeted, and that it faces an ever-increasing burden of new and complex issues.

Commission Chairman Richard E. Wiley asserts that the FCC has started to take positive steps to form policy that will meet the country's future needs in telecommunications. Among other things, says Wiley, the Commission has created a new office to manage research into policy and plans; it has given this office $750,000 for outside research into telecommunication's policy, some of which will be disbursed to the academic community.

What kind of research? For one thing, Wiley seeks information on which to base policy about this country's official posture toward international common carriers.

Looking ahead. In general, Wiley claims the FCC is starting to anticipate more and will be able to render decisions faster than before.

Bernard Strassburg, retired chief of the FCC's Common Carrier bureau and now a consultant in Washington, also expects changes in the regulatory situations. "Consumers are finding fault with the system," he says, and he anticipates more public participation in regulatory affairs.

Legislation is again pending in Congress to create a consumer advocate, who would likely intervene in regulatory matters in behalf of consumers. As the consumer becomes more conscious of the regulatory scene, the whole question of regulatory administration itself may be open to challenge. "The regulatory official is veritably untouchable," says Strassburg. But the consumer, both individual and corporate, may insist on a briefer term and liability for removal for lack of performance.

Independents fight. Almost unnoticed by the typical data communications user but active on the regulatory scene is the Independent Data Communications Manufacturers Association (IDCMA). According to IDCMA counsel Herbert E. Marks, "IDCMA takes positions before government agencies to assure policies that preserve, create, and encourage effective competition in the data communications marketplace."

These activities are, of course, basically in behalf of IDCMA's nine company members whose archrival is AT&T and its operating companies. "We are convinced," says Marks, "that competition means better and cheaper equipment and services, and if the public is convinced that competition is good, then IDCMA is also serving the public interest through its regulatory activities."

Among the issues IDCMA is currently fighting before the FCC is the disparity in rates and connection requirements between intrastate and interstate lines supplied to "other common carriers" interconnecting to Bell lines; the modem rate case, in which IDCMA believes that Bell's modem prices should be based on fully allocated costs; and Bell's "no-mix" rule on the use of "independent maker" modems on Dataphone digital service (DDS) links. ∎

Terminals to steal more host functions

Technology's latest nostrum for low-cost problem solving, the microprocessor, promises to make the cost of adding intelligence to terminals so low that terminals sold in the future will almost always be intelligent.

But don't expect price reductions for intelligent terminals very soon. To cover the heavy software development in designing microprocessors into systems, manufacturers look to hold the line on prices and try instead to provide more computing power for the dollar.

Availability of this intelligence will result in terminals taking on more data processing and communications functions. Simple data analysis, editing, formating, sorting, encryption, and error control, all may be built into the terminal of the near future.

"I see both batch and on-line editing functions being pulled out of the host computer," says Donald J. Alusic, marketing manager for the Deccomm product line of Digital Equipment Corp. "Minis or intelligent terminals at local sites will perform data preparation, entry, and validation. The host in turn will be optimized more towards handling block messages rather than being a character-at-a-time 'plunker.'" As a result, terminal makers will be selling more complete processing systems for local computing sites.

Competition. Not all terminal makers see themselves in this role. But those who are spreading out in this way may be embarking on a collision course with the minicomputer makers who have also opted to take the "total solution" approach to selling combined data processing and communications functions. With the minicomputer mainframe being threatened by the rapidly escalating capabilities of microcomputers, and with terminal makers able to add processing power at low cost, the total-solution market may soon be up for grabs.

The microprocessor also will add flexibility to terminals—a benefit to manufacturers and users alike. A common complaint among users has concerned the necessity to adjust their operations to fit standard terminal functions.

"Although there is a great degree of product standardization in this market, there is an awful lot of variability in what the user wants," notes Ron Moyer, president of Digi-Log Systems Inc. But the cost of customizing keyboards, software, interfaces, and display formats often has been a stumbling block. Before long, however, it may not be necessary to buy a standard terminal box like a model T Ford. Instead, the user will configure his terminal from a catalog of standard functions, much the way today's car buyer orders the body, decor, engine, transmission, and accessories.

Each optional function will be provided by an LSI chip or circuit board, or by a microprocessor with a read-only memory specially programed for the function. Economy will often dictate that several functions be combined in one of these types of hardware. But the variety of mix-and-match combinations will be much greater. "We see ourselves as doing more of the programing work, making the user a true user," says Richard Gorton, director of marketing support for Sanders Data Systems.

Moyer asserts that the true potential of the microprocessor is only beginning to unfold. "We are just now scratching the surface on how intelligence can be used in a terminal. The obvious way to start is to replace hardwired functions. But after that who knows?"

Whither CRTs? Recently-introduced large plasma displays already are cost competitive with graphic CRTs, but whether their prices will eventually be competitive with alphanumeric tubes is an open question, according to terminal builders. Some believe it is only a matter of time.

The advantages of plasma displays are flicker-free operation, selective erasure, translucent screens permitting low-cost rear-projected overlays, and compact size. The latter attribute makes plasma devices especially attractive for use with portable terminals.

Keyboards will have increasingly customized key functions and layouts. Programable look-up tables for customizing key functions are on the market. ∎

Communications processors to take many shapes

Today's building blocks of communications processors—minicomputers—may not be the sole basis for these units in the future. One reason is that the computing resources of tomorrow's processors will vary from a microprocessor embedded in other data communications hardware to a sizeable computer that also functions as a local host processor in a distributed network. Another reason is that conversion of standard, repetitive communications software into hardwired LSI chips and programed microprocessors will significantly change the processor's architecture.

Computer manufacturers who offer large software libraries also look forward to the next generation of communications processors in the hope that they may not have to support these functions in their general-purpose product lines, thereby removing a major component of software costs.

DEC's Nathan Teichholtz explains that, "Trying to provide network and communications support in a variety of operating systems for each computer family is becoming unwieldy. We have about 10 major operating systems for PDP-8s, 11s, and Decsystem 10s. Trying to provide comprehensive network and communications for each one of them is very expensive."

He adds, "We are looking at new kinds of com-

MULTIPLEXERS

munications processor architecture, which might be microprocessor-based. Then, instead of supporting N different types of terminals with M different types of processor operating systems, requiring N times M combinations, we could use a single communications-processor operating system and only worry about supporting N-type terminals. Another benefit is that the host will have to contend only with a single-type communications interface."

The evolution of the communications processor will be somewhat like that of the UART (universal asynchronous receiver-transmitter) over the past few years.

"Years ago," says Teichholtz, "we used a whole PDP-8 just to monitor a serial line and do character assembly and disassembly. Today instead of being a huge program that nobody can understand, the UART is a single chip.

"We're gradually moving all the things that used to be communications programs into hardware. And the user will do little or no programing of standard communications functions. Instead he will be concerned more with interfaces to the user program, resource sharing, and accessing files."

Harold Buchanan of Interdata takes a somewhat different view, at least for the near future. Buchanan does not foresee microprocessors having enough computing power to replace minis. Instead microprocessors will operate in conjunction with communications processors as they become increasingly burdened with data-processing tasks. This trend will parallel the early development of data communications in which processors were in the form of separate boxes to relieve the main computer of the communications load. ■

'Muxing' faces bigger role in data transmission

Time-division multiplexing will be another beneficiary of the proliferation of new digital transmission networks from AT&T and Datran. And more TDMs will contain "intelligence"—either as a minicomputer or microcomputer—to provide error control and dynamic channel allocation to make the most efficient use of available bandwith of the long-haul line. By 1980 it's estimated that 40% of the TDMs sold will feature dynamic channel allocation. Also by 1980, TDMs will be smaller in terms of the number (up to 10) of asynchronous input

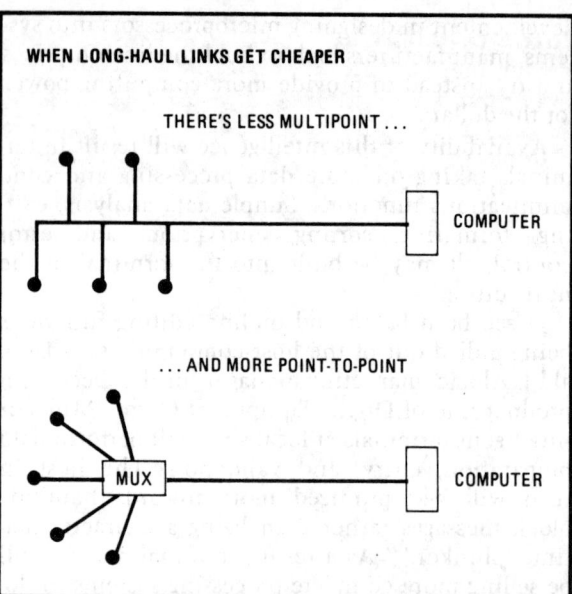

Multiplexers will obsolete many multipoint systems because of the economies of single, wideband lines.

channels they can support. Furthermore, multipoint-circuit architecture will gradually fall into disfavor and be replaced by point-to-point links.

Time-division multiplexing takes several low-speed signals and merges them into one higher speed signal for long-haul transmission, thereby taking advantage of the lower cost-per-bit of high-speed lines. AT&T's new Dataphone digital service (DDS), with its substantially lower cost for long-haul lines than present voice-grade circuits, will make multiplexing more economical and therefore more extensive.

M. Lloyd Bond, vice president for marketing of Timeplex, described the economies in this manner: To multiplex at 9,600 b/s on a voice-grade line costs about $5,000 for the multiplexer and $20,000 for two modems. On DDS, the same speed can be obtained with the same $5,000 for the multiplexer, but about $1,500 for the DDS digital service units that play the same role as modems.

A. Brooks Carll, marketing manager of Infotron Systems, says the TDMs built for DDS will be much less expensive because DDS operates with universal, synchronous clocking. In TDMs built for analog lines, a considerable amount of expensive circuitry is devoted to adaptive digital logic to accommodate speed variations on both sides of the TDM and to perform digital sampling fast enough to distinguish data from spurious noise. With the tighter clocking tolerances available on DDS, these circuits may be greatly simplified.

With these TDM cost reductions and the conversion from asynchronous to synchronous channels, the end-to-end cost of operating a typical data communications system will be cut almost in half. Further, the real cost of the TDM itself may go down by 50% in the next five years.

Point-to-point. Because DDS will make multiplexing more attractive, there will be a tendency away from multipoint links and toward small clusters of terminals, each terminal feeding into the multiplexer on a point-to-point basis. The clustered point-to-point arrangement eliminates the polling delays and overhead addressing characters that go with multipoint links. Also, because of the benefits of clustering, there will be more terminal-cluster nodes in a network, with each TDM handling fewer input lines.

The two common ways to perform line concentration now are the conventional time-division multiplexer and the remote data concentrator. Then there's the "smart mux." Charles P. Johnson of General DataComm Industries says flatly that, "Smart muxing is better than dumb muxing, and any [vendor] company not moving into smart muxing is foolish."

Some vendors, such as Digital Communications Associates and Timeplex, favor an error-control mechanism in the smart mux, an important function not available in conventional multiplexers. However, Johnson feels the improved error rates available on digital transmission systems, or on voice-grade links with suitable error-resistant modems, will make the question of error control in TDMs an academic issue. A better approach, he believes, is to put error control at the ends of the links—just in those terminals and for those applications that mandate error-free data. ∎

Speed not everything in future modems

Notwithstanding the predicted upswing in all-digital transmission systems from AT&T and Datran, analog links will not disappear, nor will the voice-grade modems associated with them. In fact, more voice-grade modems will be sold in 1980 (374,000 units, total) than in 1975 (280,000 units). But different modems will be affected in different ways. Sales of modems operating at 2,000, 3,600, and 7,200 b/s will dwindle to a trickle as terminal and multiplexer manufacturers stop producing equipment requiring these bit rates.

Meanwhile, long-held opinions that all users will opt for higher-speed equipment as they become cost effective are contradicted by the finding that more modems operating at less than 1,200 b/s will be sold in 1980 than any other speed. This supports the predictions that more "little users" will start to benefit from data communications through the use of low-speed administrative-traffic terminals, and that the sales of inquiry/response terminals will remain high through 1980. Here, the point is that some modems don't really need to be any faster than a person's typing speed.

The modems expected to enjoy the healthiest growth in sales are those operating at 1,800, 2,400, and 4,800 b/s. The upswing here will occur for two reasons. One is the increase in the number of terminals requiring these speeds. The other is that reduced costs of long-haul links will encourage the use of more multiplexers serving smaller geographical areas and fewer terminals, which can therefore operate at lower trunk speeds.

Sales of 9,600-b/s modems, the fastest now available for operation on a private voice-grade line, will remain fairly steady through the years to 1980.

Sang Youn Whang, senior vice president and technical director of the International Communications Corp., agrees that the 9,600-b/s modem will continue to be popular despite the spread of AT&T's Dataphone digital service (DDS) and Datran's digital network. "Otherwise," he says, "why would AT&T invest in the development and production of its type 209 [9,600-b/s] modem, knowing that DDS would provide services that, in effect, would compete with 9,600-b/s analog links? Why would they introduce D1-conditioning on private voice-grade lines for these modems if they expected DDS to replace such analog links?" Whang, in fact, expects DDS to open up the market for modems, including 9,600-b/s and lesser speeds, for linking multiplexers and terminals by way of analog links to DDS hub and end offices.

Not compatible. Whang believes that no company will venture into the development of a 14,400-b/s modem for use on private voice-grade lines. Technically, says Whang, such a modem could be developed, but it would be electronically complex and expensive, and probably would not perform too well, "considering the quality of the phone lines—even with D1 conditioning."

Furthermore, a 14,400-b/s speed is not compatible with the digital-channel offerings on DDS, and so there would be no market for these modems on DDS off-net analog links to distant terminals.

Robert L. Wagner, executive director of the Bell Labs' data communications division, agrees on the fate of 14,400 b/s modems. He says the cost of going to a higher-speed modem becomes prohibitive. "At 9,600 bits per second, the law of diminishing returns is operating on voice-grade channels," he says. Furthermore, such a modem would require a super-powerful automatic equalizer whose retraining time would be excessive and adversely affect the link's error performance, particularly if the link were carrying multiplexed data. In the latter case, each subchannel would suffer errors.

All this is not to say that modem developments will lag during the next five years. Far from it. On the one hand, users can expect modem vendors to exploit the large-scale digital integrated circuit in the form of the microprocessor, and, on the other hand, to apply to lower-speed modems those modulation/demodulation schemes normally associated

MODEMS

with higher-speed units. Digital rather than analog circuits will also provide better noise immunity, smaller size, and no field adjustments.

Bill Myers, marketing vice president of Penril Data Communications, anticipates more functions to become available in modems without an increase in price, because of the microprocessor. For one thing, a programed microprocessor will provide continuous fault monitoring and isolation—and automatic warning—in a troubled modem. For another thing, the microprocessor will also be used to control the modem's modulation and demodulation and to perform automatic equalization.

Such a design approach will make it easier and cheaper for a vendor to develop one basic modem package and then add a specifically programed read-only memory to accomplish the unique function required by a customer.

ICC's
Sang Youn Whang

The president of General DataComm Industries, Charles P. Johnson, has examined the technical and economic aspects of AT&T's Dataphone digital service to determine marketing opportunities for "independent" (non-Bell) companies such as his. To Johnson, DDS signals the beginning of real change in the way data communications will be carried out. DDS means low-cost intercity communications, he says, and this situation will reduce the cost of end-to-end transmissions, increase demand for data communications, and spur innovation. Furthermore, as digital transmission starts to dominate, then digitized-voice communications will piggyback on the digital links.

Can compete. Depending on speed of service, DDS will serve customers within a distance of up to eight miles from a hub or end office. Analog links will be required outside this "circle of service". But, says Johnson, speeds of 2,400 bits per second or less will remain analog because of the $1,400 per-link investment (independent of speed) required by the phone company for its DDS signaling units. This means that a full-duplex 2,400-b/s modem that sells for $680 or less would allow analog links to challenge DDS on economic grounds.

Johnson appears confident that electronics technology and improved modulation schemes will permit such a modem to be developed in the next few years. And he foresees a 300-b/s modem operating in a full-duplex mode on a single pair of copper wires that will sell for about $200.

On the other hand, DDS operates full duplex on four-wire circuits. Johnson says the telephone company spends $150 a mile to install one pair of copper wires. In all, he says the telephone companies will have to invest about $15 billion for copper circuits to service DDS customers—and copper is scarce. That's why he believes that modems will have to be designed to operate full duplex on two wires. In this light, he fully expects that 9,600-b/s full-duplex modems operating on a pair of wires will be available in five years for about $1,000 each.

If all this comes about, then full-duplex, 9,600-b/s transmission on the dial-up network with one phone connection becomes feasible.

Implicit in Johnson's cost predictions for future modems is that these modems will serve on relatively short hauls, with intercity DDS links providing the long-haul requirements. The governing principle of short distances regarding modems is that the shorter the line, the less the impact of line impairments and the fewer causes for data errors. Therefore on a dollar-for-dollar basis, instead of the modem's sophisticated electronic circuits having to "fight" the effects of excessive line impairments, the electronics can be used to perform more advanced filtering and modulation/demodulation.

Most modems makers agree that the emphasis will be on making modems more functional and less costly rather than on increasing bit rates. Ray Mazurek, director of research for Tele-dynamics, ventures the following predictions on modem prices: In the next two years, the cost of 4,800-b/s modems will be halved, but it will take a little longer before the 9,600-b/s modem price is cut in half. Medium-speed modems—1,200- and 2,400-b/s—will drop about 10% a year for the next few years, and low-speed modems will probably drop in price only slightly.

Acknowledging the presence of 56 kilobit-per second speed on digital transmission networks and the development of time-division multiplexers having trunk speeds of 56 kb/s, Maruzek predicts that suitable modems for this service will go for about $10,000 to $15,000 each, or about what a 9,600-b/s modem costs now. ∎

part 2
data data data data data data data data data

terminals

The broad array of data terminals for communications

From the supermarket of data terminals, the user selects those programable and nonprogramable functions that can best match specific requirements of different types of business applications

Roy M. Salzman
Arthur D. Little Inc.
Cambridge, Mass.

Because the remote data terminal is the only device in a data-communications system that interfaces directly with business activity, it must be subject to several levels of critical evaluation. Management must be certain that terminals satisfy functional requirements within a prescribed cost. Operators must find them acceptable in the work environment. And system planners must be sure the terminals are technically compatible with the rest of the system.

Fortunately, the user today can select from many different kinds of terminals. They range from simple, low-cost keyboard/printers—basically Teletype machines and their emulators—to rather sophisticated, high-speed communications terminals, which can also serve as stand-alone data processors. This variety is due to recent rapid advances in terminal design, the most significant of which is the addition of programable electronics to the terminal.

Programability makes a general-purpose terminal suitable for tailoring to many different business applications. It enables the terminal to take over certain data manipulations, and so both reduce the amount of data sent over the communications lines and relieve the central computer of some jobs unrelated to its basic work of data processing. That is, when needed in a given situation, programability enhances terminal performance in the business environment and alleviates substantial communications and computer losses.

Business requirements dictate the nature and form of the information to be sent to the computer, the urgency with which this information must be processed, and the nature and form of the processed information delivered back to the sending site. Therefore, the first thing to consider when selecting a remote data terminal is the input/output data requirement of the business application. This will take any of four forms:
- Enter transaction-oriented data.
- Enter batch data.
- Output batch data.
- Retrieve—or ask for—information from a data base.

Input/output classifications

In transaction-oriented data entry, data is keyed in at the place where the transaction occurs, and goes at once to the host computer to update the data base. For example, an operator at a remote branch warehouse might key in data describing a shipment and so update the central inventory-control data base immediately. (Sometimes, though, data from several transactions may be briefly stored in the computer mass memory and then processed in small batches.) Terminals suited for transaction-oriented inputs resemble those used in "interactive" or "conversational" applications, in which the operator engages in a dialog with the computer or with an operator at another terminal.

Being dependent on the operator's peak keying rate, idle time, and other performance factors, data enters the transaction terminal relatively slowly, and cannot leave it any faster. Consequently, transaction-oriented functions employ terminals that can be classified as having slow input and slow output.

In batch-data entry, a large quantity of data is gradually accumulated on magnetic tape or some other machine-readable medium, and later sent to the host computer as one long message. The accumulated data, usually in fixed record lengths, could represent eight hours of local business operations, while the total batch may take perhaps 10 minutes to transmit. This type of terminal may therefore be classified as slow input/fast output.

In batch-data output the host computer quickly transmits a quantity of data to the local site, where it is recorded as fast as possible on punched cards, magnetic tape, disk packs, or—most commonly—paper. A high-speed line-printer terminal, for example, could print out invoices generated by the computer and based on batch data that had been entered earlier. For this type of application, the terminal would be classified as fast input/fast output.

Data-base inquiry is simply the retrieval of a fairly small amount of information from a centrally situated data base for display at the local site. The classic example is an airlines reservation system. Here, the terminal has a keyboard for requesting specific flight data and a cathode ray tube (CRT) screen for displaying the answers. Often data-base inquiry requires a hard copy—for example, a completely filled-in airline ticket, which is produced by a teleprinter. The inquiry function has much in common with the transaction function.

To satisfy one or more of these basic data input/output requirements, manufacturers have developed several classes of terminals. Besides the keyboard/printers (or teletypewriters), alphanumeric CRT displays, and remote batch terminals already mentioned, they include such special units as point-of-sale and bank-teller terminals.

More distinctions

Terminals with a slow, manual input generally fall into the asynchronous class. They have a keyboard, much like that of a conventional typewriter, and generate a coded character each time the operator strikes a key. Therefore their output is normally asynchronous, in that it depends not on a fixed time base, but on the unrhythmical performance of the operator. Asynchronous terminals frame each character with a start bit and a stop bit to enable the receiving terminal (often a computer) to know when another coded character has been transmitted and is to be detected.

Machine-input terminals, and manual-input batch terminals that accumulate slow input data, generally deliver output data at a fast rate to the line, and use synchronous transmission. In synchronous transmission, long data blocks made up of a string of character codes, are transmitted, with each block—rather than each character—framed by a special synchronizing code.

Finally, remote data terminals can also be categorized as senders, receivers, and sender/receivers. Each has a controller and a buffer memory. The sender has a data input mechanism, most commonly a keyboard, while the receiver has a data output mechanism, most commonly either a character or line printer or a CRT display. The printer, of course, supplies hard, or permanent, copy, while the CRT displays the message

1. Powerhouse. Data 100's model 78 programed terminal for high-volume data-communications applications contains a stored-program processor, which enables the terminal to meet a wide variety of business requirements and support its own peripherals.

33

temporarily. The sender/receiver consolidates the send/receive functions into one physical device as, for example, in a keyboard/printer terminal.

The buffer found in an electronic terminal is a memory that can store enough bits to represent at least one character. Larger buffers store a word, a line, or even a whole message. With the development of integrated-circuit memories and their high-volume production, buffering cost has come down, so terminal makers can afford to install more internal buffering in their terminals. External buffers, too, like a disk pack, may be used to enhance a terminal's versatility.

Basic tasks

The controller is a logic device—it directs all the tasks that the terminal must perform to convert, say, a keystroke into a sequence of bits for transmission on the communications link. Depending on the particular terminal, the controller may be hard-wired and thus have been fixed in its functions when the terminal was designed, or it may be programable so that its functions and tasks can be suited to a particular application.

Whether hard-wired or programable, the terminal's controller must perform several basic operations. One of its simpler tasks is to handle the few signals that govern the input or output device. Here, the controller sends or receives signals to start up or shut down such devices as keyboards, card readers, CRT screens, character and line printers, magnetic-tape drives, magnetic cassettes, and punched-paper-tape drives. A control signal also initiates translation of data from human-readable form to machine-readable form, and vice versa.

As for buffers, the controller in a sender fills this memory with bits from the input device and then empties it bit by bit in proper sequence onto a communications line. The reverse occurs, of course, when the terminal is a receiver. Buffer control may be quite complex, as when the controller simultaneously manages separate buffers for send and receive functions in the same terminal, or when the terminal employs double buffering to improve terminal speed. (In double buffering, one buffer is being filled while the other is being emptied; then the buffers reverse roles, and so on.) Control of simultaneous and double buffers, while complex, is a relatively fixed function and well within the capability of hard-wired logic.

A third task is code translation. The controller converts a character from the code form in which it was sensed by the input device into another code form, one suited to transmission and computation, or vice versa. In a sender, for instance, it might translate Hollerith (punched card) code into ASCII code (American Standard Code for Information Interchange), one of several standard communications codes. Most modern teleprinters and CRT displays contain internal logic that directly accepts ASCII-coded information sent by the computer and translates (or generates) the coded information into characters.

In the generation of error-detection codes, the controller adds a 0 or 1 checking bit, called a parity bit, to a coded character before it is sent to the receiving terminal. The controller there scans the received character, including the parity bit, and its logic tells whether an error has occurred during transmission. (Error control is the subject of another article in this book.)

The character (or block, or message, depending on the particular requirement) must be stored in the sender's buffer until it is told by the receiving terminal the character has been received correctly. Otherwise, the message is repeated until it has been correctly received. Because the buffer cannot be filled with a new bit sequence until it is emptied, and time elapses while the receiving terminal evaluates the correctness of the received character, terminals often increase their data rate by using the double buffering mentioned earlier.

Automatic answering and transmission enable a sending terminal to be used with no operator present. With its aid, data can easily be telephoned in by an unattended remote batch terminal in response to a request made during the evening, when low-cost night-rate telephone tariffs apply. The automatic answering equipment is not a part of every terminal's controller, but can usually be obtained as an option.

The host computer's overhead

However much is done by the controller of a terminal, the host computer is still left with the job of handling terminal signals. This is overhead as far as the computer is concerned, since any time the computer spends other than in processing data reduces its efficiency. As will be discussed in the articles on remote data concentration and programable front-end processing, equipment intervening between the terminal and the host computer can alleviate the host computer's data-communications losses due to overhead. However, regardless of whether the terminal interfaces with a remote data concentrator, a programable front-end processor, or a host computer, one of these devices will have to provide the data-manipulation services required to interpret the hard-wired terminal's output. In the following discussion, it will be assumed the terminal feeds directly into the host computer.

For character-by-character (asynchronous) transmission, the computer must continually observe the lines coming in from the terminals. It can do this by

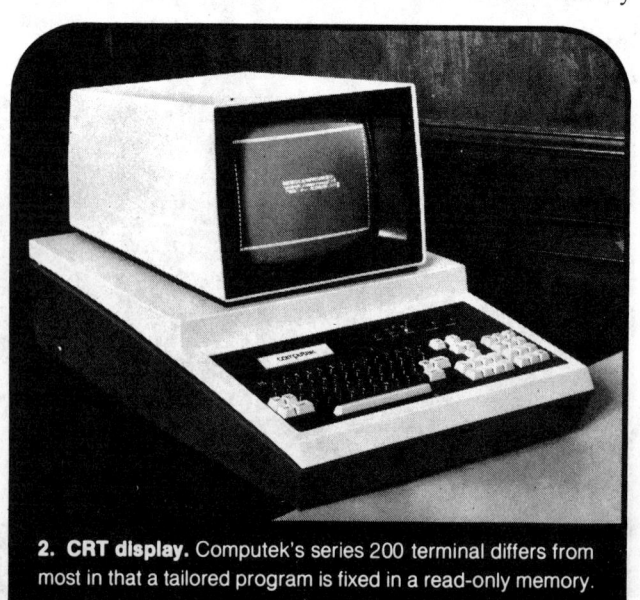

2. CRT display. Computek's series 200 terminal differs from most in that a tailored program is fixed in a read-only memory.

frequent polling—that is, by asking each terminal if it has any messages to send—or by watching for an interrupt signal, which forces the computer to read the interrupting terminal's message.

After connection has been established, the computer must assemble each sequence of incoming bits into a full character, strip off the start and stop bits, translate the data from transmission code into computer-processable code, test the parity bit to see if an error has occurred, and place the character in the correct location in its memory for assembly into words and messages. Furthermore, the computer must determine whether the received character is a message character or a control character. A control character sets up a different level of activity within the computer. For example, it could indicate the end of a message, which allows the computer to disconnect the terminal and go on to other communications or data-processing tasks.

In character-by-character transmission, the computer must perform each of the above activities from six to 10 times each second for each terminal in the network in turn. Comparable activity occurs for outgoing characters, which have to be handled at up to 30 times each second for each terminal. All this is a considerable burden on the computer.

Batch terminals with hard-wired controllers also place a data-manipulation load on the computer. Because such terminals produce blocks of bits and operate in a synchronous mode, they do relieve the computer of some of the burden of individual character detection, validation, and removal of synch bits. Even so, the computer still has much to do to support their operation and communications requirements.

Such batch terminals as punched-card readers and magnetic-tape readers transmit information on a record-by-record basis. For example, one punched card may constitute a record. The computer must read the record, test it for accuracy, and either accept it or ask for a retransmission. For a good record, the computer must then perform necessary code translation, interpret card data for special commands, and position the data correctly in its memory.

Note that the entire record of a punched card, usually 80 characters, must be transmitted by the terminal and processed by the computer, even though the card itself may only contain perhaps 20 or 30 meaningful characters. The balance of the 80 fields on the card may be blank, contain card deck identification, long sequences of zeros, and other extraneous characters. Most hard-wired batch-terminal controllers cannot distinguish meaningful from extraneous information, and thus when the computer sends a record to a card-punch terminal, it must transmit all 80 fields—even though some are meaningless.

Now a large computer is designed for data processing, and not for such data-communications tasks as terminal control, character and block analysis, and message assembly. Electronic programability has been added to terminals for one main reason: to take much of the specialized work away from the general-purpose computer and do it instead at the terminal site. As a side benefit, the programable terminal—sometimes called a "smart" or "intelligent" terminal—also reduces the data load on the communications links.

Benefits of programability

What has made this development economical and therefore feasible is the advance in semiconductor memory technology. It is now easy to build special memories, called read-only memories (ROMs), program them for a fixed application, and insert them into the terminal. The ROM is basically an expanded equivalent of hard-wired logic which adds significantly to the functions the controller can perform.

Some terminals even contain a stored-program computer—a minicomputer with its functions programed by software just as in the large host computer. Such a field-programable terminal, with its own mass memory or other peripherals, then becomes very flexible and powerful. It can serve as a stand-alone data processor in a local environment, besides communicating with the host computer. It also permits the development of distributed data processing/data-communications networks.

The stored-program terminal, therefore, depending on the amount of electronic equipment it contains or on the creativity of a programer in setting it up, can perform a wider or narrower selection of the following functions:

FORMATTING INPUT AND OUTPUT. Many normal business applications require information to be displayed or entered in a highly structured form. Cases in point are invoices or insurance applications, that could best be prepared on preprinted forms, and customer records that could do with a standardized form for display on a CRT screen.

From a careful description of the form, a receiver-terminal controller can be programed to generate the necessary spaces, tab stops, carriage returns, line feeds, and other device control characters to properly position the carriage on a character printer or a cursor on a CRT display. Similarly, with suitably described field formats for a sender-terminal an operator simply enters the necessary information in each field. If the information does not fill a fixed-format field, the operator then keys in a delimiter symbol, and the controller fills in the field automatically with characters that are meaningless in the context—dollar signs, zeros, or spaces. The computer

3. Guidance. IBM's 3735 programable terminal uses data on magnetic disk (corner) to tell operator what to do next.

is programed to ignore field-filling characters, but as a check will count all characters to make sure the field is filled to its prescribed length—no more and no less.

DATA COMPRESSION. Instead of transmitting entire field-filling and other extraneous character sequences, the controller in a programable terminal may simply send a two- or three-character code to indicate how many filler characters were removed from the data stream. The receiver then skips that many characters. Or the terminal can send a short code giving the position (or address) in the record at which the next meaningful character or sequence starts. Removing sequences of extraneous zeros or spaces is a simple application of data compression.

More significant compression can take place if the terminal stores several fully expanded (English language) text messages to be displayed or printed on receipt of a short code corresponding to each piece of text. Such texts might be the specific payment terms to be printed on an invoice, an account credit status to be displayed on a CRT terminal, or a full product description to be printed on an inventory status report. Since some of these texts may be quite long—perhaps 50 characters—the ability to store them at the terminal, rather than at the computer, and call them up with a short address code reduces line traffic significantly.

CONTENT-ORIENTED ERROR DETECTION. Though all receiver-terminal controllers check parity bits to determine whether an error has occurred during transmission, parity checking cannot tell whether the correct data was sent. However, programable logic at the terminal, particularly if it is based on a minicomputer, can undertake a rather more thorough check for errors. It relies on the content and context of the data being transmitted. The programable terminal can make sure that received data conforms to prescribed format rules. It will, for example, prevent a field that should contain alphabetical characters from accepting numerical characters. It will count characters to make sure a field is filled and, if the field is not filled, will ask for a retransmission. It will perform quantitative checks, making sure a received number does not fall outside a prescribed minimum-maximum range.

BUFFERING MULTIPLE BATCHES. An effective use of the terminal's buffer storage is the accumulation of character-by-character input data, as from a manually operated keyboard. This accumulation may last for as little as a line or as much as a page of information. The terminal assembles characters into messages, inserts redundancy-check characters where appropriate, interprets control characters, and responds as necessary to the display or printer to let the operator know his input has been correctly received at the computer. Here, each time a line or a page is completed, it is immediately sent out as one long message to the computer. In short, the programable terminal performs those functions that a central computer might otherwise do for character-by-character transmission, but much less time is taken on the communications link and the computer's overhead is substantially reduced.

At a more complex level, one programable terminal can handle batches accumulated from several input devices (keyboards). In this situation, it operates as a gathering place for what amounts to multiple keypunch operations being performed away from the central computer site. The input from each keyboard may be assigned its own field on a disk pack, with each keyboard operator perhaps performing a different processing task. When all work is finished at the remote site, or on demand from the computer, information stored on the disk is transmitted at high speed to the computer.

LOCAL EDITING OF DATA. When the programable terminal contains a large local buffer, a complete line or page can be reviewed by the operator, who then edits the message to correct keying and other errors. When he spots an error in a displayed message, he simply backspaces the cursor to the error position and keys in the correct data. Once satisfied with the message's accuracy, he causes the buffered data to be transmitted. Local editing thus reduces the communications link and computer overhead that would otherwise have been used to handle wrong data.

HANDLING SIMPLE COMPUTATIONS. With programability in the controller, the terminal can perform computations on accumulated data locally, rather than having to send all raw data to the host computer. Typical examples of such computations are price extensions from unit prices and quantities, quantity discounts, and tax amounts. With such local computations, the terminal can then prepare an invoice locally.

APPENDING LOCAL CONSTANT DATA. During a business day, constant data is added to each document or business transaction serviced by the terminal. The controller can be programed to include terminal location and identification, operator identification, date and time, and security code to allow only authorized access to the terminal. Such data is automatically added to the variable data being inputted by the operator. In a sense, appending local data is a form of data compression.

CONTROL SEQUENCE NUMBERING. To avoid confusion within a business operation, it often is important to consecutively number messages, invoices, and other documents. If the terminal is programed to assign and print numbers in sequence and to use a number only when it produces a valid document, no skips will occur in numbering sequences.

AUTOMATIC RESTART AND RECOVERY. From the standpoint of system integrity, one attractive use of a programable terminal is to accumulate a local log, with

4. Getting around. Mobile Teletype machine, using acoustic coupler to the line, serves as input/output to computer.

5. Matrix. Burrough's TD 500 keyboard terminal forms each character by a matrix of dots, and displays 256 characters.

checkpoints along the way, of all transactions passing through it. Then, should the data-communications/ data processing system malfunction, operation can be backed up to the last checkpoint and restarted, and the logged data automatically transmitted up to the point at which the malfunction occurred. This eliminates the need for the operator to find source documents and re-key information, because the terminal's "audit tape" simulates the operator's actions. When recovery transmission has been completed, the operator continues the transmission with new manual inputs.

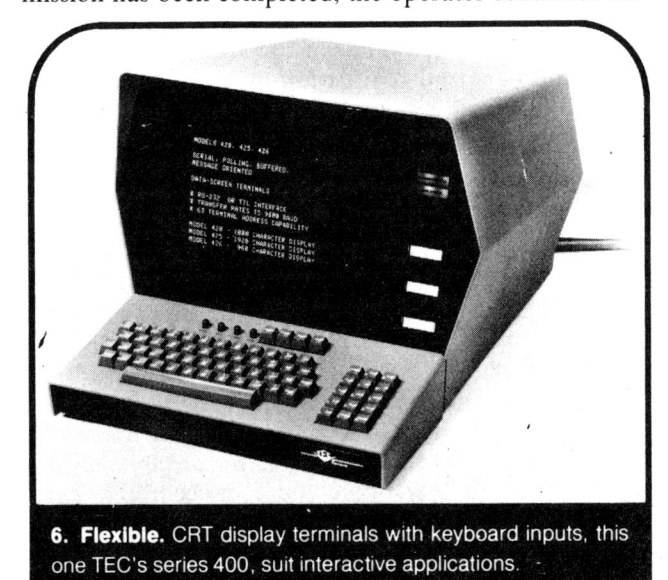

LOCAL OPERATOR GUIDANCE. Some programable terminals are designed to handle a single type of operation, but others, particularly ones based on a minicomputer, can be changed by the operator to handle several different assignments during the day. In either case, the input has to be entered in a structured form. Often, the operator will make mistakes either in keying in good data or in entering the wrong kind of data for a particular field in the prescribed format. To circumvent these situations and thus to improve over-all efficiency, terminals can be cleverly programed to display a specific message on a CRT screen that will tell the operator what to do next. And if he does it wrong, the program can even lock the keyboard, preventing any further input, and flash a message telling the operator what he did wrong and what to do instead.

Comparative popularity

Though programable data-communications terminals are available, the most popular terminals are still the essentially unbuffered Teletype senders and receivers and other makes of terminals that emulate this type of teletypewriter or keyboard/printer. An Arthur D. Little Inc. study estimates that about 60,000 to 70,000 teletypewriters were shipped in 1971 for use as true terminals to a host computer in data-communications systems (that is, not including teletypewriters used for straight message communications in TWX and Telex networks and elsewhere, or those used simply as consoles to minicomputer installations).

Shipments of unbuffered keyboard/printers may continue to rise slightly in 1974, but then taper off. However, the shipment of programed keyboard/printers will increase steadily from about 11,000 in 1972 to upwards of 22,000 a year by 1976.

Remote batch terminals are just beginning to be installed in substantial numbers, about 2,000 in 1971. Their greater power and their considerably higher cost may never make them the most widely used data-communications terminals. However, the survey does estimate that 15,000 to 20,000 of them will be shipped in 1976, making for a total installed base of about 60,000 remote batch terminals—in addition to the many full-fledged data processing systems with communications facilities being used for some terminal functions.

The alphanumeric CRT display terminal for data-communications applications has the greatest growth rate and may well overtake the keyboard/printer within five years. About 30,000 CRT terminals were shipped in 1971, and this will increase to 50,000 to 60,000 by 1976. If these estimates are correct, then over 150,000 new terminals will go into operation in 1976.

Even right now, though, users have a wide range of terminal types, functions, and costs to balance off in making a selection for a particular site or application. For a simple application, getting the terminal on line may be no more complicated than mating a few connectors. For complex applications, programming will probably have to be done by an in-house or a vendor-independent software group. But, whether his requirements are simple or complex, the user will find a terminal available that suits his needs. □

6. Flexible. CRT display terminals with keyboard inputs, this one TEC's series 400, suit interactive applications.

Facsimile terminals send the documents with the data

With 100,000 facsimile terminals already in intracompany and intercompany service via the dial-up network, fax makes data documents available remotely for entry into the computer system

George M. Stamps and Peter S. Philippi
Magnavox Systems Inc.
Fort Wayne, Ind.

Facsimile transmission may fulfill many requirements of a business better than conventional low-speed data terminals. In many instances, fax can get information to the destination faster and at lower cost. Whether permanently connected or acoustically coupled, a fax transmitter can send as many as 550 words on a sheet of paper to a fax receiver anywhere on the dial-up telephone network in as little as three minutes. The receiver faithfully reproduces all images on the transmitted documents, including alphanumerics, graphics, formats, signatures, and handwritten notations. The received hardcopy document is easy to handle, file, and reproduce.

Although a faster terminal that transmits images digitally will soon be introduced, existing fax machines require separate conversion to digital form by an operator who keys information onto a data terminal when the information must be processed by a computer. A company's evaluation of fax transmission should be governed by many factors, chiefly traffic volume and how much data processing is necessary on message content.

Because facsimile machines require little skill to operate, they provide an economical way to transmit source documents that are free from operator errors and with a minimum of line-induced errors from many low-volume outlying stations to one central point. At the central terminal site, the documents can be converted to digital coding by a skilled keyboard operator for processing by a host computer.

Although fax can't compete with high-speed data terminals, it can frequently be more cost-effective and faster than such low-speed terminals as teleprinters. For example, if a branch office must send an eight-page set of specifications and request-for-quotation to a central sales office for review and pricing, the complete set of specs can be in the sales office in less than an hour. And it is easily reproducible. If a teleprinter hookup were used, the job would probably take four hours, and if sent by mail, at least 24 hours. Furthermore, when a user transmits over his leased line or WATS (Wide Area Telecommunication Service), he is not bound by the minimum charge by the common carrier.

Fax may save money

Although fax delivery of a document costs more than mail delivery, the difference in speed may be translated into savings in a business. Possible measurable advantages include additional time to prepare comprehensive plans and answers to requests for quotations and the faster release and utilization of inventory, vehicles, and other assets to profit-making opportunities.

What's more, two companies can rapidly transmit documents by facsimile to each other merely by dialing the telephone network, whereas it is highly unlikely that their data-communications systems—unless coordinated beforehand—would be able to intercommunicate because of differences in techniques, format, and operational structures.

For example, many purchasing departments use fax to communicate rapidly with vendors' sales departments. Since a vendor will receive documents with different formats from its many customers, an experienced operator can interpret the customers' variations to conform to the vendor's requirements.

Thus, the customer may request via fax a price on a ¼-inch-diameter rod, but the vendor's data base only recognizes decimal values, so the operator keys in 0.250. The vendor's computer immediately sends back to the sales department a price-and-delivery quote, including transportation charges, and the sales operator faxes this information back to the customer. This information, in turn, may be keyed into the customer's data base.

Fax adapted to phone set

The use of the telephone network to carry acoustically coupled signals from facsimile machines began in earnest in 1968, about two years after Magnavox developed the first acoustically coupled facsimile machine, the Magnafax 840, for business-office applications. Marketed by Xerox as the Xerox-Magnavox Telecopier, it transmitted all the information on an 8½-by-11-inch sheet in six minutes.

Since that time, several other companies have developed and marketed facsimile machines. And new faster units, with such features as continuous rolls of paper, tab stops to set paper length, and unattended operation, have been made available. Present facsimile terminals deliver a page in six, four, or three minutes. Under de-

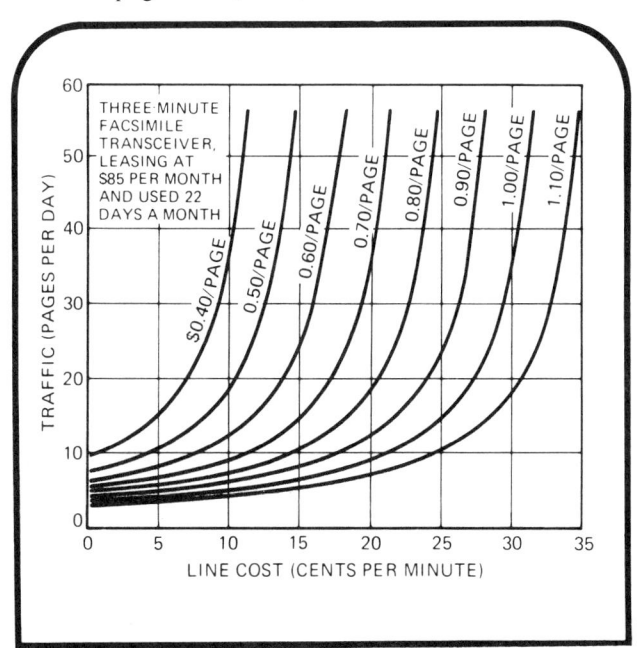

1. Cost factors. The delivery cost of a page of material sent by facsimile is the sum of the machine rental and the line cost, per page, shown here for a three-minute fax transceiver.

velopment and expected to be on the market within a year is a one-minute machine which will lease for about $150 a month. In comparison, the six-minute-a-page transceiver costs about $45 to $50 a month and the three-minute transceiver about $85 to $90 a month.

When analysis of business needs indicates that facsimile may prove useful for electronic mail delivery or for transmitting data for subsequent input to a data-communications/data processing operation, or both, then facsimile costs must be considered. These transmission costs, rated in delivery price per page, are a function of traffic volume in pages per day, the machine rental cost in dollars per month, and the telephone line charges in cents per minute. While operator time may also be included in a cost analysis, usually facsimile-operation time is incidental to a person's main assignment.

The cost profiles for facsimile usage of a three-minute transceiver are shown in Fig. 1. They are based on a rental cost of $85 a month, with a usage of 22 working days a month, and the applicable line costs. Thus, the cost per page is the sum of the rental cost per page and the communications cost per page.

Line cost compared

The line cost shown in Fig. 1 can be the dial-up station-to-station rate applicable between the transmitting and the receiving locations. More likely, though, the fax terminals will be communicating on a private line, or the user will subscribe to telephone company's WATS, a bulk-tariff arrangement.

In general, bulk-usage telephone costs are averaged to cost per minute per call independently of distance. But even if distance is included for each call, it still comes down to cost per minute. Thus, these cost profiles are usable for a station-to-station connection, an "averaged" arrangement, or a time-distance assessment.

As Fig. 1 shows, at a line cost of 20 cents a minute and with an average traffic of 10 pages a day, the delivery cost for fax is $1 per page. Normally, the rental cost and the line charges are assessed against the sender.

Even though the communications cost per page is double that for a three-minute fax machine, a six-minute machine may actually provide a lower cost per page because of its lower lease cost. For each pair of machine speeds, there is a certain volume of break-even traffic, in pages per day, for which the delivery cost per page is the same. Based on 22 days of use a month, the break-even boundaries and resulting zones are shown in Fig. 2 for a six-minute machine leased at $45 a month, a three-minute machine at $85 a month, and the projected one-minute machine at an estimated $150 a month. This set of curves shows that, at 10 cents a minute for line cost, the six-minute machine is more economical when the traffic is five pages a day, the three-minute machine is more economical at 10 pages a day, and the one-minute machine, when available, will be more economical at 20 pages a day.

Technologies compared

The faster the machine, the lower the line cost per page will be. Facsimile design engineers have known this for many years and have kept up a continuing development program to reduce page-transmission time.

One approach is to avoid transmission of white space containing no information and send only the black areas containing information. To accomplish this, the image must be coded.

This technique, known as information (or data) compression, requires sophisticated real-time calculations and manipulation of the facsimile signals to compress them, packing the coded signal for transmission, and extracting the coded signal at the receiver. While engineers have long understood the concepts and logical procedures to effect data compression, they needed a small computer to do it—impractical from a cost viewpoint. The recent availability of low-cost large-scale integrated circuits makes possible the production of a facsimile transceiver that can deliver a page in one minute—and rent for about $150 a month.

The one-minute fax machine will have certain distinctive characteristics that differ from its slower counterparts. For one thing, it will have many of the properties of the conventional data terminal—producing digital, rather than analog, outputs. For another thing, it will not have a gray scale, but print only in black and white. The information images on transmitted documents will be converted to a sequence of digital codes that will essentially indicate the coordinate position (or address) for each scanned area, and tell whether that area is black or white.

The resulting data stream can be handled in much the same manner as that from any other high-speed digital terminal. Some one-minute fax machines will accept a message header, probably a prepared card that the machine will read first to "seize" the receiving fax machine. Then, the message will proceed through the digital data-communications network just as if it were a message sent through a conventional data terminal. When the receiver is busy, digital messages will be stored in a mass memory and forwarded when the receiver becomes ready to take a new input.

Fax vs TWX costs

The most common terminal in data communications is the keyboard sender/receiver. Often, the user leases these terminals and connects them into an in-house data-communications network using either narrowband channels or voice-grade lines leased from the telephone company. Another alternative is the TWX network, in which the user leases the terminals and pays time charges to the common carrier. It turns out, though, that a three-minute fax machine can often deliver messages at lower cost than TWX.

The monthly rental cost for a TWX machine is about the same as that for a fax terminal. Therefore, the cost comparison between the two approaches is based on the actual communications cost per word. A typical sheet of paper, typed single-spaced, can contain at least 550 words. To transmit this amount of information by fax takes three minutes. The maximum daytime station-to-station telephone rate for a phone call is $1.35 for three minutes, even for distances exceeding 2,000 miles. Thus, via fax the cost is less than 0.25 cent per word.

By TWX, the cost per word includes both the labor cost to convert the source document to punched tape and the 2,000-mile communications cost of 70 cents per minute. To dispatch a 550-word TWX message requires about 40 minutes of operator time—17 minutes to punch a tape, 6.5 minutes to verify it, 10 minutes to rerun and correct errors, and 6.5 minutes to transmit. At $2.50 per hour for the operator, this direct labor is $1.66, or 0.3 cent per word.

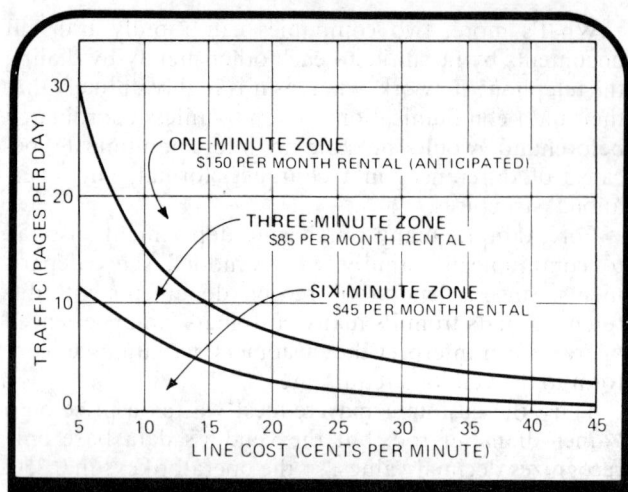

2. Break even. Balancing off traffic in pages per day against line cost shows that even when rental cost goes up for faster fax transceivers, delivery cost per page can go down.

Although the TWX system is nominally rated at 100 words a minute, about 15% of the time is occupied by control signals, such as those needed for carriage return on the terminal, which reduce the actual message rate to 85 words a minute. Therefore, at 70 cents a minute, the communications cost for delivering a 550-word message is 0.82 cent a word—for a total cost of 1.1 cent a word by TWX, compared with 0.25 cent by three-minute fax. Similar calculations will show that for any distance, the cost of delivering a 550-word message is lower by fax than by TWX. □

A new read-only terminal: your television set

The long-awaited home computer terminal moves closer to reality with teletext systems

Timothy Johnson
London, England

Round up

Unbeknown to the general public, a small but significant portion of the electromagnetic signal used in putting a picture on the home television screen is going to waste, dissipating into thin air. Before long, however, a stockbroker may turn to the same commercial TV channel on which a soap opera is being played, press a decoding switch, and be presented with data on current market conditions. Or, on another commercial channel, a homemaker might consult a data bank for a recipe, or an interested citizen might keep an eye on wire-service dispatches.

More intriguing is the possibility of advertising on extremely short notice, for spare seats on charter flights, for example, or for cut-rate space on return trips in trucking operations. Another idea is the rental of time slots to businesses for internal information purposes. A company could, by renting a given time slot each day, distribute an assortment information like price changes, stock shortages, or general memoranda—more quickly and cheaply than by telephone. Programable decoders on the TV sets would assure that only authorized receivers could read out the information.

Far from being futuristic visions, the means for these and any number of other such services are available now. The resources are here, the technology is at hand to use them, and, in fact, pilot projects are already in operation utilizing the principles involved. The questions that remain are mostly a matter of economics.

The so-called "teletext" systems emphasized in this article (Table on the next page) are quite similar to the new investor-information service begun by Reuters Ltd. But an important distinction is that the Reuters service operates over a dedicated TV cable. By contrast, the newer, more experimental teletext systems transmit pictures and text by precise injections of digitized information along segments of the same signals that carry over-the-air TV programing. At the receiving end, a special decoder on the typical TV set can convert the TV from its usual role into a read-only terminal.

The process is feasible because the electron beam scanning of the normal TV set is temporarily inhibited each time it is shifted from the bottom of the picture to the top. It is during this brief "vertical blanking interval" that a digitized data signal can be received at the set, decoded, and generated into

COMPARISION OF TELETEXT SYSTEMS

SYSTEM	U.K. UNIFIED STANDARD	HRI ADD-ON	NBS TV-TIME	REUTERS IDR
LEADING APPLICATIONS	PUBLIC BROADCASTING, CAPTIONING	CAPTIONING, MESSAGES	CAPTIONING	CABLE SUBSCRIBER SERVICE
TRANSMISSION METHOD	2 LINES IN EACH VERTICAL BLANKING INTERVAL	EACH ACTIVE LINE IN PICTURE VIDEO SIGNAL	1/2 LINE IN ALTERNATE VERTICAL BLANKING INTERVAL	DEDICATED CABLE CHANNEL
PEAK DATA RATE (BITS/S)	7 MILLION	23,600*	1 MILLION	5 MILLION
AVERAGE DATA RATE (BITS/S)	36,000	21,600*	780	5 MILLION
PAGE OR CAPTION CAPACITY	960 CHARACTERS (24 ROWS, 40 CHARACTERS PER ROW)	NOT FIXED	96 CHARACTERS (3 ROWS, 32 CHARACTERS PER ROW)	768 CHARACTERS (12 ROWS, 64 CHARACTERS PER ROW)
ACCESS TIME PER FULL PAGE OR CAPTION	0.24 SECOND	NOT FIXED	0.15 SECOND	0.0015 SECOND
SPECIAL FEATURES	GRAPHICS, COLOR SELECTION		FREQUENCY STANDARD ALSO INCLUDED	CENTRAL CONTROL OF SUBSCRIBER ACCESS

*ASSUMING 3 BITS PER LINE PAIR

meaningful characters on the screen—all without interfering with the regular TV signal. The British television standard, for example, specifies 625 scanning lines for a picture, split into two consecutive fields of 312½ lines each. In actuality, only 287½ lines of each field appear on the screen, with the remaining 25 lines being produced during the vertical blanking interval. It is along these empty scanning lines that a different signal can be imposed. In effect, about 8% of the British scanning cycle is available for other purposes, and, as in America, some of it is used for internal broadcasting information.

A somewhat different operating principle is being tried by the Hazeltine Research Corp., Chicago, Ill. Here, bits of digital information are imposed directly on the active video signal, but in a manner intended to produce negligible interference with the normal picture.

The RCA Corp. attempted in the mid-1960s to put the vertical blanking interval to good use, but RCA employed an analog signal. This worked, but the system was expensive, slow, limited in its information carrying capacity, and never got beyond the experimental stage.

Today, digitized text can be implemented with low-cost, compact, integrated circuits that can greatly reduce overall costs, a critical factor in the mass consumer market. Because digitized information can be transmitted in the megabit-per-second range, a given message can be transmitted during a shorter segment of the vertical blanking interval than can an analog transmission.

In the United States, an experimental program being run by the Public Broadcasting System involves superimposing captions for the deaf on the regular TV picture, but of course only decoder-equipped sets can read out the captions.

Probably the most highly developed operational system of its kind today is the Ceefax service, developed jointly by the British Broadcasting Corp. and Britain's Independent Broadcasting Authority, and operated on a test basis for a limited but nationwide audience. Both the British and the American PBS projects share the signals of regular TV channels, inserting their special digital programing during the vertical scanning intervals.

The Ceefax service is organized into "pages," a page being a block of information filling the TV screen. The user is initially offered an index of subject matter and the corresponding page numbers (Fig. 1). He may then punch the desired page number on his decoder. The pages are continually being transmitted in serial form; the decoder waits for the requested page to come along, locks onto it, and stores it in a memory register. The register drives a character generator to display the text for as long as the user likes.

Countries using television standards different from Britain's will doubtlessly have somewhat different teletext standards, but probably will adopt systems that are the same conceptually. Under what is known in England as the Unified Standard, teletext is imposed on two of the 25 lines in the vertical blanking interval. And since the overall British TV standard calls for a scan rate of 50 fields per second, there are that many vertical scanning intervals each second. (A significant detail to keep in mind in this regard is the standard technique of interlacing a scanning sequence to overcome the flicker effect. In this, the odd lines are scanned first, then the even lines, two interlaced scans for one complete picture. This fact is exploited in the Hazeltine experiment, but more on that later.)

The British teletext standard also specifies that no more than 40 8-bit characters can be included on one scan line, and that the characters must be coded in the ISO-7 format, a version of ASCII. The standard allows a maximum data rate of 6.9375 million bits per second. Each scan line carries the data for one row of characters on the screen. A full-page display (including the page "header," which controls selection of pages) takes 24 lines, requiring 12 vertical blanking intervals. This could accommodate 960 characters, but fewer are actually used because some empty screen space is needed for legibility. The character generator drives the electron gun to produce the readouts.

Under the British standard, the address field in a header can identify any of 800 separate pages. But because each page requires a fraction of a second to transmit, the viewer of an 800-page service might have to wait as long as three minutes for the selected page to arrive. In practice, the BBC and independent television companies are considering a maximum of 60 to 100 pages for each channel. To satisfy special interests, the British are also considering the use of special pages—broadcast at particular hours of the day rather than continuously. As they would not be part of the continuous service, the special pages would not necessarily increase the overall access time.

Moderate investment

The Ceefax project cost the BBC $120,000 for studio equipment. The hardware consists of a minicomputer with 16,000 bytes of memory, a small disk memory, four video displays for editing, a teletypewriter for developing the text characters, and special circuits for inserting the text data into the television signal. A functional diagram of the system is shown in Figure 2.

The British standard requires digital-data-acquisition circuits capable of processing 7 megabits per second, address circuits to recognize the selected page and organize the data in the memory, a 7-kilobit direct-access memory, a character generator, and circuits to decode and implement special capabilities such as color and graphics. All circuits must be relatively high-speed and provide precise switching times to lock onto a short run-in code on each line and to maintain accurate timing during the transmission of each row. The diagram in Figure 3 indicates the functions required in a decoder.

The cost of decoding equipment may be the biggest obstacle in the way of large-scale acceptance of teletext, at least for early versions. But decoders should already be affordable in the business community. Retail prices of first-generation decoders for the consumer market might be around $300 for systems similar to that of the British standard. This does not include the cost of a television receiver.

However, historical experience with integrated circuits suggests a rapid decline in costs as the mar-

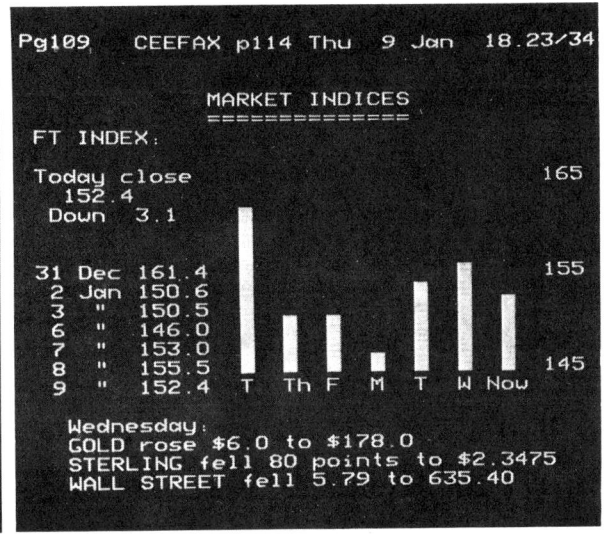

1. With the BBC's Ceefax system, the user is first shown an index of pages and then he selects the desired item, such as stock averages or news bulletins, by entering the page number on the decoder.

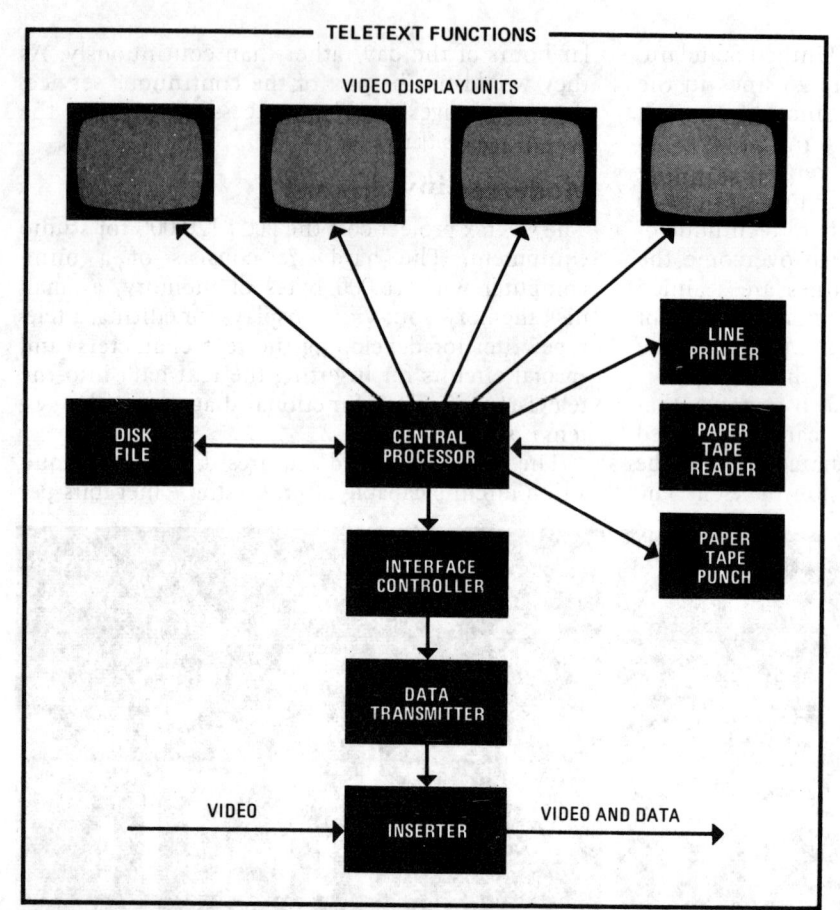

2. In a truly low-cost teletext system the central computer could be a microprocessor and the line printer, paper tape reader, and paper tape punch could all be part of a conventional teletypewriter terminal.

ket grows and circuit functions are further integrated. Within a few years decoder prices might be cut in half. They would then become affordable to a large percentage of viewers. But for business applications today, $300 plus the cost of a receiver is usually less than the cost of a terminal.

The BBC experiment, begun in 1974, is scheduled to end in the fall of 1976, while independent British television companies will begin broadcasting their own teletext project in 1975. Decoders for the British market may be produced by TV set makers at mass-market prices as early as 1978. Other European countries also are interested in a teletext system.

Translating the standard

In the United States and other countries which use a 525-line TV standard, teletext systems would be different in certain respects, but operating principles would be the same.

The initiative in providing a commercial teletext service in the U.S. was taken early this year by the Reuters-Manhattan Cable project, which began late in 1974 on a dedicated cable TV channel. A small number of subscribers pay upwards of $650 a month for access to one or more of the Reuters business news services.

Commodity traders, for example, can receive transactions on the Chicago Board of Trade, as well as prices and news from elsewhere. For now, Reuters is concentrating on the "professional" market. It also plans to offer racing results and other sports news to bars and betting parlors for about $200 a month—if the company can find a way through some legal tangles.

Reuters editors compile text with the aid of two minicomputers that format, store, and forward the coded data to the appropriate subscribers. A customer can elect to receive all or part of the financial news service. A "packet" of data carrying 32 characters is imposed on each scan line of the television signal. Each display page is made up of 12 rows of up to 64 characters each. The channel can carry up to 650 full pages per second, so even if a full service is provided, the viewer need wait no longer than a few seconds before a selected page appears. The speedy access is possible because of the dedicated channel; the system is not confined to a fraction of the total number of scan lines, as with a combined service like Ceefax. There is also an advantage in a service sold only on a selective basis. The minicomputers can control which decoders can receive which pages, and they can discontinue service without needing access to the customer's premises.

Other teletext services based on a dedicated cable are bound to follow. Reuters already has a contract with one of the major cable networks in Toronto, and is studying some possibilities in Chicago. One ambitious scheme—that of providing the equivalent of a complete newspaper by teletext to subscribers in Toronto—was dropped, but could well resurface elsewhere.

The central marketing problem in all this is to get decoders into the viewers' homes and offices at a reasonable price. At present, as in the case of Reuters, this is done by offering a specialized service at a fairly high price, which makes the cost of the decoder a relatively small part of the whole. Another route is to plan a multimillion dollar operation in which equipment can be mass-produced from the start, as planned by the British networks. Once the base of mass production is established, teletext services will spread rapidly.

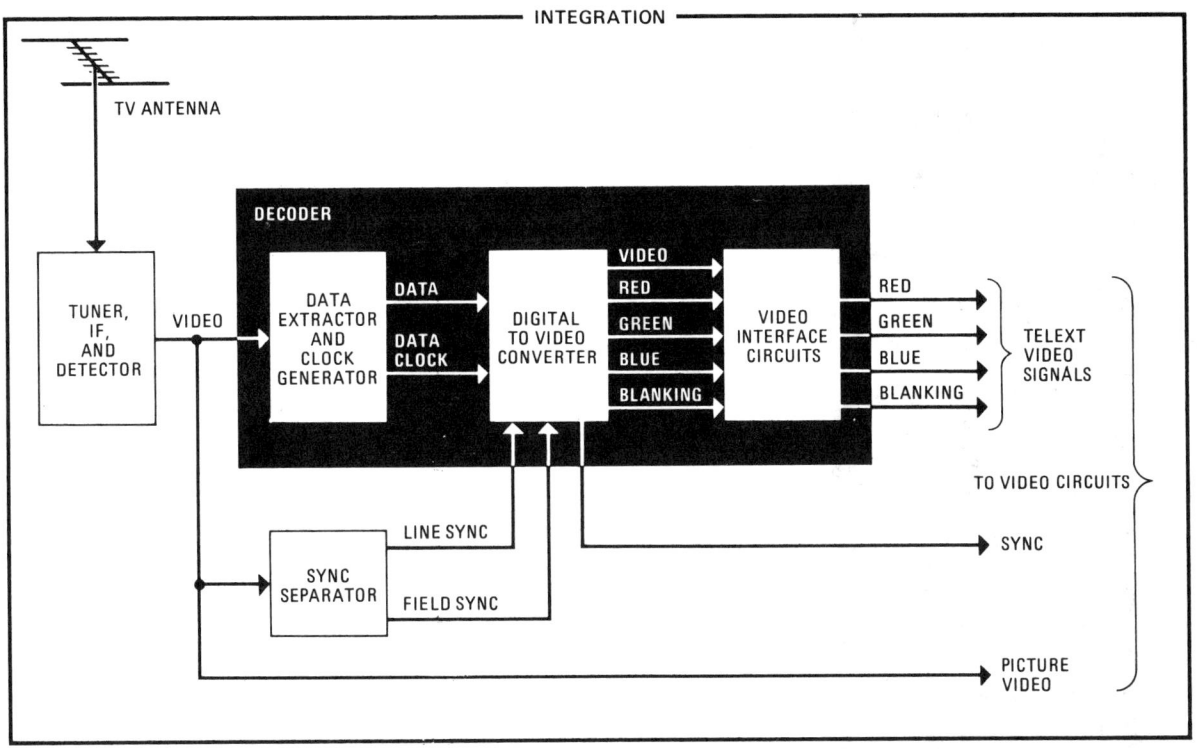

3. A decoder can be integrated into a conventional TV receiver, as shown. The digitized information is changed into video by the digital to video converter. The video interface circuits add signals for special color and graphic effects to enhance readability.

A broadcast service with a more limited capability is the National Bureau of Standards' TV-Time. Originally developed for broadcasting highly accurate time and frequency readouts. TV-Time also uses a spare line in the vertical blanking interval and imposes on it a highly stable 1-megahertz sine wave as a frequency standard, followed by 26 bits of data on alternate fields. This arrangement provides for a net data rate of 60 characters per second, or about 600 words a minute, plus addressing information. Such a capacity is far too low for a comprehensive news service, but quite ample for program captioning or news bulletins.

The PBS has been testing the TV-Time technique as a way of providing captions for the deaf. The captions are displayed only on receivers equipped with the proper decoders. If the PBS formally institutes this system, programs would be captioned in advance and the data recorded on the master videotape. Some signal modifications would be needed to indicate that the signal is not a valid frequency standard.

Trial PBS showings of news and entertainment programs to hard-of-hearing audiences in various parts of the country have won an enthusiastic response. Over 90% of the viewers said they could not have understood the programs without the captions, and would like to buy a decoder when they became available. One advantage of the low data rate of the TV-Time system is that decoders should be fairly cheap from the beginning—probably about $100. PBS hopes, if its proposals are accepted, that captioned programs will appear on a regular basis in 1977 or 1978.

PBS has also been looking into the system developed by Hazeltine Research. This approach does not use the vertical blanking interval, but relies on the actual picture lines. In its simplest form the HRI Add-On System, as it is called, imposes one bit of information on each member of each odd-even pair of television lines, each line coded oppositely to its twin, so that the visible effect of the data on the screen is substantially reduced. Special circuits are needed to strip the data from the received signal so that the text can be displayed. HRI claims that up to three bits of data can be imposed on each line pair without causing unacceptable interference with the viewer's picture. With three bits per line, the system has an information-carrying capacity comparable to system using a high data rate on one or two blanking-interval lines. It is also more resistant to multipath interference than systems like TV-Time and Britain's Unified Standard. But engineers outside HRI criticize this approach as being unnecessarily complex. Also, it is said to be difficult to remove the text signal once it has been added to a program recording, but HRI says it is working on this. ∎

part 3
data data data data data data data data
acoustic couplers and modems

Acoustic couplers lend portability to data terminals

Most often found with low-speed terminals, the acoustic coupler is not wired into a line but can make use of any telephone handset to send and receive data

Richard J. Indermill
Anderson Jacobson Inc.
Sunnyvale, Calif.

Until 1965, a user had to bring his data physically to the digital computer. But late that year, some manufacturers announced time-sharing computers, which could be fed data from a distance, through remote terminals. Moreover, the telephone network was available to link terminals and computer—provided two handicaps could be overcome. First, at that time telephone companies prohibited hard-wiring of "foreign" devices directly to their lines. Second, the data-pulse signals generated by terminals and computers are not electrically compatible with the voice-signal requirements of telephone lines. The solution, developed in 1966, was the acoustic data coupler.

The coupler is attached to a terminal, and converts its pulsed outputs into audible frequencies or tones. The tones are sensed by an ordinary telephone, which is inserted into the coupler and processes the tones as though they were voice signals. This simple approach is still in wide use today, even though hard-wired connection is now permitted.

The device, first produced commercially by Anderson Jacobson Inc., and now available from several other companies, consists of two main sections:
- Electronic circuitry that converts binary pulses produced by the terminal's keyboard into two audible tones, one representing a binary 1 and the other a binary 0. Equipment at the computer converts the two tones back into digital-computer code.
- Microphone and speaker, mounted in cups that surround the telephone handset (see photograph).

To send data to the computer, the terminal operator dials the computer's number and, when he has established the connection, nests the telephone into the coupler's cups. The coupler's speaker mates with the telephone's transmitter, and the coupler's microphone mates with the telephone's receiver. Thus, as with a keyboard/printer, the acoustic coupler serves the terminal when it is sending or receiving. Furthermore, when not in the acoustic coupler, the telephone can be used for conventional voice traffic.

The acoustic data coupler is similar in function to the modem, or dataset, which also converts pulses to tones (or some other type of modulation of the audio signal) and tones to pulses. Technically, though, they are different. The modem (see next article) is hard-wire-connected to a telephone line, and is therefore liable to special telephone company tariffs, but can operate at faster bit rates. The acoustic coupler, because it need not be attached to any telephone line, is suitable for portable or mobile terminals, is not subject to special tariffs, and

Coupler. Using a sound path, acoustic data coupler interfaces low-speed terminals to telephone line.

is generally used with relatively low-speed terminals.

Usually a terminal comes already equipped with such a coupler, though the user may sometimes buy and install one of his own choosing. For most keyboard/printer terminals, the 30 characters per second transmitted by most data couplers is a more than adequate transmission capability. (The necessary data rate here is not determined by the varying speeds of the manually operated keyboard, but by the printer, which can be driven by a steady data stream, as from a computer.) The most popular printer today, the Teletype unit and its emulators, prints 10 characters per second. The IBM Selectric mechanism, the foundation for many independently made keyboard/printer terminals, operates at 15 characters a second. Some new keyboard/printer terminals operate at the full 30 characters a second, and are gaining in popularity.

The fastest data coupler on the market can handle up to 120 characters a second. Made only by Anderson Jacobson Inc., this coupler is suited to interfacing telephone lines with CRT terminals, as well as such keyboard/printers as the Memorex 1240, which can print up to 120 characters a second (1,200 bits a second). At and beyond this data transmission rate, however, terminals are generally fixed in place and can be better served by a hard-wired modem. □

Modems marry data terminals to telephone lines

Because most data communications flow over telephone lines, modems must be connected between lines and terminals to convert data pulses to analog frequencies—and back

Robert L. Toombs
International Communications Corp.
Miami, Fla.

The telephone system is the backbone of data communications today. Yet digital pulses cannot be carried over its voice-oriented network unless they are processed through the sophisticated electronics package called a modem.

As a result, virtually every digital data terminal linked to the network requires a modem. In addition, every port or access connection in a multiplexer—which concentrates signals from many terminals into a single, fast data stream—requires a modem. And every data communications port at the host computer site needs a modem.

"Modem" is actually a contraction of the two main functions of the unit: modulation and demodulation. (The word "dataset" is a synonym for modem.) But the user's concern in less with how modulation and demodulation are achieved than with specifying the modem's data rate and choosing from among the innovations that have improved modem performance, convenience, and reliability over the past few years.

If modems are classified by the kind of lines they serve, they fall into three categories, as shown in the table on the next page. By far the most important to the business and industrial community—and therefore the subject of this article—is formed by the voice-grade units. But the wideband and hard-wired modems also deserve mention.

Wideband modems, supplied only by the telephone companies and subject to specific tariffs, provide synchronous data transmission at data rates from 19,200 bits per second (b/s) to 230,400 b/s. They operate with special wideband lines (actually groups of voice-grade lines) provided by the telephone companies, and are primarily used in computer-to-computer applications.

Hard-wired modems operate on dedicated solid-conductor twisted-pair or coaxial lines, and can perform well over distances up to 15 miles. Operating at speeds up to 1 million b/s, they are particularly useful for in-plant and on-campus installations of terminals and computers. Also, in comparison with the other two types of modems, they are relatively inexpensive, costing from $900 to $3,000 each. The exact price depends on data speed, operating features, the distance over which the modem can operate, and the quality of the unit.

Voice-grade modems

As for the voice-grade modems, they operate on the telephone companies' voice-grade lines, whether these are private (dedicated) or part of the dial-up switched network. Such modems can be obtained from three primary sources: common carriers, independent makers (that is, not related to the telephone companies), and terminal and computer makers, who generally buy their modems from one of the independents.

The price of a voice-grade modem (or its equivalent monthly lease value) increases almost in a straight line with respect to its data rate. A 2,400-b/s modem will sell for about $1,200 to $2,000, while a 9,600-b/s modem will sell for about $10,000 to $12,000.

Independent makers will sell as well as lease units, but a telephone company will only lease them. Moreover, when an independent's modem is connected to the line, the telephone company installs a protective device called a data-access arrangement (DAA) and charges the user several dollars a month for it. Generally, however, an independent modem compensates for the added expense by costing less than the telephone company's to start with, or by offering more features or better performance. In view of the number of modems that may be needed for a data-communications system, all the economic choices should be investigated—in addition to those technical innovations and performance benefits that distinguish one maker's modem from another maker's.

As shown in the table, voice-grade modems can be grouped according to their speeds. The slowest operate at 1,800 b/s or less, use an asynchronous-code format, and interface low-speed asynchronous terminals, like keyboard-printers, to the line. Such low-speed units are simple and inexpensive because they do not require a clock—the elaborate timing circuitry included for bit synchronization required for synchronous medium- and high-speed modems.

Medium-speed synchronous modems, operating at

MAJOR APPLICATION FEATURES OF MODEMS FOR DATA COMMUNICATIONS

	VOICE-GRADE			WIDEBAND			HARD-WIRE
	LOW-SPEED— up to 1,200 b/s; or 1,800 b/s on conditioned line	MEDIUM-SPEED— 2,000; 2,400; 3,600; 4,800 b/s	HIGH-SPEED— 7,200; 9,600 b/s	HALF-GROUP BAND— up to 19,200 b/s	GROUP-BAND— up to 50,000 b/s	SUPER-GROUP BAND— up to 230,400 b/s	SUPER HIGH-SPEED— up to 1,000,000 b/s
DEDICATED OR DIAL-UP	■	■	(1)				
DEDICATED ONLY							▧
AVAILABLE FROM INDEPENDENT MANUFACTURERS	■	■	■				▧
AVAILABLE FROM COMMON CARRIER	■	■	■	▧	▧	▧	
ASYNCHRONOUS	■						
SYNCHRONOUS		■	■	▧	▧	▧	
HALF-DUPLEX AVAILABLE	■	■	■				
FULL-DUPLEX AVAILABLE	■	■		▧	▧	▧	▧
FAST TURNAROUND AVAILABLE FOR USE ON DIAL-UP LINES	■	(2)					
MULTIPLE-SPEED TECHNIQUES AVAILABLE	■	■					▧
REVERSE CHANNEL AVAILABLE	■	(2)					
MANUAL EQUALIZATION AVAILABLE		(3)	■				
AUTOMATIC EQUALIZATION AVAILABLE		(3)	■				
SUITABLE FOR MULTIPOINT NETWORKS	■						▧

(1) For dial-up, must be switched to 4,800 b/s (2) Except 4,800 b/s (3) 4,800 b/s only

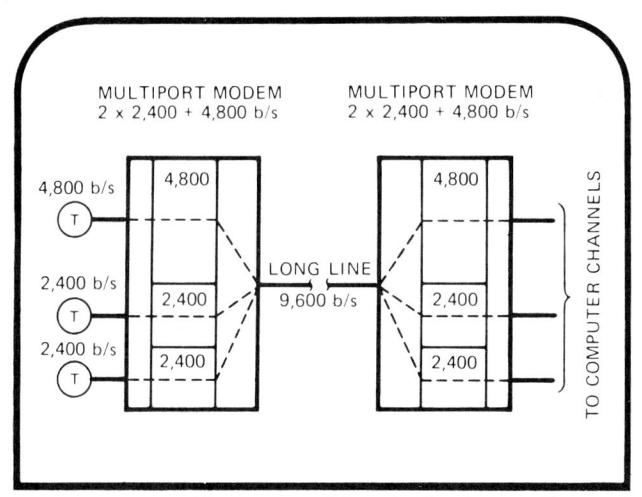

1. Multiporting. When several terminals at one site must communicate with terminals at another site, a multiport modem having individual units of appropriate data rates can cut costs in this point-to-point hookup.

2,000, 2,400, 3,600, and 4,800 b/s, are used with faster terminals and also multiplexers. High-speed modems work at 7,200 and 9,800 b/s, again with both terminals and multiplexers.

Technological advancements in data communications over the past few years have resulted in the development of several useful new modem features. Among these are: dial-up capability on the switched network; dial backup for dedicated installations; automatic equalization of electrical parameters to match the line's characteristics; secondary channel utilization; voice/data communications; built-in self-diagnostic techniques; and multiport capabilities.

The dial-up capability

Most modems rated at 0 to 1,200, 2,000, 2,400 or 3,600 b/s, and some at 4,800 b/s, can operate on the public switched (dial-up) telephone network. A line in this network is two-wire, and when it carries traffic in both directions, is said to operate in half-duplex mode. Here, traffic can flow in only one direction at a time. To turn the line around and reverse the direction of traffic, time must be allowed for the echo suppressors to reverse themselves. This takes about 150 milliseconds.

In the error-control method most commonly used for two-wire half-duplex transmission, the terminal sends a block of data, the data is analyzed for accuracy at the receiver terminal (or computer), and the receiver sends an acknowledgement signal—either an ACK if the block contains no errors or a NAK if the block is erroneous and must be retransmitted. This procedure is called ARQ, for automatic request for repeat. Implicit in this discipline is that the line must be turned around twice for each block. That is, one echo suppressor must be disabled in one direction, and another suppressor in the other direction, the total cycle taking 300 milliseconds.

To put this into perspective, a 1,000-bit block at 4,800 b/s would take about 200 milliseconds to transmit and about 300 milliseconds to turn around—which may well seem an inordinate overhead. Some recently developed modem features help reduce or circumvent the line's turnaround time, among them the use of a reverse channel, a go-back-two ARQ scheme, and a fast-turnaround technique.

Clever error control

The most efficient of these, the reverse-channel method allows the short ACK or NAK signal to be returned over a low-speed asynchronous channel that shares the line with the main synchronous data channel. The effect is to completely eliminate turnaround of the data channel during consecutive block transmissions in one direction. But while some slow and medium-speed modems contain a reverse-channel feature, it is not often exploited because it requires a special interface of limited availability at the host computer and the data terminal.

A clever variation on this idea is the go-back-two ARQ method. Here, redundant bits for error detection are formatted inside the sending modem. The receiving modem analyzes the received block for errors, and sends back an ACK or a NAK over the low-speed reverse channel. Essentially, the modem sends blocks continuously until it receives a NAK, and, when that happens, it must repeat, not one, but the last two blocks. In the proper error environment and using an appropriate block length, the go-back-two modem can increase the net data throughput substantially over that of the conventional modem used in the standard ARQ situation. However, high error rates reduce the effective throughput. This "continuous" modem contains the special interface equipment for the reverse channel.

The third way of increasing data throughput in half-duplex operation is to eliminate the need for repeatedly reversing the echo suppressor. This fast-turnaround method is implemented in certain 2,000-, 2,400- and 3,600-b/s modems. A short signal of a specified frequency initially disables the suppressors and then changes to a continuous out-of-band signal of any frequency, which keeps them disabled during connection time but does not interfere with the data channel. Thus delay time is reduced to that required for modem synchronization—as little as 8.5 milliseconds, or 17 milliseconds for each block when operating in a half-duplex ARQ mode. By way of comparison, at typically used block lengths and the typical error environment found on the dial-up network, a fast-turnaround modem rated at 3,600 b/s may have a net data throughput similar to that of a "continuous" ARQ modem at 4,800 b/s.

Modems featuring dial-up capability work in conjunction with a telephone equipped with an extra switch function besides on and off. This special switch, called an exclusion key, converts the line from voice to data communications. To initiate a manual data connection, the sending operator dials the receiver's number, and discusses with the receiving operator any necessary coordination before they both activate their exclusion keys. The two modems then go into what is called a handshaking routine, which tells both the terminal and the computer that a connection has been established.

With proper ancillary equipment and data-access arrangements, the modem-to-modem data connection can be accomplished automatically, permitting unattended transmission and reception at terminal and computer;

the complete capability is called automatic answering/automatic disconnect.

When a modem is used on a dedicated, or private, line, the data connection is established permanently. This, among the other benefits of dedicated lines, saves time that would otherwise be spent on busy signals and handshaking. But dedicated lines may fail completely or suffer such a high error rate as to be useless. When either of these situations occurs, modems rated at 4,800 b/s or less and connected to the dedicated line can also be used in a dial-backup mode. Furthermore, high-speed modems—the 7,200- and 9,600-b/s units normally only used on dedicated lines—can also be made to perform in a dial-backup mode, provided their speed can be manually switched to 4,800 b/s.

The why and how of equalization

Equalization of a modem is a way that manufacturers compensate for the inconsistencies of a transmission medium—the telephone line, amplifiers, and other equipment making up the link—that was designed for voice rather than data. Its purpose is to match line conditions by maintaining certain of the modem's electrical parameters at the widest practical set of marginal limits to take full advantage of data rate capability of the line. Operating too near or outside of these margins increases the error rate. The main parameters affected by equalization are amplitude-and-frequency response and envelope delay.

The faster the modem, the greater becomes the need for equalization and the more complex the equalizer. Most 0-1,800-, 2,000-, 2,400-, and 3,600-b/s modems incorporate fixed, or nonadjustable, equalizers with characteristics that match the average line conditions found from a statistical survey of the dial-up network. That is, for a normal and randomly routed call between two locations, modems with fixed, or statistical, equalizers should work well. If the error rate does rise to an unacceptable value, as may be indicated by a signal-quality light on the modem, the operator can often clear the difficulty up simply by dialing a new call and rerouting the connection.

Some 4,800-b/s modems employ manually adjustable equalization. The parameters are preset, or tuned, at installations to match the line. Once set, no further time is required to equalize a modem—except for a slight retuning once every three months or so—unless the line is reconfigured.

Other 4,800- and all 7,200- and 9,600-b/s modems feature automatic equalization, which continuously compensates for all the static and most dynamic anomalies found on four-wire lines. With automatic equalization a certain initialization time is taken to train, or adapt, the modem to instantaneous line conditions. Thereafter the modem continuously tracks and compensates for line conditions that will change, for example, due to temperature variations during the day and night.

The minimum initialization time for automatic equalization is about 50 milliseconds, during which no useful data can be transmitted. Because a terminal/modem combination is operating full-duplex on a four-wire line, each two-wire line always transmits in one direction so that there is no need for the conventional turnaround time to disable echo suppressors. Therefore, the initialization time for an automatically equalized modem occurs only when a "permanent" connection is being established in a point-to-point configuration, and hardly reduces net data throughput at all.

On the other hand, the total turnaround time of automatically equalized modems may preclude their efficient use in four-wire multipoint polled configurations or in two-wire systems with an ARQ protocol in which turnaround occurs often and equalization would have to occur as often. However, a special form of manual equalization, called forward equalization, makes 4,800-b/s operation practical for multipoint configurations on dedicated lines.

In a multipoint polled system, several terminals are connected to the host computer along the same physical line, but only one terminal can be active at a time. Because the distance from the computer to each terminal may vary by hundreds and even thousands of miles, the line condition from one terminal to the computer will be different from the line condition of every other terminal to the computer. If in these circumstances automatic equalization were used by the computer-site modem, the delay arising out of polling each terminal and receiving an answer would reduce data throughput. Instead, with forward equalization, the modem at each terminal is manually equalized, being adjusted for its own line length and other electrical conditions.

The secondary channel

Often a sending site may contain a fast, synchronous batch terminal and a slow, asynchronous terminal, such as a Teletype sender/receiver, both connected to the same receiver site and both operating in a full-duplex mode. One way to handle this situation is to install a dedicated four-wire voice-grade line and a four-wire low-speed line. But as a separate slow line costs about half as much as the fast one, it's far more economical to add a secondary-channel feature to the modem.

The secondary-channel technique, usable only on four-wire systems, derives two channels from the same physical line, a wide one to carry synchronous speeds of 2,000-, 2,400-, 3,600-, or 4,800 b/s, and a narrow one for the 150-b/s Teletype data stream. Actually, the secondary-channel option can even provide two slow channels and one fast data channel on the same line. Furthermore, it can be made to function as a reverse channel in two-wire half-duplex installations, but doing this requires special interface equipment.

Voice and diagnostics

Oral communication between two terminal locations is often essential for the coordination of data processing and data communications. With the addition of a voice adapter, available from most modem manufacturers, a dedicated four-wire point-to-point link installed for data communications can also be made to carry conversation. For example, the adapter permits the same line to be used during the day as a private-voice line and at night for transmission of fast batch data.

One of the more valuable developments in modems has been the inclusion of diagnostic techniques to im-

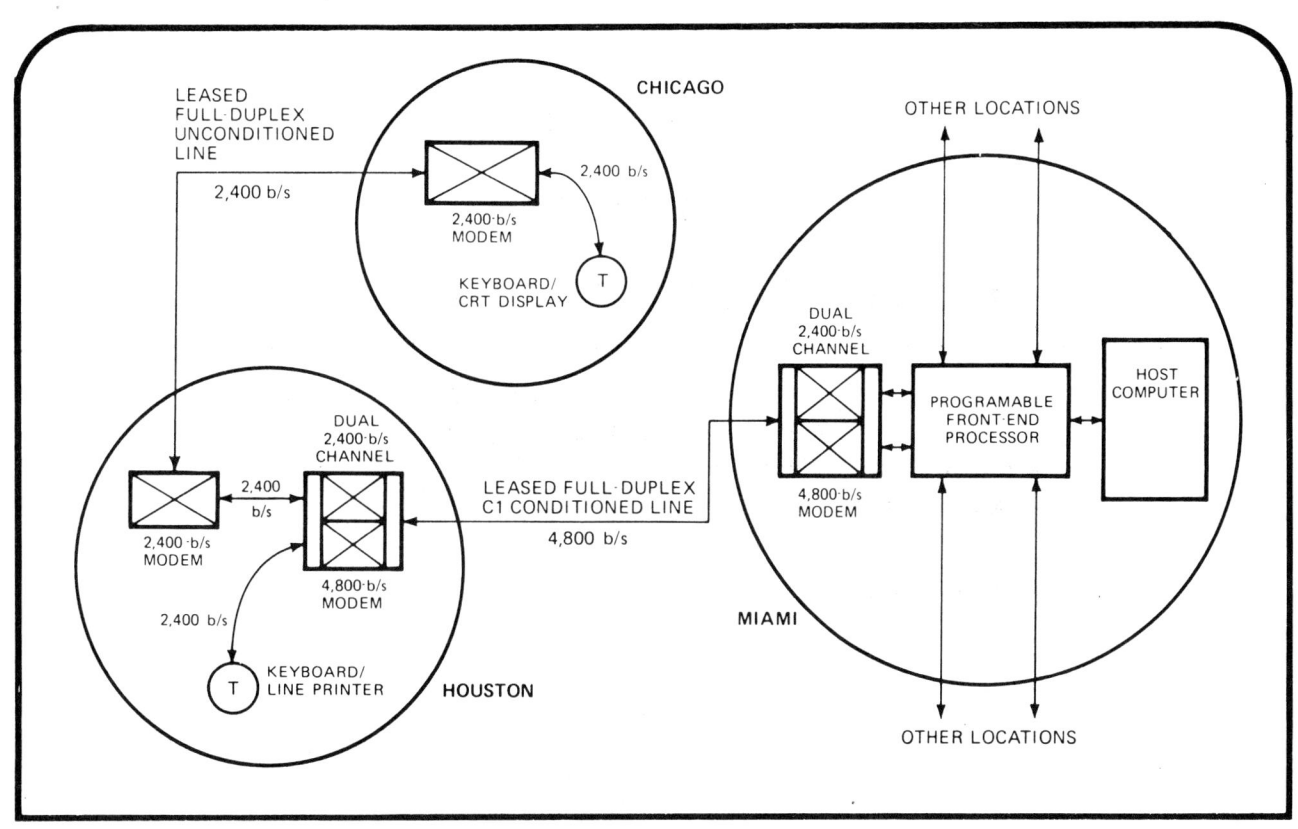

2. **Multidropping.** Multiport modems, as in Fig. 1, can also be used to drop off to terminals at an end-city and an intermediate city. Here, a multiport modem in Miami drops off data at Houston. The data then goes on to Chicago.

prove system integrity and permit rapid fault isolation. The two major diagnostics are self test and remote test. In self test, a built-in pseudo-random test pattern exercises the modem's circuits. Using local loopback connections to its own analog and digital interfaces, the modem indicates when the random pattern received is exactly the same as the one sent. That is, from just a simple light indicator an operator can tell conclusively whether or not the local modem is operating properly.

If the self test shows the local modem is in a GO condition, then, with the help of a person at the other modem, a qualitative check of the line can be made. Here, the remote operator actuates a loopback switch so that the random pattern goes down the line and returns to the local modem. If errors occur on the line during transmission, the NO GO diagnostic light comes on.

With remote test the local modems checks the remote unit and the line without the aid of anyone at the other end. Here, switches on the local unit send signals to the remote one to activate its loopback switches. The remote-test capability is meaningful for installations in which the remote terminal must operate unattended, say during the night.

Normally one modem is needed for each terminal. However, a development called multiporting allows several terminals at the same site to be serviced by one high-speed modem on a point-to-point basis over a single four-wire line, as shown in Fig. 1. For example, one 9,600-b/s modem can carry four data channels, each rated at 2,400 b/s, between two point-to-point locations. Several makers of 4,800-, 7,200-, and 9,600-b/s modems include multiporting in their product line. Currently available data rate combinations include 2 × 2,400, 3 × 2,400, 4 × 2,000, 2,400 + 4,800, and 2 × 2,400 + 4,800 b/s.

Although basically used for multiterminal point-to-point installations, multiported modems can also be used in a multidrop network. As an example of how a multipoint modem may be used to economize on line charges, consider the configuration in Fig. 2. In Miami a communications front-end processor interfaces with a 4,800-b/s modem through two independent 2,400-b/s ports. Data from the two ports is bit-multiplexed by the modem into a single carrier signal for transmission to Houston, where the signal is demultiplexed to recover the two 2,400-b/s channels. One channel serves the terminal in Houston, and the other is retransmitted to Chicago through a conventional 2,400-b/s modem.

Multiporting can save a lot on line charges, because it does away with the additional lines that would otherwise be needed if several point-to-point lines were to be used from a central site to the remote terminal locations. To cite figures, eliminating one 500-mile voice-grade line saves almost $800 a month in line charges, whereas a modem with a multiporting capability may be bought once and for all for only about $500 more than the price of a conventional unit of the same total speed.

To derive these multiple channels, modem makers include time-division multiplexing circuits inside the modems. Multipoint modems do not compete with multiplexers. Multiplexers concentrate low-speed channels, whereas multiport modems concentrate high-speed channels.

Transmitting data pulses over short distances

Limited distance modems can save money when using twisted-pair lines

Gerald Lapidus
Associate Editor
DATA COMMUNICATIONS

Hardware

The shrewd system designer can save thousands of dollars on initial investment and hundreds of dollars in monthly telephone charges for an institutional or corporate private line network. These savings can be realized by using hardwired, twisted-pair lines driven by limited distance modems (LDMs). However, choosing the most economical way to drive data pulses down these lines involves tradeoffs in cost, performance, and quality of transmission. Although LDMs lack capabilities found in conventional modems, they are often chosen because of their lower cost for localized systems such as office complexes, universities, and factories.

Typical point-to-point, multipoint, and repeater configurations are shown in Figure 1. Since such applications often have large numbers of terminals, the savings that can accrue by selection of the most economical transmission devices can amount to a substantial portion of the total system cost. However, in evaluating the possible savings, the cost of wire and its installation should not be overlooked, since these expenses can be considerable. If wiring costs are prohibitive, the alternative is to use limited distance modems in conjunction with leased telephone lines, but there may be problems in selecting units that meet telephone company standards, which will be covered later.

The four methods of transmitting data, each more expensive than the preceding one, should be carefully evaluated:

- Direct connection by ordinary copper-wire pairs.
- Line drivers, which reshape distorted pulses.
- LDMs, which are simpler versions of conventional modems.
- Conventional telephone-line modems.

The range of distances that can be achieved by the various interconnection methods are shown in Figure 2. As might be expected, the greater the distance, the greater the cost.

The RS-232-C and CCITT V.24 standards limit direct connections to 50 feet for data rates to 20 kilobits per second (kb/s). For distances greater than 50 feet, first consideration should be given to line drivers.

A single line driver generally provides adequate performance for hundreds of feet (depending on the unit) at data rates up to 9.6 kb/s. Beyond the

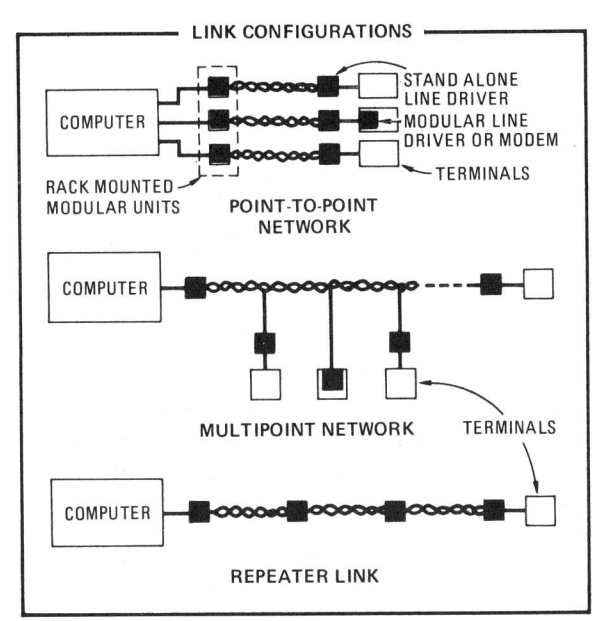

1. In the various link configurations, whether to use direct connection, line drivers, LDMs, or conventional telephone modems depends on the distances.

distance normally specified for the line driver, signal attenuation and line distortion become significant, and tradeoffs between line drivers and LDMs must be considered. However, line drivers can be strung out as repeaters (Fig. 1).

The distortion that results from long line lengths is caused by electrical characteristics of twisted-pair cables, which round off the leading edges of pulses and displace them in time. Pulse rounding, in turn, causes pulse sensing to be delayed, adding to existing delays. The effect of increasing attenuation as distance increases sometimes makes signal levels fall below the thresholds of pulse sensing circuits, thereby causing bits to be dropped.

The factors affecting the usable distances of line drivers and LDMs for long wire runs are transmission rates, distances between points, and wire types. By using line drivers, low data rates may be transmitted over several miles of wire. For instance, at speeds up to 300 bits per second (b/s), data can be transmitted over 5,000 feet of #22 twisted-pair wire without excessive distortion. But to minimize errors, high data rates require squarer pulses than low data rates. Therefore, data rates much higher than 300 b/s suffer from reduced range. For example, 2.4 kb/s is limited to several hundred feet between line drivers.

When to use LDMs

If line drivers cannot do the job, the next consideration should be LDMs. The basic elements of an LDM are shown in Figure 3. These devices are simpler and less expensive versions of conventional telephone modems. The cost advantage of an LDM increases with the data rate because three major functions performed by conventional modems can be eliminated or relaxed. At low data rates, the cost differences are not great. At high speeds, the differential may be thousands of dollars.

Among these unnecessary modem functions are multilevel modulation schemes, which compress high data rates into the narrow voice grade channels supplied by telephone companies. Another function that can be eliminated in LDMs is immunity to frequency offset. This effect often occurs in the modulation and demodulation processes in long distance telephone circuits. If the modulation and demodulation carriers are not precisely at the same frequency and phase, then the received signal is offset, causing demodulation errors.

A requirement that can be relaxed in LDMs is noise rejection. This factor often plays an important role in dictating the modulation technique of a telephone modem and tends to make it costly.

Another major reason for using LDMs is to solve the problem of servicing by vendors. Some suppliers are unwilling to maintain their equipment unless it is linked to other equipment by lines that meet AT&T standards. A pair of LDMs can satisfy this requirement, but they should be selected carefully to assure that they meet these standards.

Settling for less

In return for low price, LDMs lack some capabilities, which could limit their usefulness. Therefore, in considering tradeoffs between LDMs and conventional modems, the buyer must decide whether his system will tolerate the deficiencies.

First, LDMs are used mostly for private line, hardwired links. Some also can operate over local-loop voice grade telephone lines, provided that: there are no loading coils between the two points; the total distance the signal must travel can be guaranteed to be within the range of the modem; the bandwidth is within the range of the telephone

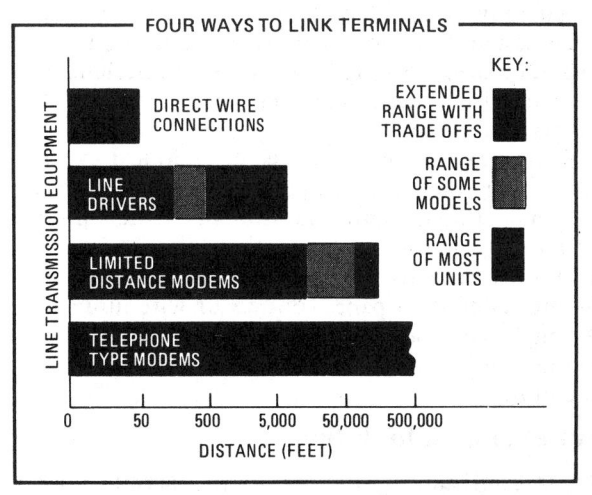

2. The typical usable distances indicated for the various short-haul transmission methods are based on medium-speed operation of 2,400 bits per second and using twisted pair copper wires.

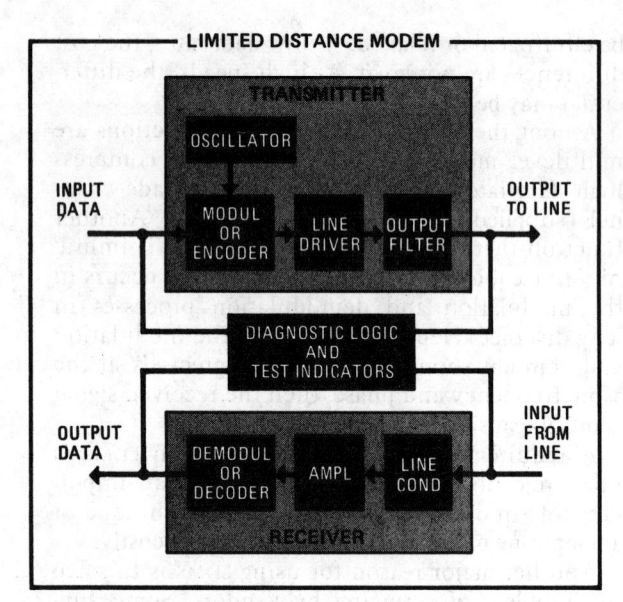

3. A limited distance modem contains a transmitter and a receiver. The input data either is modulated (analog) or encoded (digital).

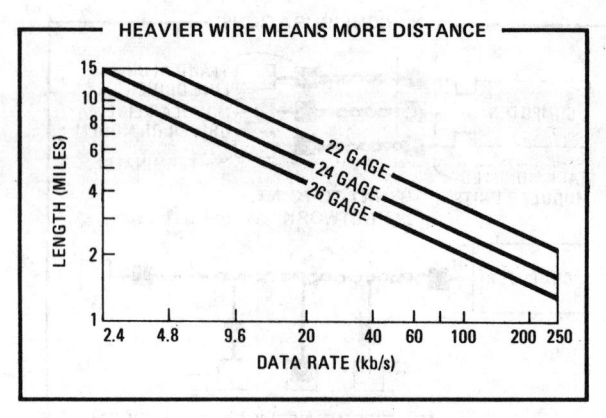

4. Maximum line length depends on both data rate and wire gage. Graph shown above applies to Computer Transmission Corp.'s Intertran.

lines; and the power level meets standards for metallic circuits (Bell System Technical Reference 43401). Although manufacturers may promise that telco bandwidth and power requirements can be met for some models, when the LDMs are to be used over telephone lines, it is best to determine in advance of purchase whether or not the local telephone company will find them acceptable.

Another problem is that some LDMs have modulation techniques with inherent defects, so that certain bit patterns cause temporary loss of synchronization. The buyer should find out whether such patterns affect the LDM being considered.

Ease of adjustment should be given serious attention, especially for new systems. Several manufacturers note that customers often select a given speed, only to find that by the time the system has been tested and optimized, they really require a different speed. Similarly, changes in line lengths may require equalization and power level adjustments on the units.

Some LDMs require a degree of technical expertise and certain test equipment to change speeds and make line matching adjustments. Others must be returned to the factory for these jobs. However, a number of LDMs do facilitate these adjustments by means of front panel controls or wire jumpers. As might be expected, the ease with which adjustments can be made often depends on the price of the unit.

What else to look for

At short distances, the error rates specified for most units are about 1 error in 10^8 bits, but these rates increase rapidly with respect to distance, and the user should check with the manufacturer on the conditions under which the LDM has been tested for the specified error rate. He should also find out what the error rate may be for the LDM/line combination he is considering and the noise pickup expected.

Many LDMs provide for alternative clocking modes of operation. For example, Northern Electric's STE-2 offers master-master, master-external, and external-slave modes. In master-master, a crystal clock circuit in each unit controls its own transmitter, but the receiver is slaved to the unit sending data. In external-external, an outside clocking signal, such as that from a terminal, synchronizes the transmitters of the units on line, but the receivers are slaved to the sending units.

For the external-slave mode, an outside clock drives the transmit circuit of one unit (external mode) and the receiver circuits of both the external and slave LDMs are synchronized to the received clocking pulses. In addition, the transmitter of the slave unit also is synchronized to this clock, so that only one master clock is needed to control the link.

Diagnostic capabilities vary from one LDM model to another. Increasingly, manufacturers are turning to self-testing methods to permit the user to easily find out if his unit is defective or if trouble exists elsewhere, with one or more lights to indicate equipment status and alarm conditions.

The self-testing features usually involve some form of loopback testing. An example of an extensive set of diagnostics is found in the Codex 8200. The simplest diagnostic is a local loop-back test, which returns the transmitter output to the receiver of the same LDM so that the transmitter signal can be checked for errors. Another test, dc busback, transmits the received data and clock back to the LDM at the other end of the line to provide an end to end test. Still another test, remote loopback, triggers dc busback at a remote site to test the elements bypassed in the preceding test. In all of these tests, an internally generated message may be used instead of existing data in order to fully exercise the system. If an error appears in the received data

CAPABILITIES OF LIMITED DISTANCE MODEMS

MANUFACTURER	SPEEDS	MAXIMUM DISTANCE (#22 twisted pair wire)	OTHER FEATURES
Astrocom Corp. 1502 Minnetonka Industrial Rd. Minnetonka, Minn. 55343	400 series— 10–96 kb/s 200 series— 1.8–19.2 kb/s	400 series—5 miles @ 10 kb/s 200 series—13 miles @ 2 kb/s	400 series—current interface compatible with Western Electric 301 and 303 modems. 200 series—RS-232-C voltage interface
Codex Corp. 15 Riverdale Ave. Newton, Mass. 02194	Model 8200 2.4–19.2 kb/s	15 miles @ 2.4 kb/s	Transmit power is strap selectable from -16 to +3 dBm. Comprehensive self-testing.
Computer Transmission Corp. (Tran) 2362 Utah Ave. El Segundo, Calif. 90045	Intertran— 2.4–250 kb/s Directran— 0–2.4 kb/s	Intertran—15 miles @ 2.4 kb/s Directran—15 miles @ 300 b/s	Intertran—Modulation scheme makes unit independent of line length, eliminating equalization adjustments. Directran—Simulates the presence of dialup function of Western Electric 113A and 113B data sets for use with equipment needing dialup protocol
Gandalf Data Comm. Ltd. 15 Grenfell Crescent Ottawa, Ontario, Canada K2G 0G3 U.S. distributor: Penril Corp. 5520 Randolph Rd. Rockville, Md. 20852	LDS 200D 2.4–9.6 kb/s LDS 100 0–9.6 kb/s LDS 250 series— 9.6 kb/s–100 kb/s	LDS 200D—13 miles @ 2.4 kb/s LDS 100—10 miles @ 1.2 kb/s LDS 250—4.25 miles @ 50 kb/s	LDS 200D-Output levels, equalization, gain, carrier (controlled or continuous), and clocking are selectable. LDS 100–DC continuity required. 105 version designed for private, asynchronous computer exchange. LDS 250—Multipoint capability
General Electric Co. Data Communications Prod. Dept. Waynesboro, Va. 22980	Dignet–300 2.4–9.6 kb/s	5 miles @ 2.4 kb/s	Switch selectable speeds. Automatic equalization. Can be connected to 8 remote terminals.
International Communications Corp. 7620 N.W. 36th Ave. Miami, Fla. 33147	Com-Link II— 2.4–19.2 kb/s Modem 1100— 4.8–230 kb/s	Com-Link II— 10.5 miles @ 2.4 kb/s Modem 1100— 8.2 miles @ 4.8 kb/s	Com-Link—Switch selectable speeds. Line adjustments by selectable strapping. Modem 1100—Speeds up to 1 Mb/s can be ordered.
Northern Telecom Inc. 140 Federal St. Boston, Mass. 02110	STE-2 2.4–9.6 kb/s	Greater than 6 miles @ 2.4 kb/s	Switch-selectable data rates. Does not require dc continuity.
Syntech Corp. 11810 Parklawn Drive Rockville, Md. 20852	LDM 2496 0–9.6 kb/s	5 miles @ 2.4 kb/s	Multipoint operation in polled mode is possible either in 2 or 4 wire configurations.

pattern, the signal detect indicator is extinguished for 200 milliseconds.

Hardware and software interfacing usually present no special problems. However, since inspection of vendor literature will not always indicate whether such problems may exist for the particular hardware and software environment in which the modem is to be placed, it is advisable first to borrow or rent a pair of units.

The ways of building line driving units into a network depend on the most economical packaging available. Most line drivers and LDMs can be purchased in modular or stand alone form. Some terminal equipment also may have these units built in. If several units can be located in the same place, the best approach is to order them in a rackmounted enclosure as shown for the private line network in Figure 1.

In a polled system, only one driving unit, either modular or stand alone, is needed at the primary site. For systems in which the wire runs exceed the limits of the line driving units, repeater stations may be needed. If they are found to be necessary, it is a good idea to rethink the selection process, as the most economical approach may be to use a combination of units with different ranges or perhaps even substituting line drivers.

Comparing prices

The point at which LDMs begin to cost less than telephone modems of the same speed usually begins at 2.4 kb/s rates and rises dramatically with the data rates. For example, Gandalf Data Communications Ltd. offers a simple 2.4 kb/s LDM for a little more than $500, while the same company will sell a simple unit with three times the speed (9.6 kb/s) for about $200 more. Products that contain a full complement of features and operate at high speeds sell for $1,000 and up. In contrast, conventional telephone modems may sell for $900 or more for 2.4 kb/s speed, but 9.6 kb/s capability costs 10 times as much.

Prices of LDMs begin at about $250. Most units provide for simplex, half duplex, and full duplex configurations; synchronous or asynchronous operation; and built-in self testing. The lower priced units generally operate at the low data rates, offer meager self checking features, few if any adjustments for matching the lines' electrical characteristics, and make on-site data rate changes difficult or impossible. The more expensive units do not suffer from these deficiencies but instead offer extensive local and remote diagnostic capabilities; switch or strap selectable speeds and clocking arrangements; higher data rates; equalization, amplitude, and other line conditioning adjustments; and provide a variety of interface arrangements.

Data rates of most LDMs range between 2.4 and 9.6 kb/s. However, some units offer speeds down to 300 b/s or less and others reach data rates up to 1 Mb/s. The major features of various LDMs are listed in the table. Note that since most applications do not require data rates exceeding 9.6 kb/s and that distances rarely exceed 5 miles, the range usually is not a problem.

Hooking up the system

An important consideration in designing a system containing LDM's is selection of the most economical form of wiring. If the system is wholly situated on private property, cables can be strung, or telco lines can be leased. It is illegal to connect to unused carrier-owned lines strung within the premises without permission from the telephone company. Similarly, stringing private lines along public property is usually prohibited.

If the decision is made to use private lines, there may still be some legal problems, as indicated in the following incident, related by one LDM manufacturer, on problems of installing an internal telephone system.

A plant had installed private PBX equipment for internal telephone service. The building was built with troughs recessed in the concrete floors to house cable installed by the telephone company. At first, the plant management asked the telephone company for a quote on purchasing the cable, but the price was far beyond what had been budgeted for wiring the system.

The company then considered installing its own cables in the troughs, but found out that this, too, is forbidden because the telephone company has exclusive use of these troughs, even though it owns only the cable in them. In the end, the company had to resort to stringing the wire through drop ceilings, bypassing the cable troughs intended for telephone lines.

Wire sizes and types deserve special attention. Twisted-pair wire serves for relatively short distances, and coaxial cable may be needed for longer runs. For example, International Communications Corp. specifies that for its Model 1100, a data rate of 4.8 kb/s may be transmitted over 15 miles of #19 twisted-pair wire, but only five miles for #24 wire. At speeds near 1 Mb/s, these distances drop to 10% of the 4.8 kb/s distances. If, instead, 1100s are interconnected by video quality coaxial cables, the signals can be reliably transmitted as far as 17 miles at 4.8 kb/s and 2.4 miles at 1 Mb/s. Each manufacturer can provide a set of curves showing suggested bit rates for wire lengths for various wire sizes. An example is shown in Figure 4.

The manufacturer also should supply information on what signals appear on the various connector pins, the connector types, and the loading requirements. While most of this information is spelled out in the standard interface specifications such as RS-232-C or CCITT V.24, the user should make sure that all required signals actually are brought out to the connector. ∎

part 4 data data data data data data data data data

processors

Multiplexing cuts cost of communications lines

By concentrating traffic from many low-speed lines onto one higher-speed line, users can take advantage of the cost variations in transmission-line tariffs

Robert S. Smith
General DataComm Industries Inc.
Norwalk, Conn.

Data communications users usually multiplex their high-capacity traffic on leased voice-grade lines between major data-accumulation centers so that they can transmit large amounts of data at reasonable prices. By means of multiplexing, one high-speed link can carry the same amount of traffic that several low-speed ones could handle without it.

By taking advantage of the rate structures for the various types of leased lines available from common carriers, the user can lease one high-speed link for much less than a number of low-speed lines would cost. More than likely, the savings realized from multiplexing—concentrating—traffic into one high-speed line will pay for the multiplexing equipment.

How much the user can save depends on how common carriers have constructed their tariffs for leased lines. Intrastate tariffs will, of course, vary from state to state, but interstate tariffs—notably AT&T tariff FCC #260—demonstrate how multiplexing can cut costs.

The table lists specific #260 charges for three types of commonly used lines operating at different speeds. They represent the kinds of channels available to accommodate the slowest to the fastest terminals found in normal business applications. (The prices, subject to change, were in effect as of Oct. 1, 1972.)

For data rates up to 75 bits per second, the charge for 1005-type lines ranges from $1.58 per mile per month for the first 100 miles down to $0.34 per mile for distances over 500 miles. In addition, at every city serviced by the circuit, the telephone company charges $30.25 per month for the first service terminal—which consists of equipment and facilities providing the subscriber line—in a given exchange area and $9.08 for additional service terminals in the same exchange area.

The 1006-type lines provide 150-b/s service. Here, charges run from $1.93 per mile per month for the first 100 miles down to $0.39 per mile for more than 500 miles—the first service terminal costs $34.38 per month, and each additional one $10.32. A station arrangement—electrical interface equipment—at each termination costs an additional $13.75 per month per station.

The type-3002 voice-grade line is usable at data rates up to 9,600 b/s, the actual data rate depending on the modem and terminal equipment connected to the line. The telephone company can condition the 3002 line, which extends the line's bandwidth, and thus make it suitable for higher speeds and more subchannels.

For an unconditioned line, commonly called a C0 line, charges start at $3.30 per mile per month for the first 25 miles and are reduced to $0.82 per mile per

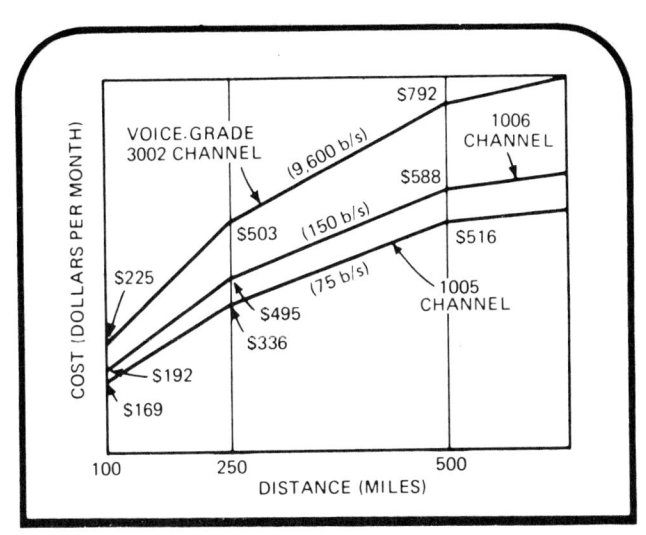

1. Tariffs. A voice-grade private line, which can handle up to 9,600 b/s, costs less than twice as much as either of the two lower-speed lines, thus offering savings with multiplexing.

month for more than 500 miles. Service terminals cost $16.50 a month for the first terminal and $11 for each additional one in the same exchange area. C1 conditioning adds $5 per month per station; C2, $19 per month per station, and C4, $30 per month per station.

Multiplexing saves line costs

Another view of the relative costs of lines is given in Fig. 1. Note that the voice-grade line always costs less than twice the price of the slower circuit, yet it can carry many more than twice the bit rates of the slower line. Thus, if more data can be added to a voice-grade line and some slower lines eliminated—the purpose of multiplexing—there is an excellent prospect for trading off the reduced line costs against the monthly cost of multiplexing equipment.

Consider an application in which a computer center in New York City communicates with branches in the western part of the country. As shown in black in Fig. 2, a nonmultiplexed configuration—which may have evolved as the company established more branches—uses four point-to-point 1006 lines. This requires eight service-terminal/station arrangements at a cost of $352 a month. The mileage cost for the four lines is $5,354 per month, bringing total communications cost to $5,706 a month.

The geographical disposition of these western locations indicates that multiplexing could be beneficial by, in essence, bundling all four direct lines into one multidrop 3002 line running from New York to Denver, to Las Vegas, to San Francisco, and to Portland, as shown in color in Fig. 2.

The cost for five service terminals (instead of eight) is now $83 per month, and the line mileage cost is $4,113, for a total of $4,196 per month. The net communications-link savings is thus $1,510 per month, probably more than enough to pay for the monthly cost of the multiplexing equipment needed to concentrate the four low-speed direct lines into one higher-speed line.

If a preliminary analysis indicates that multiplexing may save money, consideration must then be given to selection of either frequency-division multiplex (FDM), time-division multiplex (TDM), or a combination of FDM and TDM. As the names imply, FDM achieves its concentration by dividing the telephone-line frequency band into smaller frequency segments, and TDM allocates time segments to the various traffic channels. Generally, FDM is used when one line must service many terminals in a local area through a multidrop arrangement. In contrast, TDM is usually used in long-distance point-to-point configurations, typically between major cities.

FDM divides band

In FDM, a voice-grade line with a certain bandwidth, typically 3,000 hertz, is split into narrower channel segments, often called derived channels or data bands. Each channel, tuned to a specific frequency, has a transmitter at one end and a receiver at the other end. Binary characters are indicated by some type of presence or absence of signals at each data channel's assigned frequency. The width of each frequency band determines the data-rate capacity of the subchannel. For example, the band can be made narrow enough to pass only 75-b/s data streams or wide enough for 150 b/s, or even as many as 300 b/s, if required by the terminal speeds. Guard bands separate the data bands to prevent data on one channel from interfering with data in the adjacent channel.

FDM is code-transparent; that is, once the data band is set, for example, to carry up to 150 b/s, any terminal device operating at 150 b/s or less can be run through that channel without concern for code format. Therefore, the channel can be operated at 110-b/s Teletype speed for part of the day, and then be switched to an IBM 2740 communications terminal at 134 b/s for another part of the day. Generally, FDM is used for low-speed asynchronous terminals.

Typical data-band spacings are shown in Fig. 3. The CCITT (International Consultative Committee for Telegraph and Telephone) series provides 17 75-b/s channels on an unconditioned (C0) line, 19 channels on a C1 conditioned line, 22 channels on a C2 line, and the full 24 channels on a C4 line. These data bands were established many years ago by the common carriers and governmental agencies so that the channel spacings were

```
INTERSTATE LINE COSTS PER MONTH
AT&T TARIFF FCC No. 260

1005 LINE (UP TO 75 b/s):
  Mileage..................... $1.58 per mile to 0.34 per mile
  Service terminal ............ 30.25 (first one)
                                9.08 (each additional unit)

1006 LINE (UP TO 150 b/s):
  Mileage..................... $1.93 per mile to 0.39 per mile
  Service terminal ............ 34.38 (first one)
                                10.32 (each additional unit)
  Station arrangements ........ 13.75 each, per station

3002 LINE (UP TO 9,600 b/s):
  Mileage (unconditioned) ..... $3.30 per mile to 0.82 per mile
  Service terminal ............ 16.50 (first one)
                                11.00 (each additional unit)
  Conditioning (per station) .. C1  $ 5.00
                                C2   19.00
                                C4   30.00
```

61

geared then—as now—to keyboard terminals.

For higher data rates, the CCITT standards also allocate 12 and 6 channels from a voice-grade line (Fig. 3). These channels are simply frequency multiples of the 24-band standard. These CCITT frequencies, although in international service, are also used to derive channels on voice-grade lines within the United States. However, with the development of terminals operating at speeds that don't match CCITT derivations, new families of FDM equipment have been developed (for example, General DataComm Industries' Chan-L-Stak) to split the voice-grade line into data rates required for the 110-b/s and 134-b/s terminals in common service today. Wider FDM subchannels now provide more efficient use of voice-grade lines than the older CCITT standards.

Functionally, FDM consists of a section of common equipment and a channel set for each derived channel serviced by the multiplexer (Fig. 4). The channel sets are independent of each other. Thus, if only five 75-b/s channels are needed on an unconditioned voice-grade line, only five channel sets, not the full 17, are installed. The only cost for later expansion of the system is for the addition of necessary channel sets—without additional line charge. Typically, a computer center houses the common equipment and the channel sets needed to get in and out of the computer. An FDM channel set would also be located with each terminal.

FDM operates on two or four wires

FDM normally provides full-duplex operation on a four-wire circuit. Each channel set has a transmit and a receive section. All transmit tones go out on one pair of wires, and all receive tones come back on a second pair. However, FDM can also operate full-duplex on a two-wire system. For example, with 24 channels available on the line, one channel set is tuned for line-channel 1 to transmit, and channel 13 receive; another for channel 2 transmit, and channel 14 receive; and so on. Here, the number of data channels is halved, but this technique will save the difference in costs between four-wire and two-wire lines in a network servicing a small number of full-duplex terminals.

TDM uses full bandwidth

In contrast to FDM, time-division multiplexing utilizes the full bandwidth of the line, but divides it into a sequence of time segments, with each segment assigned to a character (or a bit) from each low-speed terminal. The fundamental operation of TDM is shown in Fig. 5. Here, each low-speed terminal is assigned a channel register that accumulates bits representing, say, a character in a message. The TDM's common equipment generates a synchronizing pattern, and then a scanner circuit interrogates each channel register in time sequence, packing the bits in each character into a continuous, synchronous bit stream that then goes through a high-speed modem down the voice-grade line to a demultiplexing equipment at the other end. This cycle repeats continuously so that the second, third, fourth, and subsequent characters from each terminal appear at the receiver. In this manner, the TDM equipment handles data from many low-speed terminals and concentrates it into one long high-speed synchronous bit stream.

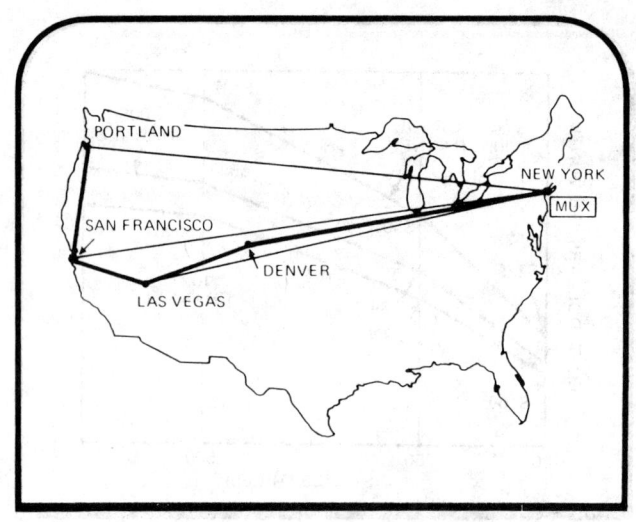

2. Alternative. Connecting four outlying cities to a host computer can take four direct private lines, shown in black, or, at less cost, with one multiplexed, multidropped line, in color.

At the receiving end, similar equipment works in reverse to demultiplex the bit stream and assemble the characters into coherent messages from each sending terminal. The multiplexers at each end of the leased line are essentially identical, and each channel has data registers to transmit and receive data, as shown in Fig. 6. This multiplexing technique is called character interleaving. An alternative, called bit interleaving, scans one bit from each register in sequence.

Compared with FDM, TDM requires more equipment in common, as shown in Fig. 6. Further, operating in a digital mode, TDM requires a modem to interface with the line, whereas FDM, already operating in a frequency mode consistent with the voice line, does not need a modem. Often, TDM makers will include the modem as an option, particularly for operation at 1,200, 1,800, and 2,400 b/s. At 3,600 b/s and faster, the modem must be selected as a separate stand-alone unit.

Needs dictate choice

While FDM and TDM are often thought to be competitive techniques, the choice of one or the other is usually dictated by different parameters; in fact, most data-communications systems contain both types. The main considerations in multiplexing are channel capacity, equipment costs, and line configuration.

CHANNEL CAPACITY. With FDM a C4 conditioned line can be split into 24 channels, each carrying 75 b/s, for a total line capacity of 1,800 b/s. And other channel/speed combinations can be obtained. For example, the same C4 line can carry 16 channels at 134 b/s each, for a total of 2,144 b/s. Thus, the maximum throughput of the voice-grade line is in the range of 1,800 to 2,100 b/s. With TDM, the capability of the modem sets the maximum line data rate. Thus, on the same C4 voice-grade line, the maximum data rate can be 9,600 b/s. And, with special wideband lines available from the carrier and with an appropriate modem, lower-speed channels can be multiplexed up to 40,800 b/s.

EQUIPMENT COST. For preliminary planning purposes, the following prices can be used to estimate FDM

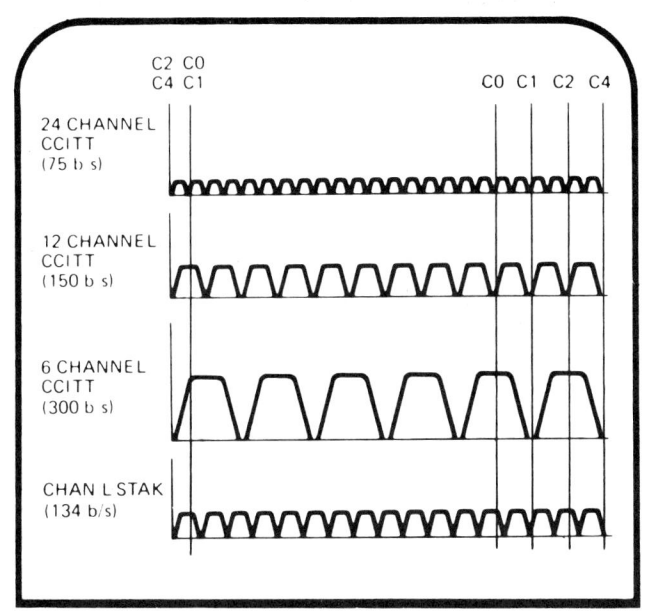

3. Conditioning. With proper adjustment of the line's electrical characteristics, called conditioning, the voice-grade line can (with C4 conditioning) provide up to 24 75-b/s channels.

and TDM equipment costs. In FDM, the common equipment capable of handling from nine to 16 low-speed channels costs about $600. Each channel set at each end of the line costs about $500. As mentioned earlier, FDM entails no modem cost.

In TDM, the common equipment to handle about 25 or 30 low-speed channels costs about $3,000. Each channel end, with full diagnostic and network-control features, requires about $175 worth of equipment. In addition, each end of the high-speed line requires a modem, the cost depending on the composite bit rate. For planning purposes, a 2,400-b/s modem costs $1,500 to $2,000; a 4,800-b/s modem, $4,000 to $5,000; and a 9,600-b/s modem about $9,600. Note that these are equipment-purchase costs. The equivalent monthly cost can be determined by dividing the purchase cost by 28, the actual divisor depending, however, on the specific terms of a lease contract or how the buyer chooses to amortize capital investments.

LINE CONFIGURATION. FDM equipment is nested at the host computer location. One voice-grade line can be multidropped to serve many dispersed terminals when there is a geographical relationship like that shown in Fig. 2. The basic application of time-division multiplexing is to provide high-speed data rates between two widely separated points on a leased line—for example, between New York City and Los Angeles. However, as shown at the top of Fig. 7, TDM can also be used in a form of multidrop. Here, a relatively uneven split in number of channels exists between a central city (Chicago) and the two end-point cities.

As shown, 30 channels are multiplexed onto a 4,800-b/s line in Los Angeles. They enter a multiplexer in Chicago, which splits out seven channels for distribution to local terminals, and the remaining 23 channels are forwarded on to another TDM in New York City. The only thing the Chicago TDM does to the channels going to New York is to buffer and retime the data. This concept can be extended to include several more midpoint drop-off locations, but the added cost of TDM common equipment and modems may obviate this approach for more than three or four locations.

TDM can be split

Another way of TDM multidropping, shown at the bottom of Fig. 7, is to install a "split" TDM at the computer center in Los Angeles, run to Chicago at, for example, 4,800-b/s, and drop off about half the channels through a smaller multiplexer. The rest of the channels run on a 2,400-b/s line to New York City where, again, a smaller multiplexer distributes the channels to appropriate terminals. Because of the lower speed on the Chicago-New York City link, less expensive modems can

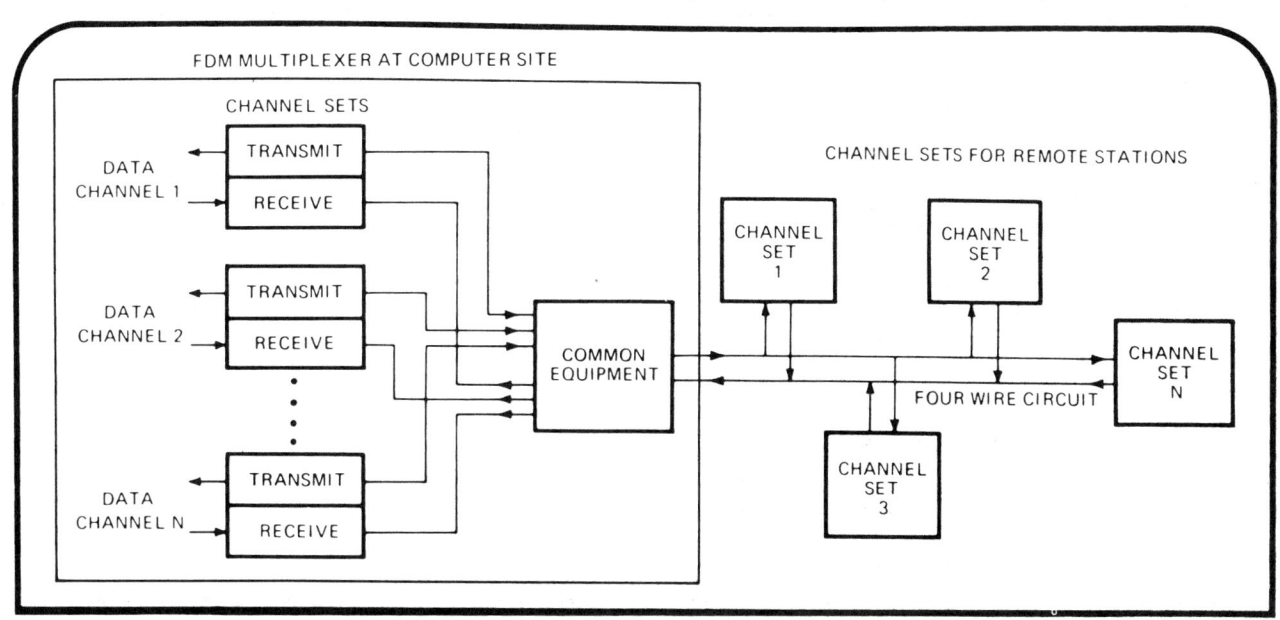

4. Multidrop. Frequency-division multiplexers use a common equipment, along with individual receive or transmit channel sets, at the computer site, and the individual channel sets at each remote site are dropped from one four-wire link.

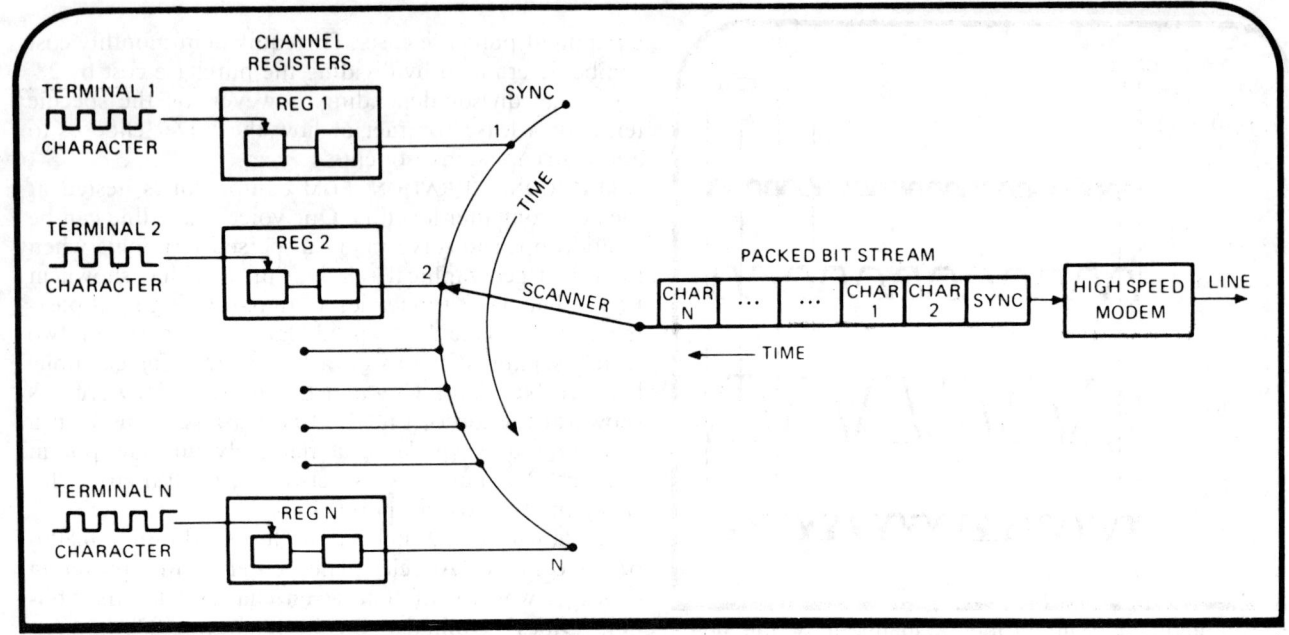

5. Packed. In time-division multiplexing, a channel register assigned to each terminal accumulates bits from each character, and then the characters are scanned and transmitted in sequence by the multiplexer as a high-speed packed data stream.

be used. Costs of the line and TDM equipment remain about the same for both methods of multidropping.

In summary, FDM has the advantage of greater multidrop capability, it is code-transparent, and, for systems of about 10 or fewer channels, will probably cost less than TDM. Thus, even when TDM is used for transmission at high data rates between two widely separated cities, FDM is often used in the same system to multidrop channels to clustered terminals in the drop-off areas. TDM achieves more efficient utilization of a voice-grade line because of the multiplexing technique. It can mix asynchronous and synchronous channels and can handle higher terminal rates. On large systems, the over-all TDM equipment costs will be less than those of FDM. However, since TDM actually restructures the code, and often strips out start-stop bits from asynchronous terminals to raise net data throughput, the channel equipment in TDM is code-sensitive. Thus, TDM requires programing for a particular code and channel data rate.

An actual user-owned on-line data-entry network (Fig. 8) employs a combination of TDM and FDM equipment operating on 3002 voice-grade lines. The computer center is in Dallas, with time-division multiplexing links fanning out to regional areas served by Orlando, Washington, Chicago, and Los Angeles. Extending from these regional centers are FDM circuits to various local points where FDM channel sets serve IBM 2740 communications terminals used by the operators.

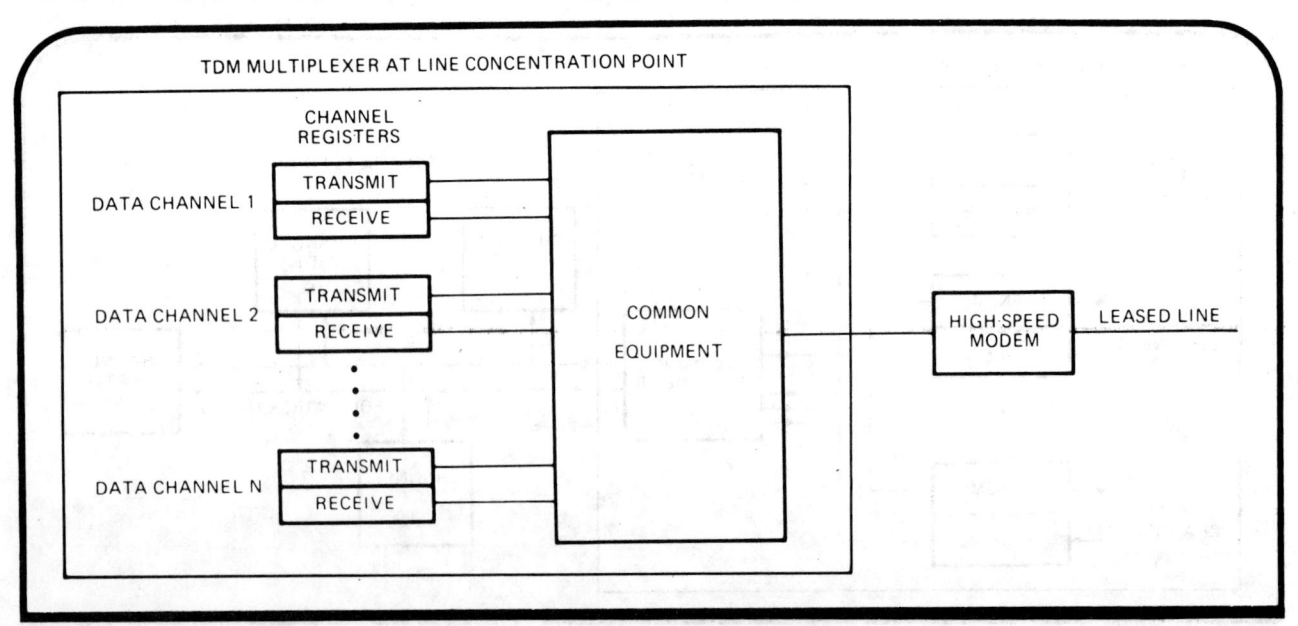

6. More. Compared with FDM, the time-division multiplexer employs more equipment in common at the location at which the low-speed lines from the terminals are concentrated, and each data channel requires a pair of channel registers.

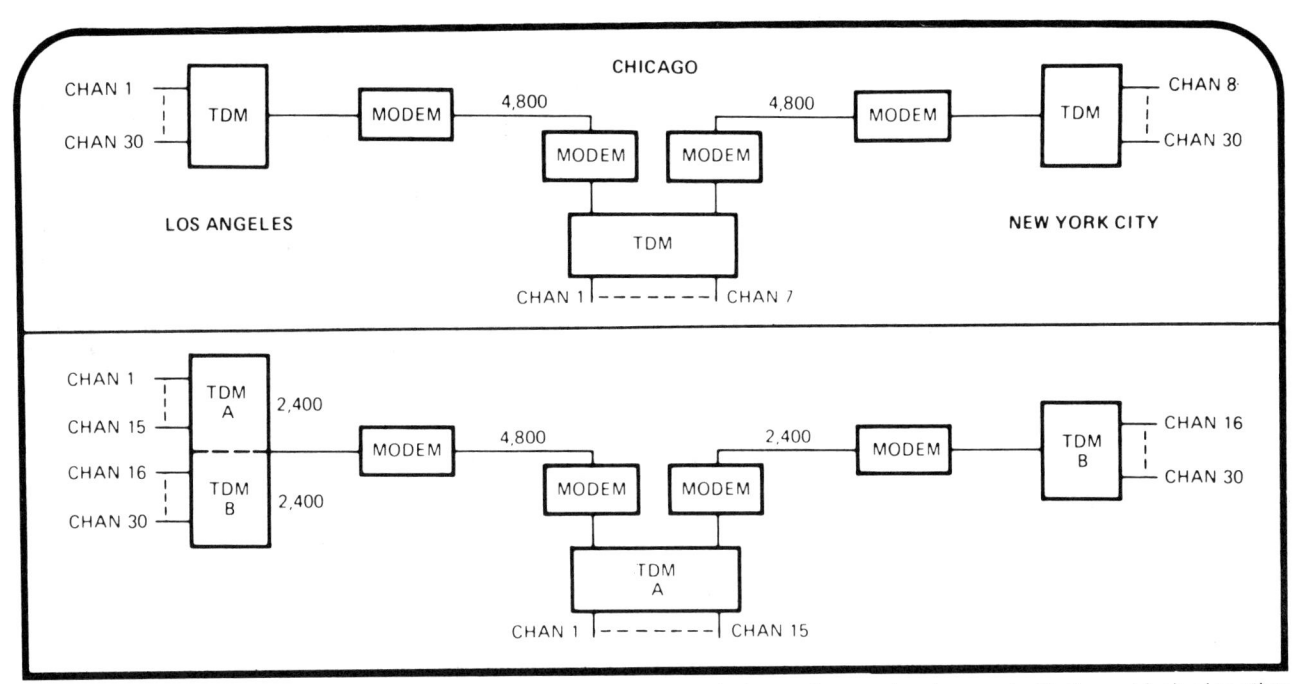

7. Channel splitting. In the top multidrop configuration, several channels may be dropped off at a mid-city location with the same data speed forwarding the balance of the channels—but, at bottom, the split multiplexer permits use of lower-speed modems.

This configuration is typical of many user-owned, private communications networks. Because neither the numerous equipment vendors nor telephone companies can be responsible for total system maintenance, the company's communications manager must supervise maintenance to keep the data-communications system operating. Since technical expertise tends to be located at the computer center, this appears to be the logical place to establish maintenance management. And to manage maintenance from one location, each major piece of equipment should include remote control and diagnostic capabilities.

Diagnostic capability cited

Typical of diagnostic capability is that contained in General DataComm's multiplex equipment and in modems from many vendors. Take as an example the Dallas computer center, the high-speed link to Orlando, and the Orlando terminal concentration, as shown in Fig. 9. The communications group in Dallas can, first, call for a loopback—denoted by curved arrow—around the Dallas TDM, to check the integrity of that equipment. If all is well, the Orlando TDM outputs can be checked channel by channel from Dallas, looping each channel back, without affecting operation of other channels. Also controlled from Dallas is the channel-by-channel checkout of the Orlando FDM equipment, as well as the FDM subsets right at the interface of the IBM 2740 terminals. Included in the discipline of fault diagnosis is the ability to determine from Dallas when any line is not working properly, even if it's within the jurisdiction of the phone company in Florida.

Related to remote fault diagnosis is the capability to repair the equipment. Often this is merely a matter of replacing, from a set of spares, a module isolated by the diagnostic routine. This replacement can be easily accomplished by a company employee—with very little technical training—at the site. More difficult repairs may require an in-house service group or assurance that equipment vendors have properly trained maintenance men suitably deployed to reach the trouble site quickly. Since a data-communications system is an integral part of a company's operation, down-time can be costly. □

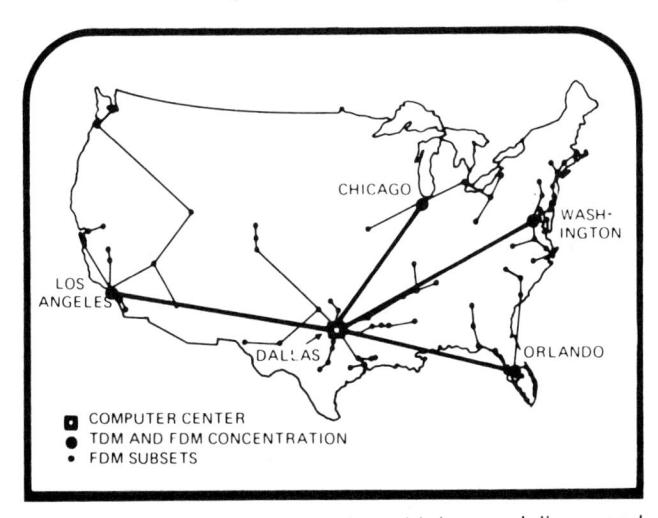

8. Nationwide. Network employs high-speed lines and TDMs between regional centers, FDM subsets locally.

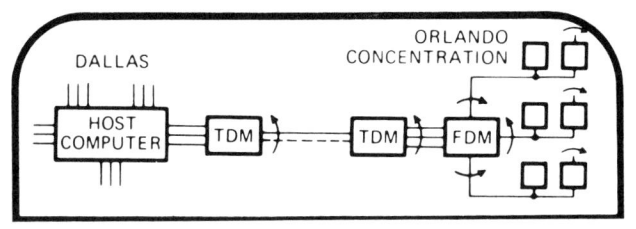

9. Big brother. At console in Dallas, diagnostic routines can be initiated to check out the Dallas and Orlando equipment.

Programable front-end processors

Based on a minicomputer, the programable front-end processor serves as a line controller and relieves the host computer of most of its data-communications overhead

David Stackpole
*Digital Equipment Corp.
Maynard, Mass.*

When computer-based data-communications systems began coming into use in the 1960s, the preferred interface between the host computer and the communications lines was the hard-wired transmission controller—notably IBM models 2701, 2702, and 2703 (or 270XJ). Occasionally even then, small computers, called data-communications preprocessors, were installed at the front of the host computer by such large-computer makers as Burroughs and Digital Equipment Corp.

Then, several minicomputer makers began to advocate actually replacing the hard-wired units with programable front-end processors. The idea took hold so well that in early 1972 IBM announced its own model 3704/5 series of programable communications processors.

A programable front-end processor (PFEP) often costs less than a hard-wired controller. Its real advantage, however, is its ability to free the host computer's internal memory, software, and execution time of much of the burden of data communications. First, compared with the hard-wired unit, it makes much less demand on the host computer for line control. Second, being far more versatile than the hard-wired unit, it can take over such data-communications tasks as polling, code conversion, formatting, and error control.

In a well-designed data processing system, with the computer operating in a batch-data processing mode, average computer utilization is about 75% to 80%. The balance of 20% to 25% has to be left available as a reserve for servicing peak loads. But adding a data-communications mode increases the average load by perhaps another 20%, eliminating the reserve capacity.

In this situation, if the extra load is handled by upgrading the host computer to a larger mainframe or more memory or both, thousands of dollars will be added to the lease cost of the system. But if retaining a peak-load reserve is handled by adding a PFEP, it costs relatively little to buy the PFEP, develop suitable software and pay the manpower to install, test, and start it up. A cursory analysis of computer utilization and comparative costs will quickly reveal whether it is economical to opt for a PFEP.

While only the simplest data-communications systems will in the future employ hard-wired controllers, those controllers now in operation are candidates for replacement by a PFEP. But the line-control discipline implemented by the hard-wired controller is so thoroughly embedded in the technology of computer-based data communications that the PFEP will, for some time to come, be initially programed to emulate the hard-wired

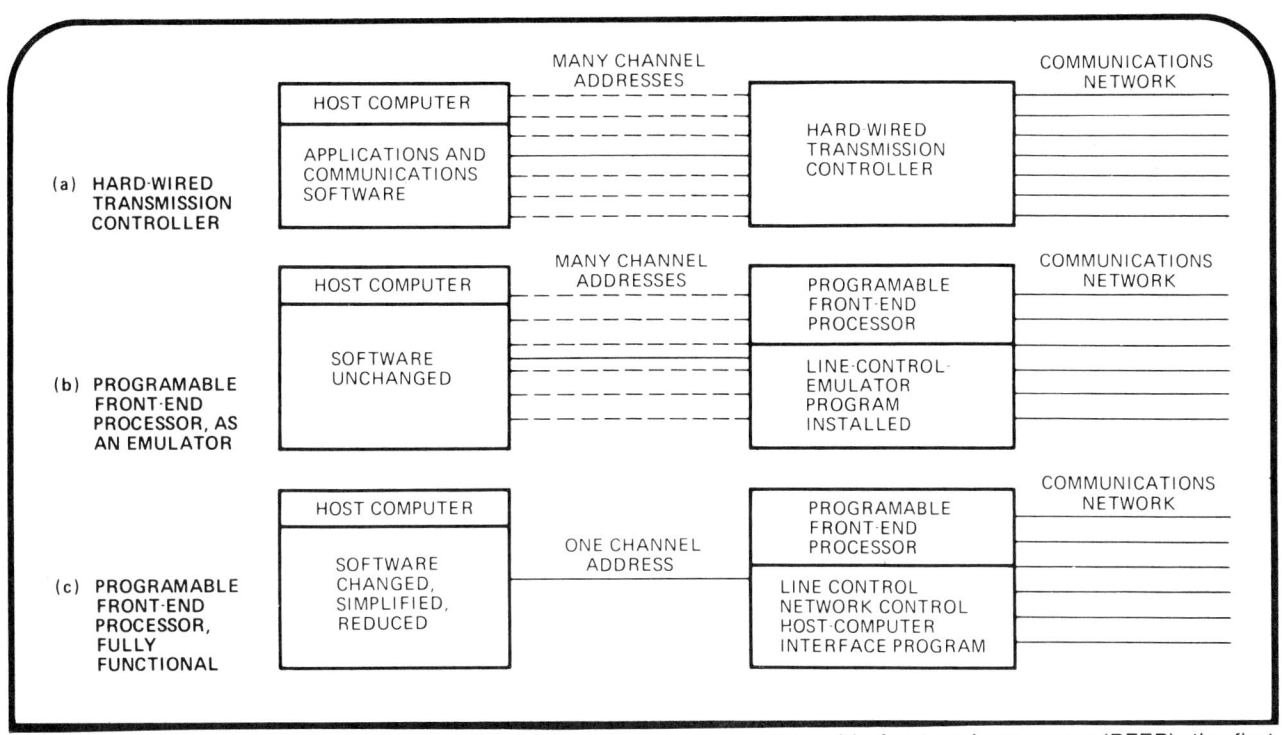

1. Evolutionary. When a hard-wired controller (a) is replaced by a programable front-end processor (PFEP), the first step (b) is to make the PFEP emulate the controller, and the second step (c) is to move communications programs from the computer to the PFEP.

controller. Thus the installation of a PFEP occurs in two major steps. The first is the programing of the PFEP with the line-control functions of the hard-wired controller (Fig. 1a and b). The second is the transference from the host computer to the PFEP of much of the software related to specific data-communications functions (Fig. 1b and c). The result is to make the network managed by the PFEP appear to the host computer almost like a single input/output device that is ideally compatible with the design and operation of the host computer.

As a preliminary to discussing the major project management and software aspects of getting a front-end processor on line, it is instructive to review the complex situation at the interface between the transmission lines and the host computer. A typical data-communications system includes several types of terminals as well as multiplexers and perhaps a remote data concentrator, operating asynchronously and synchronously and with several codes and speeds. Thus, when all bits in all characters in all messages converge at the central computer site, the host computer is confronted with a random, interleaved, and intermittent data stream from all on-line terminals and other devices.

The host computer is not designed to directly process these diverse inputs. The bits comprising a character are entered one after the other at the terminal and continue as a series along the transmission link to the computer. But the host computer takes in all the character bits in parallel, at one time. In computer terms, this bit-parallel character is called a byte. Moreover, the host computer can accept bytes at a rate at least 1,000 times faster than characters come off a high-speed voice-grade line. It would be wasteful for such a machine to spend its time, worth perhaps $5 to $20 a minute, on slowly accumulating bits and converting them into bytes.

Instead, this job is done by hardware registers in the interface. There, each serial-bit character coming from a given terminal is directed into that terminal's interface channel and converted into a parallel byte for the computer. The computer in turn temporarily stores that byte in a dedicated address space in its internal memory, from which the byte, or character, moves either to another area in internal memory or to a secondary memory that accumulates words and messages. One address space is needed for each line.

All this the hard-wired controller does under control of communications programs stored in the host computer (Fig. 1a). When the PFEP is used only as an emulator (Fig. 1b), it operates under control of the same communications software. And it too has as many channel addresses between the host computer and the PFEP as there are lines into the PFEP. But note that in the true front-end processor (Fig. 1c), just one channel is used by the PFEP to communicate with the host computer. This means that only one address space is needed in the host computer, leaving the more space for use by other peripherals. Also the access-method program, which "connects" input lines to the application program, is simplified, since the host computer interfaces with only one line, not many, to the PFEP.

Acting solely as an emulator, the PFEP offers no real economic advantage since the data-communications overhead still burdens the host computer. This overhead consists of extensive manipulation, or preprocessing, of data that must be accomplished by the host computer before it can actually operate on the messages. These preprocessing programs require host-computer execution time and internal memory, both of which

67

may be in short supply in a given application. Depending on installation size, the amount of internal memory dedicated to data-communications tasks will range from 20,000 to 60,000 bytes.

When most of the overhead is transferred out of the host computer and into the PFEP, the front-end processor will, among other things:
- Poll the input and output devices to determine whether an information transfer should take place.
- Restructure incoming data to more compact forms to increase host-computer input efficiency.
- Convert data to the code most suited to the host data processing computer.
- Check incoming data to make sure that errors are not present, and reject data blocks if errors are present.
- Route messages from one terminal to another, that is, perform message switching, without data having to enter the host computer at all.

When these communications chores are performed within the PFEP, only about 10,000 to 20,000 bytes, out of the original 20,000 to 60,000 bytes, remain in the host computer to service a vestigial form of control program. Thus, 10,000 to 50,000 bytes of the host computer's internal memory, together with related instruction-execution time, are freed for data processing. Though about the same amount of memory will be needed in the PFEP, removing these many bytes in the host computer may obviate the necessity of leasing more core memory which, depending on the particular host computer, may have to be obtained and paid for in increments as large as 256,000 bytes. On the other hand, more memory for the PFEP's minicomputer can be added in increments as small as 8,000 or 16,000 bytes.

No channel limitation

Conserving the host computer's internal memory and execution time is not the only reason for preferring a front-end processor to a hard-wired transmission controller. Another is that the PFEP's programability makes it easier to change the system—when, for example, substituting one type of remote terminal for another—without having to make actual wiring changes at the host computer. A third reason is that a minicomputer, the kernel of the PFEP, is designed to handle more, and more varied, types of input and output devices than is the large computer, which is primarily designed to rapidly process large batches of data.

Thus, in its data-communications mode using a hard-wired controller, the host computer may be restricted in the number of lines it can handle either by the limit of the computer's channel addresses (address space) or by the fixed number of channels available on a particular hard-wired controller. In either case, the installation is channel-limited, not throughput-limited.

The minicomputer, on the other hand, is not channel-limited but throughput-limited. By way of example, if the minicomputer has an instruction execution time of 1 microsecond, it can perform 1 million instructions each second. Suppose that sampling a line to tell if a pulse is a 1 or a 0 takes 25 instruction times and that, to define the start and stop edges of a bit's pulse, each bit is sampled eight times. Thus, each bit uses 200 instruction times. Therefore, the minicomputer can service this throughput as eight (input or output) lines at 600 b/s, 30 lines at 150 b/s, 45 lines at 110 b/s, or some combination of speeds that does not exceed 5,000 b/s. More than likely, assuming a random traffic pattern, the PFEP could handle even more lines. Moreover, if systems analysis indicates that the instantaneous data rate may exceed 5,000 b/s, then the peak data can be stored at the PFEP, in a mini disk pack, for inputting (or outputting) when traffic slows down. This small mass memory can also augment the host computer's larger memory.

However, most present minicomputers use inexpensive hardware interfaces to perform serial-to-parallel conversion of bits to bytes. Depending on the particular minicomputer, throughput can thus be increased by 5,000 to 20,000 bytes, or characters, per second.

The whole business of a programable front-end processor unburdening a host computer is easy to grasp. The practical problems lie in performing a technical

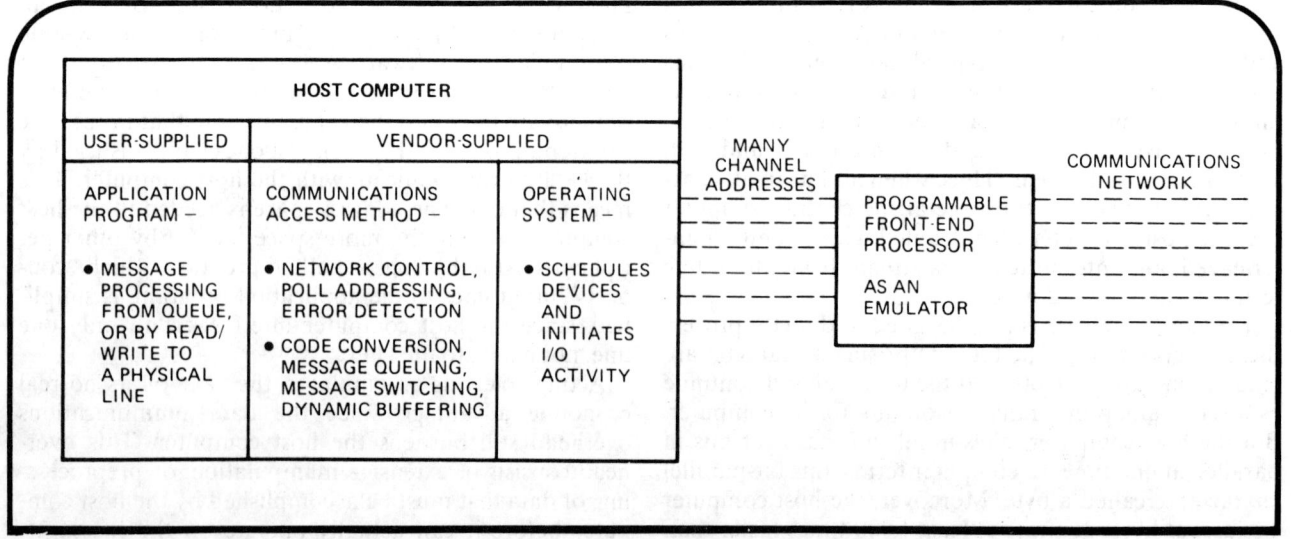

HOST COMPUTER				
USER-SUPPLIED	VENDOR-SUPPLIED		MANY CHANNEL ADDRESSES	COMMUNICATIONS NETWORK
APPLICATION PROGRAM — • MESSAGE PROCESSING FROM QUEUE, OR BY READ/WRITE TO A PHYSICAL LINE	COMMUNICATIONS ACCESS METHOD — • NETWORK CONTROL, POLL ADDRESSING, ERROR DETECTION • CODE CONVERSION, MESSAGE QUEUING, MESSAGE SWITCHING, DYNAMIC BUFFERING	OPERATING SYSTEM — • SCHEDULES DEVICES AND INITIATES I/O ACTIVITY	PROGRAMABLE FRONT-END PROCESSOR AS AN EMULATOR	

2. Software. When connection to a PFEP used as an emulator, the host computer contains the operating system and the communications access method, both supplied by the host-computer vendor, and the applications programs, developed by the user.

HOST COMPUTER				PROGRAMABLE FRONT-END PROCESSOR		
APPLICATION PROGRAM	SEQUENTIAL ACCESS METHOD	OPERATING SYSTEM (OS)	ONE CHANNEL	HOST-COMPUTER INTERFACE PROGRAM	NETWORK CONTROL PROGRAM	LINE CONTROL PROGRAM
MESSAGE PROCESSING	MESSAGE ACCESSING	I/O CONTROL		MAGNETIC-TAPE EMULATOR	MESSAGE CONTROL	

3. Magnetic tape. One approach to having the PFEP serve as a single peripheral to the host computer is to make the PFEP seem to be a magnetic tape unit, with the host computer using the sequential access method suited to reading and writing on tapes.

and economic analysis to assure that a PFEP of adequate size, including all necessary peripherals, is specified, and in managing the software development.

Installing a front-end processor

Assuming the data-communications system is already operating with a hard-wired controller, the first—and easier—step is to remove the present hard-wired transmission controller and immediately replace it with a front-end processor programed as an emulator. Fortunately, minicomputer vendors who seriously pursue the PFEP market have developed their own software packages to emulate the functions of hard-wired controllers. This line-control program can be loaded into the PFEP and completely tested at the vendor's site.

Thus, about all that has to be done at the user site is to physically transfer communications wires from the transmission controller to the PFEP—a matter of perhaps four to eight hours—and connect power to the minicomputer. Essentially, then, the PFEP is plug-to-plug compatible with the transmission controller. A few hours of tryout will prove whether the emulated controller works as well as the hard-wired unit.

The balance of the conversion—the development of the programs that will transfer data-communications tasks from the host computer to the PFEP—will take about one to six man-months of programing effort. Implementing the conversion during nonproduction hours is preferable, since it reduces operational interference and means the hard-wired controller can be returned to the supplier, resulting in savings of lease cost.

However, if there are no nonproduction hours—that is, if the system must remain on line all the time—then the hard-wired transmission controller must also remain on site for the balance of the conversion period. Then, one way to handle the conversion project is to connect the input and output lines to channels in both the PFEP and the transmission controller. Each unit is assigned its own set of channel addresses in the host computer. During normal operation, the computer addresses the channels assigned to the transmission controller, and during the software development it addresses the channels assigned to the PFEP.

But if the host computer is out of addressing space, the PFEP could be assigned the same address spaces as the transmission controller. Doing so requires frequent transfer of cables—or a complicated switching arrangement—from the hard-wired unit to the PFEP and back again as the project progresses and new pieces of programs need to be installed, tested, and debugged. This approach is considered unsatisfactory by most data processing managers because of the added lease cost and the possible interference with the data processing peripherals and operations.

Improving systems software

After the line-control program emulating the transmission controller has been installed in the PFEP, the host computer still contains three major kinds of software—the operating system, the application program, and the communications access method (see Fig. 2).

Figures 3 through 6 show various software tradeoffs between the host computer and the fully functional PFEP. These alternatives obey two pragmatic rules: vendor-supplied software is not to be modified, and user application software will have to be changed.

The operating system (OS) serves as a master program that dictates when data processing and data-communications programs become operational, and that schedules peripheral and initiates input/output activities. The OS, a massive undertaking of the host computer maker, serves hundreds and perhaps thousands of computer installations. It is subject to changes and improvements, called releases, that must be compatible with all OS installations. Therefore, the OS should not be modified by the user because such changes may not be compatible with future OS releases.

The application program relates to the specific needs of the business implemented via a data-communications link. It is usually developed by the user, but it must be compatible with the access-method and interface programs in the rest of the system.

The access-method program, however, which serves as a traffic director between the network-control program and the application program, offers a real area of savings in both memory and execution time. A host

[Diagram: Host Computer 360/370 containing Application Program, Graphic Access Method (GAM), Operating System (OS), Message Processing, Message Accessing, I/O Control — connected via ONE CHANNEL to Programable Front-End Processor containing Host Computer Interface Program, Network Control Program, Line Control Program, IBM 2848 CRT Control Unit Emulator, Message Control.]

4. Graphic access. Another way to interface the PFEP to the computer employs the graphic-access method (GAM), in which the PFEP's interface program emulates an IBM 2848 CRT display control unit, which can generate an interrupt to flag the computer.

computer may receive inputs from one or more hard-wired controllers, each servicing as many as 176 channels. The access method for the hard-wired controller, or an emulator, will have to look at each input/output channel and will require as many special modular programs as there are different types of terminals connected to the controller. This modularity and the multiplicity of channel addresses reduces programing efficiency and consumes more memory.

Interfaced by a true PFEP, however, the host computer benefits from a special-purpose access method that—generally—reads and writes data from only one input/output device. Thus, going to a true PFEP simplifies the access-method software, and reduces the number of address spaces and amount of required memory.

The PFEP behaves like one peripheral—a magnetic-tape unit, for example—as far as the host computer is concerned. It concentrates the messages from all incoming terminals into a data stream fed into the host computer. Thus, the special-purpose access method has to continuously interrogate the concentrated data stream from the PFEP to determine the source and address of the individual messages from different terminals. Therefore, the access method has to block and unblock messages, just as it would do if the input peripheral were a magnetic tape.

Magnetic-tape emulator

In fact, the most popular special-purpose access method used in data communications is actually a sequential access method derived from a non-communications application. As Fig. 3 shows, the PFEP contains the programs for line and network control, as transferred out of the host computer. In addition, the PFEP mates with the host computer through an interface program that emulates a magnetic tape. As far as the program for that sequential access method in the host computer is concerned, it is receiving information from and sending information to an emulated magnetic tape. In this case, the application program in the host computer now has to be changed to interface with a program that delivers data from one high-speed peripheral (the emulated magnetic tape) instead of from many different physical lines, as was the case with the transmission controller.

Graphic access method

When the host computer is an IBM model 360 or 370 operating under OS, the best non-communications access method to use is GAM (for graphic access method). GAM was designed specifically to access one input/output device, the IBM 2848 CRT display control unit for serving several IBM 2260 CRT displays. Thus, as shown in Fig. 4, the PFEP's interface emulates a 2848, while the host computer contains GAM. Emulating the 2848 as the host-computer interface is fairly simple and efficient, since this display control unit requires a small command set. Two of these commands are READ and WRITE, the only ones actually needed to transfer data into and out of the host computer.

In addition, GAM is the only access method supplied by a host-computer vendor that can accept the ATTENTION interrupt from the front-end processor. That is, whenever the PFEP has a message for the host computer it "raises" an interrupt signal. With such an interrupt available, the PFEP does not have to be periodically polled by the host computer, so that the polling overhead in the host computer is reduced.

Furthermore, GAM is actually transparent to data codes and message length. Therefore, the application programs do not have to make the host computer think the interfaced device is an emulated display with a fixed-size buffer. Avoiding emulation means that the application program in the host computer can interface directly and efficiently with the network-control program in the PFEP. In all, GAM requires fewer than 5,000 bytes of internal memory in the host computer.

Subset access method

Figure 5 illustrates an accessing approach that contains several subsets of a common communications access method, yet requires only one channel between the PFEP and the host computer. Here, the access method uses subsets of the IBM 360 BTAM (for basic telecommunications access method). For example, if the network includes such low-speed asynchronous termi-

HOST COMPUTER 360/370				PROGRAMABLE FRONT-END PROCESSOR		
APPLICATION PROGRAM	BTAM ACCESS METHOD	OPERATING SYSTEM (OS)	ONE CHANNEL	HOST-COMPUTER INTERFACE PROGRAM	NETWORK CONTROL PROGRAM	LINE CONTROL PROGRAM
MODIFIED FOR BSC*	BINARY SYNCH ONLY			IBM 2703 TRANS-MISSION-CONTROL EMULATOR (BSC* ONLY)	MESSAGE CONTROL	
MESSAGE PROCESSING	MESSAGE ACCESSING	I/O CONTROL				

*BINARY SYNCHRONOUS COMMUNICATIONS PROTOCOL

5. Subset. Another way of connecting the PFEP with the host computer is to make the PFEP emulate an IBM 2703 transmission controller, and to use subsets of the basic telecommunications access method (BTAM) in the host computer.

nals as Teletype Model 33/35 automatic sender/receivers, and IBM 2741 communications terminals, as well as high-speed, synchronous, block-oriented, terminals operating under binary synchronous communications (BSC) protocol, then the access program requires a BTAM software module for each of these terminals. The PFEP's interface program to the host computer emulates only that portion of a 2703 transmission controller related to the BSC protocol.

In this approach to front-ending, the BTAM program thinks it has to receive and send data, on only one binary synchronous line, through the network control program in the PFEP. This approach is not as desirable as those of Figs. 3 and 4 because its software modularity leaves many inefficiencies.

Front-end access method

The most efficient access method is not a variation of existing ones as described above, but software designed specifically to interface with the front-end processor itself (see Fig. 6). This avoids many of the inefficiencies inherent in emulating some other device. Such a PFEP-oriented access-method program is installed in the host computer, and may be supplied either by the host-computer vendor or the front-end vendor.

Using an access method customized by the host-computer vendor is perhaps the safest thing to do, but it may introduce some inefficiencies if the host-computer program has to be forced to conform to PFEP requirements. Using an access method customized by the PFEP vendor is more efficient. The only possible disadvantage is that the operating-system interface with the access method might be changed by new releases from the host-computer vendor, a most unlikely event since such a change would impact too many access-method programs installed in other host computers.

In sum, though making all these software transfers is fairly complex and can be fraught with difficulties, the way to minimize problems, as mentioned, is to install, test, and debug each software element in succession. The first step is to install and prove the emulation program for line control. The next is to try out a demonstration program that uses the new access method and the true front-end software as supplied by the front-end vendor. Last, the user can make and debug changes in the application programs. □

HOST COMPUTER				PROGRAMABLE FRONT-END PROCESSOR		
APPLICATION CONTROL PROGRAM	PFEP ACCESS METHOD	OPERATING SYSTEM (OS)	ONE CHANNEL	HOST-COMPUTER INTERFACE PROGRAM	NETWORK CONTROL PROGRAM	LINE CONTROL PROGRAM
MESSAGE PROCESSING	MESSAGE ACCESSING	I/O CONTROL		DESIGNED FOR PFEP	MESSAGE CONTROL	

6. Preferred. Probably the most efficient interface and access method, in terms of resource utilization, is one designed specifically for the PFEP, with the host computer's access method developed either by the PFEP maker or by the host-computer maker.

Programable front ending beats computer upgrading

Off-line simulation studies prove that—dollar for dollar—a programable front end can be more efficient than a larger main frame in a computer based network

Charles J. Riviere
Telcom Inc.
McLean, Va.

Economics

VICE PRESIDENT OF OPERATIONS: "All our district managers are complaining about the long time it takes to turn around their batch processing work. We're thinking of installing a remote job entry RJE terminal in each of the eight outlying district headquarters, and tying them on-line to our main computer. Can our IBM 360/50 handle the added communications load? What will the impact be on getting our local batch processing done if we give priority to RJE batch communications? And if the 50 can't handle the added communications throughput, what choices do we have to work with?"

DATA PROCESSING MANAGER: "My latest operations statistics show we have some reserve throughput capacity in the 50's CPU, but I'm not sure if it's enough to service the RJE terminal traffic. I guess it will all depend on how fast they can transmit. Perhaps the safe thing to do is upgrade to a 360/65, which operates three times faster than the 50 and thus can triple throughput. Then the communications group can do whatever seems best. We'll probably have plenty of CPU reserve, too. But upgrading to a 65 will increase rental by about $12,000 a month."

DATA COMMUNICATIONS MANAGER: "Well, we know a few things for sure. One is that we'll have to install private lines from each district to the central site. Another is that we can choose one of several customary transmission speeds, from 2,400 bits per second to 9,600 bits per second. If we go too slowly, complete jobs may take too long to transmit. Transmitting at 9,600 b/s raises the line cost somewhat, because of more expensive modems, but the real problem would seem to be the impact of high-speed communications on the CPU throughput. There may be no capacity to process local records. But if we need more CPU throughput, there's another way to go. We ought to look into using a programable front-end processor to see how much CPU capacity it can really release. I've read a lot about front ends. Now let's find out. Adding an IBM 3705—or some other vendor's front end—might do the trick, and it will increase rental by only about $4,000 a month."

VICE PRESIDENT: "So we can upgrade at $12,000 a month or front-end at $4,000 a month. That's almost $100,000 a year difference. Too big to make a snap decision. It's worth spending some time and money to make a thorough study of all reasonable alternatives before we make a final commitment on configuration and line speed."

DATA COMMUNICATIONS MANAGER: "Right. Both groups, data communications and data processing, ought to invest in a joint study using a computer simulation to test out our different ideas. Modeling and simulation will help us arrive at the right answer for this specific project. And the overall analysis should give all of us a more general and more complete understanding of what really happens, in terms of number of communications messages and data processing records that can be processed, as we change line speeds and revise system configuration. Getting this insight will help us zero in much more quickly and accurately on answers for the next project we hit."

Designing and reconfiguring a computer-based data communications system is a major undertaking: wrong decisions can be very costly. To reach an acceptable design requires so many cost-performance tradeoff calculations that designers need a methodology and a tool. The methodology is systems modeling and the tool is computer simulation models. Then, by selecting certain operating parameters and making them "drive" the model, the designer can have the off-line simulation computer produce a set of answers rapidly and accurately. And he can adjust the parameters and come up with a new set of answers. In short, computer simulation allows a designer to investigate system performance without having to make any investment in on-line computer or communications equipment—until he's satisfied with a particular configuration's performance and cost.

What a designer finally obtains from computer simulation runs is a set of curves that permit easy comparison of the effect of different system conditions on performance. Furthermore, if in the first place the model is sufficiently general, then specific results can be extrapolated to provide valuable insight to performance of a general class of problem—here, upgrading versus front ending.

Consider the situation discussed in the foregoing dialog. Computer simulation, using Telcom's Data Communications Analyzer Model (DCAM), proved that front-ending the host central processing unit

BASELINE AND UPGRADE

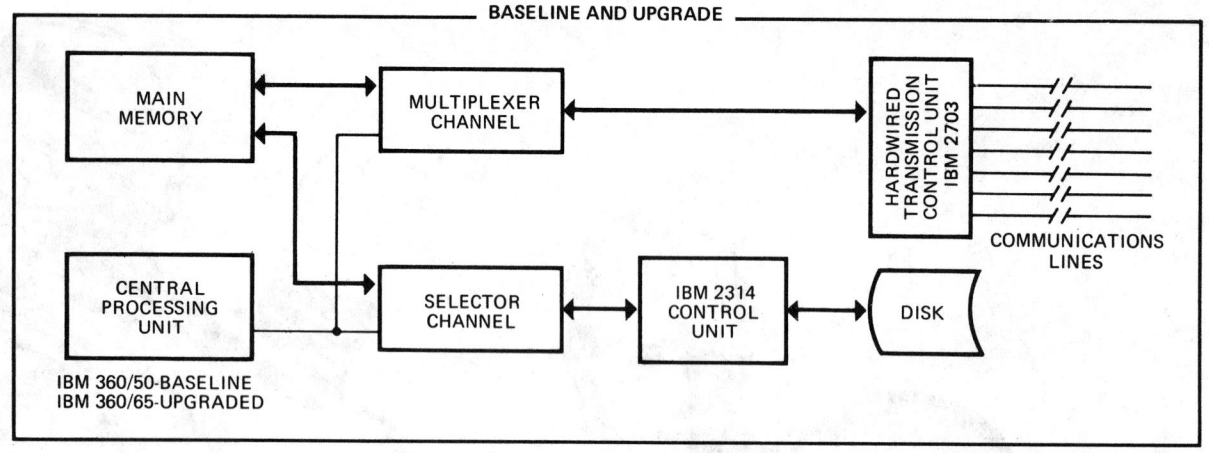

1. The upgraded computer-based configuration contains the same equipment as the baseline configuration except that the host computer is an IBM 360/65, not a 360/50.

paid off better than upgrading the CPU. The results of a portion of the actual overall system analysis are given here. They shed light on the interaction of data communications inputs on the performance of batch data processing computers.

The user company has an IBM System 360 Model 50 which mainly performs batch processing but also services messages from eight remote cathode-ray tube (CRT) keyboard/display terminals for operator-computer interactive applications. The CRT terminals, each on a private line, operate at 2,400 b/s. Thus, the baseline configuration already has a data communications environment (Fig. 1).

This baseline configuration typifies about 50% of the medium and large computers installed in the United States that have some sort of remote communications applications. And of these, most center around an IBM 360 or 370 central processing unit. DCAM simulates computer-communications networks using the IBM System 360 line operating under OS 360, IBM's multiprocessing disk operating system. Because of the upward compatability of the 360 to 370 line, by extension DCAM also models System 370 installations. (With modifications, DCAM can simulate other manufacturers' systems.)

As discussed, the company wants to add eight remote job entry terminals—a heavy increase in communications traffic on the CPU. DCAM will be used to investigate the baseline configuration and two alternatives:

- Upgrade overall capacity by going to a faster CPU, so that throughput increases correspondingly. Retain the presently installed hardwired IBM 2703 transmission control unit. The configuration is the same as in Figure 1, except that the 360/50 becomes a 360/65.

- Add a programable front-end processor (PFEP—here, an IBM 3705—to relieve the host CPU of most of its line handling chores, which frees a substantial percentage of CPU capacity for more batch and teleprocessing tasks. Using the 3705, the 2703 can be removed. This front end configuration is shown in Figure 2.

The main objectives of DCAM are to simulate the data processing and data communications operations, emphasizing the several contention points in the system. Contention points, those resources which may be simultaneously requested by several system components or activities, include main memory, data channels, central processor, logical and arithmetic units, and disk memory and other peripheral devices. Contention points also occur in application, supervisory, and load-module elements of system software.

Simulation runs measure the system's present and reserve capacity, resource utilization, utilization efficiencies, sensitivities, and configuration reliability—and reveal bottlenecks.

Bottlenecks imply saturation or degraded performance. Saturation refers to a system's inability to process specific demand requirements. For example, data communications links and interfaces can only handle a given maximum amount of traffic; the link saturates above that maximum. Similarly, a given CPU can only handle a given throughput; above that it saturates. Obviously, both the communications links and the CPU should have compatible throughput capacities.

Implicit in the present study is that the links and interfaces from the RJE terminals to the host computer will be chosen, later, to have the required capacity, or transmission speed to handle the traffic and avoid saturation. That is, finding out whether the CPU can service the combined throughput resulting from local record processing and the high-speed message processing is the primary point of the study. Selecting the best line speed is a later consideration.

Thus, what has to be determined is how many records and how many messages can be processed

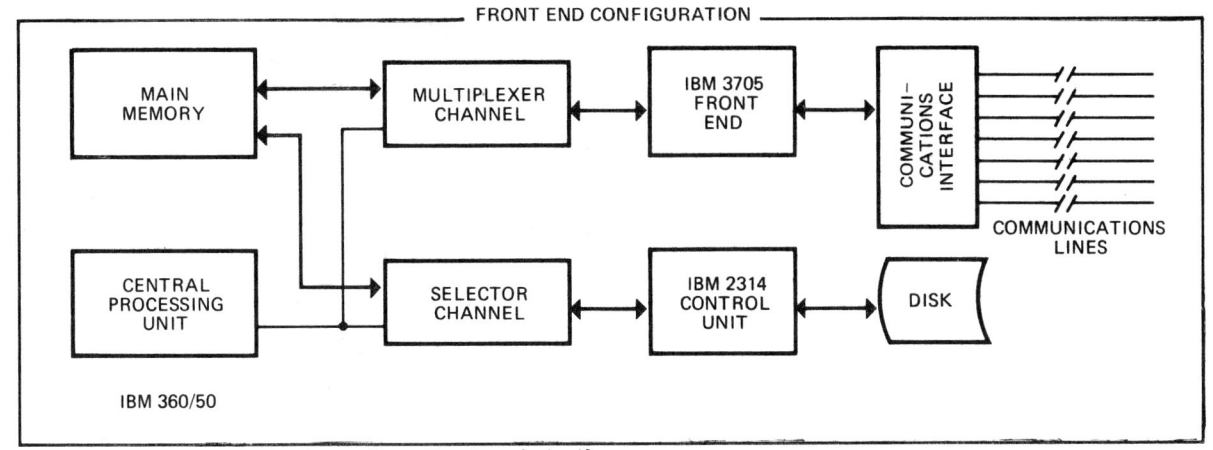

2. In the front end configuration, the hardwired control unit is replaced by an IBM 3705 communications controller, but still uses the IBM 360/50 as the host.

at any prescribed RJE line speed before the CPU saturates or the system performance degrades.

Because such an overall analysis involves comparing several configurations, each with various sets of operating conditions, DCAM was purposely designed to be highly parameterized and table driven. Furthermore, DCAM contains two submodels: a background submodel representing the processing of local batch records and a foreground submodel representing the processing of the high speed communications messages. Implicit here is interaction with another functional model, the computer's operating system. The background and foreground submodels compete for system resources under control of the operating system and they also compete for operating system services. In simulation, as in real operation, the operating system (OS) is given highest priority, the foreground submodel the next highest priority, and the background submodel the lowest priority.

Records and messages

In the background submodel, records are called from main memory, processed, and stored back into mass storage. A record in this instance is a block of 512 characters.

The communications-message submodel includes a selectable dynamic human delay of about 5 or 10 seconds that will occur when the operator interacts with the CRT display information. (The impact of the CRT traffic on the CPU in this study is relatively small, compared with that from RJE terminal traffic, so it will be neglected in further discussion. However, for the sake of clarity and completeness, it is pointed out that each buffered CRT always transmits at 2,400 b/s, an inbound message is 50 characters long, and an outbound message is 300 characters.)

In the foreground submodel for the RJE messages, a message is four blocks of 256 characters each. A decision block requests that four characters (the block size handled by the IBM 2703) be entered into the buffer, and the buffer accepts such blocks until an END OF BLOCK control character is recognized—at which time the operating system starts up the task.

Before describing how DCAM provides information for the systems-alternatives analysis, it may be worthwhile to detail its construction. The simulation runs presented later were produced on an IBM 360/65. DCAM is programed in GPSS (General Purpose Simulation System), a digital simulation language that is easy to use, facilitates model formulation, provides good error diagnostics, and applies to a wide range of problems. And it can simulate single or multiprogramed operations.

Choose the parameters

Data Communications Analyzer Model employs over 70 different variables. The most important ones include the characteristics of the central processor hardware, the operating system, the communications access methods, the front-end processor hardware, the peripheral devices and channels, the various types of terminals, the data link control (line protocol), the buffering schemes, as well as the response time requirements, the number and speeds of the communications lines, and the number of files and file architecture. In a given study, most variables are, in fact, parameterized—that is, set at some fixed value related to operation of a particular piece of hardware: the average access time for a disk memory is a typical parameter.

Developing the model, converting it to a simulation program, and entering appropriate parameters costs hundreds of man-hours of professional-level effort. But even when a model is complete, debugged, and available, a system designer would be rather foolhardy to call for excessive sets of runs, trying out the impact of one variable after another, on some performance criterion, say CPU utilization.

Succumbing to the appeal of simulation can be costly, since each run—that is, set of conditions—could cost from $5 to $10 worth of computer time just for tabular printouts. Graphic plots would cost more. Furthermore, the indiscriminate use of simulation would soon result in so many data points the designer would then be hard pressed to extract meaningful designs from the morass.

In practice, the system designer can find out quite a bit about performance by making just a few runs on just a few variables. Selecting the variables to run depends on the designer's knowledge, experience, and intuition—attributes that become sharper as he gains simulation experience.

In the present example, it has already been decided to compare three different configurations. What has to be determined next is the effect of different RJE communications loads (line speeds) on each configuration, and the cost effectiveness of each alternative configuration and speed. The model will be driven with all high-speed RJE lines operating at the same speed—and at 100% line occupancy so there is no interarrival time between messages. Operating a line at maximum message handling capacity means that not only does the system simulate peak loads, but it also eliminates message distribution statistics as a system variable. Thus a design derived from subsequent analysis will be conservative.

During the runs, lines will be simulated at four common terminal and modem speeds: 1,200, 2,400, 4,800 and 9,600 bits per second.

To be measured at each of these four different line speeds are:
- The number of high-speed messages received
- The number of background records processed
- Foreground (message) CPU utilization (CPU_F)
- Background (record) CPU utilization (CPU_B)
- Operating system (OS) utilization (CPU_{OS}).

Once the models and the appropriate parameters are loaded into the simulation computer, a 20-second snapshot run is taken at each speed for each of the three alternative configurations; resulting data is plotted. (Fig. 3).

Communications first

Here, the model first takes care of processing the message (communications) input, which is a direct function of line speed; if any CPU time remains then as many records as can be handled are processed until the CPU is 100% utilized. Thus, except for a negligible amount (1% to 3%) of idle time, $CPU_F + CPU_B = 100\%$ at each message volume (line speed) condition.

By comparison with the busy foreground pro-

3. Sets of curves derived from simulation runs reveal significant differences in CPU utilization, record processing, and message throughput for the three configurations.

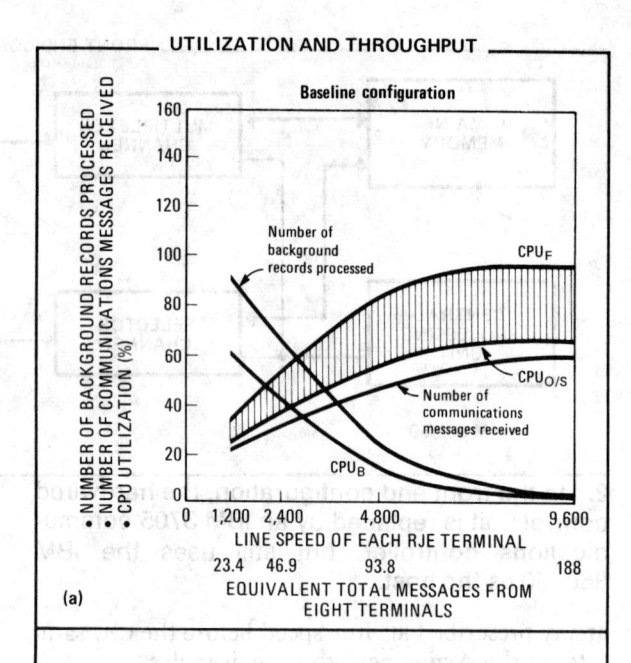

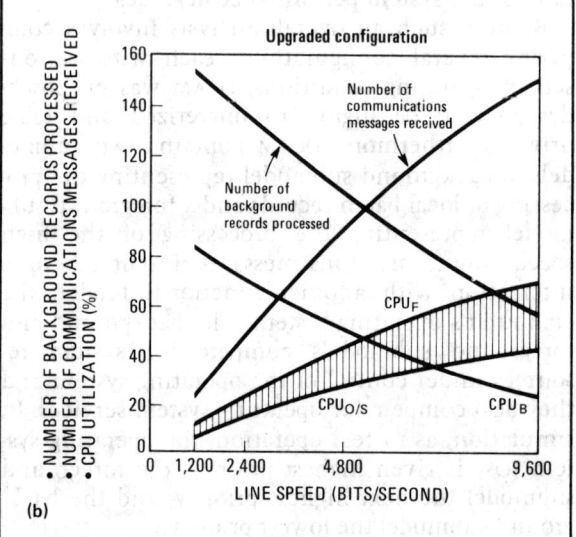

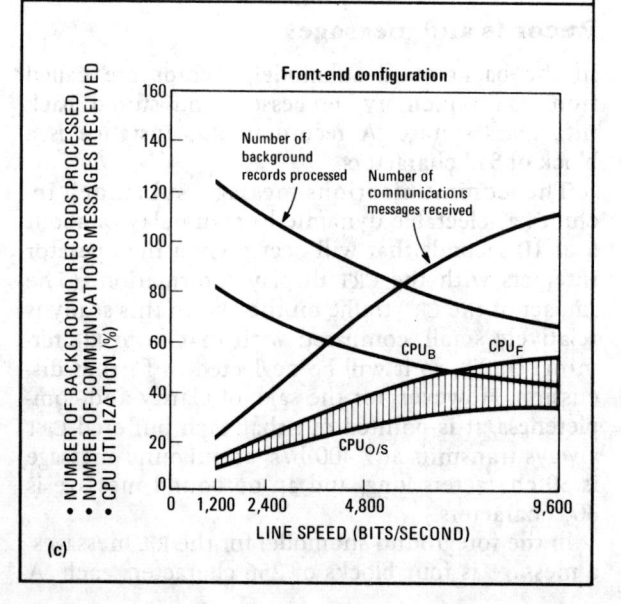

UTILIZATION AND THROUGHPUT

(a) Baseline configuration

(b) Upgraded configuration

(c) Front-end configuration

cessing, background processing involves few interrupts and little intertask switching. It can be safely assumed then that all the operating system's CPU utilization can be assessed against the handling of foreground processing. As can be seen, even at the lowest level of message rate (1,200 b/s) the operating system overhead for servicing communications traffic for the baseline configuration is 25%. This 25% is included in the 35% for foreground utilization at the same speed. In other words, only 10% of the CPU was employed for "true" communications throughput, at the expense of a 25% overhead. This "true" communications throughput, the difference between CPU_F and CPU_{OS}, is shaded in the figures for emphasis.

Besides wanting to know the communications message throughput and record processing situation at different sets of conditions, the designer also needs insight as to performance degradation and saturation.

Performance degradation means that the amount of system overhead—that is, the amount of operating system utilization of overall resources—increases faster than communications throughput. Consider Figure 4. As long as the shaded area widens as line speed increases, then system utilization is improving: initially, CPU is going up faster than CPU_{OS}. But at about 4,800 b/s the area stops widening. Performance has started to degrade.

Saturation, a special class of performance degradation, occurs when the CPU cannot process any more records or messages as the number of inputted messages (proportional here to the line speed) increases. Figure 3a shows that communications-message saturation occurs at about 7,200 b/s and higher. While it may be pragmatic to operate in a region of degraded performance under certain conditions, operating in a saturated region is wasteful of resources and self defeating.

At this point it is well to relate the line speed from each RJE terminal, all transmitting at the same line speed, to the equivalent total messages to be processed as communications throughput by the CPU. A message is defined as four 256-character blocks (1,024 characters), each character eight bits long. There is no interarrival time between blocks or messages. Therefore, the eight terminals operating at 1,200 b/s each produce 23.4 messages during the 20-second snapshot. Doubling the line speed doubles the total input messages. This information is carried on the horizontal axis of Figure 3a and other plots.

Plenty of capacity

Note in Figure 3a that at 1,200 b/s the communications throughput is about 22 messages—just about equal to the theoretical throughput of 23.4 messages. That is, even though the foreground processing requires a 25% overhead to handle this traffic, the CPU is easily capable of servicing input traffic. But at 2,400 b/s the communications throughput is increased to only about 28 messages, not the theoretical 46.9. What has happened here is CPU's relatively slow execution time prevents the system from handling all incoming messages. Some messages will disappear, or terminals may shut down or slow down automatically. And, as Figure 3a demonstrates, even at 2,400 b/s both CPU foreground and OS utilization go way up. Performance has degraded. For all intents and purposes the postulated System 360/50 CPU configuration has limited utility at line speeds exceeding 4,800 b/s. The system has become saturated. And that's what the data communications and data processing people wanted to find out.

Figure 3b represents the performance of the alternative configuration that uses the same IBM 2703 hardwired transmission control unit as in the baseline (Model 50) configuration, but with the CPU upgraded to a three-times-faster Model 65. Figure 3c depicts performance of the Model 50 baseline configuration but with the hardwired control unit replaced with a Model 3705 programable communications controller. Compared with the baseline configuration (Fig. 3a), either alternative configuration has a significantly better performance. In particular, saturation has disappeared; in one case because the upgrading has provided a 300% increase in CPU capacity; and in the other case because the programable front-end processor eliminates most line handling tasks within the CPU and thus greatly reduces CPU_F utilization and CPU_{OS} overhead.

Vast improvement

Upgrading to a Model 65 increased communications throughput by (48 - 32)/32, or 50%. Background processing increased by (135 - 70)/70, or 93%. Similarly, for front ending, communications throughput went up 25% and background processing went up by 62%.

Even more insight can be obtained by comparing plots of the CPU utilization of the baseline configuration with the two alternative configurations and of the CPU utilization of the two alternatives compared with each other (Fig. 4).

Figure 4a clearly demonstrates the improved utilization of the Model 50 CPU when using a programable front-end processor instead of the more traditional hardwired transmission controller employed in the baseline configuration.

For one thing, the front-end configuration reduces CPU utilization so much it does not permit the CPU to saturate. In other words, the front-end and CPU combination could handle RJE terminals operating as fast as 9,600 b/s. For another thing, even at 9,600 b/s, CPU_F only rises to about 50%, basically because of the CPU_{OS} overhead has decreased since the front end introduces fewer and less frequent contentions for the CPU, the multiplexer channel, and the mass storage. Therefore considerable CPU capacity is left over for handling background records.

From the viewpoint of CPU utilization, the upgraded configuration will also fit the bill (Fig. 4b). It can operate at all speeds between 1,200 and 9,600 b/s without saturating.

The cost-benefit performance of CPU utilization in either alternative configuration is best illustrated in Figure 4c. This comparative plot shows that the front end provides better CPU utilization of the Model 50 (at a net rental increase of $4,000) than does upgrading the Model 65 (at increase of $12,000).

More messages, fewer records

Priority is given to communications messages. The essential point is how many input messages can be processed through the CPU. The throughput comparisons of the baseline configuration versus the front end configuration are shown in Figure 5a, baseline versus upgrading in Figure 5b, and front ending versus upgrading in Figure 5c.

In Figure 5a, the number of communications messages and the number of background records handled at any speed by the front-end configuration is significantly greater than for the baseline configuration. Even so, the front end cannot service all traffic coming in at the higher speeds: at 9,600 b/s the front-end configuration can handle about 120 messages, not the theoretical input of 188 messages at that speed. Messages will get lost or the system can signal one or more RJE terminals to stop sending, thus reducing throughput and slowing down job turnarounds.

Figure 5b reveals that the communications and records throughput of the upgraded CPU is superior to that of the baseline configuration since the Model 65 is three times faster than the Model 50 and thus can handle three times the input load.

Again, the cost-performance comparison between the two alternative configurations becomes the important consideration (Fig. 5c). The question is: Is it worth a net difference of $8,000 a month for the upgraded CPU to provide a relatively small improvement in throughput at any RJE transmission speed? Probably not.

What options does the design team now have in determining CPU configuration and related data communications link design on overall system performance and cost? (Remember, the simulation results shown here include conservative, worst-case, conditions.) As examples, these further questions can be asked:

■ What is the lowest transmission speed at which the RJE terminals can be operated? A collat-

4. Comparisons of CPU utilization for all configurations show that the baseline configuration saturates at lower speeds than either upgraded or front end configurations.

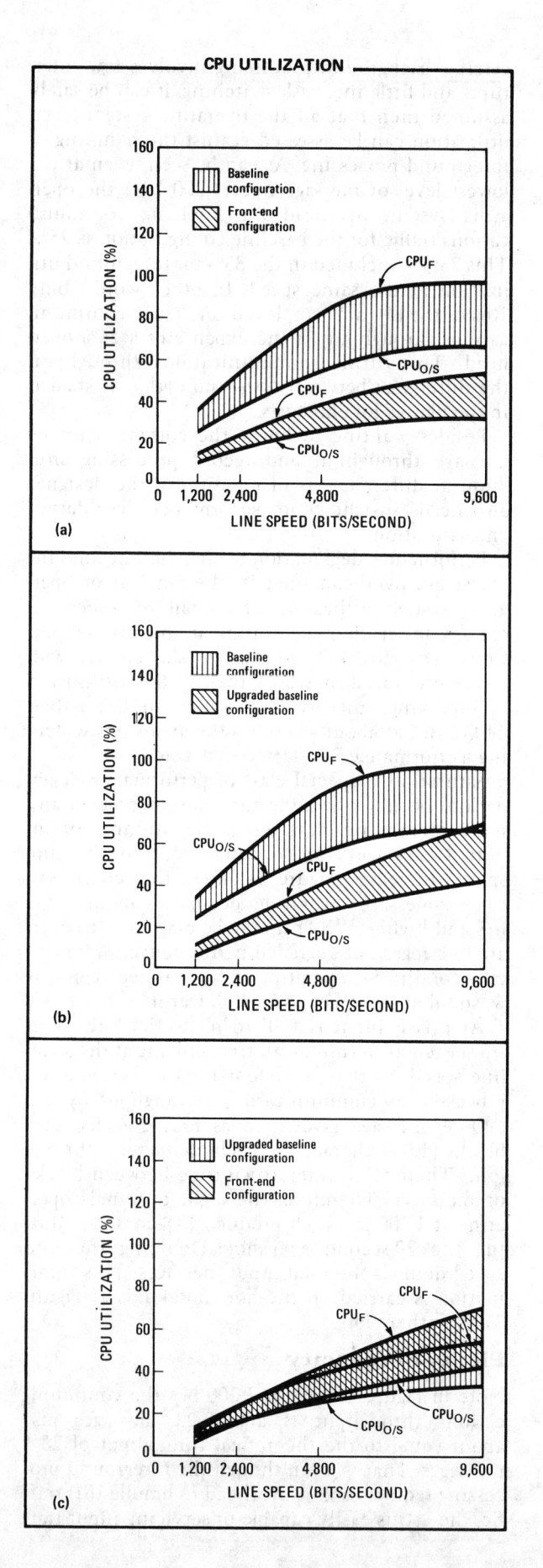

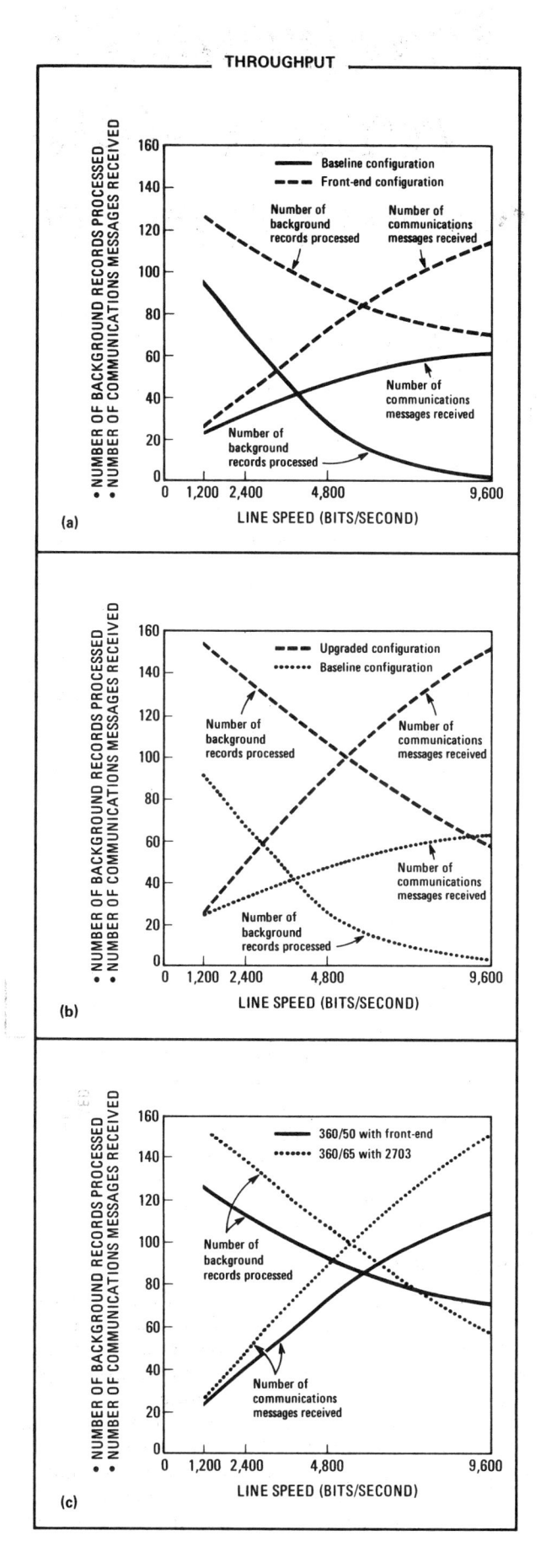

5. Comparisons of message throughput for all configurations reveal that both front end and upgraded configurations handle more traffic than baseline configuration.

eral question is: How long will it take, from the first character to the last, for a remote batch job of a specified volume (number of characters) to be transferred from a district office to the central CPU site? If this time is too long at a specified line speed to suit a minimum job turnaround time, or if the communications load ties up the CPU too long, then the transmission speed will have to be increased. But in general the lower the speed, the better—primarily because of reduced modem costs.

■ If high-speed transmission, say 9,600 b/s, must be used to reduce remote-batch transit time, then another realistic question that can be asked is: Will all eight RJE terminals have to be transmitting simultaneously? Probably not, due to time zone differences. Furthermore a reasonable operating strategy could be to prevent no more than, say, four RJE terminals to input to the computer at any instant. If so, then even though each on-line RJE terminal transmits "flat out" at 9,600 b/s, the net total message throughput is the same as if all eight RJE terminals were transmitting at 4,800 b/s. And, as Figure 5c indicates, at 4,800 b/s the cost-performance benefit favors front ending.

■ Furthermore, transmitting at 100%, or flat out, line occupancy at any speed may not be realistic in practice. When an RJE terminal is on-line, there may be some interarrival delays between blocks and messages. This can happen, for example, when operating in a half-duplex mode, which then requires modem-line turnaround delay and ensuing reduction of net character throughput (but not rated line speed). Therefore, the design team could ask the questions and find the answers to: What is the probability distribution of message arrivals and message lengths at the central site? How many terminals can operate simultaneously without choking the CPU because of data messages?

Using the basic insight to system alternatives presented here, the data communications and data processing specialists can specify one operating condition after another, often making slight compromises from optimum conditions. For example, for the given application, front ending does seem to be the best configuration for the least added cost. Then new answers for postulated operating conditions, say line speeds and number of terminals on line, can be obtained rapidly and at little cost by performing several more simulation runs only on the front-end configuration. Using simulation, it's simple to find out, for example, the effect on communications throughput of the front-end configuration by driving the model first by four terminals, then five, and so on until throughput starts to appear unsatisfactory—if at all. ∎

Programable remote data concentrators

Serving both as a programable time-division multiplexer and as an outpost of the host computer, the remote data concentrator reduces line costs and improves host-computer usage

E. Walter Brown and Alex C. Latker
General Electric Co.
Lynchburg, Va.

A data-communications system may grow so complex that it needs a remote data concentrator (RDC) to interface clusters of terminals with the long line to the host computer. Used instead of, or in conjunction with, time-division and frequency-division multiplexers (see preceding article), the RDC relieves the communications lines of unnecessary traffic and unburdens the host computer of all those chores that assure a fast flow of virtually error-free messages. It therefore serves as a super-multiplexer, mainly because it is built around a small, specialized computer.

As a multiplexer, the RDC not only combines many low-speed lines into a high-speed line or lines, as conventional time-division multiplexers do, but also handles more terminals that a TDM can—and does so without increasing data speed on the fast line(s). With its computer architecture, the RDC takes over such jobs as line servicing, code and speed conversion, traffic smoothing, and error control—in short, all the data communications overhead necessary for improving line and host computer utilization.

In hardware form, remote data concentration can be implemented with a general-purpose stored-program minicomputer or with a communications-control computer designed specifically for the purpose. The difference between them, essentially, is that a minicomputer includes certain hardware, like an arithmetic multiplier or floating-point hardware, not needed in data communications, and it usually uses software to service communications lines. Conversely, the communications-control computer contains no superfluous equipment, and employs hardware for multiplexing, logical programing, and data manipulation at the line interface. Either, however, can be programed to implement specific functions required in a particular system, and can be readily modified to suit system changes and growth. And each contains extensive memory in which input data can be temporarily stored.

Line servicing

The illustration gives a good idea of the RDC's location in a data-communications system and its extensive and varied capability for line concentration. Assume that the concentrator shown is located in Denver. Inputs reach it from terminals and multiplexers situated around Denver and to the west and south, including some low-speed inputs dialed in over the telephone network. Several high-speed lines leave the RDC, and go directly to ports of a large host computer in Cleveland.

The line servicing of such a setup involves tasks like

link establishment, terminal identification, recognition of speed and type of service required, and polling. All of these take time and memory, and if performed by the host computer in Cleveland, would reduce both its and the high-speed lines' performance. Instead, the RDC handles all overhead tasks, and sends only essential message information on to the host computer.

Consider polling in more detail. Many private data-communications systems have a polling protocol for addressing individual terminals that share the same line but that can only communicate one at a time. Here, transmission does not start until a terminal receives its unique address that tells it to start transmitting if it has a message. This address could be sent from the distant host computer. Instead the RDC can do it locally.

Furthermore, any message transmitted by an addressed terminal contains a message header designating the recipient. If the recipient is the host computer, the RDC merges the message with other computer messages and sends out a composite bit stream on the high-speed line. However, many data-communications networks must include the ability to send messages from one terminal to another. The message header would be read by the host computer and the message routed back to the addressed terminal, again loading up the high-speed line. Instead, the remote data concentrator can perform the message switching when the terminals are within the geographical area serviced by the RDC. Finally, because the RDC will usually have an extensive amount of bulk memory, usually disk packs, it can store the message until the addressed but busy terminal is free to accept it.

Code and speed conversion

Any fairly large data-communications network contains a variety of terminals operating at different speeds and in different code formats. Transmitting this intermix over a high-speed line to different ports in the host computer again places a severe overhead on both line and computer. Instead, the programable RDC can convert both speeds and codes at a relatively local level.

Such RDCs contain a hardware interface that accommodates such standard terminal speeds as 110, 134.5, 150, and 300 b/s. This hardware, in combination with suitable software, detects the incoming speeds prior to reading the data, and executes code and speed formatting. For example, it will strip out start-stop bits from asynchronous codes—a procedure that by itself improves line utilization by about 20%—and will arrange mixed codes into a synchronous bit stream, with a fixed format suited to the host computer. In this manner, one computer port can service a variety of terminal speeds, instead of a separate port being needed for each speed.

The RDC's code- and speed-conversion capability has several advantages. It allows users in remote cities to dial up the computer, through the RDC, and insert data from various terminals operating at any common speed. It permits the RDC to undertake message-switching between terminals operating with different speeds and codes. Furthermore, it allows for system expansion without disturbing the host computer's software or hardware, and permits a new, more efficient, type of terminal to be substituted for another without any modification at the host computer installation.

When implementing line servicing and code and speed conversion, the RDC uses a special form of memory called a buffer. Each input channel to the RDC has its own buffer. This memory accumulates one or more characters, which are then read into the RDC's fast internal core memory, which performs immediate processing of the data streams.

The RDC also contains mass memory—usually disk packs or drums—in which it holds data for longer time intervals as may be needed for traffic smoothing and to save data in case of temporary outages of the line. The RDC will also have output buffers to interface the accumulated high-speed data stream onto the line between the concentrator and the computer. Often, these output buffers serve instead of software to implement error-control procedures. Consequently, how much and what kind of memory is needed depends on two major functions of remote data concentration—traffic smoothing and error control—both of which have to be carefully matched to user requirements.

Traffic smoothing

Conventionally, systems planning will include an analysis of traffic to be carried from the terminals to the

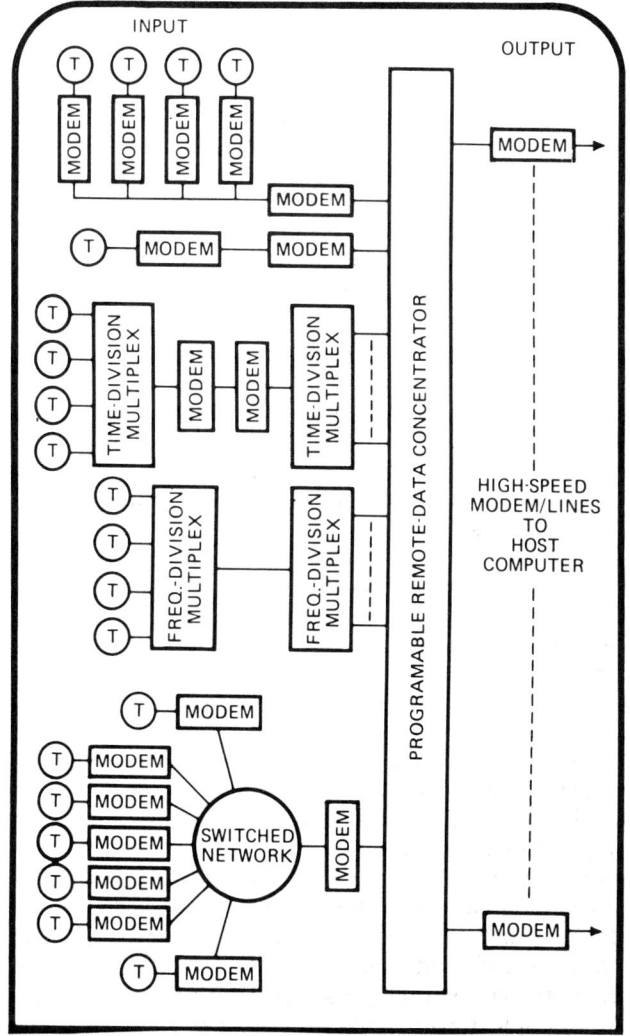

Pulling together. Remote data concentrator consolidates inputs of many codes and speeds into one high-speed output data stream.

computer and to other terminals. Peak loads will be found to occur at certain times of the day, and will determine the number of high-speed lines needed to carry the peak traffic. But if this peak-load determination assumes all terminals will operate simultaneously at their full rated speed, the system may be overspecified. Many terminals, particularly the manually operated low-speed keyboard types, may be connected to the line, but there will be scattered intervals during which operators will not be hitting the keys. Thus, the actual data throughput of the system will be less than rated capacity. Therefore, with the traffic-smoothing function made available by adding mass memory to the RDC, the fully utilized average speed on the line to the computer can be less than the sum of the rated speeds of the terminals going into the concentrator.

For traffic smoothing, the mass memory acts as a temporary reservoir for bits entering the data concentrator when the total input rate is higher than the output rate to the line. When traffic slows, stored messages leave the memory for their destinations. That is, properly designed traffic smoothing takes care of the effect of random variation in terminal traffic on the system, and assembles the message completely in the memory before transmission to assure full utilization of the high-speed line.

Even though the memory size may be economically selected for some assumed peak traffic load, this load may be exceeded from time to time. In this case, the RDC, under software control, will have to be able to raise a busy signal to incoming terminals to prevent them from transmitting.

Besides providing maximum utilization of the line to the computer during peak load periods, the mass memory can aid system integrity. It can store one or more consecutive outgoing data blocks as may be needed for a request-for-retransmission error-control procedure, and it can store incoming terminal messages during an outage on the line between the concentrator and the computer.

Outages occur for many reasons. For instance, the common carrier may have to switch the primary leased line to a back-up line when trouble arises. A Bell System document defines two types of outages: a hit, lasting less than 300 milliseconds (as might happen from a lightning strike) and a dropout, lasting 300 milliseconds or more. A dropout of 300 milliseconds on 4,800-b/s lines means a loss of about 1,500 bits. The loss of so much data, which could have originated at many different terminals, could upset company operation—unless the bits are retained within the RDC's memory for retransmission in case of an outage.

The mass memory can also be made large enough to implement a store-and-forward feature, particularly desirable in networks handling considerable traffic between terminals. With this feature, the sending terminal does not have to wait for a busy receiving terminal to finish a call and hang up. The store-and-forward memory acts as a temporary receiving terminal. Then, for example, the line protocol can allow each "free" terminal to request messages addressed to it, if any, that are stored in the RDC's mass memory. The feature may also be desirable for data-communications systems requiring especially high operational certainty. Here, the need may be not only to forward messages but also to store all traffic for up to 24 hours, in order to provide repeats of messages that may have gotten lost or to retain data until an audit of the day's work has been completed.

Thus, the size of the mass memory depends on several factors. As a minimum, the memory should be able to retain a line of text from each terminal.

Because the concentrator can actually deliver composite data messages to a high-speed port at the computer, there is no need for demultiplexing equipment to split the received data into slower channels. Further, the number of lines between RDC and computer can often be reduced since dedicated but underutilized time or frequency slots—as in conventional TDM and FDM—are not required to handle each terminal's output.

Error control

One of the main benefits of using a remote data concentrator is that it can check data coming in from all terminals for errors and add a checking code to traffic between the RDC and the host computer. That is, if the concentrator detects an error, it can request retransmission of a message over a short link, without involving the host computer. Moreover, the corrected messages can be sent in long, economical blocks over the high-speed line to the computer.

Error detection and control can be implemented with several clever procedures, as explained in the article on error control. One thing all these procedures have in common is that extra coding bits are added to the data block. These redundant bits, which can range from a simple parity bit to perhaps one-third redundant bits in a forward error-correction method, require memory to store messages temporarily while they are being analyzed for proper coding and decoding. And extensive memory is available in remote data concentrators at a relatively low cost per bit.

Error control can be implemented by software or hardware. For the commonly used cyclic redundancy checking, for instance, special dedicated hardware is assigned to each line. Although such hardware raises the RDC's installed cost, its advantage is that checking is done in real time with no software overhead required at the data concentrator. Software implementation reduces the RDC's cost, but adds a burden to the computer within the concentrator. For example, a single 4,800-b/s link controlled by a concentrator with a 1.8-microsecond memory cycle can use up to 10% of available real time for performing the checking function by software. The decision whether to use hardware or software error control will fall out as part of the over-all technical and cost analysis of the network requirements.

In a complex data-communications system, the RDC's many functions each require an applications program, usually prepared by the user. The preparation and verification of these programs is a major task, but one that is made easier if the concentrator vendor supplies standard support software. A competent vendor should at least furnish an assembler program, and, desirably, an operating system. □

Estimating buffer size and queuing delay in a remote concentrator

Simple models in graphic form aid the system designer to understand the interactions of equipment in a multiterminal inquiry/response network without getting involved in probability and queuing theory

Ronald I. Price
Interdata Inc.
Oceanport, N. J.

HARDWARE

In the typical interactive inquiry/response system, the human operator of a teletypewriter or other keyboard terminal transmits his inquiries in irregular bursts of characters. His output might occupy a line leading ultimately to the host computer only about 3% of the connection time, although momentary spurts might reach as high as 20% line occupancy. Such a situation exists when the use of a terminal is essentially incidental to the person's main job assignment, as in law-enforcement applications.

A much more uniform and thus more efficient use of the long-haul line can be achieved if the random inputs from many such terminals are not sent directly to the host computer but are collected in the memory of an intervening data concentrator. There the messages can be sorted and queued into a much smoother data stream that can be transmitted more steadily, with no waste of connection time.

The remote data concentrator (RDC) is usually based on a small computer and includes some main memory for storing the programs that operate it and some for temporarily storing (buffering) messages to and from the terminals and the host computer. While its main advantage is in savings of line costs, the random nature of the traffic in a multiterminal interactive inquiry/response system permits a further extension of this advantage.

For one thing, the speed of this "backdoor" link to the host computer, in characters per second, need not necessarily equal or exceed the sum of the maximum input-line speeds, as happens in time-division multiplexers, another type of line-concentration equipment. In fact, the backdoor speed may more closely equal the average traffic rate of the (mostly idle) lines, with buffering provided to smooth any instantaneous traffic peaks.

For another thing, the concentrator can compress incoming data from many lines, so its backdoor rate can be kept even smaller. This permits the use of low-speed, hence lower-cost, datasets (modems) and may obviate the need for costly line conditioning.

Typically, not all the "active" terminals will be on line during a given time interval, and even of those that are on line, not all will be sending or receiving characters at any given instant. That is, many inquiry/response terminals are normally placed on and off line and transmit and receive under control of the user, not of any fixed polling or selection procedure. Thus, the system is "freewheeling." Characters and bits comprising characters are not assigned to dedicated time slots in the over-all multiplexing cycle, as is the case for synchronous multiplexing. Within this context, then, the remote data concentrator operates, on the terminal side, as an asynchronous time-division multiplexer.

Thus, in an inquiry/response system messages arrive at the RDC at random intervals and have varying lengths which can be described in terms of statistical distribution. A common buffer pool is best for this type of operation. The buffer pool should be large enough to assemble and queue messages from the terminals, and also to disassemble messages coming from the computer for each terminal. If sized correctly, this common buffer pool will have a small probability of overfill, and will therefore have little influence on queuing delay. Further, the buffer size should permit as slow a backdoor transmission rate as possible. But queuing delay is extremely sensitive to backdoor rate, so the designer must perform a tradeoff analysis to properly balance over-all cost and performance.

An estimate of the correct size for a common buffer pool can be readily obtained from generalized system models, presented here as a set of graphs. These graphs derive from previous mathematical studies of actual multiterminal systems (see "Where the RDC models come from," next page). By using them, the system planner does not have to resort to extensive and exhaustive mathematical investigation of a system that is characterized by multiple streams of statistically distributed, intermittent characters.

Later, the graphical models will be used in a numerical example. Although these graphical models are simple plots, their valid use does require some definition of terms and an understanding of the characteristics of the statistical nature of intermittent data traffic.

Data—and data about data

First, consider the character-by-character traffic on a typical low-speed terminal going to the computer and returning (Fig. 1). The activity shown here represents characters, not bits, and assumes—for the moment—that the RDC is not in the system. The user sends a message, a group of characters, to the computer. This group is sequence 1 in the illustration. Note the extensive amount of idle time, due to the user's nonrhythmic keying speed, necessary thinking periods, and physiological response lags.

When the user signals the end of his inquiry message by sending some type of control character, the host computer, after a slight delay for retrieving data from its own memory, responds with sequence 2. The message from the computer to the terminal may be completely assembled beforehand so the return transmission occurs as a burst. The only idle periods during computer transmission may be, for example, to provide time for carriage return of the terminal's printer mechanism. In general, the computer's transmissions occupy the terminal-to-RDC line about 30% of the time during a typical receive-transmit session.

Implicit in the activity graph in Fig. 1 is the assumption that the terminal is operating in a half-duplex mode—the usual situation in inquiry/response systems. That is, the terminal can either send or receive, but cannot do both at the same time.

When all terminals feed into the RDC's front end, the over-all data traffic is much more complex (Fig. 2). Although this illustration shows only four low-speed lines, in an actual system the number of low-speed terminals into one RDC may well exceed 50. The section at the left of Fig. 2 shows typical traffic patterns on these lines. Note that no characters are time-related to other characters, either on the same line or on any line. That is, each terminal is freewheeling, and each character is asynchronous.

The middle section of Fig. 2 represents the common buffer pool, divided into two sections (figuratively): one for assembling characters into blocks that correspond to whole messages from the user, and the other for temporarily storing the message blocks in queue until their turn arrives for them to go out over the high-speed link, as shown at the bottom of Fig. 2.

Block assembly from a terminal starts when the terminal initiates a START OF MESSAGE sequence. User (terminal) identification characters are appended to the message. Message characters are then accumulated as they come in at random. The user signals END OF MESSAGE by sending an EOM control character. If a message has more characters than can be handled by one block, then—of course—more than one block must be used. But each block must contain message characters and necessary appended identification characters. Depending on system operation, the delineating EOM character (say, CARRIAGE RETURN) may actually represent only one message segment of a total inquiry.

All this is rather elementary. The essential reason for reviewing it is to distinguish between message-character rate and transmission-line character rate. In other words, for transmission purposes, message blocks contain appended characters which affect both the required

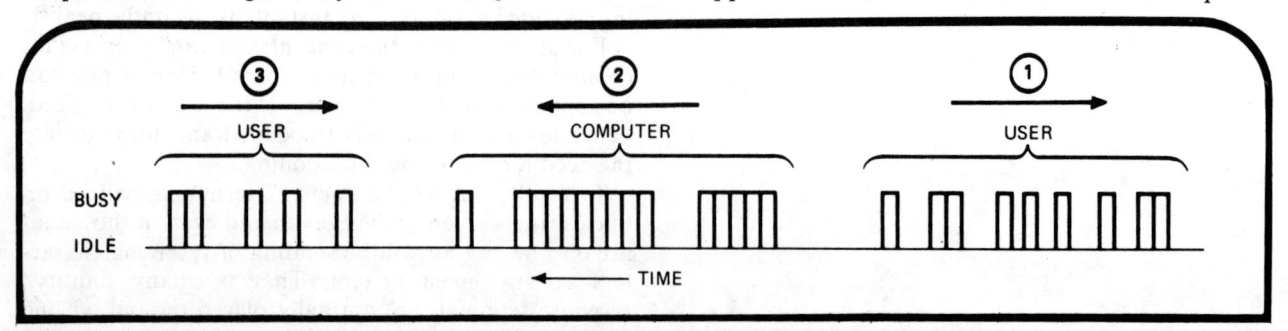

1. Nonrhythmic. When a human operator sends a message on a keyboard terminal, there is extensive idle time because of erratic keying speed and thinking periods, but the computer can send back longer messages in steady bursts.

size of the common buffer pool and the actual "message" throughput over the backdoor high-speed line.

Assembling and queuing

Although the common buffer pool will not actually be partitioned into areas for message assembly and areas for queue storage, in estimating over-all size it is simpler to assume that the partitions exist. Separate estimates yield a conservative design: that is, buffer size will be rather larger than necessary.

The block-assembly pool estimate can be obtained from the model in Fig. 3. Here, buffer size depends on the total number of unbuffered terminals in the system. This graph takes into account the statistical aspects of typical inquiry/response data communications systems. Not all the terminals are connected to the RDC at the same time. Of those that are on line some are sending data, while the others are receiving.

The model is based on the statistical probability that only one message (or block) out of a million—coming from the terminals actually on line—will be barred from entering the pool because at that instant the pool is fully utilized. Such overflow messages, if they occur, will be lost within the system. When the user gets no response or an error return, he will have to retransmit.

The buffer size for assembling blocks is given as multiples of average message length. Thus, to determine the actual assembler-buffer size, in bytes, the planner will have to survey actual traffic conditions to obtain the average length of the messages in his system.

Note that a condition stipulated in Fig. 3 is that the system have identical lines. By this is meant that each line have the same distribution of message length and hence the same average message length. Thus, the model in Fig. 3 is independent of rated terminal speed, provided message length follows the prescribed distribution. Therefore, the model is equally valid for unbuffered terminals of different speeds.

Furthermore, Fig. 3 also applies when buffered terminals are used. Such terminals send or receive data in a burst of characters representing one or more blocks. For this case, the buffer pool's size can be estimated by taking the size of buffer pool indicated in Fig. 3 for the unbuffered terminal and multiplying it by the average occupancy percentage of lines connecting the buffered terminals to the RDC. Average line occupancy, whether buffered or unbuffered, is the ratio of the traffic rate at the RDC's front end to the sum total of all terminals operating at nominal maximum speed. Hence, when buffered terminals provide operational or economic advantage, they result in a smaller RDC common buffer pool.

As shown in Fig. 2, once blocks have been assembled, they wait in the queue buffer pool until they can be sent on to the host computer through the RDC's back door. The estimate for this part of the common buffer pool can be obtained from Fig. 4. Here, the size of the queue buffer pool, again in multiples of average message length, depends on two major factors: the average line occupancy percentage, and an important parameter designated "R" which simply characterizes a remote data concentrator in terms of its input-to-output speed reduction or expansion ratio.

The defining equation for R is contained in Fig. 4. The denominator represents all the characters that

---

Where RDC models come from

An extensive study by personnel at Bell Telephone Laboratories of four different non-Bell inquiry/response systems offers useful and significant support for simplifying modes of on-line computer communications processes. Study results were reported by E. Fuchs and F. Jackson in "Estimates of distributions of random variables for certain computer communications traffic models" in Communications of the ACM, December, 1970, published by the Association for Computing Machinery.

In summary, Fuchs and Jackson found that the discrete variables in a data stream model appear to be represented quite well by a geometric distribution and therefore this distribution is used to characterize message lengths. Furthermore, character interarrival time is random. Such random arrivals, also known as Poisson arrivals, can be characterized by a negative exponential function.

These two types of distribution function held true for each of the four systems, even though the four systems contained different computers and terminals and operated in different applications. That is, the findings were robust: the stated distribution functions gave the best fit for each of the four systems. Robust results means that the distribution functions can be confidently assumed for other interactive systems. Then only certain parameters—for example, average message length—differ from system to system.

The results are also amenable to the modeling process if use is made of standard analytical techniques and a large body of probability and queuing theory. The first-order-approximation models presented in this article derive from the application of this kind of mathematical procedure to the results obtained by Fuchs and Jackson and also by W.W. Chu ("A study of asynchronous time-division multiplexing for time-sharing computer systems," AFIPS Conference Proceedings, Fall Joint Computer Conference, 1969). The outcome is a set of graphs that can be applied to a particular system configuration once several easily determinable parameters of that system is obtained.

One important parameter is the average message length, in characters. The general relationship in a geometric distribution between average length of message and the probability of occurrence of a message n-characters long is:

$$P_n = \frac{1}{\bar{n}} \left(1 - \frac{1}{\bar{n}}\right)^{(n-1)}$$

where $\bar{n}$ is the average message length, in characters, and n is 1,2,3,..., etc.

---

would enter the RDC if all terminals in the system were on at the same time and transmitting at maximum rated character speed. The factor K in the numerator modifies the actual backdoor transmission rate to account for appended characters which must be transmitted but are not part of the actual message content. For example, if a transmission block contains 20 actual message characters to which are appended four control characters, then K is 20/24.

To review, the size of the message-assembly pool can be determined from Fig. 3, and the size of the queuing pool from Fig. 4. In practice, though, each pool may have to be inflated somewhat to account for such factors as the use of fixed-size blocks in the buffer space and of block linking. Discussion of this inflation factor will be deferred, however, till after introduction of the queuing delay model (Fig. 5).

Queue service time delay

The delay suffered by a message while stored in the concentrator and awaiting transmission of its first character depends on the average line occupancy and the backdoor transmission rate. (Such delay within the RDC is only part of the total round-trip delay, since an additional delay occurs within the host computer, and perhaps even at intermediate RDCs and multiplexers.)

In Fig. 5, the delay is normalized to character periods of the backdoor line rate, expressed as multiples of average message length. Note that the delay also depends on the concentrator parameter R.

Consider now the use of these models in a typical inquiry/response system. Coming into the RDC are 100 active terminals operating at 10 characters per second (c/s). The planner can assume that messages arrive in a random manner and that their lengths are geometrically distributed. What he needs to find out is the length of the average message. Suppose a study indicates the average message length is 20 characters with each message requiring four control characters appended by the RDC before transmission to the host computer. Assume each character fills one 8-bit byte of storage in the pool. The backdoor transmission link operates at 2,400 bits per second.

The estimate for the assembly buffer pool is found from the model in Fig. 3. Here, 100 terminals need a buffer pool of 78 average message lengths. Thus the assembly pool is 78 × 20 = 1,560 bytes.

To find the size of the queue buffer pool first requires a determination of average line occupancy, which should allow for peak-load conditions. For example, if all 100 terminals were to send at an average peak of 2 c/s simultaneously, the line occupancy (LO) is (2 × 100)/(10 × 100) = 20%. The other value needed to estimate the buffer size is R. Here:

R = (20/24)(300)/(100)(10) = ¼

Figure 4 shows that at a 20% line occupancy and an R of ¼, the required queue buffer pool is 65 average message lengths. Thus, 65 × 20 = 1,300 bytes is required for the queue buffer pool.

These two models thus indicate the required common buffer pool is 1,560 + 1,300 = 2,860 bytes. In practice, this value will have to be inflated, as will be explained.

As Fig. 5 shows, the average queue delay at R = ¼ and LO = 20% is 3.8 character periods multiplied by the average message length. Thus, the average delay within the RDC is (3.8/300)(20), or a quarter of a second.

The design options

The models for the queue buffer pool and queue delay provide insight into alternatives open to the system designer. For example, Fig. 4 shows that, for the same 20% LO, the amount of required queue buffer can be cut from 65 to 20 by going from R = ¼ to R = ½; that is by raising the backdoor transmission rate from 2,400 b/s to 4,800 b/s. But then a more costly dataset (modem) may be required.

Sometimes traffic intensity (that is, line occupancy) may increase slightly because of system growth or tem-

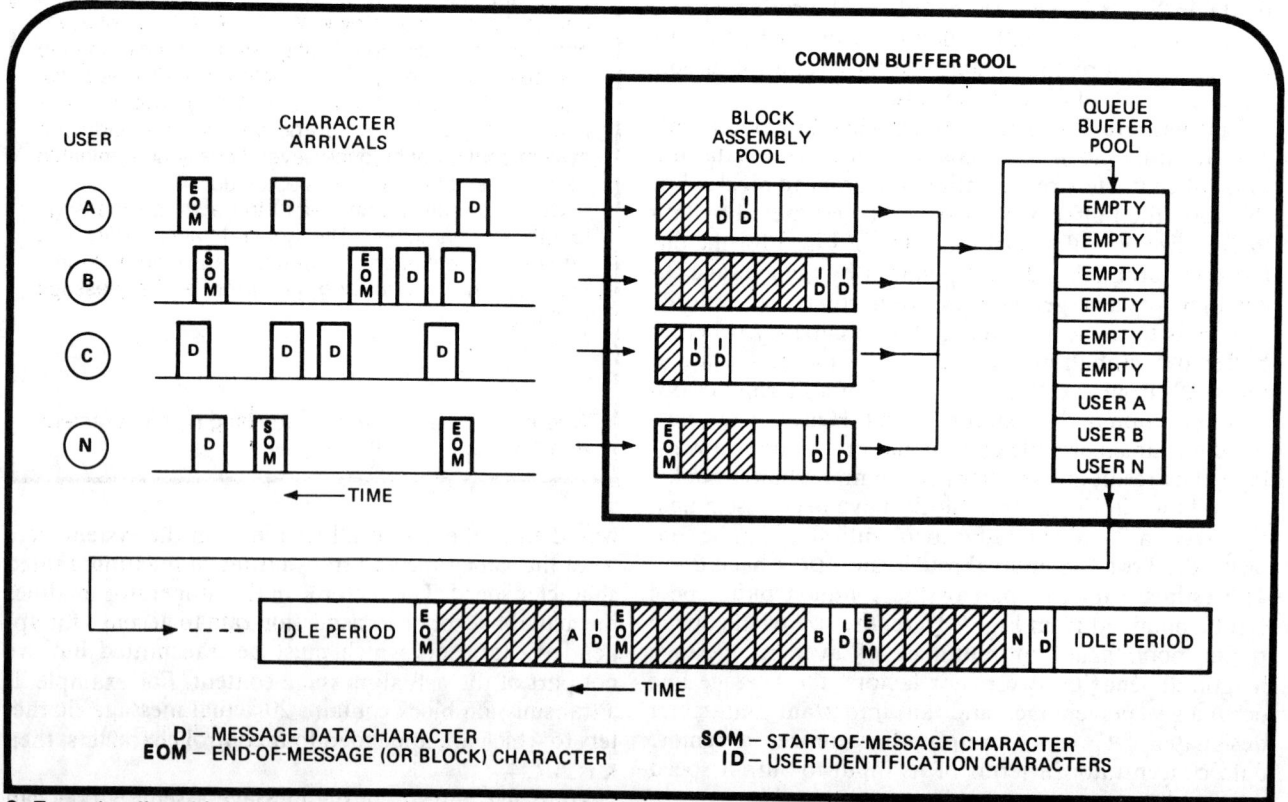

2. **Freewheeling.** Lines coming into a remote data concentrator are asynchronous, and characters can start and stop at any time, but the common buffer removes idle times and assembles and queues messages or blocks for transmission.

porary overload. In that case, though the extra traffic can easily be accommodated by adding to the assembly and queue buffer, the real problem is the impact on queuing delay. As Fig. 5 indicates, at R = ¼, raising the line occupancy even a small percentage over the 20% design point will increase the average queue delay to 8 or 10 character periods. In fact, delay quickly approaches infinity in the model. Obviously at 20% LO, R = ¼ is a very poor design choice.

One solution is to increase R, since queuing delay decreases as R increases. Thus, at R = ½, the delay drops to 0.7 holding time, compared with 3.8 at R = ¼. This is about a fivefold improvement. Moreover, doubling the transmission rate halves the character period, so that, overall, the average delay time is cut to about a 10th. Delay decreases even more dramatically at R = 1 and R = 2, and buffer size, too, goes down, but not as fast.

However, the game of doubling transmission rate cannot be played without penalty. For one thing, the costs of the datasets and communications links increase. The needed transmission rate may also tax the acceptable state of the art. Multiple backdoor links could be employed, but at a cost of more complicated control programs, and the models presented here would then not apply. For another thing, the faster the backdoor transmission rate, the greater the software and hardware burden on the RDC, and this overhead reduces the number of terminal lines (throughput) a given RDC can service efficiently and properly.

The example inquiry/response system, with its 100 terminals rated at 10 c/s, results in a backdoor transmission rate of 9,600 b/s at R = 1. Suppose the system designer wants the small delay as well as the traffic overload protection available at R = 1 for a given line occupancy, but must operate at a backdoor rate of 4,800 b/s. He will get this kind of performance by halving the number of terminals that enter the front end of the concentrator. Using these 50 terminals and a backdoor rate of 4,800 b/s (or 300 c/s) also yields R = 1. In other words, he will need two RDCs (or two backdoor links) to meet performance specifications.

Determining just how many concentrators to employ for one distributed system involves much more than merely selecting appropriate queuing delay and common buffer pool size. A thorough network configuration and optmzation study is involved. However, the models can indicate when conditions within just one RDC violate or satisfy operational criteria.

The inflation factor

Buffer space in a remote data concentrator is normally provided in blocks of bytes (or characters) that are fixed in size. Each block may need to include space for characters that have nothing to do with message content, such as those used for control characters and linking address. Furthermore, although messages can be characterized for design purposes in terms of average message length, in fact some messages are short and others long. Therefore, the last block in a chain of character-blocks making up a given message may not be filled with meaningful characters. This fill waste—added control characters, linking characters, and the like—expands the actual amount of fixed-block-size buffer as determined from the assembly and queuing buffer graphs. The amount of expansion, or inflation factor, is contained in the set of curves in Fig. 6. The curves are based on a calculation of the average number of blocks consumed by geometrically distributed messages.

Suppose that, as in the preceding numerical example, the average message length is 20 characters and the fixed-block size is selected as 10 bytes. The ratio of buffer block size to average message length is 0.5. This value is used to enter the graph and intersect at a given value of P. Here, P is the percentage of characters added to the buffer block for such things as control and linking

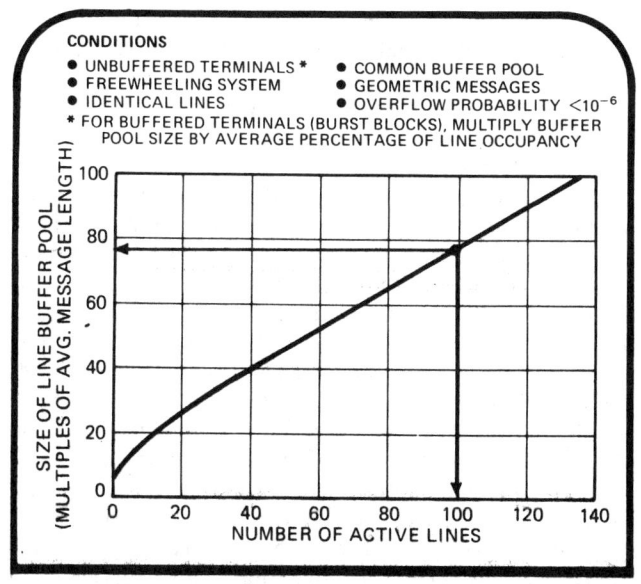

3. **Assembly sizing.** The amount of buffer needed to hold characters depends on the number of lines occupied according to a geometric distribution.

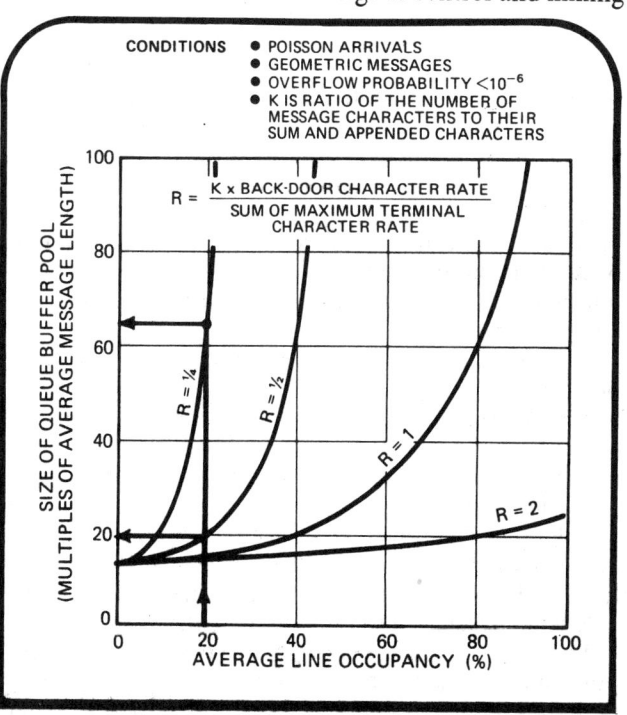

4. **Queue-buffer sizing.** The amount of memory needed for queue buffering depends on average line occupancy and the RDC's input and output data speeds.

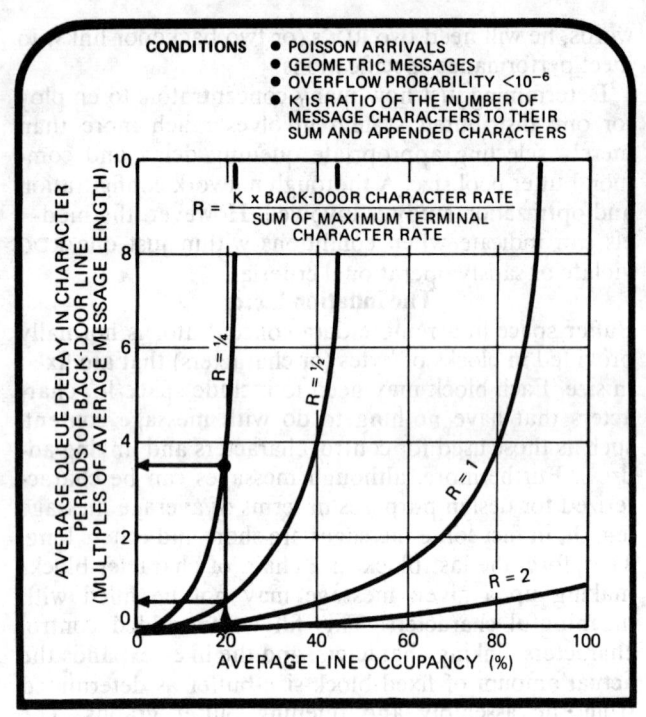

5. Queuing delay. The amount of time a message is delayed in the RDC depends on average line occupancy and the relative speeds of the RDC's input and output.

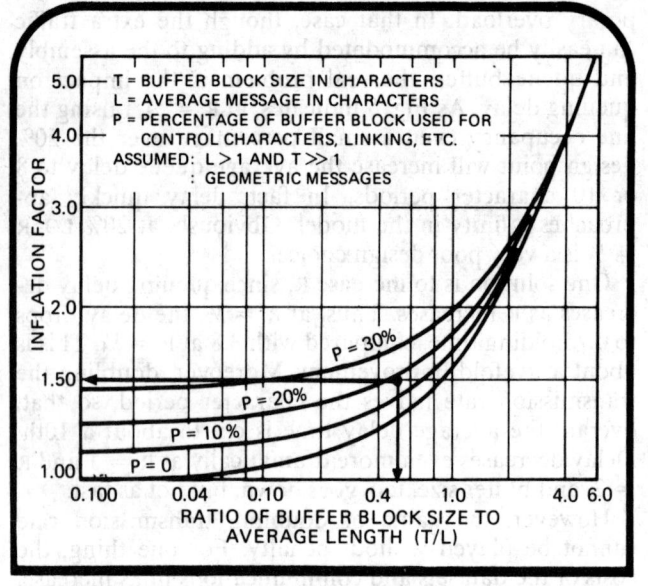

6. Inflation. Fill waste—added control characters, linking characters, and the like—increases the amount of buffer, as determined from assembly and queue models.

characters. If two characters are added to the 10-byte block, then P = 20%. Therefore, as the colored lines in Fig. 6 indicate, the required inflation factor is 1.52. Hence, the common buffer pool of 2,860 bytes computed earlier would have to be multiplied by 1.52, and a pool of about 4,350 bytes would be required to assemble and queue the traffic.

Return journey

The discussion so far has centered on the sizing of the common buffer pool when the concentrator multiplexes incoming characters on the terminal lines. However, the computer returns messages which must be demultiplexed and forwarded to corresponding terminals. With certain provisos, the common buffer pool sized for the multiplexing mode should also suffice for the demultiplexing mode. One proviso is that the messages coming back from the computer not be excessively long. Another proviso is that messages be evenly distributed amongst terminals.

In an inquiry/response system, traffic from the computer may be 10 times more than from the terminals. If one very long message must go to one terminal, the queue buffer may be too small and will overflow. Such message traffic is more typical of a batch remote job entry system—and the models are not valid. One way to handle long messages that might otherwise fill the queue buffer is to program the host computer—or a front-end processor if there is one in the system—to divide long messages into shorter segments, or blocks.

These blocks, along with similar block segments from other messages for other terminals, can then be transmitted to the RDC's queue pool, and then disassembled and sent to the appropriate terminals. (Implementation of message segmentation is actually more subtle than simply dividing messages into pieces; host-program scheduling is involved.) The host computer programing must make the computer appear like many freewheeling low-speed buffered terminals sending short blocks that can be characterized by random arrivals and geometric message-length distribution. If so, then the model in Fig. 4 still gives a good first-order estimate of adequate queue buffer pool.

Another approach for servicing host-to-RDC traffic is to install a control program enabling the computer to send messages to the concentrator until the latter's queue buffer is full. At that point, the concentrator signals the host computer to stop transmission temporarily and tells it to store the messages in its buffer until such time as the RDC can again handle the traffic.

An alternative control scheme which may have better queue performance for some applications is where the RDC requests succeeding segments of a segmented message on a per-terminal basis when needed. Here, in a well-coordinated system, the risks involved in providing adequate buffering for the host-to-terminal traffic via the RDC are not ordinarily as critical as for the terminal-input situations.

In summary, the models yield a first-order approximation of the size of a common buffer pool required by a remote data concentrator to provide prescribed performance in an inquiry/response system. Like most models, they offer rapid solutions and give insight into the effect of interdependent system parameters. But, also like most models, they are less accurate at extreme values. While these models can aid the designer, their simplicity may be disarming. They should not be employed without full awareness of such other important aspects of data communications systems as the consequences of line transmission rates and excessive queuing delays, the value of buffer space in a main frame, the number of terminals that can be serviced by an RDC, and the impact of RDC implementation on the scheduling algorithms in the host computer or programable front-end processor.

Message switcher links diverse data services, speeds

Store-and-forward computer enables user to intermix line speeds, formats, and hardware, optimizing network efficiency and throughput

Walter J. Heide and Patrick J. Hennelly
McGraw-Hill Inc.
New York, N.Y.

Application

By redesigning its news gathering and dissemination network around a message-switching computer, and dual-buffered video editing terminals at the busiest locations, McGraw-Hill overcame some significant data communications limitations.

The resulting improvement in the speed and efficiency of its worldwide operations was accomplished by building in a high degree of flexibility in transmission services and speeds, including code- and speed-conversion capabilities; almost instant delivery of messages to more than 135 remote sites; improved error control; dynamic polling of high-traffic locations, and unattended communications.

Now, any of those offices—in the U.S. and abroad—will be able to send copy and administrative messages directly to any single location or group of locations anywhere in the world through the IBM System/7 message switcher. Located at McGraw-Hill's communications center in New York, the switcher can handle communications between points having quite different line speeds, at rates up to 1,200 bits per second.

For nine years, the company had relied on a low-speed leased-line network to handle the traffic between its 17 most active domestic locations. The AT&T 83-B-3 system consisted of five 10-character-per-second half-duplex circuits, each terminating in New York City. The 125 less-active sites were on dial-up Telex or TWX service. They still are, with a few modifications, such as adding Data-Phone units to the TWX locations, so that they also can be contacted through the telephone network.

The redesign's main thrust was to revamp the network that connected the 17 most active locations. At about 100 words per minute per circuit, it had become too slow for a large news-oriented organization. In addition, a Teletype terminal could only communicate directly to other terminals within its own circuit—with New York the only common denominator. For example, Chicago could contact Detroit directly, but not Washington, a pivotal city for a news-gathering company. Instead, a time consuming procedure was followed: The message was sent to New York, where a paper tape was produced, hand-carried to the tape reader of the multipoint circuit serving Washington, and transmitted to the nation's capital.

Other problems with the old arrangement were

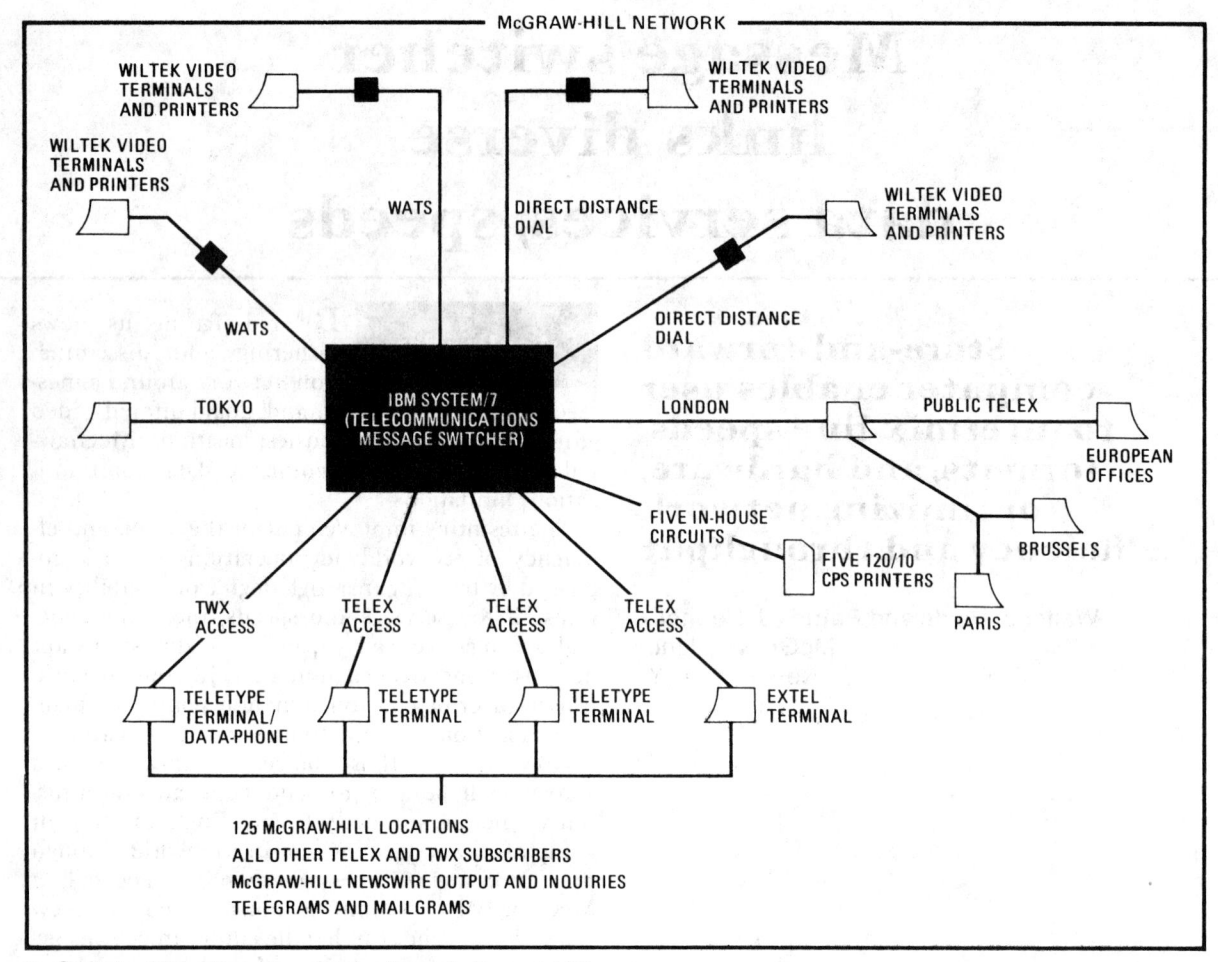

1. Switcher handles communications between 137 company sites, and also provides access to the entire Western Union Co. Telex and TWX networks.

that the leased lines were expensive, Telex and TWX units incurred time and usage charges, and there was no error-checking ability, nor even a hard copy to proofread prior to transmission. Someone also had to stand by at the sending location to make sure the paper tape didn't jam or tear during transmission. Since the majority of sites don't have terminal operators, the local personnel had to double as keyboarders and tape watchers.

The new approach clears up these problems and gives a 15% annual savings. And the system can be expanded to support five times the present number of lines and 10 times the number of terminals. Upgrading the modems to 2,400 b/s would double throughput on the high-speed lines.

The heart of the system (Fig. 1) is the Telecommunications Message Switcher (TMS), which interfaces two outward-bound, nationwide WATS lines that are used exclusively for this network; two direct distance dialing (DDD) lines that link with the dedicated voice-grade, in-house Centrex system and its tie-lines; and three Telex access lines and a TWX access line to the full Western Union Co. (WU) networks. There are also five in-house lines,

along with two 75-b/s, full-duplex lines and a half-duplex, multipoint circuit for international use. The entire network is run by CCAP/7, a Communications Control Application Program from IBM that was modified, under contract, to meet McGraw-Hill's specifications.

All 15 lines in the system are connected to either of two multiplexer modules in the TMS (Fig. 2). The multiplexers have a one-character buffer—with a serializer/deserializer, activity timer, and alert register—for each line. Consequently, each line is treated as an independent entity. Simultaneously, one line may be sending, a second receiving, a third initiating a polling, another may be idle, and so on.

Sequence of events

For data coming into the TMS, here's what happens: The central processor unit (CPU) accepts data on a contention basis—that is, first come, first served. When a character is complete in its buffer, it is moved into main memory. The CPU knows at what speed each line transmits data and what format it uses. Thus, it can handle bits that arrive at the multiplexer at different intervals. After a specified number of characters are in memory, the data is written onto the TMS's disk. A given message is not placed contiguously on disk, since its segments are handled on a contention basis. Instead, it is written on dynamically assigned sectors of the disk as segments are received (Fig. 3).

Two special message-control fields are added to each segment. A field added at the beginning indicates where the previous segment of the message is stored. Another field at the end specifies which disk sector will hold the next segment of the message. Depending on how many individual messages are contending for space, the pieces of the message may be close together or widely separated on the disk.

On the output side of the TMS, the procedure is reversed. Again, because the CPU knows the speed and format of each line and terminal, the message is sent at the right speed. Data is taken off disk in sectors, put into main storage, and presented, character by character, to the multiplexer's buffer.

The 17 most active locations are equipped with Wiltek Inc. dual-buffered video terminals and printer terminals, and only those sites are polled. In all, there are 44 terminals at the key locations. Concurrently, the TWX, Telex, and overseas circuits are continuously monitored, calls answered, and the messages put through the multiplexer and stored for subsequent forwarding.

Polling is done on a programed timing schedule and sequence, adjusted by the TMS every hour, 24 hours a day. The adjustment reflects historic peak-load periods for each location. The sites with the heaviest traffic are generally queried approximately every 10 minutes, the others every 20 minutes. But this changes. At lunch time on the East Coast, for example, eastern terminals are polled less frequently than at other times during the business day, although sites in other time zones may continue to send as much copy to the east as during any other period of the day.

Once entered in the central memory, messages from any location on the network are forwarded at the first opportunity—usually within seconds, but never longer than a few minutes. With messages being delivered that quickly, there is no heavy buildup of traffic for a given location.

Dynamic polling

At the beginning of the day, there is an initial polling sequence, but that sequence changes rapidly. The frequency of contacting each station is the critical factor. For example, once a station on a 10-minute polling schedule has been contacted out of turn so that it could receive a message, it is automatically rescheduled for polling 10 minutes later.

Let's assume that Los Angeles, San Francisco, Chicago, and Detroit are the next stations due for polling, and each one is to be contacted every 10

Building a network

Installing video and printer terminals at 17 locations around the U.S., bringing them on-line, and training operators can create a lot of problems. But when those terminals, plus 120 existing Telex, Teletype, and TWX locations, are all going to be hooked up to a new message-switching system, the data communications manager can expect many more headaches.

One way to be certain the entire network runs smoothly as soon as it goes into operation, is to put the elements together on a building-block basis, making sure each step is of manageable size. First, get the video/printer terminals up and running well. Then check out the other locations for compatibility with the proposed system. Finally, prove out the message-switcher portion. Only then integrate the pieces.

That's exactly what McGraw-Hill did. The Wiltek Inc. terminals were installed and plugged into a proven message-switching computer at Wiltek's offices in Norwalk, Conn.

Once the terminals were in, operators were trained and the units debugged. The procedures for moving information to the other terminals in the network were the same as those that would be used once the company's IBM System/7 message switcher was installed in New York.

After several months of testing terminals, procedures, programs, and people, attention turned to installing the System/7. For checking out the switcher and its capabilities, the group of terminals in New York were designated as the guinea pigs, using the following procedure. All the Wiltek terminals, including those in New York, would remain working through the computer in Connecticut. But all messages originating in New York would also print out on a terminal in New York. After the messages went out, the send buffer would be reversed to a given point and stopped. The System/7 would call the local terminal, take all the data, perform all its functions, and then deliver the message to a second terminal in New York. Only when everything was working to management's satisfaction, would the other 16 locations be moved onto the system.

The same approach was used to prove-out the Telex, Teletype, and TWX locations.

minutes. Should a polling of Los Angeles pick up a message for Detroit, delivery of that message is tried as soon as it is on disk. When Detroit is on the line, it sends any messages it has to the computer for storage. It then gets the message from Los Angeles—as well as any other messages that may be stored for it—and the next Detroit polling is rescheduled for 10 minutes from that point.

In the same example, Detroit might have had a message for Chicago. It, then, would trigger the immediate-delivery routine. On completion of the delivery, Chicago would be rescheduled for polling 10 minutes later. Calling times and schedules are, thus, dynamically updated. If Detroit has a message for a Western Union location, such as a Telex or TWX site, or service, such as Mailgram, no reshuffling of calls is needed since those locations are not polled.

The video terminals and the TMS have automatic answering, so no operator intervention is required. However, when the destination terminal cannot be contacted, the TMS retries the line four times in rapid succession before messages for the called terminal are put on "hold," and the console operator in New York advised of the line problem.

Manual override

Because of McGraw-Hill's news-dissemination needs, there often is last-minute copy for a publication about to close. In that case, the console operator at the communications center can manually override the polling routine by keyboarding some special instructions, which take effect as soon as the present call is completed. For example, a station can notify the operator that it has closing copy and, starting at a given time of day, wants to be polled every five minutes, or every 10 minutes in the case of sites normally polled every 20 minutes. The operator can put this change into effect by hitting a few keys.

When a WATS or DDD line becomes overburdened, the console operator can instruct the TMS to move part of that line's polling over to one of the other high-speed lines. When the overburdened channel falls back within acceptable parameters, he can switch the polling back to its normal line. The operator can determine when this problem arises by asking the computer how many messages are in queue for a given site, or line.

In case the TMS itself becomes inoperative, the network can still function. The lines are taken out of the TMS and connected directly to the terminals at the communications center via a backup device that simulates the computer. A remote site is then manually dialed, and messages sent straight through from terminal to terminal.

Minimizing costs

Having a variety of lines not only makes the system efficient, but offers the flexibility needed to minimize transmission costs. The tie-lines, for instance, are relatively unused for voice calls during certain hours of the day—primarily early morning, lunch time, and late in the afternoon. At those times, the TMS transfers some of what would normally be WATS-line polls to the tie-lines, which cost the same monthly rate no matter how much they are used. As a result, the WATS lines, which carry a flat-rate charge for up to 240 hours per month usage, are cleared for communicating with TWX/Data-Phone locations. This juggling of lines eliminates charges that would have been incurred had the sites been called over a WU line, as they are during most of the day.

16,000 addresses are stored

The TMS has 40,000 bytes of main memory. Part of that capacity stores the addresses and phone numbers of all the video terminals, a four-character identification code for all other McGraw-Hill (M-H) locations, and frequently called WU numbers. In addition, it has the destination identity codes for 150 membership, or group, lists, each of which can consist of up to 100 different telephone numbers, and all the required code- and speed-conversion tables.

The disk memory has an additional 4.9 megabytes of storage, of which 4.3 megabytes are for temporary storage of messages. The remaining 600,000 bytes are overlay memory holding the phone numbers, answerbacks, and addresses of WU terminals, as well as complete data on the less frequently called M-H locations and the hundreds of outside subscribers to the company's three daily newswire services.

Messages within the system cannot exceed 25,000 characters. Under the old setup, a long message took 45 minutes to send. Now, it is completed in under three minutes because of the 1,200-b/s modems that link video terminals with the line. That long a message, however, is highly unusual. Since a keyboarder can't type that fast and most of the terminals are polled roughly every 10 minutes, long messages are generally broken into shorter "pages" for pick-up on the next several pollings of that site.

The average message, in fact, is 1,900 characters long, and a terminal may have a number of messages to send. In all, the system handles about 1,000 different messages per day. Many messages, however, go to multiple locations, making daily throughput about 1,800 messages. Stated another way, 2.3 million characters are sent a day, plus the variable volume entered for outside subscribers to M-H's daily newswire services. This load still leaves plenty of room for expansion because the system can sustain a throughput of 450 c/s, or in excess of 38 million characters per day.

Using the average length, the disk can store over 2,000 messages, or twice the present total. The aver-

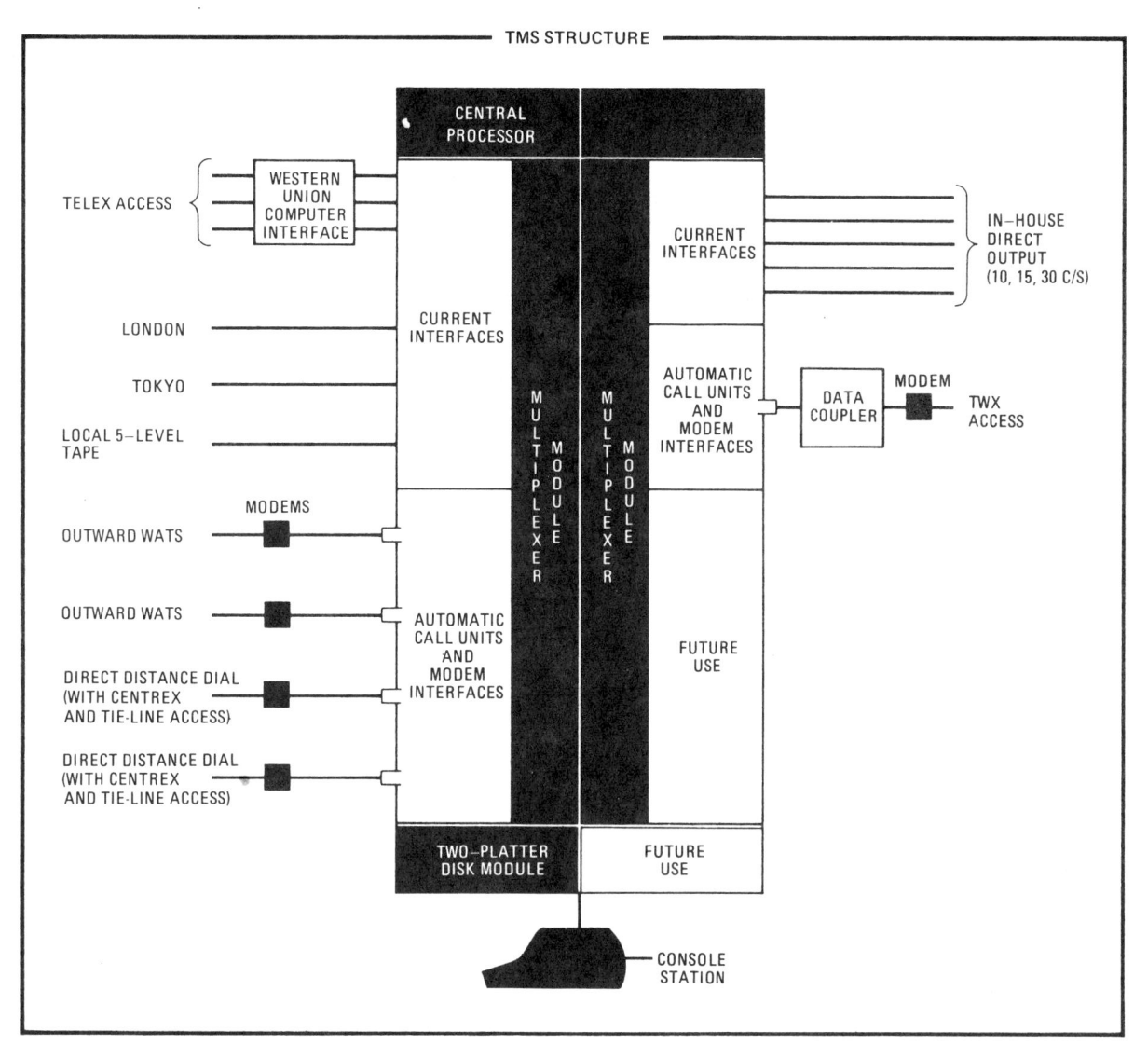

2. To simplify network expansion, the switcher's disk can store twice its present volume, and the unit's line capacity is five times today's total.

age time an undelivered message is resident on disk is less than 2 minutes. Once stored on the two-platter disk, any message can be retrieved up to 36 hours later.

Message handling

A message entering the system has a header made up of local message number and the code address of the terminal to receive the data. Before being placed on disk, the TMS code-converts any message to EBCDIC, the computer's internal code. Coming off disk, the language is again translated, this time to ASCII for transmission to the video terminals, national or international baudot for Telex, or TTS for five-level-tape teletypewriters.

Before the transmission is completed, the TMS assigns an internal number to the message. The number can later be used for tracking the message should someone say it was never received. Each terminal's messages are numbered in order from 0001 through 9,999 for a month, and then the cycle starts again.

The hard-copy printout at the originator's end contains an accepted or rejected code, the date and time, to the second, that the message was accepted by the system, and the number of characters sent. If a message is rejected by the computer, the reason why is given.

At the receive end, the header contains the same data as on the sender's copy—including who else is receiving the same message. This last part is considered particularly important because some locations that should be informed may have been forgotten. What's more, knowing exactly who else is receiving the same message allows editors to coordinate efforts between field offices, and eliminates duplication.

At the end of the message appear the date and

MESSAGE FLOW

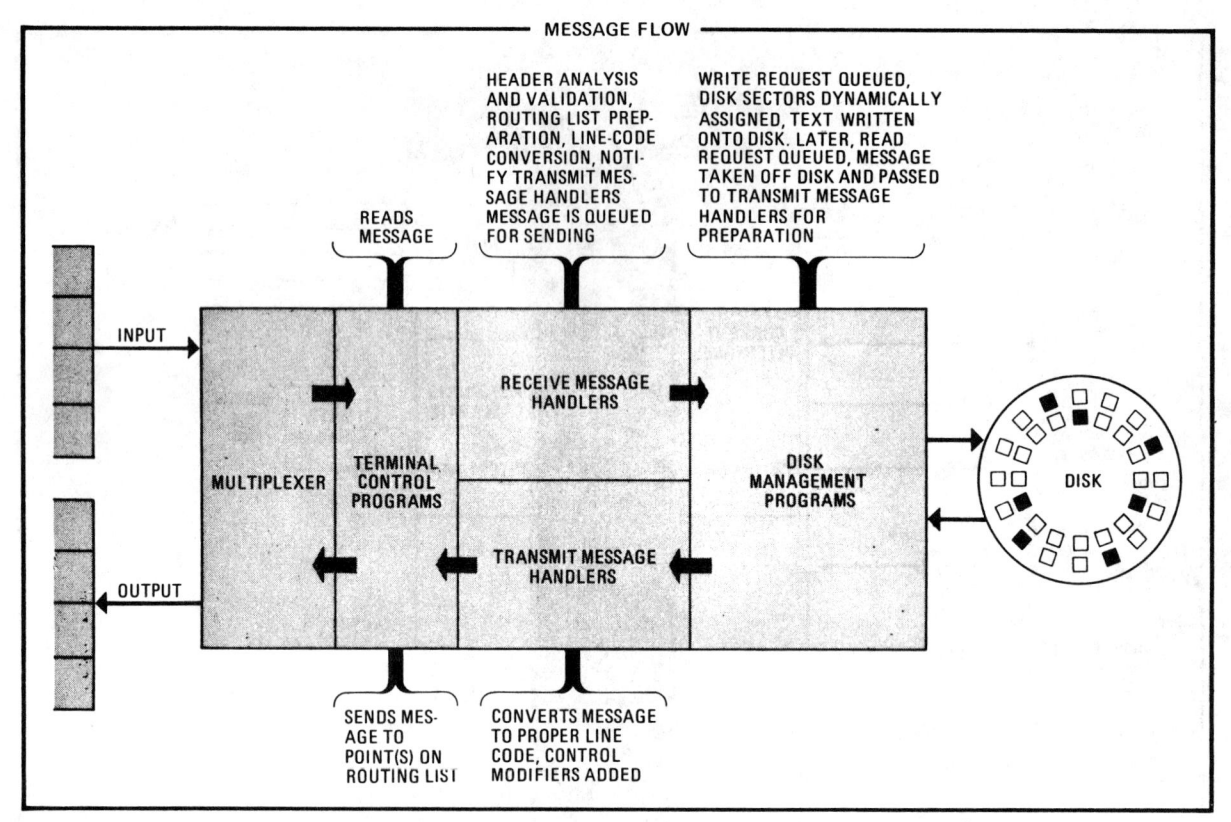

Diagram labels:
- INPUT → MULTIPLEXER → TERMINAL CONTROL PROGRAMS → RECEIVE MESSAGE HANDLERS → DISK MANAGEMENT PROGRAMS → DISK
- DISK → DISK MANAGEMENT PROGRAMS → TRANSMIT MESSAGE HANDLERS → TERMINAL CONTROL PROGRAMS → MULTIPLEXER → OUTPUT
- READS MESSAGE
- HEADER ANALYSIS AND VALIDATION, ROUTING LIST PREPARATION, LINE-CODE CONVERSION, NOTIFY TRANSMIT MESSAGE HANDLERS MESSAGE IS QUEUED FOR SENDING
- WRITE REQUEST QUEUED, DISK SECTORS DYNAMICALLY ASSIGNED, TEXT WRITTEN ONTO DISK. LATER, READ REQUEST QUEUED, MESSAGE TAKEN OFF DISK AND PASSED TO TRANSMIT MESSAGE HANDLERS FOR PREPARATION
- SENDS MESSAGE TO POINT(S) ON ROUTING LIST
- CONVERTS MESSAGE TO PROPER LINE CODE, CONTROL MODIFIERS ADDED

3. Multiplexer has a separate one-character buffer, with serializer/deserializer, for each line. CPU then accepts characters from the different buffers on a contention basis—first come, first served.

time the delivery was made, as well as a repeat of the date and time it went into the system. This allows the recipient to quickly determine whether the TMS is handling messages promptly.

Dual-buffered terminals

The buffered editing terminals at the 17 high-traffic locations play a critical role in moving copy quickly. Each location has at least one cathode-ray-tube (CRT) terminal with full editing capabilities. Instead of having to compose a story at a typewriter and then enter it into the terminal by keyboarding—which he might have to do when no keyboarders are available—an editor can start right out at the terminal and see what he is typing displayed on the screen. To help him along, the keyboard has special function keys for deleting a word or line, adding or deleting spaces, and for overprinting any previously entered word.

When a short message (not more than a screenfull) has been entered, the operator inserts the end-of-message code, moving everything from the CRT's memory to the send buffer for storage until transmittal. The terminal keeps an internal count of the number of messages it has stored at any given moment, so that only complete messages are sent when polling takes place.

Long messages are handled with a slight variation. As the CRT is filled, a different key is hit. This also moves the data to the send buffer—at the same time, wiping the screen clean. But, the data is stored there and not sent until the balance of the message has been entered and the end-of-message characters added.

A second major plus for the terminals is the dual-buffer arrangement. The separate 50,000-character send and receive buffers can be used simultaneously, and each buffer has its own read and write heads. When the terminal is polled, it answers the call, performs the handshaking routine, and starts to transmit the contents of the send buffer. Typically, while sending through the read head of the send buffer, that buffer is also taking input from the CRT memory. If that message is completed while the terminal is still sending, the new message will be sent right then.

With dual-buffers, this entry of data by the keyboarder can continue even when the terminal turns the line around and starts to accept messages from the TMS. The incoming data is stored until it is printed out on the hard-copy printer, which starts up as soon as the first character hits the endless-loop magnetic tape buffer.

At the medium-to-heavy-load locations, the printer churns out message traffic at 300 words per minute. At low-traffic points, it's printed at 100 wpm. A hard copy of outgoing material is made via a loopback, built into the terminal controller, between the send and receive buffers. To get an extra copy or to retransmit a message, the operator backs up the appropriate buffer to the beginning of the message he wants.

Error control is provided in two steps. There is a parity check that takes place between the CRT and send buffer. In this case, any error appears on a printout as an asterisk, alerting all parties. The error can then be corrected at the next polling. Second, a cyclic redundancy check of each block is made at the send buffer and the TMS and later by the TMS and the receive buffer at the ultimate destination. When an error creeps through three times, the balance of the message is scrubbed and the call placed again. Transmission is picked up at the point where the error was made.

To hurry the flow of copy in New York, home of most of the company's magazines, printers can be located in the editorial offices of the various publications. All copy addressed to a publication would print out both there and on one of the print-only terminals in the communications center. Eventually, as the system expands, the copy at the center will be eliminated. Without printers in editorial offices, messages have to be hand-delivered or sent by pneumatic tubes to the proper floor for hand pick-up, which results in an unnecessary time delay. A printer's speed depends on how much copy is forwarded to that specific publication. Heavy traffic calls for a 30-c/s unit. Other printers can be slower, down to 10 c/s.

International copy

Time-zone differences, especially on an international basis, were a headache under the old system. The Tokyo office, for example, couldn't communicate directly with the international head office in London because the two cities were on different circuits. Everything had to go through New York. But, during the work day in Tokyo (13 hours ahead of New York) it was night in New York. The message for Europe had to sit in New York until the next morning. By the time it was delivered to London (five hours ahead), it was already afternoon in England. From there, the message was relayed to other M-H offices on the continent.

Although this operation still runs over a four-wire leased line, the TMS immediately forwards the message to its destination, no matter what time of day or night it comes into the computer. The London office then distributes the messages—to Paris and Brussels by leased wire and to other European bureaus by the public telex network.

Newswire service

The availability and distribution of news to subscribers to any of M-H's three daily newswire services has been enhanced by the TMS. Until the unit was installed, finished copy, plus a list of several hundred subscribers, would be transmitted on paper tape from the communications center to a Western Union computer every afternoon. WU then sent the material across its Telex and TWX networks to the subscribers. Today, newswire copy is put on disk and distributed to subscribers right from the TMS. This saves both the time and costs previously involved with first transmitting the data to the WU computer. Since three separate newswires are involved, the saving is substantial.

What's more, an important additional service is possible because of the power of the TMS. A subscriber need no longer wait until the end of the day to get his news. Up-to-the-minute news can be put on disk as soon as it is written, instead of everything being batched for entry late in the afternoon. Thus, a subscriber would be able to dial a special number at any time during the day and, once verified by the TMS as a valid subscriber, have access to all the news stored in the computer.

Current proposals for expansion call for an even greater improvement in service in the near future. The TMS may be programed to combine the news from the three newswire services—metal, chemical and oil—and sort items into specialized subcategories, such as financial, plastics, particular types of petroleum, and so on. A subscriber to this new service would be able to call in and get the news from just those categories that interested him.

Keeping up to the minute

Several statistical reports for management are prepared regularly by the TMS. The purpose is to give an over-all view of any problem and help in analyzing traffic and load patterns.

One report, which covers all the video terminal locations, lists (to the second) the total amount of dialing time for a given period, transmission time in total, and the average transmission time, as well as the number of connected and disconnected calls by site. The report can be generated at a moment's notice, but is usually made every 24 hours. It is used to see how efficient the system is and how the most recent day's performance compares with another day.

Another daily report logs all parity errors for each individual site by specific character, time made, and message number. Among other things, should a particular error crop up too often, this log helps in alerting management to possible circuit or terminal problems.

Line errors are detailed in a separate daily report, which totals how many blocks were retransmitted because of problems on the communications lines, and how many calls were terminated as a result. A fourth report specifies how many calls were placed over the WATS lines, broken down by location. One use for the report is to monitor the telephone company's monthly statement. ∎

Making the most of simultaneous voice-plus-teleprinter techniques

Operating with voice and data channels on the same private line can lead to significant economies

Stephen A. Dalyai
Quindar Electronics Inc
Springfield, N. J.

Hardware

When a private voice-grade line is equipped at each end with a voice-plus-teleprinter unit, one or more teleprinter lines along the same route can be eliminated. The longer the voice line (above a certain minimum), the greater the monthly savings in lease costs, since these rise with distance.

Basically, the voice-plus-teleprinter unit (V+TU) uses frequency-division multiplexing to divide the voice channel into a not-as-wide voice band and up to four narrow bands for low-speed teleprinter data. The derived voice band, despite the loss of some frequencies, delivers speech that is fully intelligible to most listeners. What's more, it can be used also for high-speed data and facsimile transmission, though the maximum possible speed may be a little less than on conventional voice-grade line.

The simultaneous voice-plus-teleprinter technique is not to be confused with such line-sharing methods as alternate voice/data and interpolated voice. And it is also different from conventional data communications, where frequency-division multiplexing derives up to 24 low-speed channels from a voice-grade line, but no voice channel.

The technique is not new. Also known as speech-plus-data and voice-plus-data, it enjoys a certain popularity in Europe, South America, and Asia, and it is offered as a tariffed service by several international record carriers (for example, RCA Global Communications Inc.) and by at least one domestic specialized carrier (Southern Pacific Communications Corp.). However, neither Western Union nor domestic telephone companies provide such tariffed offerings. Furthermore, despite potential savings and technological maturity, simultaneous voice-plus-teleprinter has been little used in private networks in the United States—perhaps because many communications managers and planners aren't aware either that it exists or that equipment for implementing the technique is available from several domestic manufacturers.

The principal reason for using the voice-plus-teleprinter technique is to eliminate the need for one or more separate Type 1005 (75 bits per second) or 1006 (110 b/s) teleprinter lines between two or more locations that already have private-line connections for voice or high-speed data traffic. Thus, the amount saved each month depends on the difference in monthly cost between one or

PAYBACK PERIOD AND MONTHLY SAVINGS

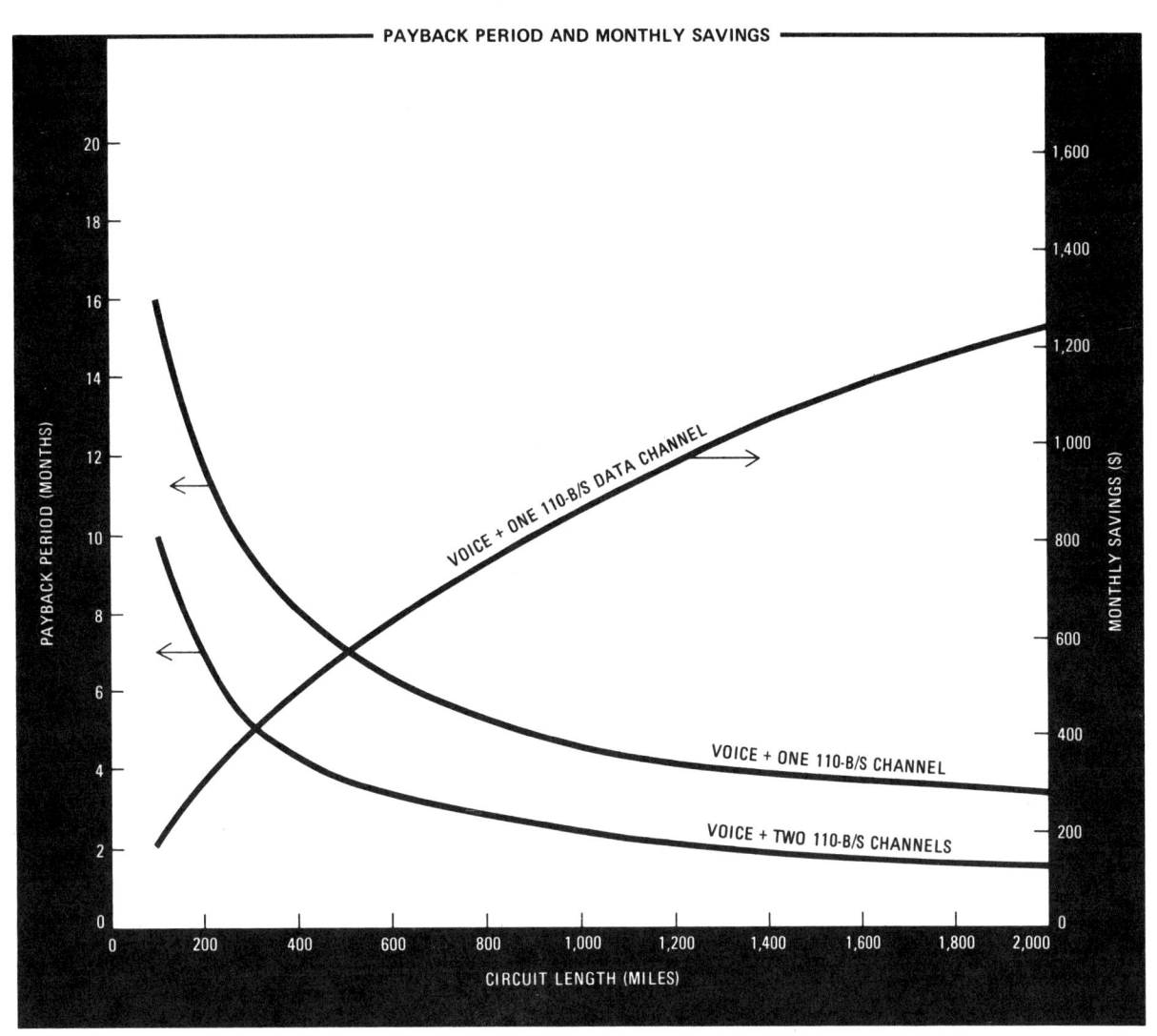

1. Voice-plus-teleprinter techniques have favorable payback period and monthly savings economics, particularly for long distances.

more teleprinter lines and the equivalent monthly cost for the V+TUs.

The cost of a Type 1005 or 1006 line, found in FCC No. 260 tariff, increases with line length. While the figure for a V+TU will vary somewhat depending on requirements, a pair of them should cost at most about $4,000. The colored curves in Figure 1 show how long it will take, by not having to pay for a Type 1006 line and related service termination units, to recoup the equipment cost in two applications. One application requires a voice band and one 110-b/s (or baud) teleprinter channel, the other requires a voice band and two 110-b/s channels. For example, eliminating one 1,000-mile Type 1006 line and its service termination units (upper colored curve) will pay for two V+TUs in about 4½ months. Eliminating two 1,000-mile low-speed lines (lower colored curve) will pay for two V+TUs in about 2½ months.

The black curve in Figure 1 takes two other factors into account—maintenance cost and installed lifetime—to arrive at the monthly savings that result from not leasing one Type 1006 line plus its service termination units. Here, again, the cost of the V+TUs is estimated at $4,000, the maintenance is, at most, $400 a year, and installed lifetime is assumed to be a conservative 10 years, making the over-all cost of the equipment come to $67 a month. The black curve then shows that eliminating a 1,000-mile teleprinter line will save more than $800 a month during the V+TUs' lifetimes.

Variety of networks

By using the voice-plus-teleprinter technique with both private and tie-line voice circuits, a variety of low-speed data networks can be produced, including point-to-point, remote computer access, and message-switching types.

A four-wire point-to-point private line, with a V+TU installed at each end, can be made to pro-

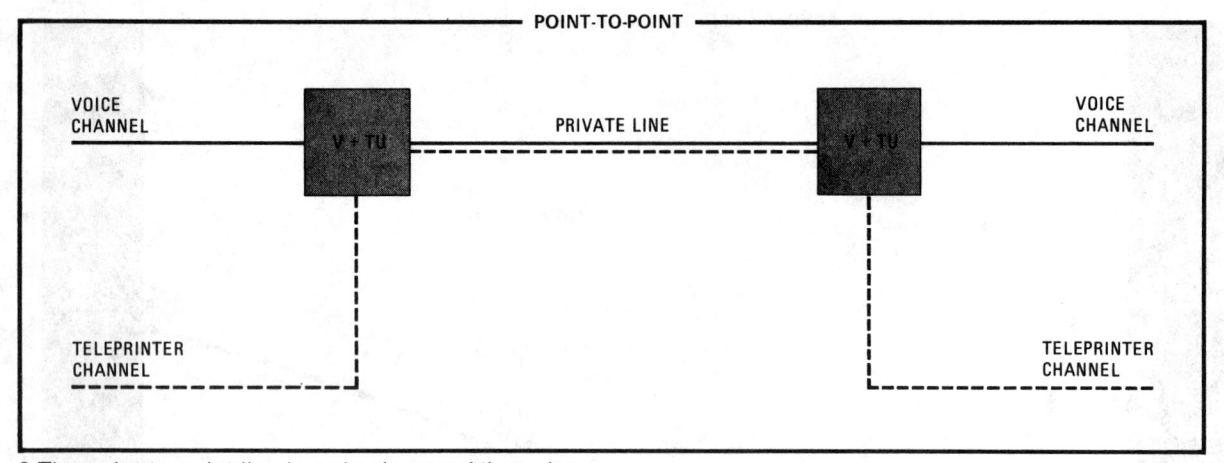

2. The point-to-point line is a simple use of the voice-plus-teleprinter technique.

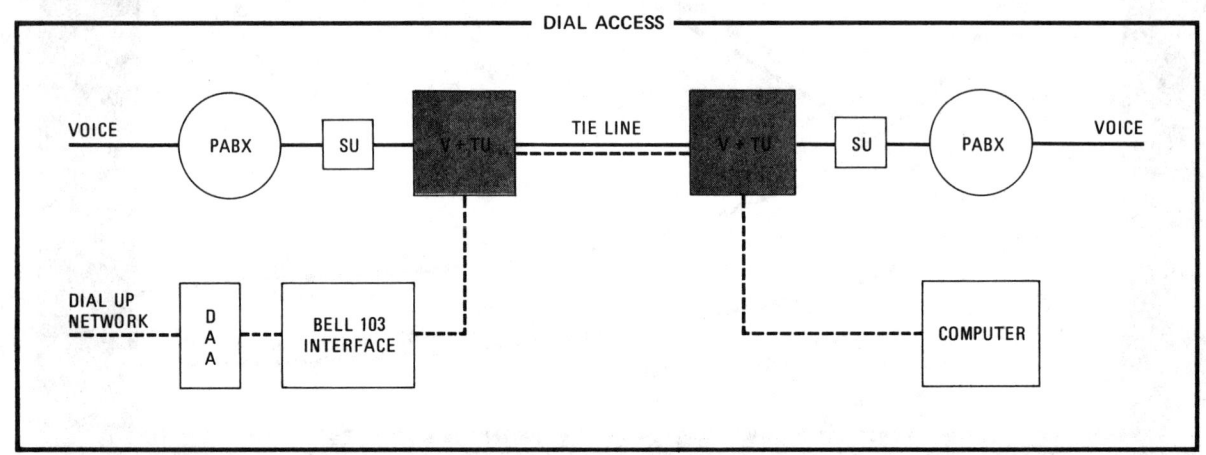

3. Voice-plus-teleprinter channels can be dialed up and go over a private-branch-exchange line.

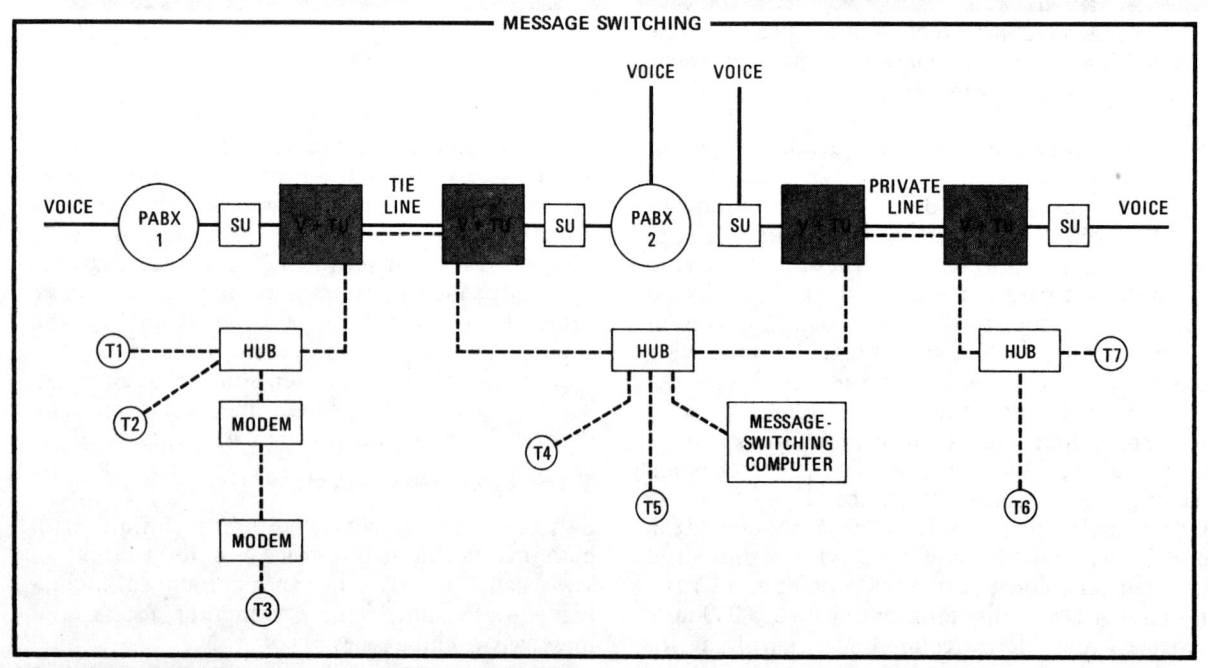

4. A variety of ways to access channels can be used in a message-switching system.

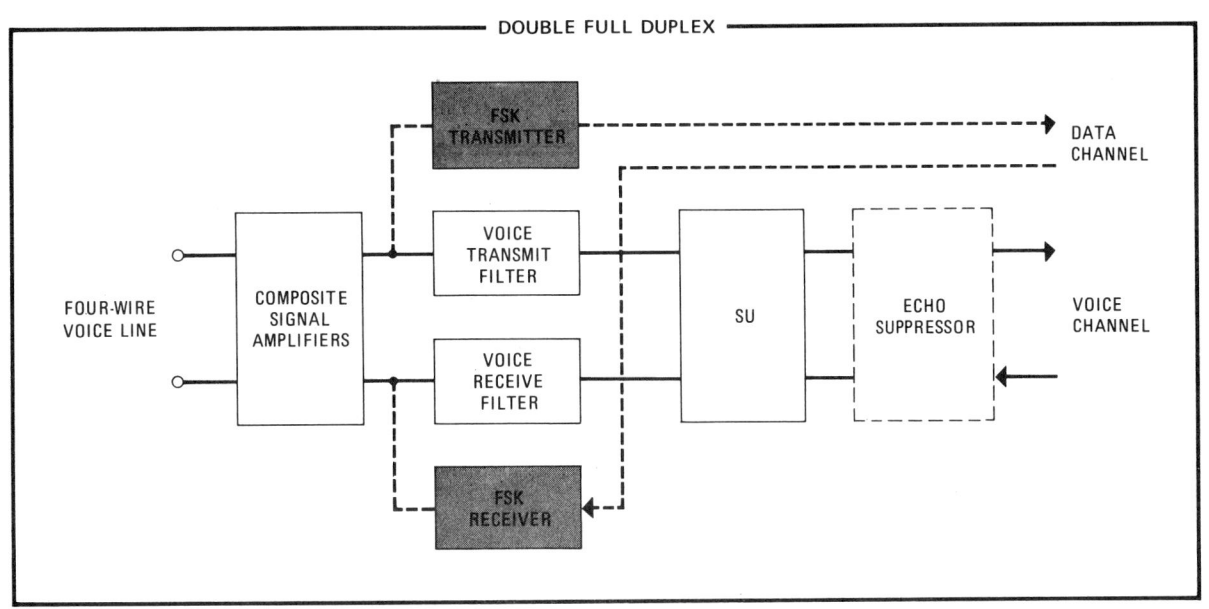

5. The basic electronic elements in the voice-plus-teleprinter unit are amplifiers and filters.

vide full-duplex voice transmission and full-duplex teleprinter transmission (Fig. 2). Here, the V+TUs break out the four-wire voice circuit (solid color) from the four-wire teleprinter channel (dashed color). If the V+TUs had, say, provision for two teleprinter channels, then two separate, private, four-wire, low-speed channels would be available.

Organizations with geographically scattered installations often use private automatic branch exchanges (PABX) to reduce voice-line costs (Fig. 3). Here, a tie line connects each PABX to one or more remote PABXs. (Signaling units (SU) must also be installed.) Therefore, point-to-point teleprinter service can be obtained on tie lines, too, more or less as in Figure 2.

However, the arrangement in Figure 3 goes a step further in the application of the V+TU technique. Here, a remote terminal uses a data access arrangement (DAA) and an interface to dial up the teleprinter channel of a V+TU so that, as shown, any terminal connected to the dial network can access a remote computer (or other terminal) connected to the other end of the tie line. The terminal could, for example, operate at 300 b/s, so that the voice-plus-teleprinter unit would derive one channel with a 3,000-Hz carrier frequency with the necessary ±120-Hz frequency shift.

When a company needs extensive voice communications and extensive message switching over essentially identical routes, the voice-plus-teleprinter technique has great economic advantages. Figure 4 shows a message-switching network employing both private and PABX tie lines.

Here, the message-switching computer selectively polls any terminal on the network either to forward data from another terminal to the addressed terminal or to ask if the selected terminal has a message for any other terminal. A hub, or bridge, permits any of the computer-selected terminals to communicate one at a time with the message-switching computer. Note that the hub is linked to nearby terminals with current loops and to remote installations by modems. The terminals can operate at any standard speed from 75 b/s to 300 b/s, and the bandwidth of the teleprinter channel in the V+TU is selected accordingly. Standard bandwidths for low-speed data channels are given in the table.

How the V+TU works

These low-speed channels can be squeezed onto the conventional voice-grade line only because human speech contains most of its power in frequencies below 2,500 Hz. In fact, tests have shown that intelligible conversations can be held with the highest frequency at 1,500 Hz—although this situation is not normally exploited by the voice-plus-teleprinter technique. Using FDM, sharp-cutoff filters limit the band for voice and prevent this band from interfering with filter-derived low-speed data channels. At the center of each derived data channel is placed an audio tone that carries the data stream. This carrier is shifted up and down in frequency by the binary 1 and 0 bits in the transmission to yield binary-modulated audio signals. Modulation of a carrier frequency by a signal frequency is called frequency-shift keying (FSK).

Figure 5 shows the basic elements of a voice-plus-teleprinter unit. Amplifiers in the receiving and transmitting paths control the level of the composite signal. The voice channel interfaces with manual or automatic switching systems and incorporates an SU. An optional echo suppressor can be inserted for long-distance circuits. The FSK transmitters and receivers, used to derive the data channels, are connected to the composite-signal amplifiers. The interface of the data channels to the terminals, not shown, can accommodate current (20-milliampere or 60-mA) or voltage (RS-232-C) signals.

Deriving data channels

Theoretically, the narrow-band channels for teleprinter traffic could be trimmed from the low, middle, or high frequencies of the normal bandwidth available on a voice-grade line. The three alternatives are shown in Figure 6.

In the top drawing, a high-pass filter cuts off the lower portion of the voice spectrum, and the data carrier tone (or tones) are placed at the low-frequency end of the spectrum. This scheme has two major disadvantages. Firstly, filtering out the low frequencies would degrade voice transmission the most because they are the most responsible for its intelligibility and quality. Secondly, power line and other 60-Hz noise can cause intense interference at the lower frequencies.

The middle drawing in Figure 6 shows the center of the voice spectrum "notched out" and the data carrier placed in the notch. This scheme is popular and very effective in applications where only one data channel is needed and the wideband channel is used only for transmission of speech. However, a data carrier in the middle of the spectrum would cause interference if the "voice" channel were used to transmit medium-speed data or facsimile.

The arrangement in the bottom drawing of Figure 6 is best for most purposes and is used in most voice-plus-teleprinter equipment. Here, a low-pass filter with a voice cutoff frequency of F_{VC} band-limits the channel so that all the low- and mid-frequency components of the voice are retained but the upper frequencies are greatly attenuated and create a spectrum for the data channels.

If F_{VC} is kept above 2,000 Hz, both voice quality and intelligibility will remain good to most listeners. Certain systems have even employed an F_{VC} of 1,500 Hz with reasonable success. But when F_{VC} is 2,500 Hz or more, the average listener cannot tell the difference between the band-limited voice channel and normal telephone quality. The low-pass filter may also be delay-equalized, to permit medium-speed data transmission in an alternate voice/data mode.

For effectiveness, then, the data carriers should be placed in the portion of the spectrum above the voice cutoff frequency. Each data channel requires two guardbands to protect it from interference by adjacent channels. The width of each channel plus its guardbands is a function of the maximum transmission speed of the channel. Therefore, the number of data carriers that can be placed above the derived voice channel depends on the voice cutoff frequency (F_{VC}), the cutoff frequency of the facil-

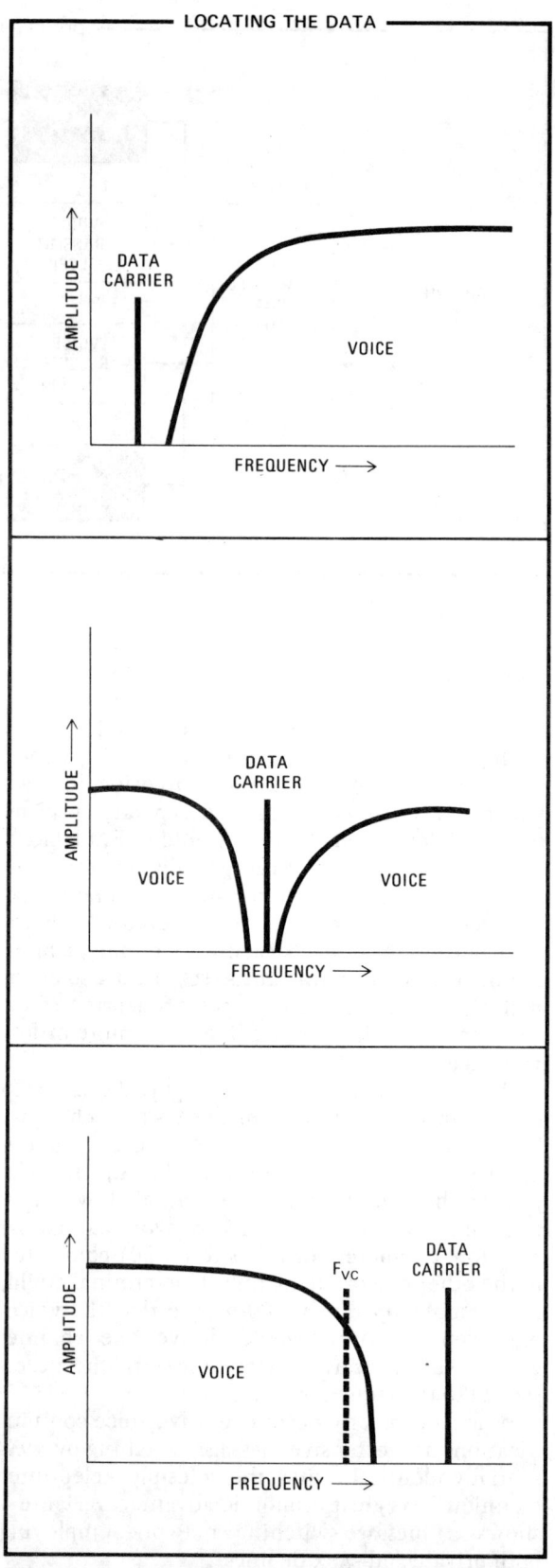

6. The carrier frequency for the data channel can be located below, in, or above the voice channel, and experience has shown that the best place is above, so as not to pick up electrical interference.

STANDARD LOW-SPEED CHANNELS

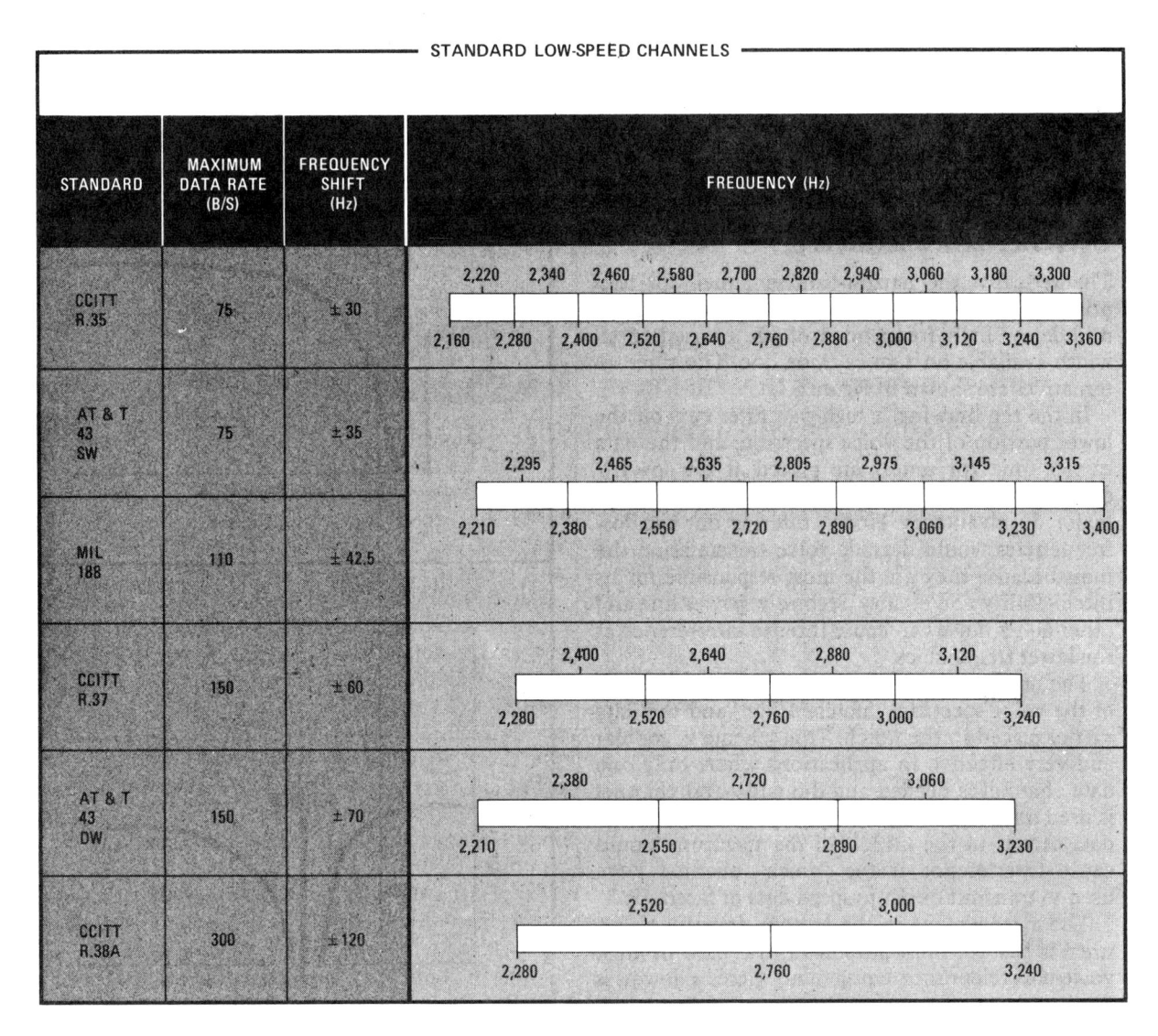

STANDARD	MAXIMUM DATA RATE (B/S)	FREQUENCY SHIFT (Hz)	FREQUENCY (Hz)
CCITT R.35	75	±30	Upper: 2,220 2,340 2,460 2,580 2,700 2,820 2,940 3,060 3,180 3,300; Lower: 2,160 2,280 2,400 2,520 2,640 2,760 2,880 3,000 3,120 3,240 3,360
AT&T 43 SW	75	±35	Upper: 2,295 2,465 2,635 2,805 2,975 3,145 3,315; Lower: 2,210 2,380 2,550 2,720 2,890 3,060 3,230 3,400
MIL 188	110	±42.5	
CCITT R.37	150	±60	Upper: 2,400 2,640 2,880 3,120; Lower: 2,280 2,520 2,760 3,000 3,240
AT&T 43 DW	150	±70	Upper: 2,380 2,720 3,060; Lower: 2,210 2,550 2,890 3,230
CCITT R.38A	300	±120	Upper: 2,520 3,000; Lower: 2,280 2,760 3,240

ities, or voice circuit (F_F), and the maximum transmission speed of the data channels.

Conditioning increases capacity

The spectrum made available for low-speed data channels, or data bandwidth (D_{BW}), is therefore:

$$D_{BW} = F_F - F_{VC}$$

That is, the higher the cutoff frequency of the circuit (or facilities) and the lower the cutoff frequency of the voice channel, the larger the number of available data channels of a given speed. Unconditioned lines supplied by telephone common carriers have an F_F of, at best, 3,000 Hz, but on microwave and many international circuits F_F is usually 3,400 Hz. Conditioning the common carrier lines can thus increase D_{BW}.

The table contains a range of channels that can be derived from voice circuits, including conditioned ones, according to various domestic and international standards. If a 2,100-Hz cutoff frequency is assumed for the voice channel, C0 conditioning will provide one 110-b/s (or baud) channel; C1 conditioning provides two such data channels; C2 conditioning, four channels, and C4 conditioning, five channels.

For example, appropriate line conditioning enables one arrangement, typical of those used on international circuits, to squeeze in four teleprinter channels above a voice cutoff frequency of 2,700 Hz. Here, the data carriers have a maximum transmission capability of 75 b/s each and are placed 120 Hz apart according to CCITT (International Consultative Committee for Telegraph and Telephone) Recommendation R.35 (see table).

The voice cutoff frequency of 2,700 Hz, besides providing excellent speech quality, also allows a standard 2,600-Hz tone to be used for signaling—a distinct asset, since the presence of a signaling tone in the voice channel means that the system can be operated with manual switching systems or automatic switching systems like PABXs. Finally, the equalized low-pass filter used to create the voice band also enables that channel to transmit facsimile or medium-speed data (up to 4,800 b/s) in an alternate voice/data mode. ∎

Fundamentals of software for communications processors

It's the choice among 10 software modules that makes a minicomputer into a remote data concentrator, a front-end processor, an intelligent terminal, or a message switcher

John F. Chyzik
Data General Corp.
Southboro, Mass.

TUTORIAL

Communications processors play four different roles in up-to-date data communications systems. They serve as programable front-end processors, as message switchers, as remote programable data terminals, and as remote data concentrators. For the most part, all are based on general-purpose equipment—the minicomputer, together with necessary interface hardware and main and secondary memory.

What distinguishes one system from another, essentially, is the software invoked to make the processor perform the detailed operations required by the processor's application. And even though the over-all software will differ from one type of system to the other, all of them use many software modules in common. Thus, a good understanding of data communications systems can be obtained by reviewing the roles of the various software functions, together with some implications of module relationship to memory utilization, and to programing difficulty.

The sharing of a data communications processor by many terminals (or stations) is accomplished, at any given instant, by one of 10 software modules. Each module is an intricate program of explicit instructions within the processor that implements a specific communications function. The 10 software modules are:

- **Dial-up,** used when the processor communicates through the public switched telephone network
- **Dedicated line,** used when leased lines connect terminals to the processor
- **Polling and selecting,** used with dedicated multipoint lines to call up the remote terminals
- **Code conversion,** used when the network equipment—including the terminals, the processor, and the host computer—employ different codes, as is usually the case in data communications systems
- **Message control,** which determines routing and further action to be taken on a particular message
- **Buffer and queue,** which temporarily stores incoming characters and organizes them into messages or blocks
- **Message switch,** which directs particular messages onto their corresponding lines
- **Peripheral control,** a standard software package related to a particular peripheral, such as a line printer or card reader, and usually supplied as part of the operating system
- **Host-computer link,** which organizes the processor to be compatible with a local host computer
- **High-speed synchronous link,** which organizes the

processor's high-speed input/output data stream to be compatible with a remote host computer

Certain software modules handle all types of communications, while others are specific to one or a few types, as shown in Figs. 1 through 4. For example, the remote data concentrator, Fig. 1, gathers inputs from many lines—dedicated, dial-up, or both—and merges them into a high-speed synchronous data stream for transmission to a remote host computer. Thus, remote data concentration requires dedicated lines, together with poll-and-select, dial-up, code-conversion, message-control, buffer-and-queue, and high-speed synchronous-link software modules. As a programable front-end processor, though, the minicomputer would employ the host-computer-link module instead of the high-speed synchronous link module (Fig. 2).

Operating system supervises all

Once the appropriate software modules have been determined by the processor's application, they are coordinated in time and by event by another software package, the operating system (OS). The OS is usually supplied by the minicomputer vendor. For example, when the time comes to perform message buffering and queuing of an incoming message, the OS calls in the buffer-and-queue software module, actually a program of perhaps hundreds of machine instructions that executes the detailed operations.

The detailed programing of a communications processor is complex and time-consuming, and the amount of over-all effort depends on the number and types of software modules, the processor architecture, the programing language, the amount of hardware-interface equipment, and the size of the data communications network. Often, programing effort can be decreased by using, where appropriate, dedicated electronic hardware that can do the same job as the software, but more quickly and with less complexity—for example, error control, as explained later. If hardware can be used, then the OS task is simply to bring the hardware into play at the proper instant.

Programing will not be discussed further here. It is a task well understood and implemented by professional programers who execute the activities specified by data communications systems planners. However, one major aspect of programing does depend on the system application and thus is determined by the planner: If the application must or can be implemented totally within the minicomputer's fast-access main memory, a real-time operating system is used, but if the application requires extensive memory and can tolerate slower-access, but less costly, disk systems, a real-time disk operating system must be used.

Line programs get the messages in and out

The first software module encountered by a message is either a dial-up-line program if the message comes in through a dial-up port, or a dedicated-line program if the message comes in through a dedicated, or leased-line, port. The two line-program modules differ in two ways: How control is exercised to connect a terminal to the processor and how electrical signals actually make the connection. These differences will be explained in the following discussion of the five subroutines that both types of line program have in common:

- **Link connection,** which establishes electrical connection between the terminal and the processor and gets the voice line ready to carry data
- **Receive data,** invoked when the processor acts as a receiver of data sent over the connection
- **Transmit data,** invoked when the processor sends data to a remote terminal
- **Error control,** which detects and can correct erroneous data when the communications processor is in its receive-data mode
- **Link control,** which is an agreed procedure on how message data passes between the terminal and the processor, and which insures message integrity by including an error-control discipline

The essential elements of establishing a link connection between the datasets for the terminal and processor is shown in Fig. 5. Here, the processor dials up the phone number associated with the terminal's dataset (modem). The dataset sends back a RING signal, which the processor dataset answers by sending back a DATA TERMINAL READY (DTR) signal, or does not answer by letting the RING signal continue until shut off by the terminal. The terminal dataset responds to the DTR by sending back a DATASET READY signal. Then the terminal dataset sends a CARRIER DETECT signal which, if detected by the processor dataset, means that compatible datasets are voice-mode connected. To shift into the data mode, the processor dataset then raises a REQUEST TO SEND, which tells the terminal it wants to transmit data, and, after a short delay while the telephone line echo suppressor is disabled to permit data transmission over the lines, the terminal dataset responds with CLEAR TO SEND. Now, the link is ready to carry data.

When dedicated lines are used, neither the RING report nor the DTR are needed, since the two datasets are permanently connected and can begin communications at any time. Further, in full-duplex links, the REQUEST TO SEND is raised permanently, while in half-duplex links, the direction of the REQUEST TO SEND signal determines whether a terminal is to send or receive data.

Connecting terminals on multipoint lines

Although, a connection in a dial-up network is made by dialing the telephone number associated with a terminal's dataset, in dedicated multipoint (or multidrop) systems, all datasets for all the terminals and the processor are permanently connected and ready to send or receive data. In this case, the communications processor sends out a bit sequence, called an address or identification (ID). When the addressed terminal senses its own ID, it will respond to the processor.

Two alternative ways to "connect" with a terminal on a multipoint line are polling and selecting (also called polling and addressing). In polling, the processor sends the ID which means: "Do you have any data to send me?" In selecting, or addressing, the ID is interpreted by the terminal as: "The processor has data for you. Are you ready?" The poll-and-select software module consists of one or more tables similar to the one shown in Fig. 6. Here, the IDs (1A, 2A, etc.) are listed in the order required by the application. That is, the IDs do not have to be in numerical sequence. For example, if terminal 1A generates a high volume, it may be polled several times during the cycle, as indicated in the figure. The ID

FIGURE 1 REMOTE DATA CONCENTRATOR

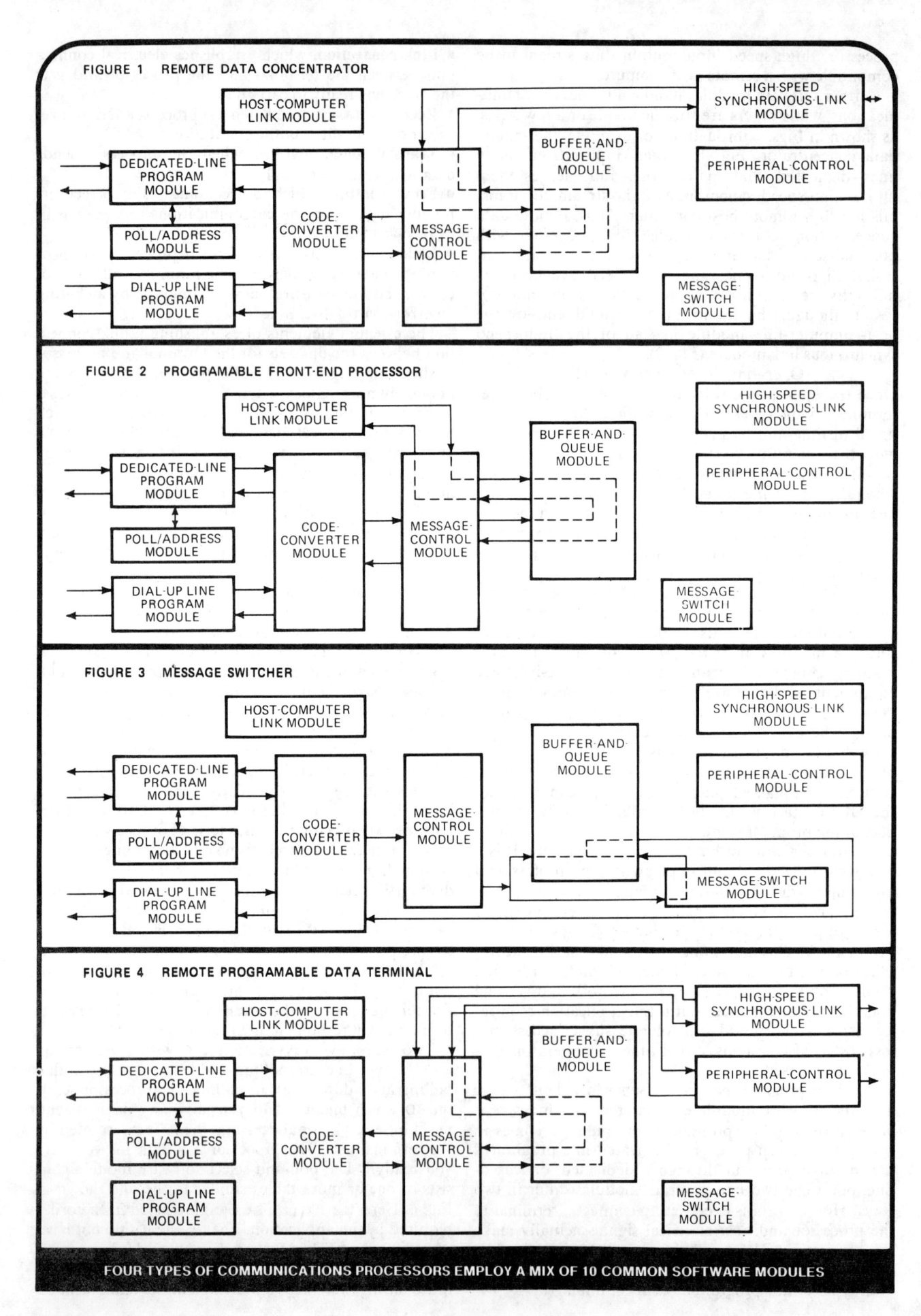

FIGURE 2 PROGRAMABLE FRONT-END PROCESSOR

FIGURE 3 MESSAGE SWITCHER

FIGURE 4 REMOTE PROGRAMABLE DATA TERMINAL

FOUR TYPES OF COMMUNICATIONS PROCESSORS EMPLOY A MIX OF 10 COMMON SOFTWARE MODULES

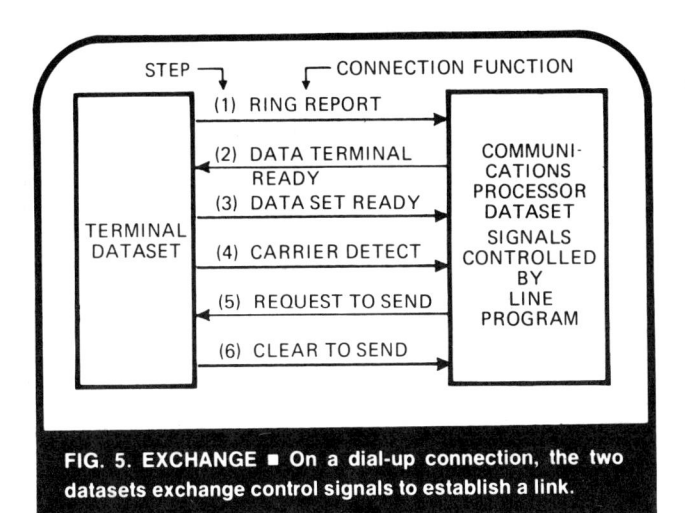

FIG. 5. EXCHANGE ■ On a dial-up connection, the two datasets exchange control signals to establish a link.

may include peripheral-device identifiers, such as A for a keyboard/CRT-display terminal and D for disc. An indicator (the check mark) tells if a particular device or terminal is inoperative; if so, it is skipped.

Poll-and-select tables can take many variations, such as alternating between polling and selecting, as it runs through its ID chain. Depending on the application, once a terminal is polled or selected by the processor, the line is held until the total message is completed. Then polling or selecting continues on to the next terminal on the list. Another alternative is for the list to scan terminals, pull out part of a message, (such as a data block), from the terminal's buffer, go on to other terminals, then return to the first terminal for another block, and so on. Even when the poll-and-select list scans for short blocks, the communications processor is so much faster than line data rates that the list can be stopped for a period before the cycle starts again. During this built-in delay, the processor performs its other routines.

Dynamically alterable polling and selecting

A data communications system that exhibits a change in traffic pattern during the day should have several poll-and-select modules. For example, if a system with the processor in the East has terminals on both East and West Coasts, at 8 a.m. Eastern time, the heavy volume of traffic comes from the Eastern terminals. Therefore, one poll-and-select table calls up the Eastern terminals more frequently than the Western ones. At 11 a.m., Eastern time, traffic starts to come in from the West, so that another table has to give about equal priority to both Eastern and Western terminals. At 5 p.m., Eastern time, though, local traffic slows down while heavy Western traffic continues for another three hours. Therefore, a third table is needed to weight polling and selecting in favor of Western terminals.

The number of tables and the organization of the polling list within each table depends on the actual traffic pattern for a given application. If three poll-and-address tables are needed, then three software modules must be programed in advance. The appropriate one is called in at the right time, automatically by program or manually by the system operator through a hardware switch or a simple instruction at the console. This approach is known as using dynamically alterable poll-and-select software.

Because of its high speed, the communications processor can sequentially scan the connected lines from all on-line low-speed terminals. Using its receive-data routine, the processor accepts incoming data from these terminals. One method of scanning the connected lines is through a tumble table, a temporary buffer (Fig. 7). Under control of the software program, it picks up a single character from each line, stores it in an empty cell, and transfers the characters to the buffer-and-queue module, which in turn scans the tumble table and organizes the individual characters back into messages from each line.

Each incoming line is scanned until an END OF TRANSMISSION (EOT) character is recognized, or a predetermined number of bits has been counted, and then the link-connection modules disconnects the particular line. The receive-data subroutine also appends additional characters to each message character to identify the line, terminal, and code.

The transmit-data module is called in when the processor wants to send data to the remote terminals. Its operation is essentially the reverse of the receive-data module. That is, the transmit-data modules takes messages from buffer-and-queue storage, sends them back through the tumble table, and out on connected lines to specified terminals.

Assuring link discipline and integrity

During a send/receive transmission, the data stream is managed by the link-control module. Synonymous terms for this module are line discipline and line protocol. The data stream not only contains data characters but also special characters, called control characters, that tell the processor and terminal such things as when a message or data block starts or ends and whether a transmitted message has arrived correctly or contains an error. The link-control module is interrelated with the link-connection module; for example, in half-duplex links, the link-control module tells the connection when to reverse the direction of transmission.

In most systems based on communications processors, the link-control module can be quite complex. It must, as an example, distinguish between data and control characters. If a control character is detected, link control must tell the software and hardware what action to take next: for example, when a transmitted block contains an error and requires a retransmission. The actual programing of a link-control module depends not only on the adopted protocol for handling data streams, but also to some extent on the hardware approaches used within the terminals and the communications-processor interfaces. Therefore, the detailed programing of the link-control module can be one of the most time-consuming and exacting tasks in setting up the software modules for a given application and network configuration.

An alleviating factor of link-control software is that the more popular protocols, generally written by host computer manufacturers, have been in existence for some time, are highly refined, and have become more or less recognized as industry "standards."

The operation of a link-control procedure is shown in a most simplified manner in Fig. 8 for two terminals connected on a multidrop line to the communications processor. Here, the processor polls the first terminal by ENQ ID1. Terminal ID1, having no data to send, responds with its ID and the control character EOT, for END OF TRANSMISSION. Sensing the EOT under this circumstance, the processor returns to its poll list and sends out ENQ ID2. The second terminal, having data to send, responds with ACK ID2, which tells the processor to start sending its message. Transmission continues until an EOT indicates that the transmission has ended, or, as shown, the terminal—using its error-control facility—senses an error in a block of data and sends a NAK ID2, for negative acknowledgment. The processor retransmits the erroneous block, and the message continues, until the terminal either sends an ACK or a NAK, and finally an EOT.

Calling in the error-control module

Because errors will occur during transmission, an error-detection-and-control procedure is embedded within the link-control module and implemented by a processor (or terminal) when it is in the receive mode. In essence, each character or block of data is checked at the receiver for errors. Two common techniques use parity-bit checking. These are vertical redundancy check (VRC), which checks for error in each character in a manner similar to that used in checking data coming from a magnetic tape, and longitudinal redundancy check LRC, which checks the parity of all characters in a fixed-length block of characters. Often, VRC and LRC are used together to detect and correct errors.

A more powerful method is called cyclic redundancy checking (CRC), which can be used effectively for long data blocks. The message block is divided by a prescribed divisor. The remainder existing after the division, called the checking block, is transmitted along with the message block. The receiver performs a similar division and compares the received remainder with the expected remainder. If they are the same, the receiving terminal sends an ACK; if not, the receiving terminal generates a NAK and sends it back to the transmitter to initiate retransmission of the erroneous block.

Error checking often is implemented by software. But because of recent decrease in electronic-hardware cost, it has become practical for error control to be implemented in special hardware registers. In CRC, the register counts message characters, performs the division, compares the received remainder with the expected remainder, and provides one signal if the block is good and another if bad. In this way the error-control programing is eased, since all it must do is sense the status of the register. And error-control execution time within the processor can be substantially reduced.

Converting to correct codes

In any fairly large data communications system, several codes may be involved from the terminal through the communications processor to the host computer and back. In fact, some terminals may use one code and other terminals, another. Thus, when a message clears the line program, it carries with it information that tells the communications processor its own code structure. For example, one terminal may use a five-level Baudot code and another terminal the seven-level ASCII (American Standard Code for Information Interchange) code. The minicomputer communications processor may itself employ the ASCII code structure, and when it communicates with a host processor may have to convert to eight-level EBCDIC (Extended Binary Coded Decimal Interchange Code).

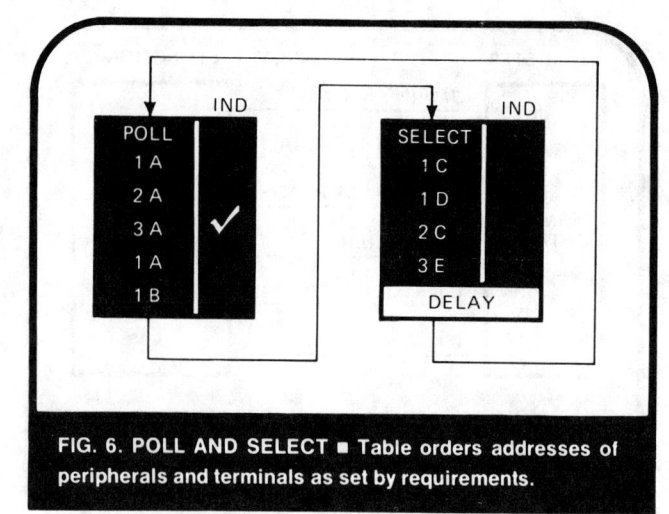

FIG. 6. POLL AND SELECT ■ Table orders addresses of peripherals and terminals as set by requirements.

Thus, whenever data in one code structure passes to another part of the system that employs a different code structure, the message must undergo a code conversion. Such code conversion is usually based on a direct lookup table. One table could, for example, convert characters sent in Baudot to its equivalent bit pattern in ASCII. And another table would perform the reverse conversion. In general, direct look up code-conversion tables are relatively short: an ASCII-to-EBCDIC table takes 128 bytes, but more important, direct lookup is fast and simple to program.

Certain *a priori* knowledge is required—for example, whether or not the seven-level ASCII code is actually being transmitted within a 10-bit block, with the extra bits being used for parity and for start-stop character-framing bits. These extra bits must be accounted for—that is, ignored—during the conversion.

Suppose the communications processor knows it is to convert seven-level ASCII to eight-level EBCDIC. The terminal sends a lower-case m in ASCII. The conversion module recognizes the pattern 1101101 (ASCII for m), looks up the table, and changes it to 10010100 (EBCDIC for m). Often an extra bit position is appended to the left of the left-most bit in ASCII. This appended bit will always he a 0, unless this bit position is used for a parity-check bit. Such an "eight-level" ASCII code can also be stored in a standard 8-bit-byte register common to the EDCDIC structure and memory-cell size.

In a similar manner, each incoming data or control character is converted through the direct lookup table. However, a separate table is required for each type of code conversion. Because of the repetitive nature of table lookup, fast conversion is a necessity. Thus, the tables are stored in main memory. However, if many tables are required, a considerable amount of main memory would be permanently occupied, yet only one table is converting data at any instant. One way to reduce this code-conversion load on the main memory is to store the tables on disk and have the operating system call the needed table out of disk and overlay it on the space allocated in main memory for code conversion. In this way, all tables will share the same smaller block in main memory.

After the message has been converted to an appro-

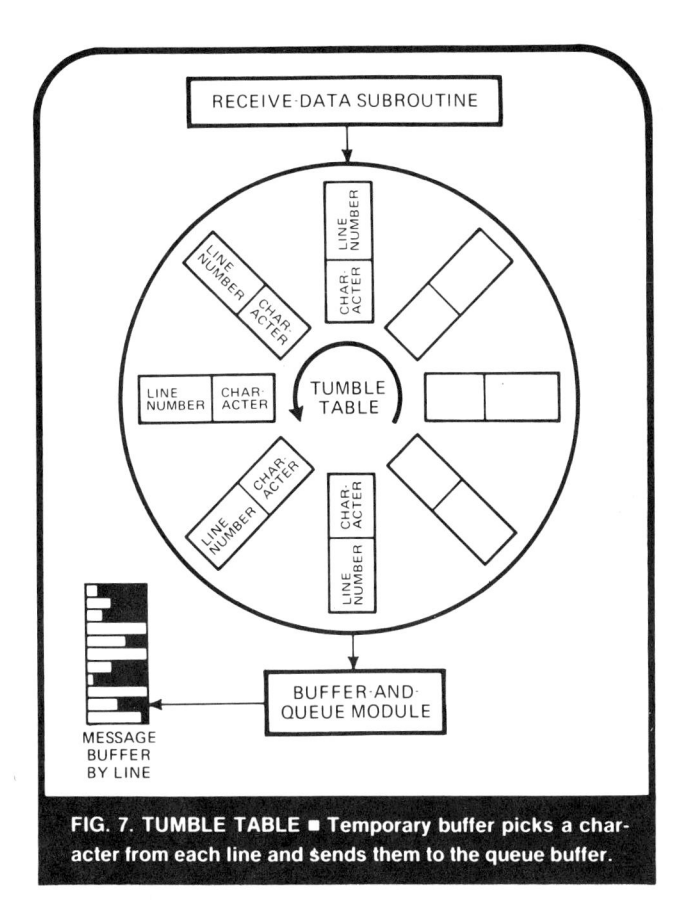

FIG. 7. TUMBLE TABLE ■ Temporary buffer picks a character from each line and sends them to the queue buffer.

priate code, the message-control software module analyzes the initial characters in the message, called the header, and determines message routing. This software module first determines whether a message is a system command, a message for another terminal, or a message to be transferred to the console or another peripheral.

A system-command message informs the processor's OS about a change in the status of equipment anywhere in the network. The command can indicate, for instance, that a peripheral device or a data terminal is inoperative for some reason and therefore should not be polled until restored to service. Whether the out-of-service time is short or long, the system commands continually update the tables that need to know status—usually the poll-and-select and line-program modules.

If the message from one terminal is simply to be switched by the processor to another terminal, the message is transferred by the message-switch module into the queue buffer and stored there until it can be switched out, again under control of the message-switch module, at the earliest moment. This is the way the communications processor operates as a message switcher (Fig. 3). If a disk is used, the message could be routed from the queue buffer through the peripheral-control module to the disk.

A message for the system console or other peripheral devices is shunted by the buffer-and-queue module to the main memory (or disk) and called out through a peripheral-control module appropriate to the device when the line to the device is available.

Finally, the message-control module contains the programs that can maintain system security. The module can, for example, deny requests from specific terminals for access to particular data-base files. Similarly, the message-control programs can permit restricted information to be transmitted over certain lines but not on others.

Storing and organizing the messages

Once data clears the message-control module, it goes to the buffer-and-queue module or, in the case of a message switcher, to the message-switch module and then to the queue module. Queue-module programs route the characters to storage locations (cells) in main memory to assemble messages (or blocks), remember where messages are located, and place the completed messages in queue so they can be recalled when needed. The queue module also manages the main memory, keeping track of empty cells available for storage and how much total memory is in use at any moment.

The three common queuing techniques used with main memory are double buffering, random storage, and sequential-and-random storage. In double buffering, a section of main memory is dedicated to each line or terminal (Fig. 9). For example, a section could consist of two 100-byte buffers serving one line. When one buffer fills up, the message-control module removes the character bytes, routes them to their destination, and inserts the other, empty, buffer to capture additional data coming in on the line. Double buffering minimizes throughput delay, is simple to program, and requires little program-execution time. However, it uses main memory inefficiently, since if a line is idle so is the memory dedicated to it. Even so, double buffering may be appropriate for systems with relatively few terminals.

Random storage employs fixed-size cells of memory. A message is divided into cells of characters, and each cell is stored at random in the next available cell (Fig. 10). Further, cells from other messages are stored in available cells anywhere in the memory. The address of the first cell of each message is put in a table stored in the message-control module. The operating system appends the storage address of the next cell at the end of each preceding cell of each message. For message retrieval, these addresses chain the cells back into a complete message. For example, Fig. 10 shows that when message 3 is to be called out, the table says start at D1, and D1 says go to D2, then to B3, and so on.

The random-storage technique uses memory more efficiently, despite the burden of block addresses, since it serves as a common buffer pool for all lines and is relatively independent of message length. Programing, however, is more difficult than double buffering, and program-execution time may be longer than it is for double buffering.

Sequential-and-random queuing uses memory most efficiently when the lengths of messages are roughly the same and the length-distribution pattern is known. In this queuing system, complete messages are deposited serially in main memory, eliminating block-addressing overhead. The queuing module creates a table that lists the first-character address and length of each message.

As messages are read out of memory and create empty cells, the table also keeps track of location and length of cells in which to put new incoming messages—that is, the queuing module dynamically updates the table. Thus, in Fig. 11, if the 80-character message, starting at the 481 address, is removed from main memory and sent to a terminal, the table is simultaneously notified that the space is vacant. Therefore, the queuing module knows it can assign one or more messages totaling up to 240 characters (80 + 160), starting at byte-address 481 and extending to 720. But programing is more complex.

If disk storage is used in conjunction with the real-time operating system, the processor employs a queuing module called journal buffering. First, a message or block is accumulated in main memory, then transferred to an empty cell on the disk. A table, or journal, in main memory keeps track of line identification, message destination, and first-character address on the disk. When the main memory journal buffer is filled, it too can be transferred to disk.

Switching the messages

The message-switch module directs traffic between the processor's main memory (or disk) and the output lines. The module maintains a message-switch table that includes line numbers and addresses of the first blocks of the highest-priority messages scheduled to go out on each line. If a line is available, the message-switch module puts the address into the status table maintained by the line programs, and the message is sent. If the line is busy, the module sets a flag in the line-status table.

When transferred to the line-program status table, the address is wiped out of the message-switch table. This action tells the message-switch module that the line can accept another message. And the module searches through queuing journals for the next-highest-priority message for that line.

In some systems, the message-switch module also contains tables and develops appropriate branching journals that designate alternate destinations for messages, should a primary line or terminal go down, and that module lists recipients for multiple transmissions of the same message.

Getting to the peripherals

When the data communications system's processor includes a peripheral, such as a disk memory or a line printer, the device is often supplied by the minicomputer maker. The peripheral-control software module also is supplied by the manufacturer—generally as a standard software package embedded in the communications processor's operating system. As a result, minimum programing effort is needed to incorporate a data processing peripheral into a data communications system. And even when the peripheral is made by another manufacturer and is in wide use, the minicomputer maker may still provide a standard software package for the "foreign" peripheral.

The main functions of the peripheral-control module are to format data going to and from the processor to the peripheral, add check characters to ensure data integrity, and issue commands to control the peripheral's operation.

Often, data transfers between a communications processor and a peripheral device are buffered into the portion of main memory dedicated to the operating system. The advantage of buffering is that the fast processor does not have to wait on the slow peripheral. For example, a data block being transferred from a disk is

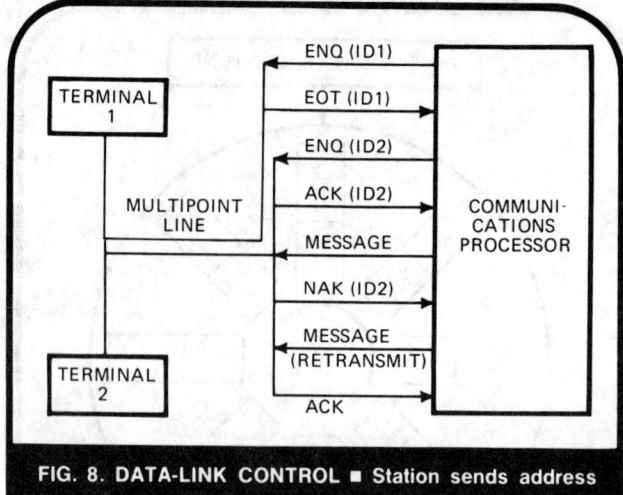

FIG. 8. DATA-LINK CONTROL ■ Station sends address and gets response that tells it to hold or advance.

read into the buffer space before it is actually required by the other software modules in the processor. This "call-ahead" method allows the processor to start transfer of a block of data, go on to other tasks, and then come back to the buffer to pick up the data after the transfer has been finished.

Talking with the host computer

The two software modules invoked when the communications processor interacts with the host computer are the host-computer link and the high-speed synchronous link. The host-computer link is used when the minicomputer, in the configuration of a programable front-end processor, is physically close to the host computer (Fig. 2). The high-speed synchronous link applies when the minicomputer, used as a remote data concentrator, is distant from the host computer (Fig. 1).

The host-computer-link module can be a relatively simple, standard software package or a complex, customized package. Which way to go depends on many system considerations over and above the processor's software ramifications. In its simpler form, the host-

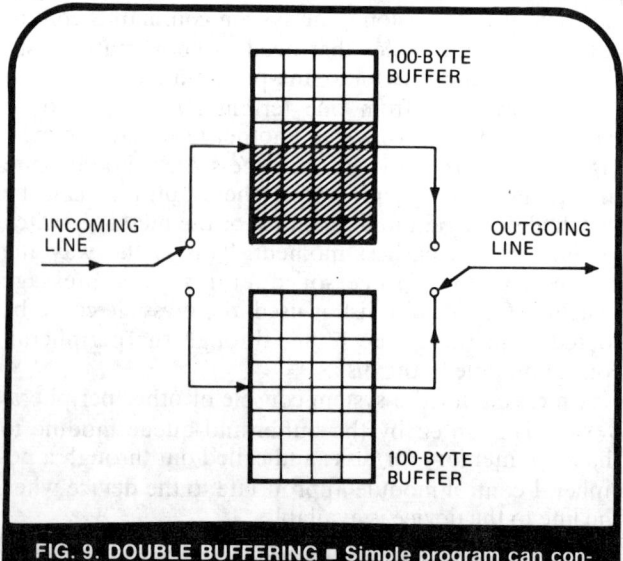

FIG. 9. DOUBLE BUFFERING ■ Simple program can control a double buffer: one section fills while other empties.

FIG. 10. RANDOM STORAGE ■ Messages are divided into cells of characters, and each cell is stored at random in the memory, then the address of the first cell in each message is listed in a table to permit retrieval of message cells.

computer link is an emulator, making the processor look like a standard peripheral device or a communications-control unit. This software link works in conjunction with the physical link that goes to one of the host computer's multiplexer or communications-adapter ports. For example, using a standard software package, generally supplied by the minicomputer maker, the processor can emulate a magnetic tape or one of the IBM 2700 series of transmission-control units. As an emulator, the processor would not include certain software modules, such as buffering-and-queuing, since these functions would be done in the host computer. And few if any modifications need to be made to the host computer's programs to support the data communications functions. However, the host computer remains burdened with execution-time and memory overhead for providing this support.

In its simple form, the standard package converts the code used by the communications processor to the code used by the host computer. Such code conversion is accomplished in the host-link module in the manner described previously in the section on the code-conversion module. The host-link module also controls the device being emulated by the programable front-end processor. The user only need enter the control commands associated with the emulated device into a command table. A status table also keeps track of the messages passing between the processor and the host computer.

More complex forms of host-computer links in the minicomputer take over many data communications tasks—such as network control—normally implemented in the host computer. Therefore more "work" is done in the communications processor, and the modules employed are as shown in Fig. 2.

When the communications processor serves as a remote data concentrator, Fig. 1, or a remote programable data terminal, Fig. 4, it connects to the host processor's high-speed data channel port over a high-speed synchronous line, and therefore it requires a high-speed synchronous-link software module. This module does much of the same work as the host-computer link.

The high-speed synchronous-link module emulates any high-speed device ordinarily connected to a host computer, such as an IBM 2780 data transmission terminal. This software package is table-driven, with the user loading the instruction code for the emulated device into appropriate blanks in the table. Filling in the table blanks permits the high-speed synchronous link to convert between ASCII and EBCDIC, if necessary, to perform CRC error detection and correction, and to establish and confirm the availability of the communications link. To do these tasks several of the modules needed to handle the low-speed input lines are included within this high-speed output module.

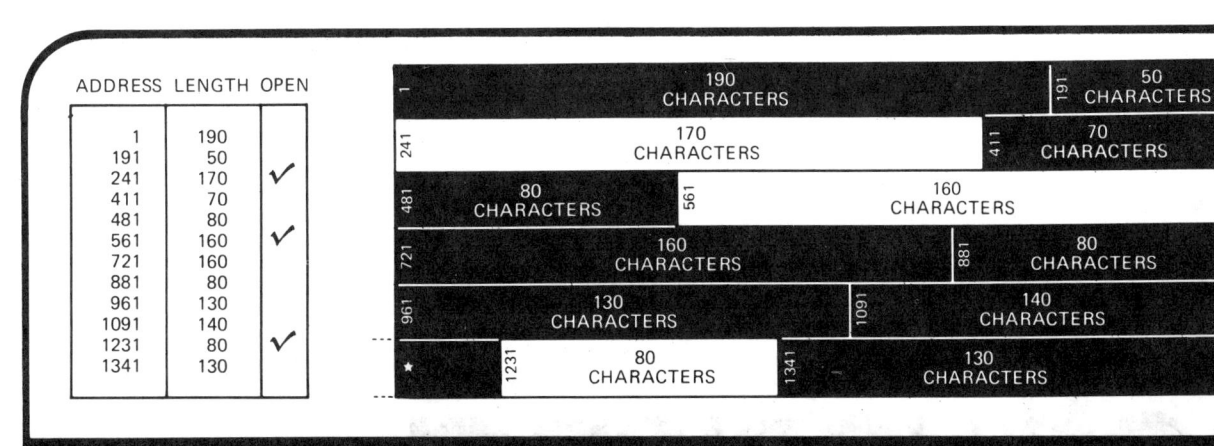

[EOT]

FIG. 11. SEQUENTIAL AND RANDOM ■ Here, complete messages are deposited serially in main memory, and the queuing module creates a table that lists the first-character address and the number of characters in each message.

part 5
data
data
data
data
data
data
data
data

networking

Computer networking-- the giant step in data communications

Connecting data-base computers together via high-speed links has been proven to be technically feasible, but the future of multicomputer networks for business and industry depends on an understanding of what these networks can do, how much they will cost, and who will own them

Dixon R. Doll
DMW Telecommunications Corp.
Ann Arbor, Mich.

OVERVIEW Many large companies can increase their operational efficiency and perhaps even improve their organizational structure by redistributing their large centralized data bases into smaller functional data bases to serve specific regions. The relatively new technique of computer networking will also have a profound influence on the traditional methods of handling data processing applications. Moreover, large computer networks are likely to effect major cost reductions and become more productive, largely because of improved efficiency in use of data communications facilities. For a summary of practical applications, see "Using multicomputer networks," next page.

Although commercial computer networking is still in its infancy, it has attracted the attention of many forward-looking company managers and systems planners. Coincident with this deep interest are the divergent opinions about computer network configuration, operation, technical alternatives, and economic benefits:
- What constitutes a multicomputer network?
- What is the present and future state of network tariffs and technology?
- What are the key application possibilities?
- How is the feasibility of a computer network evaluated?
- Who will own and operate the network?

A computer network can be defined as an interconnected group of host computers that automatically communicate with one another and that can share such resources as programs, data bases, memory space, and long-haul links with each other. In a true network situation, therefore, each constituent computer system can either operate in a local mode under its own operating system or participate in network activity under the direction of a higher-level network-supervisor program, or it can do both.

This definition does not necessarily imply a geographic separation of the data processing resources and terminals. An entire computer network may be confined within a room, floor, building, or office complex. But, since most computer networks will cover large geographical areas, they will require high-speed voice-grade and wideband common-carrier circuits and equipment to employ and share these circuits. The added communications costs and intricate technical considerations must be weighed against reduced data processor cost and improved organizational performance. In fact, some present distributed teleprocessing

> ## Using multicomputer networks
>
> When the computers in a network belong to different organizations or to different divisions of the same company, then linking the computers via the network offers many exciting prospects:
> - Reservation computers at different airlines may exchange information to eliminate lost revenue or overbooking caused by multiple flight reservations by same customer.
> - In-house corporate computers can be linked to computer-service firms' data processors for peak-period load-leveling and for direct access to proprietary data files and application programs.
> - Computers belonging to retail stores, banks, and credit-authorization companies can be networked to provide each user with better financial control and with access to expanded and more timely information about their customers.
> - An interbank computer network can quickly and accurately clear checks and implement funds transfer.
> - Shipping and transportation companies' computers can expedite interline settlements and waybill exchanges and keep track of the location of trailers, freight cars, and other vehicles.
> - Regional and statewide university computer networks can eliminate the duplication of data processing resources and thus reduce the competition for diminishing state funds by campuses and agencies.
> - When a company operates over a wide area, it may choose to distribute its data base and link the smaller data bases using the network. For example, a manufacturing firm with a large customer file can place records for its West Coast customers in a western regional computer center and records for eastern customers in a compatible system on the East Coast. Such a network linkage probably will reduce communications costs below those of a completely centralized data base, since access to regionally located records will not require transcontinental message transfers.
> - The multicomputer network can serve as a marketplace in which buyers and sellers of computer resources may carry on their business.

systems have virtually all the earmarks of being multicomputer networks.

The table summarizes the major characteristics of the better-known systems built since the genesis of computer networks in 1968. Note the dominance in the table of university and research-oriented networks, primarily because they were built under sponsorship of "public" organizations. The proprietary nature of existing corporate computer networks often precludes disclosure of their details. This lack of disclosure has been a major deterrent to acceptance of the networking concept. Even so, it is unmistakably clear that major commercial organizations are now thoroughly evaluating the possibilities of benefits through computer networks for geographically dispersed operations.

ARPA, the pioneer network

Much basic work already done will prove meaningful to the configuration of future systems. The Advanced Research Projects Agency (ARPA) network, generally recognized as the pioneering effort in computer networking, interconnects about 40 geographically distributed computer complexes. And each complex has generally different operational characteristics. One initial objective of the ARPA network was to serve as a pilot plant for exploring the technical problems of a network whose goal was to permit people and programs at one Department of Defense research center to interactively access data and programs existing in any other remote computer center. Started in 1968, the ARPA network has consumed tens of millions of dollars and hundreds of man-years of development.

In the ARPA and other computer networks listed in the table, each local computer complex, or node, interconnects to a communications subnetwork that provides rapid, reliable, and cost-effective transfer of programs and data files between the network's nodes, or sites. A computer network is homogeneous if all host computers are functionally compatible, as in the TUCC (Triangle University Computing Center.) and IBM TSS (Time Sharing System) networks using IBM computers. Other examples of heterogeneous networks are the Cybernet, Merit, and Octopus systems. Achieving full-fledged networking capabilities in the heterogeneous environment is a major technical problem that has been only partially solved.

Basic network configurations

The common network configurations are centralized (star), distributed, and loop (ring) structures. Most nets constructed so far employ a distributed form of organization (Fig. 1). One notable exception is the DCS (Distributed Computing System) network, which employs a loop structure (Fig. 2). And the Network/440 and the TUCC network are examples of the centralized (star) organization (Fig. 3).

In the centralized network, such communications functions as message routing, speed conversion, code/protocol conversion, and error checking are performed at one common location. Similarly, network-control functions, relating to network job scheduling and the allocation of host resources to users, are monitored from a common location. The network-control process may be implemented in a computer dedicated to the function, or in a host computer that alternates between the network-control mode and a background-processing mode.

Because network control takes place at one node, the advantages of the centralized network are that the network-control problem is greatly simplified, and that the control hardware and software can be fully shared by all hosts on the network. The main disadvantage of the centralized structure is that each communications line applies to one host computer. Since a line cannot be shared, then alternate routing of messages can be achieved only by adding redundant links between the central (star) node and the remote hosts. The centralized structure is reasonably attractive when each remote host computer needs only to address another host that's at a higher level in the network hierarchy. In contrast,

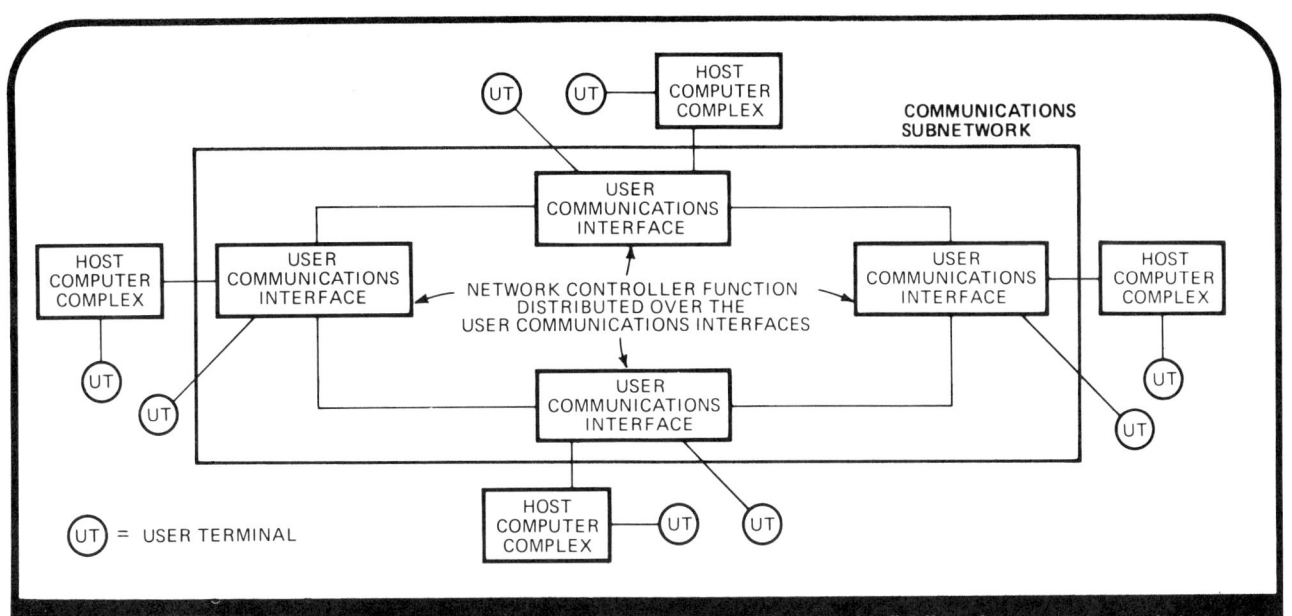

FIG. 1. DISTRIBUTED ■ Failure at one node doesn't affect the operation of the rest of a distributed-computer network, but such networks are difficult to control and require complex and costly communications-network interfaces at each node.

the distributed ARPA net permits intercommunication between virtually any node pair in the network.

If properly designed, distributed networks can offer significant reliability advantages, since a failure at one node does not affect the rest of the network. The main problems with distributed structures, though, are that they are difficult to control and that they require complex communications-network interfaces at each node. Unless the users are clustered geographically, it is quite difficult to do much, if any, sharing of interfaces. Furthermore, if a distributed network is heterogenous, then interfaces must be designed to operate with a multitude of line protocols and computers, which makes standardization difficult.

The loop configuration works very well in a network in which the remote resources are relatively close to each other, for example, where the nodes are clustered within the environs of one large city. Communications interfaces become cheaper. But the loop circuits carry the maximum traffic derived from all nodes. Therefore,

Some computer networks in operation now

ARPA. Advanced Research Projects Agency network interconnects a wide range of computer models at 35 to 40 Department of Defense and university research organizations throughout the U.S.

MERIT. Serving the three largest campuses in Michigan, the Merit network employs a mix of host computers: University of Michigan (IBM 360/67), Michigan State University (CDC-6600), and Wayne State University. (IBM 360/67).

TUCC. The Triangle University Computing Center network interconnects a mixture of IBM host computers at North Carolina University and North Carolina State University.

DCS. The Distributed Computing System, an experimental network, employs a digital megabit loop structure to interconnect small- and medium-size processors located within a relatively small geographical area: the University of California (Irvine) campus.

IBM NETWORK/440. Using a star topology, this research network at IBM's Yorktown Heights, N.Y., facility interconnects heterogeneous processors at various universities and IBM locations throughout the United States.

CYBERNET. Control Data Corp.'s homogeneous commercial computer network interconnects about 40 computers or nodes, around the U.S., using both voice-grade lines and lines operating at 40.8 kilobits per second.

OCTOPUS. An elaborate network, Octopus interconnects five Control Data Corp. supercomputers at Lawrence Berkeley Laboratories in California and uses a 12-megabit-per-second digital loop, or ring, configuration.

IBM TSS. Composed only of commercial hardware, this experimental network of IBM 360/67 computers is interconnected over switched (dial-up) voice-grade lines, and all network communications functions are accomplished within the host computers.

INFONET. Computer Science Corp.'s commercial network handles mainly remote batch and remote job-entry applications for a variety of subscribers.

MARK III. General Electric Information Services' commercial network interconnects both GE and customer computers located around the world.

TYMNET. Tymshare Inc.'s network, another commercial system, interconnects several Xerox Data Systems host processors and uses packet-switched, store-and-forward technology.

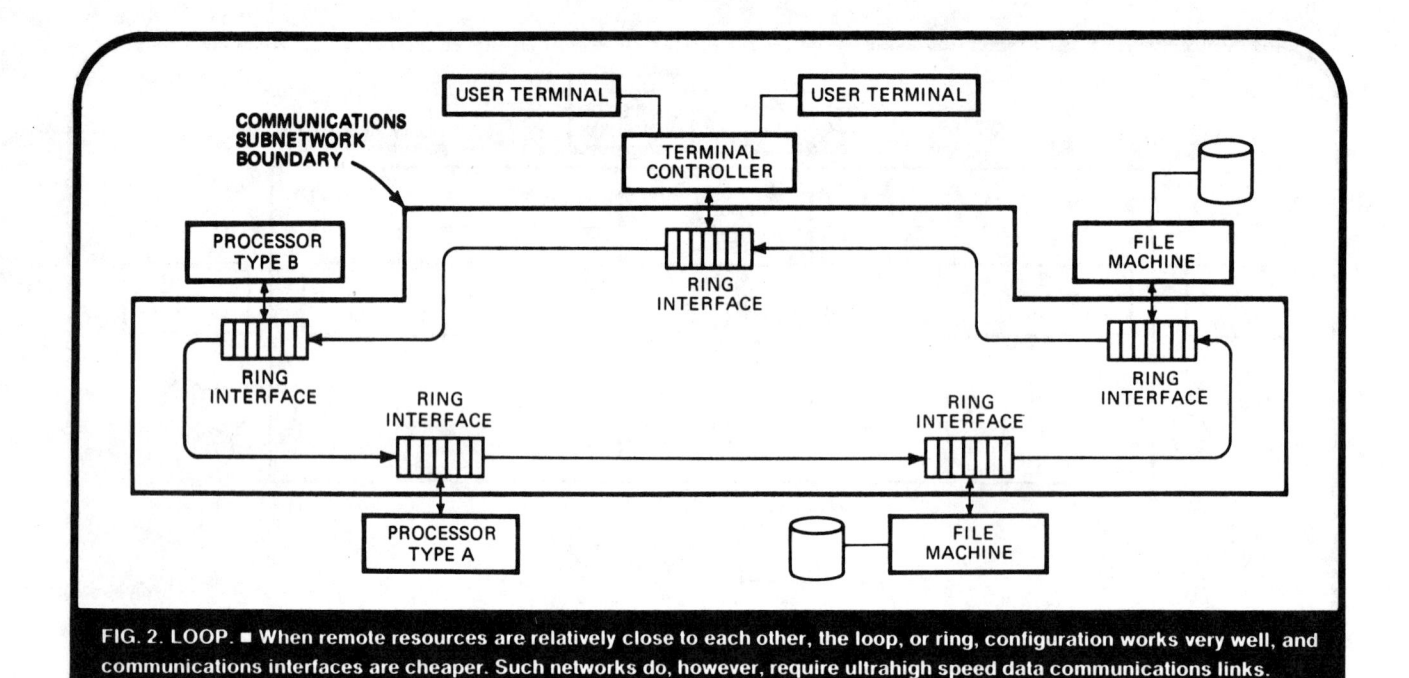

FIG. 2. LOOP. ■ When remote resources are relatively close to each other, the loop, or ring, configuration works very well, and communications interfaces are cheaper. Such networks do, however, require ultrahigh speed data communications links.

a practical problem is the high cost and lack of ready availability of ultrahigh-speed lines (operating at megabits per second) for connecting resources that are geographically dispersed.

Obviously, then, many factors must be evaluated in choosing the most suitable configuration. However, one major factor can exert a pronounced influence on this choice: the type of participation by the nodes. Any node can be a provider of resources exclusively, a user of resources exclusively, or some combination of resource-provider and resource-user. On the one hand, centralized structures are intuitively appealing when a relatively large number of the sites need only reach one or a few sites which participate merely as resource providers. On the other hand, fully distributed structures are most appropriate when all sites participate both as resource-users and resource-providers.

Five ways to operate

The five major modes of computer network usage each have distinct functional differences, unique cost justification, and can affect the choice of configuration:
- Remote job entry
- Remote batch processing
- Interactive
- Dynamic file access/transfer
- Load sharing

The remote job-entry network involves one computer simply assuming the appearance of a remote job-entry (RJE) station. Only ready-to-operate jobs (program and data) are sent to a remote host, which processes the jobs

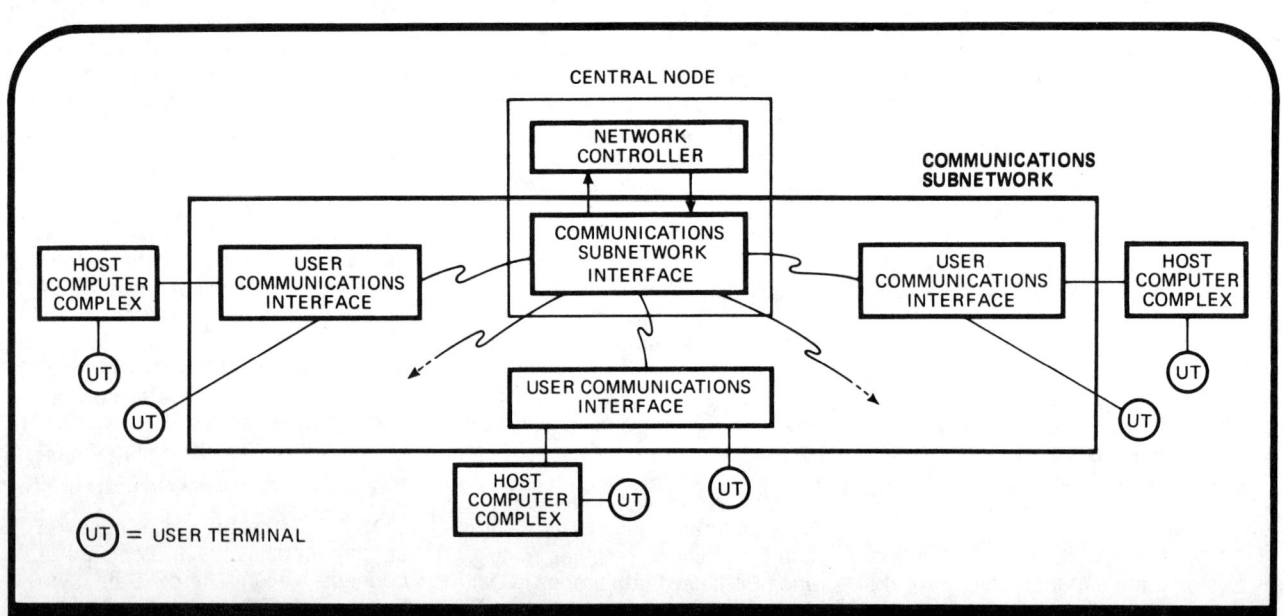

FIG. 3. CENTRALIZED ■ Because network control takes place at one node, centralized, or star, networks can share this function with all host computers on the network. Communications lines, however, cannot be shared with other nodes.

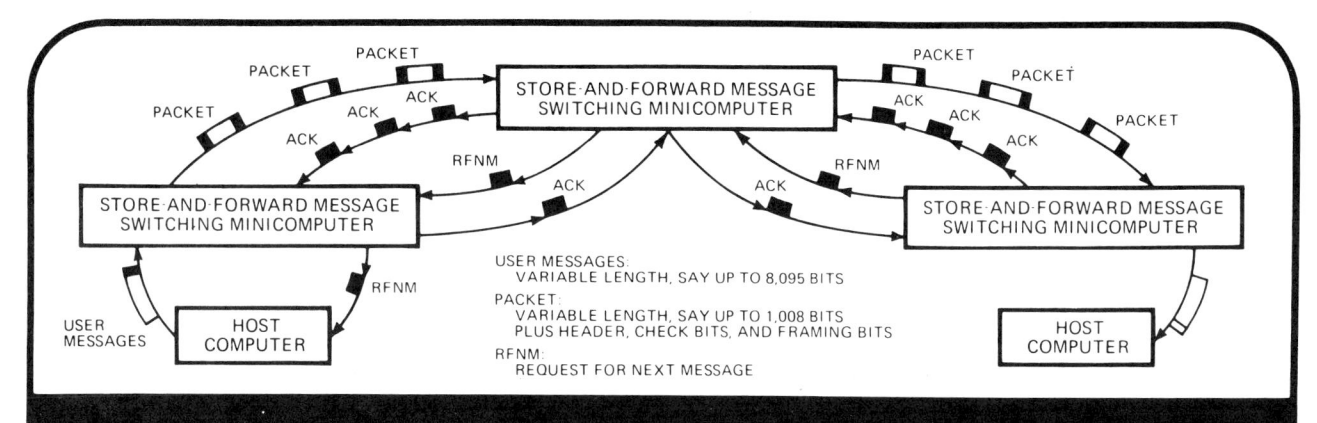

FIG. 4. PACKET SWITCHING ■ In a distributed network, minicomputers can divide long messages into fixed-length segments—called packets—and in such packet-switching networks, each packet of perhaps 1,000 to 8,000 bits is treated individually and forwarded along the best available path. Each packet is checked for errors at each node along the way, and at the destination another minicomputer reassembles related packets into complete messages for the subscriber's use.

and sends the results back to the originating host.

In the remote batch-processing mode, the local host operates in its batch mode but has access, through the network, to data files (in mass memory) at a remote resource node. Here, the network-monitor programs and the communications interfaces maintain necessary file directories and perform all required code conversions between two systems. However, only the batch mode of operation is permitted. The network-accessible remote file is a powerful concept, since it allows the use of any data file in any host computer in the network.

The interactive mode is a straightforward situation in which a user may control one or more batch processes at remote systems from the same interactive terminal.

(In the interactive, remote batch, and remote job-entry modes, it is implicit that the user specify which systems the jobs are to be assigned to, and schedule the subtasks of a complete network job in a meaningful and sensible manner.)

The dynamic file-access/transfer mode permits a user at an interactive terminal or a host computer executing a batch process to automatically request a data file to be transferred to it from another location and receive the data file in real time and on demand. With this capability, virtually any type of interaction between host computers is possible.

In the load-balancing mode, the user submits a complete job to the network via his local complex. He does not know or care which site gets the job. In effect, he tells the network-control program to find all the necessary data, construct the proper job-control sequence, and assign the job to be processed in the best way by the network's aggregate resources. The goal of load-balancing offers best—lowest cost—utilization of computer and communications resources, and hence represents the economic objective of multicomputer networking.

Basic network divisions

Besides the host computers, three types of facilities are generally required to accomplish computer networking: the user communications interface, the communications subnetwork, and the provision for the network-control function.

The user communications interface processes information sent to it from the local host and the communications subnetwork. The network controller interprets the user's command language requests for networking activity and feeds the proper sequences of instructions to the user interface to assure satisfactory execution of network jobs.

The communications subnetwork, completely responsible for reliably and rapidly transmitting messages between all participating network sites, is the major concern here, since it represents a major part of the cost of a multicomputer network. The user can build his own communications subnetwork now. Eventually, as will be discussed, he may be able to buy a communications-link service.

Most computer networks now operating implement their communications subnetworks on leased wideband circuits of, say, 50,000 bits per second, obtained from the common carriers. Obviously, the cost is not trivial, and choice of configuration becomes significant. For example, to directly interconnect all sites in an n-node distributed system requires $n(n-1)/2$ links. Thus, to connect each of 10 computers to all other computers would require the leasing of 45 private, long-haul, wideband circuits.

In centralized networks, a link runs from each remote site to the star node. Therefore, for two remote sites to communicate, their traffic must be switched through the star node. Further, in a centralized net, a reasonably clear-cut distinction exists between the user communication interface and the network-control facilities.

However, in distributed computer networks, the hardware and software extensions to each local host jointly perform both the user communications interface functions and the network-control activities.

In distributed computer networks employing leased lines, special communications processors, usually minicomputers, accept long messages from each network user, multiplexes them onto a wideband link, and, using an optimum routing, switches messages from any node to any other node.

The minicomputer divides the message into fixed-bit-length segments, called packets, for transmission through the network. In a packet-switching network,

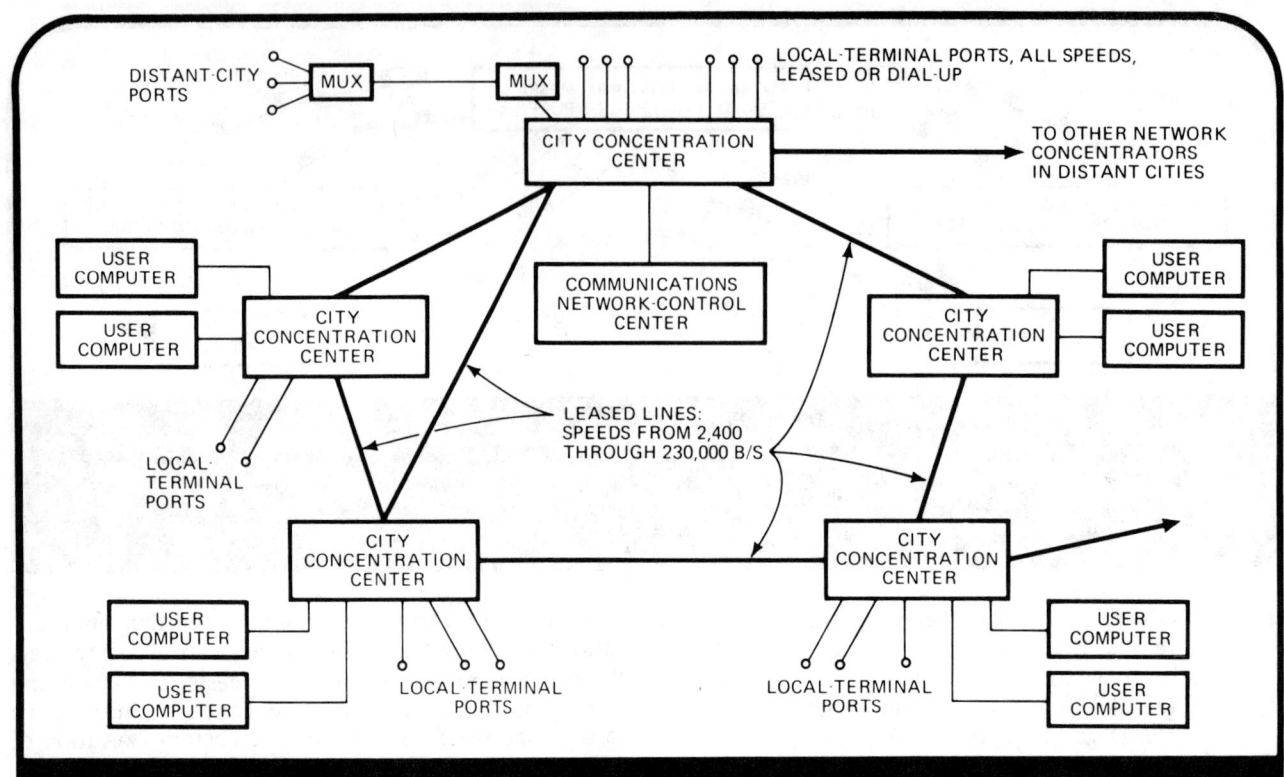

FIG. 5. VALUE-ADDED NETWORK ■ In a VAN, the services added by the vendor for the users would include error control, speed conversion, code conversion, and alternate/adaptive routing of messages between network subscribers.

each packet is treated individually (Fig. 4). Each is forwarded along the best available path at any given instant. And each packet is completely error-checked along the way—for example, each time another wideband link is traversed. At the destination, another minicomputer reassembles packets into complete messages, which are then presented to the addressee. Packet-switched, multiplexed, communications subnetworks more fully utilize the relatively expensive wideband lines, and users enjoy the resulting reduction in cost.

The ARPA network employs the packet-switching concept. User-message packets as long as 8,000 bits are typically sent through the network between origin and destination sites in less than half a second (including the time for acknowledgment—signaling back from the destination to the originating host).

Although traffic levels have increased dramatically in the ARPA network, in aggregate, the network's communications traffic capacity is still well below its maximum utilization. Thus no real experience has been obtained about the actual performance of packet-switching networks under prolonged periods of system-wide heavy traffic. Critics of packet-switching technology note a tendency for wide variation in message delays, particularly during peak traffic periods.

Value-added networks

This year, several commercial organizations have petitioned the Federal Communications Commission to permit them to provide "value-added" communications services, based, in large measure, on multicomputer networks employing packet-switching technology. Such a value-added network (VAN) would, it's hoped, allow an ultimate user to enjoy the cost benefits inherent in a large subnetwork without requiring him to build his own private network (Fig. 5). Further, a VAN would permit interconnection of computers within a company or between companies.

While some people see a VAN as the functional equivalent of the ARPA network, there is, in fact, an essential distinction. ARPA includes the host computers as an integral part of its organization. The proposed value-added networks will not include host-computer services. Each user would connect his own data processors. That is, a VAN would be functionally equivalent only to the communications subnetwork portion of the ARPA net.

The value added by the vendor will be such services as error-control, alternate/adaptive routing, speed-conversion, and code-conversion. Therefore, each user would subscribe for communications services, and also elect to be a computer-resource user, resource provider, or both. For example, the XYZ Co. and the ABC Co. might subscribe to such a VAN service, and they might also interconnect their computers via a subnetwork.

When financing and technology eventually provide full-fledged computer networking on a wide scale, network-user and network-owner management will be confronted with a multitude of administrative situations and attitudes. The user, for example, will be concerned about the security of intraorganizational administrative messages and business data.

To assess the administrative problems associated with computer networks, consider the different interest groups with stakes in the network project. Among them are the network owner or vendor, the network subscribers, and resource buyers and sellers. The network vendor will have primary responsibility for keeping the network fully available and financially self-supporting. Here, the management function includes the development and marketing of new network-related goods and services to attract new customers and retain existing ones.

For resource sellers and resource users, the computer network creates the vehicle for them to conduct business. But whether potential sellers will want to sell or potential users will want to buy depends on satisfactory arrangements for such factors as price, accounting procedures, availability, program copyrights, data security, and product support.

Another uncertain factor is the regulatory status of the value-added network concept. The network owner is clearly in the position of providing improved communications services to the public, using facilities leased from the common carrier.

Traditionally the sharing of lines by many users has been disallowed by the long-haul common-carrier tariffs, with certain exceptions. However, the Bell System has recently filed a proposed tariff change with the FCC. The carrier wants to permit value-added vendors to legitimately lease lines, add value of the types described, and sell the resultant hybrid computer/communications service to the public.

But this issue, not yet resolved, involves two basic points: first, should a common carrier be permitted to lease lines to VAN vendors under existing or possible new tariffs, and second, if they go into business, will the VAN vendors be required to file tariffs of their own?

Anticipating the cost

One intriguing point relates to the prices likely to be charged by value-added network vendors, whether tariffed or not. What is the user going to pay for the communications subnetwork portion of a multicomputer computer network? Some insight into the answer can be gained from ARPA experience and from VAN-vendor statements. The ARPA network charges its users about 30 cents for 1,000 packets transmitted over its high-speed communications network, where a packet contains as many as 1,000 bits of data from a user terminal or computer.

Present ARPA-network line payments to common carriers for nationwide 50-kilobit-per-second circuits serving 40 nodes exceed $1 million a year. That is, the circuit cost alone averages more than $2,000 a month for each node. At 30 cents per kilopacket, each user site (node) must therefore transmit over 6,000 kilopackets (that is, over 6 billion bits) per month just to pay for its average share of the total circuit cost. The 30 cents per kilopacket is essentially a subsidized price, since many aspects other than line cost must be included to determine a true cost.

The commercial packet-switching network proposals recently filed with the FCC include several pricing options. On average, the cost to user will be about $3 per kilopacket. However, it is still too early to predict future VAN rates with any exactness. Whatever the ultimate figure, for planning purposes, it certainly seems safe to conclude that the cost will not be mileage-sensitive, but probably will be response-time-sensitive. Some VAN organizations may offer fast-response (several seconds)

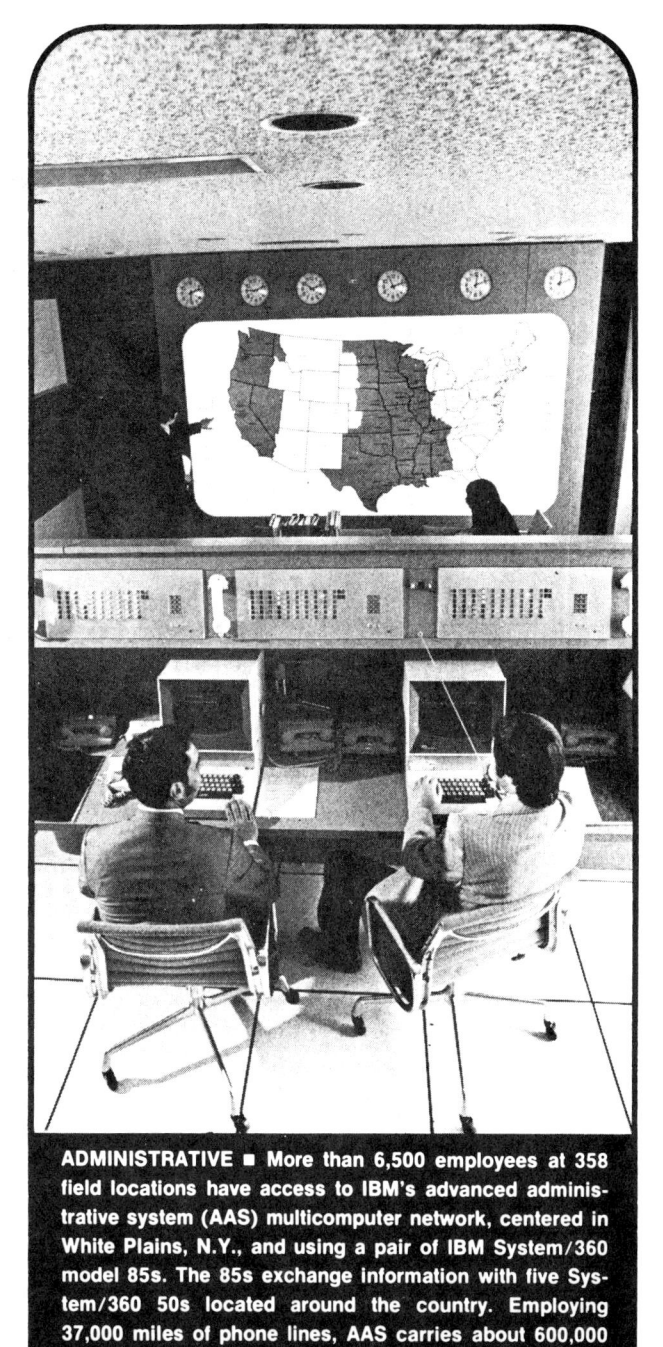

ADMINISTRATIVE ■ More than 6,500 employees at 358 field locations have access to IBM's advanced administrative system (AAS) multicomputer network, centered in White Plains, N.Y., and using a pair of IBM System/360 model 85s. The 85s exchange information with five System/360 50s located around the country. Employing 37,000 miles of phone lines, AAS carries about 600,000 messages each working day from and to 1,150 terminals.

services suited to interactive applications. Others may service batch applications, in which bulk data transfer takes hours. And others may provide both fast- and slow-response services. The price for priority bulk-transfer service is likely to be significantly less than for the fast interactive services, primarily because networks handling bulk transfer are less costly to build and more efficient than those for interactive services.

In summary, multicomputer networks are beginning to appear more feasible in the commercial environment. The forces at work include user interest and need, a change in attitude by established common carriers, the advent of specialized carriers, consideration of the VAN network concept by the FCC, and the entrepreneurship of network managers.

New pricing structures upset data network strategies

Multi-tiered tariffs and competitive carriers give systems planners more choices of routes— and more aggravation

Dixon R. Doll
DMW Telecommunications Corp.
Ann Arbor, Mich.

Economics

Data communications systems planners have always had challenging jobs, but recent and proposed tariff changes will test their ingenuity to the utmost. New pricing structures, caused by competition between Bell System companies and other common carriers, are virtually obsoleting all popular data network design procedures. Furthermore, existing networks based on yesterday's tariffs should be reexamined in the light of the new prices.

Faced with a bewildering array of options and alternatives, planners will still be able to use design rules to come up with networks that feature minimum billing and maximum performance. What the planner must do to meet this challenge, though, is to understand the major changes in the tariffs and pricing structures, consider the way optimum networks have been laid out in the past, and examine the changes that have to be made in present optimizing procedures.

The major pricing structure changes are for WATS (Wide Area Telecommunications Service) and leased voice grade lines. They have been the subject of one or more FCC filings by the Bell System for the purpose of adjusting rates. To complicate the picture, all-digital networks have finally become a reality from Bell and Datran.

In the past, it was necessary to evaluate the services of only a single carrier, usually Bell, in configuring most teleprocessing systems. In a limited number of cases, there was the option of Western Union and the independent telephone companies. However, none were geographically broad enough to be completely viable for most large networks.

But things have changed. In 1974, specialized common carriers began providing service in the major U.S. cities. Even though true nationwide service is not yet available from such carriers as Datran, MCI Telecommunications, Southern Pacific Communications, and WTCI, their competitive effect is clearly evident in the marketplace.

The Bell System responded to the challenge this spring by filing a tariff for a new all-digital network called Dataphone Digital Service (DDS).

Initially, DDS will provide channels at speeds of 2,400, 4,800, 9,600 and 56,000 bits per second.

Among other things, DDS offers substantial price

reductions and service improvements. Table I contains the DDS pricing structure. Economic benefits are most pronounced at the lowest and highest speeds: The prices are about half that for comparable leased voice grade circuits.

Since digital transmission is employed, the analog modems are replaced by digital and channel service units. Another saving, resulting from improved data throughput, is afforded by expected average error rate as low as 1 error bit in each 10^6 to 10^7 bits transmitted as long as both terminals are on the DDS network. Also, the carrier claims that, on average, over 99.5% of the 1-second transmission intervals will be completely error free.

DDS impact on network design

Dataphone Digital Service will most likely only be available in major metropolitan areas having Bell System telephone service. A user of DDS service between two points not in the DDS service area has to use analog extensions at each end of the regular DDS channel (Fig. 1). This means charges for four conventional modems (as opposed to two when using voice lines). Further, there is a by-the-mile charge for the expensive analog channels to get into the DDS network. Naturally, if only one end of the line is off-net only one extension is required.

The user, however, shouldn't expect better service quality than that afforded by conventional analog service alone. This is because each link adds to the total error rate. And, the rate on analog links is generally 100 times that expected on the DDS link (Fig. 1).

Furthermore, users will not be able to obtain alternative voice/data usage of DDS lines. Nor will they have the possibility of employing dial-up backup schemes to keep the network operational without analog modems—and modem costs were to be eliminated in digital networks.

But to gain the maximum technological and economic improvements possible, users must do some comparison shopping. There must be evaluation of DDS, voice grade (HiD/LoD), and specialized carrier and satellite tariffs on a link-by-link basis throughout the entire network.

Bell strikes again

Another development is that the pricing structure for voice grade services from AT&T recently underwent a fundamental change. The net result is that some users will be paying more, others less.

The new High Density/Low Density (HiD/LoD) tariff went into effect in 1974 but is still being evaluated by the FCC. While a single rate system existed previously, now there is a three-level structure. In order to implement the change, the 25,000 rate centers around the country have been divided into two classes—high density and low density. If only high density centers are used, there is one rate. When a low density center becomes involved, the second rate prevails. In addition, there is a separate short haul rate which supercedes either of the two for any point-to-point or multipoint circuit where the total interchange channel mileage is 25 miles or less. Table II contains the essential elements of the tariff.

The impact of this tariff depends on the geographical dispersion of a given network. As many as 20,000 users may be affected. For a network primarily interconnecting major metropolitan rate centers (usually high density rate centers), the total price should decline significantly. In such networks optimized for the new tariff by DMW Telecommunications, reductions totaled 10% to 15%.

TABLE I

BELL'S DATAPHONE DIGITAL SERVICE (DDS) TARIFFS

DATA SPEED IN BITS/SEC	MONTHLY FIXED CHARGE FOR INTERCITY DDS CHANNEL	MONTHLY RATE PER AIRLINE MILE	MONTHLY RATE PER TYPE 1 DIGITAL ACCESS LINE	MONTHLY RATE PER TYPE 2 DIGITAL ACCESS LINE — FIXED	MONTHLY RATE PER TYPE 2 DIGITAL ACCESS LINE — PER AIRLINE MILE	ONE TIME INSTALLATION CHARGE PER DIGITAL ACCESS LINE	MONTHLY RATE PER DIGITAL SIGNAL CONVERTER (DATA SERVICE UNIT)
2,400	$20	$0.40	$65	$90	$0.60	$100	$15
4,800	40	0.60	85	110	0.90	100	15
9,600	60	0.90	110	130	1.30	100	15
56,000	125	4.00	200	250	6.00	150	20

On the other hand, networks employing mostly low density centers, with just a few HiD centers involved, will have a price increase of as much as 40%. The step jump stems from the reality that not much can be done to upgrade existing or future networks revolving around LoD centers.

Indirect routing affects billing

Dedicated lines have traditionally been priced according to the straight-line distance between the user locations. Now, the HiD/LoD tariff has opened the possibility of indirect routing through specified high density homing point rate centers. Figures 2 and 3 show examples of such routing. With the new tariff, networks must be configured, by the user, on billing constraints and opportunities, not direct airline miles.

Consequently, the design and analysis tools used to configure so many of today's straight-line networks require substantial modification to be of continuing value in laying out networks where the communications lines are obtained on a route priced basis. That is, any useful minimum cost design tool must now consider billing networks made up of indirect routes which pass through high or low density locations where there are no user terminals or computers.

In the HiD/LoD route-priced tariff, station terminals are generally identical to the previous concept of service terminals. Using HiD/LoD involves a new type of termination charge—channel terminal charges are levied for each section of the billing network, as shown in Figure 3. There are two channel terminal charges for each link. Since point-to-point lines will involve either one, two, or three links in the billing network, there will be two, four, or six channel terminal charges per two-point user circuit.

Thus, with HiD/LoD, it is particularly important that all elements of a link price be included to compare one carrier or service with another. In particular the new route-priced tariffs will dramatically increase the termination related charges as a percentage of the total link charge. Table III compares the total cost and the termination charges for a 100-mile link using the old FCC 260 tariff and the HiD/LoD tariff.

A common notion about HiD/LoD is that the user is free to route his billing layout through any HiDensity rate centers to achieve lowest cost. Not true! Bell has filed a list of one or more HiD homing point rate centers for each LoD rate center. The user may only employ eligible homing points in evaluating indirect routings. Figure 4 gives several eligible and ineligible alternative routings, together with related costs.

Although in most cases an eligible homing point is the one nearest the user location, there are exceptions. There is no algorithmic way, or orderly set of rules, to determine the valid homing points

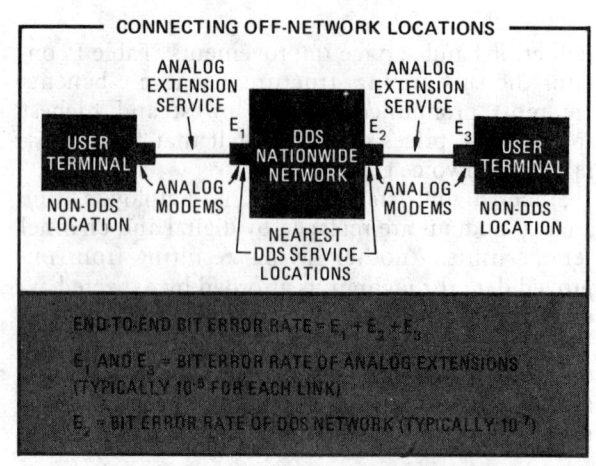

1. Each terminal off the DDS network will require an analog extension service. This means that the error rate on the over-all link will likely be the error rate of the analog extensions.

in all situations. Computer programs, table lookups, and user experience will have to be judiciously combined to produce accurate, efficient network designs.

Satellites to reduce cost

On a related front, the availability of domestic satellite channels is going to dramatically reduce long distance costs and, thus, affect network design. This, in turn, will renew interest in the centralization of computing facilities. Here, the increased propagation delay and distance-insensitive price structures will stimulate the use of remote concentrators for polling and for attaining efficient communication lines by implementing full duplex operation. Half duplex data link controls, like Binary Synchronous Communications (BSC), rely on polling and stop-and-wait error control and, thus, are likely to be very inefficient over satellite channels. New protocols, like Synchronous Data Link Control (SDLC), capable of operating efficiently in a full duplex mode will, however, have important ramifications. (Refs. 1, 2, 3).

In general, a user operating or building a large data network can now obtain a given speed data channel from any of several different vendors. He must evaluate the tremendously more complicated price structures. And he must consider whether it is worth his while to fractionate his network among several vendors, since most of the newer carrier organizations are not yet able to provide service everywhere. The cost savings associated with competitive procurements will be particularly noticeable in the major metropolitan areas where most data communications networks find their terminal traffics anyway.

Notwithstanding the increased administrative complexities of dealing with multiple vendors, the benefits of selectively employing several specialized and customized offerings for a particular networking requirement should be substantial.

In switched services, notably the WATS offering, the Bell System has been equally active in stirring up the pricing pot.

WATS revisions may hurt

The major provisions affecting WATS users are:

- Full time service will be changed to full business day service by adding hourly rates for usage in excess of 240 hours per month. The proposed overtime rate is two-thirds the equivalent hourly rate for the initial period.

- Both the full business day service and the measured time service will be more closely aligned with present long distance, station-to-station rates—the first by upping the initial period rates in the low and medium mileage categories and the second by reducing the initial period rates in the long distance categories. For measured time service, the usage rates for only low mileage categories will rise, while the others will decrease.

- In full business day service, the user will incur a utilization debit against his 240 hour per month quota of 1 minute for each minute of connection time. But if a call is less than 60 seconds the utilization debit is still one minute. This means that a user who never made any calls of more than 1 minute would be limited to a maximum of 14,400 calls per month for the basic full business day rate. (14,400 is obtained by taking 60 calls per hour times 240 hours per month)

- In the old WATS tariff there were six different service areas; in the new WATS tariff existing service areas 5 and 6 have been combined into a single service area.

- A number of one-time charges have been modified for such services as installation, moves, suspensions, maintenance of service, and adjustments of signal power.

The major consequence of these revisions is that it will no longer be possible to employ WATS lines on an unlimited basis for a fixed price. This will adversely affect the economics of WATS usage for those corporations using a large group of them for data transmissions at night and for voice communications during the day.

Under the old WATS tariffs, a corporation could justify WATS lines for voice usage alone during the day and then obtain as much as 10 or 12 hours of "free" data communications usage each night. Under the new tariffs this type of an approach will obviously become much more costly and will depend on the number of hours beyond 240 each month that each line is utilized.

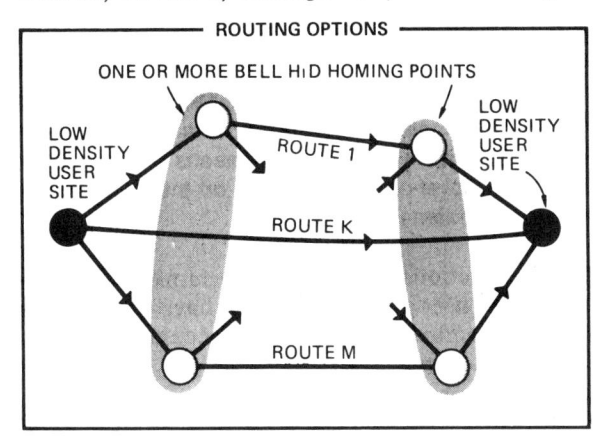

2. According to the High Density/Low Density tariff, alternative routes exist to connect to low density sites—and it's up to users to select homing points and least costly routes.

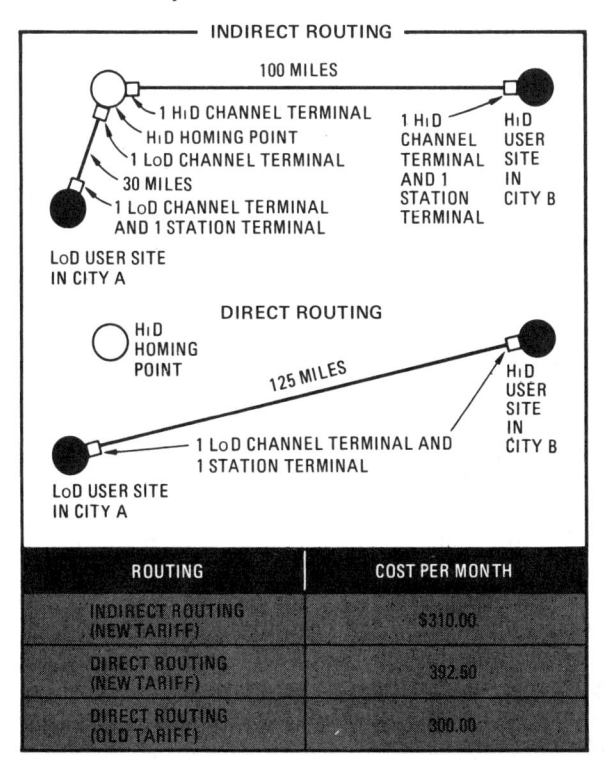

3. Differences in prices for high density and low density routes often permit indirect routes through high density sites to be less costly than low density direct routes, as shown here by example.

ROUTING	COST PER MONTH
INDIRECT ROUTING (NEW TARIFF)	$310.00
DIRECT ROUTING (NEW TARIFF)	392.50
DIRECT ROUTING (OLD TARIFF)	300.00

Overview of design techniques

Procedures used during the past decade to configure data communications systems fall into two categories: The optimization of WATS and other types of dial-up networks, and the configuration of leased line networks.

Optimization procedures for WATS generally involve linear programing packages or similar approaches executed on a computer. These determine the least-cost mixture of full and measured time lines in each service area. The inputs are the busy period calling frequencies (messages per second) for each of the locations in a network.

The design problem of optimizing a full period

TABLE II
HIGHLIGHTS OF BELL'S NEW ROUTE-PRICED HiD/LoD TARIFF

	HIGH DENSITY	LOW DENSITY	SHORT HAUL
CHANNEL TERMINAL (PER TERMINAL)	$35	$15	$3
STATION TERMINAL FIRST STATION (EACH PREMISES)	25	25	
NO FOREIGN EXCHANGE SERVICE			15
FOREIGN EXCHANGE (FX)			10
SECOND AND SUBSEQUENT STATIONS (EACH PREMISES)	3	3	3
INTEREXCHANGE CHANNEL (IXC) (PER MILE)	0.85	2.50	3

WATS network will prove to have been much simpler in the past than it will be in the future under full business day usage. Before, the price of individual lines did not depend on usage: so reasonably exact solutions could easily be determined to minimize costs. Such procedures find a minimum number of lines which satisfy a given blocking probability (the percentage of calls receiving a busy signal). The calculation of blocking probability is based on the multiserver queuing models discussed in Reference 4.

Leased line optimizations

Leased line design procedures are generally concerned with solving three types of design problems. The first involves the development of a minimum cost billing network to interconnect N user locations on the same multipoint line. This particular problem can be solved on a computer by using the algorithm known as the minimum spanning tree procedure. Developed by J.B. Kruskal of Bell Laboratories, the method assumes a knowledge of the geographic locations of each of the terminals and their point-to-point connection cost (Ref. 5).

Initially all possible links are ranked in the order of increasing cost. Each link, starting with the least costly, is considered and added to the partially formed tree if it meets certain algorithmic requirements. If not, it is permanently discarded. Then the next more expensive link is tried, and so on.

The network is obtained when a tree structure containing exactly N-1 billing sections has been formed. Figure 5 contains an example of applying the minimum spanning tree procedure. As useful as it is, the minimum spanning tree procedure does not incorporate provisions for treating transmission speed as a variable. Nor does it help locate the site for remote concentrators or consider link traffics and response times.

The second method for configuring leased line networks involves an algorithm developed by L.R. Esau and K.C. Williams of IBM that considers throughput and response times (Ref. 6). It has been implemented by a number of computer vendors, communications common carriers, and consulting firms. It accepts input data in the form of tariff information, terminal and computer locations, terminal traffic volumes, and maximum line loading or response time requirements.

The Esau-Williams algorithm assumes that a star network comprised of a point-to-point link from each remote terminal to the closest central complex is the most costly layout. Its goal is to reduce the multitude of expensive point-to-point lines to a fewer number of less costly multipoint lines. At each iteration the algorithm selects a particular pair of links. One central link is to be removed and another link is to be added so that all terminals remain connected to the central site. The algorithm produces the greatest reduction in link cost without violating the response time or loading constraints. The optimizing algorithm also assumes the fixed location of concentrators and, therefore, does not consider the use of multiple line speeds in the same network.

The third category of leased-line design techniques involves procedures for the positioning of multiplexers and concentrators. Although these algorithms are generally approximate in the sense that they do not obtain truly minimum cost networks, they produce useful solutions to such questions as: How many concentrators? What type? Where should they be located?

The basic idea implemented in most concentrator and multiplexer positioning algorithms involves the attachment or assignment of terminal locations to the nearest or least cost location which could service it. Many algorithms rank the candidate concentrator or multiplexer locations and configure a network by selecting one multiplexer or concentrator at a time. At each step a new multiplexer or concentrator is inserted only if it will produce further reductions in total networking cost. (Refs. 4 and 7).

Changes for existing design procedures

To cope with the problems of having several possible carriers, but within specified, limited geo-

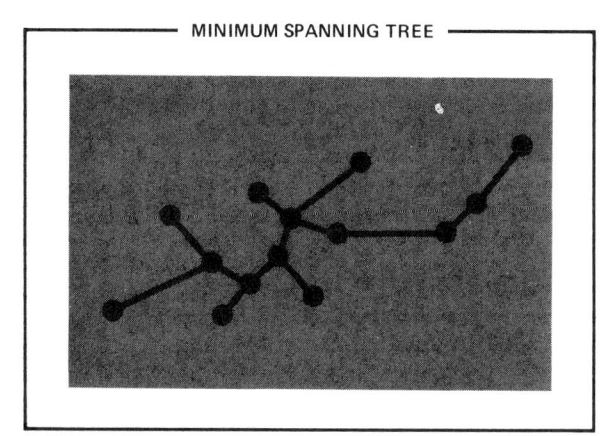

4. By use of the minimum spanning tree algorithm, all terminals are connected to form a network that results in lowest over-all prices.

graphical areas, data base management systems must be developed. The user will input his terminal locations and some new type of descriptive information to indicate which common carrier offerings are available to him in that particular geographic area. Then he will select the most cost effective service providing the desired connections between the given locations.

Network optimization problems involve most major design options open to users. There are four types of problems:
- WATS optimization.
- The minimum spanning tree.
- Multipoint network layout.
- Positioning of concentrators and multiplexers.

In the design of WATS networks, it will now be necessary to know how much total traffic is submitted by individual user locations, during the month as well as during the busy period. This tells whether aggregate volume is likely to approach or exceed the 240 hour minimum for a full business day line. Thus, WATS optimization becomes more complicated because it requires substantially more input information which is difficult (if not impossible) to obtain regularly from common carriers. And, even when the traffic information can be obtained it will usually be so tardy that calling patterns may have changed significantly before the data is in hand.

There is seldom anything a user can do to control or predict with any accuracy what his actual calling patterns are going to be for any given month. This makes network optimization particularly difficult when the tariff structures depend on usage.

Any company employing WATS lines for voice communications during the day and data transmission at night must reassess the economics of its approaches if it is near 240 hour per month or 14,400 call minutes. Perhaps the most sensible way of approaching the problem is through the development of computer programs which simulate likely calling patterns and calculate network costs for various mixtures of measured time and full business day line complements.

Simulation could allow the user to quickly investigate a large number of different mixtures for the required WATS lines. The optimum solution would be obtained by computationally finding the particular combination of WATS bands giving rise to the minimum cost of all the simulated combinations. Full blown formal simulation would not be required whenever approximate cost calculations are sufficient.

The HiD/LoD tariff complicates pricing determinations since the user has to compute direct route and indirect route costs. Even so, pricing a two-point line is relatively straightforward and amenable to computerization. But, under HiD/LoD, multipoint line layout is difficult. The user's network billing structure is permitted to include billing nodes, those locations where he does not even have terminals or computers. Therefore, any modified design procedures must investigate all possible direct and indirect routings between user locations to be connected.

Costing configurations

To accomplish this the traditional spanning tree algorithm must be modified. In the spanning tree, costs of connected pairs of locations can no longer be determined uniquely in advance. The cost functions may change as the billing network is structured. It is quite likely that the traditional spanning tree procedures for selecting user locations to be connected will work quite well if the cost entries can be changed dynamically to reflect each new configurations. Link costs can still be calculated but will change as a function of the particular homing points being used to connect the multipoint line locations.

Full blown multipoint leased line design procedures can probably be modified to reflect the new HiD/LoD and DDS pricing structures. A reasonable way of structuring the problem is to realize that most user locations will not be extremely far from

---

TABLE III
COSTS BETWEEN TWO HIGH DENSITY CENTERS

OLD FCC TARRIFF 260		HiD/LoD NEW TARIFF	
100 MILES	$232.50	100 MILES @ 0.85	$85.00
		2 HiD CHANNEL TERMINALS	70.00
2 STATION TERMINALS	30.00	2 STATION TERMINALS	50.00
TOTAL MONTHLY COST	$262.50	TOTAL MONTHLY COST	$205.00

NOTE: LINK IS TYPE 3002 VOICE GRADE IN BOTH EXAMPLES

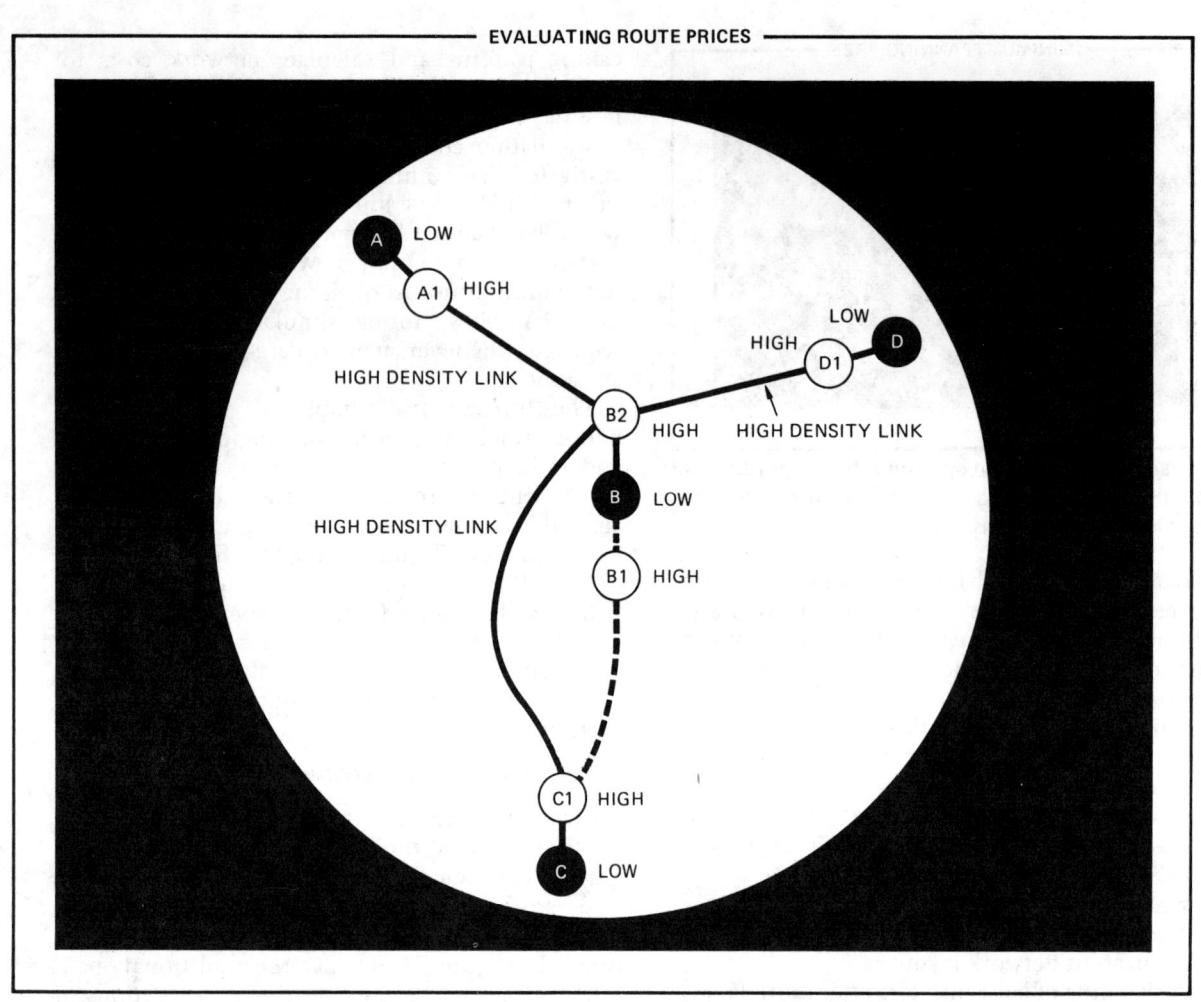

5. Although connecting all high density centers to site B2 looks to be the most economical, connecting C1 to B2 via B1 may be less costly. But this and other options must be calculated.

their HiD homing points or DDS locations.

One approach to modifying existing design procedures will assume the replacement of all user low density locations by a corresponding HiD homing point. Then the traditional design procedures for clustering and grouping the terminals onto lines can be employed.

These procedures effectively optimize the major connections spanning only the homing points and high density user locations. Next, all homing points which were substituted for user LoD locations would be replaced by their actual LoD user locations. The final solution to exact pricing would be obtained by applying the HiD/LoD minimum spanning tree procedure to each multipoint line (group of terminals) clustered by applying the Esau-Williams algorithm.

Another possibility for modifying the present line layout techniques of Esau and Williams would be to calculate the cost tradeoffs using dynamic routing comparisons of the various direct and indirect possibilities at each step in design layout. ■

REFERENCES

1. R. A. Donnan and J. R. Kersey, "Synchronous Data Link Control: A Perspective", IBM Systems Journal, Vol. 14, No. 2, 1974.
2. J. P. Gray, "Line Control Procedures", Proceedings IEEE, November 1972.
3. J. R. Kersey, "Synchronous Data Link Control", Data Communications, Vol. 3, No. 1, May/June 1974.
4. James Martin, "Systems Analysis for Data Transmission", Prentice-Hall (Chapter 31).
5. J. B. Kruskal, "On the Shortest Spanning Subtree of a Graph and the Traveling Salesman Problem", Proc. American Mathematical Society, Vol. 7, 1956.
6. L. R. Esau and K. C. Williams, "On Teleprocessing Network Design, A Method for Approximating the Optimal Network", IBM Systems Journal, Vol. 5, No. 3, 1966.
7. D. R. Doll, "Multiplexing and Concentration", Proceedings IEEE, November 1972.

How transaction cost declines as data networks get larger

Based on actual prices and tariffs, graphs vividly depict the rapid decrease in unit-transaction cost because of fuller utilization of lines and equipment as the network handles more traffic

David J. McKee
General Electric Information Services
Bethesda, Md.

ECONOMICS

When a company needs its own data communications network, its goal is to build a system that, at lowest over-all cost, will handle traffic from all locations in a satisfactory manner—and perhaps allow for growth in the number of terminal locations and the intensity of the traffic. Experience shows that the larger a properly designed data communications network is, the lower its unit cost for some defined transaction or mix of transactions. This is an instance of the well-known concept, economy of scale.

But no user would build a system larger than he actually needs just to enjoy a lower per-unit cost. Furthermore, no company by itself has a teleprocessing load large enough to push economy of scale to its limit, the point at which the cost of each transaction no longer decreases. Still, understanding the cost factors and trends in approaching the limit can shed considerable light on the future directions of data communications networks—including the impact of tariff changes and improved technology, and particularly the economic viability of multi-user shared networks.

To find out more about network economy of scale, a representative interactive-mode application was postulated, and a successive series of lowest-cost designs was computed for load (traffic) levels ranging from 0.1 million to 1,000 million transactions a month per user. The design used actual tariffs and the lowest-cost assortment of circuit options and communications equipment suited to a given load level. The results are presented here for two different situations.

In one, called the single-computer network, all traffic to and from the 50 largest cities in the United States flows to one central host computer located in St. Louis. Such a configuration typifies a private in-house network serving a single company with data from all outlying areas reaching the company's host computer. It also typifies a commercial time-sharing operation.

The second, a multicomputer network, assumes there are 10 computers located at 10 major traffic centers. Each computer is owned by a different company, and each company produces as much traffic as the single company. Further, 10% of the load from each of the 50 cities flows to each computer. The companies share the data communications network, but not their processing resources—there are no links between the computers.

In either case, the design assumes the use of asynchronous terminals operating in a half-duplex mode at a rate of 30 characters per second (or 300 bits per second). That is, the terminals can either send or receive, but not do both at the same time. The terminals are connected to the network through a dial-up facility, and enough interface equipment is available to assure that fewer than 5% of terminal dial-up attempts during the peak-load period will receive a busy signal. The duration of the average transaction—here, an interactive session from dial up to hang up—is two minutes. The traffic load from any city is proportional to the city's population, not the number of terminals in the city. Specifically, the costing of the data communications network for any load level does not include assessments for terminals or host computers.

Costing the networks

The cost per session as a function of the number of sessions per month is contained in a set of smoothed curves (Fig. 1). This set of cost results applies only to the specific, postulated application. Thus, while the results provide useful insight into general cost trends, this data will not apply to other applications, say, one designed to handle bulk transfer of data with sessions lasting up to several hours.

Figure 1 shows the cost for using only direct distance dialing (DDD) or only wide-area telecommunications services (WATS). They are clearly prohibitively expensive at these load levels, and their curves are shown simply to afford a contrast to the economically viable single-computer and 10-computer networks.

As will be shown, the major reason for the decline in session cost as the number of sessions increases, for either the single- or multicomputer case, is the lower unit cost of circuits. This is because of improved utilization of the leased circuits and the use of higher bandwidth circuits. A secondary decline in cost results from the ability to use larger, more efficient equipment.

Although new facilities are being built by the telephone companies and the specialized carriers to improve the handling of data, data communications for the most part is still the technological stepchild of voice communications. The designer can specify the use of such public, switched, telephone services as DDD, WATS, and foreign exchange (FX). He can lease dedicated telephone circuits of various bandwidths. And he can install such line-concentration equipment as frequency-division and time-division multiplexers and remote data concentrators at each communications node, to take maximum advantage of each leased circuit's transmission rate capabilities.

The mix of telephone services and leased lines will be different at each major load level, but appropriate to provide some lowest-cost configuration. Figure 2 contains a part of the single-computer network. It shows the kinds of services and communications equipment that would be used in the network. (For simplicity, this figure omits both the distribution datasets or modems required to interface each dial-up terminal to the network, and the transmission modems needed at each end of a leased line. But datasets are included in the cost calculations.)

Load level determines configuration

At low loads, the network might have 10 remote data concentrators in the cities with most traffic and some multiplexers in those with less traffic. Multiplexers or concentrators would be dialed up by terminals in outlying cities through a considerable amount of DDD, WATS, and FX. Local loops would be used in cities hosting a multiplexer or concentrator. In other words, at low loads, it would be cheaper to make greater use of switched telephone services. As the loads increased, concentrators and multiplexers would be installed in additional cities and more leased lines and local loops would be used, thus reducing the use of expensive DDD, FX, and WATS services. At highest load levels, the network would contain a concentrator in all 50 cities, and each would be accessed only through local loops. In summary, at every major load-level increment, advantage would be taken of decreasing line charges by reducing the amount of DD, WATS, and FX, and by increasing the bandwidth—that is, the transmission speed—of the leased lines.

In both the single- and 10-computer networks, geographical coverage remains the same and fixed. Only the traffic load per computer (or user company) varies. The 10-computer network, however, handles 10 times more traffic than the single-computer network. Since the cost per session is plotted against sessions per month per computer site, Fig. 1 distinctly shows the reduced cost per session made available by building a larger network and sharing it among many companies.

For example, the cost per session for 1 million sessions a month on a single-computer network is about 12 cents. But if 10 companies share a network 10 times larger, then the cost per session per company (again 1 million sessions a month) drops to about five cents.

Further, comparison of the two networks in Fig. 1 shows that a multicomputer network affords an even greater ratio of savings at lesser monthly traffic, and that any cost differential between single- and 10-computer networks essentially disappears at very high traffic intensities. In essence, for each user on the multi-

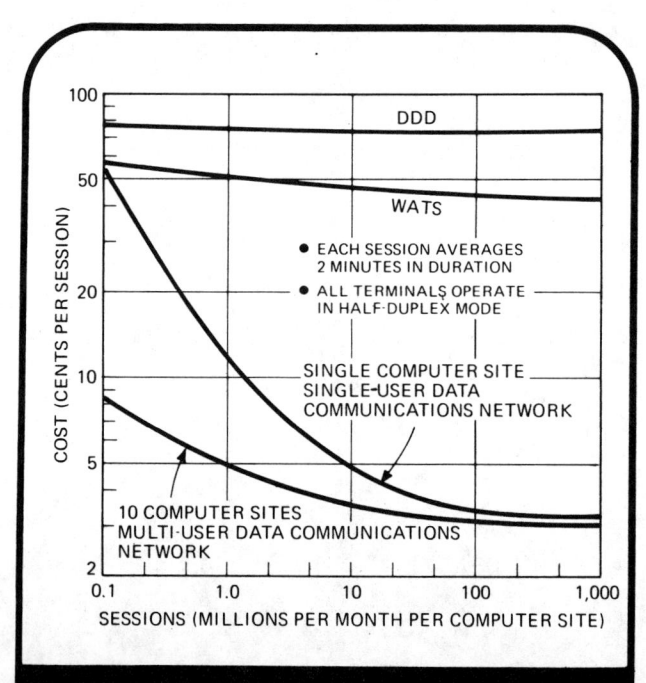

FIG. I. DECLINING COSTS ■ Transaction cost in private networks goes down rapidly as the load level increases.

computer network, the benefits of increasing line utilization, including better balance of load, occur at lower per-company traffic levels.

Send more, save more

Circuits are the major cost item in these networks, as in most data communications networks. And they yield the greatest percentage savings as traffic level increases (Fig. 3). Here, the curve labeled circuits applies to leased lines of various bandwidths and includes the DDD, WATS, and FX circuits needed at the lesser traffic intensities. The curve labeled communications equipment covers the cost of multiplexers, remote data concentrators, a programable front-end processor, and transmission datasets, each in the amount needed at each level of traffic volume. The local loops and distribution modems, the costs of which are also shown, are attached to multiplexers and remote concentrators.

The leased lines covered in the monthly circuit costs include voice-grade lines capable of operating at 9,600 b/s, as well as 50,000-b/s lines, and 230,400-b/s lines, with each line able to transmit at its rated speed simultaneously in both directions. That is, in telephone parlance, the lines are full duplex.

For each type, the longer the line, the more it costs, and higher-speed lines cost more than lower-speed lines. For a given type of line, the cost per mile decreases with increasing distance. The relationships between cost, distance, and speed, based on present interstate tariffs, are contained in Fig. 4.

The actual traffic intensity between any two widely separated points is determined by the application, so a significant design parameter in developing a least-cost network is to ask, for example: is it less costly to lease three 9,600-b/s lines or one 50,000-b/s line for a given distance? A related question is: is it cheaper to reroute lines to reduce cost? Some insight into the answers to such questions can be gained by examining the cost per bit per second (per month) as a function of distance for the three types of leased lines. This information is readily obtained by dividing the cost of each line by its own rated transmission (Fig. 5).

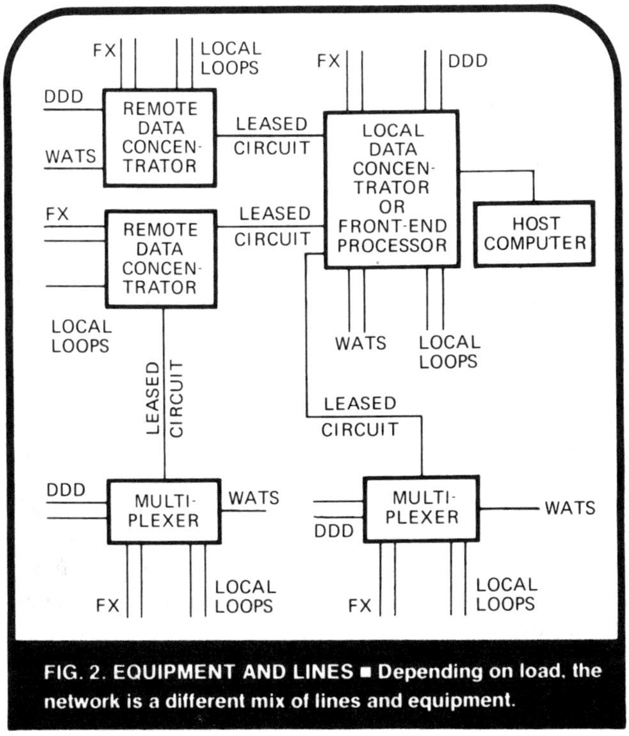

Intuitively, one might expect this cost-per-bit parameter to be less for higher-speed lines than lower-speed lines. And this is true–up to a point. But at about 500 miles or more, as indicated in Fig. 5, using five 9,600-b/s lines is' less costly than one 50,000-b/s line. A similar anomaly occurs at 2,000 miles with respect to the 230,400-b/s line.

Circuits come only in discrete transmission speeds, so smaller networks may not have enough traffic to fully utilize the capacity of a particular leased line. But when networks become larger, traffic volume for a given geographic configuration will increase and be different on different links. Then the network design can take more advantage of the significant economies available for each type of line. That is, on any link in the network, the designer can pick the lowest-cost circuit capable of handling the traffic.

Large networks exploit full duplex

In small networks, particularly ones dedicated to an inquiry/response application, the inquiry messages are generally shorter, by perhaps a factor of 10, than the response messages. Thus, there is a tendency for more traffic to flow from than to the host computer. In such a case, the capabilities of a full-duplex line are not being exploited. Although the full-duplex voice-grade line operating at 9,600 b/s can, for example, carry 9,600 bits in 1 second in each direction, one direction might only carry 960 bits in 1 second when the load is not balanced. In other words, an unbalanced load means one side of the full-duplex line operates at low line utilization.

However, in a multicomputer network with the computers geographically distributed, the messages from one computer could "pass" the messages from another computer going in the opposite direction on the same full duplex circuit. This would tend to balance the load of these circuits. Hence, load balancing allows full use to be made of the full-duplex capacity and thus reduces the required number of lines for a given traffic level.

To review, the cost curve for the two networks in Fig. 1 includes the reductions in cost obtainable from the appropriate choice of lines, multiplexers, concentrators, datasets, and load balancing as the data communications network becomes larger. The two designs apply to terminal-to-computer networks for man-machine interactive sessions. Studies have shown that similar economies of scale can be obtained for data communications networks of which the main task is to transfer bulk information from one computer to another or to switch messages from one terminal to another terminal. Regardless of the type of transaction on the data communications network, the cost per unit transaction will decrease as the network becomes larger.

At present, however, the benefits of economy of scale can be enjoyed only on a private network. With certain exceptions, shared networks are virtually prohibited by the regulatory situation.

Private and shared networks

A private data communications network is defined here as a network owned by one company and employed for

FIG. 2. EQUIPMENT AND LINES ■ Depending on load, the network is a different mix of lines and equipment.

its own use and that of any divisions or subsidiaries. On a private network, the company can implement any applications it needs and can justify. It can connect one or more computers to the network to perform remote job entry and batch data processing. It can use the system for message switching. And it can do both data processing and message switching on the same network. The economic incentive to a private company is to merge individual small networks into one larger network to obtain the economy of scale. (However, reasons other than economic may diminish a company's enthusiasm to commit itself to a larger network.) Perhaps the most important restriction on a private network is the tariff prohibition against resale of a portion of the network's capabilities to another party.

A shared network is defined here as one that can be employed by more than one company. Two shared-network alternatives are the joint-use network, which is permitted by present tariffs, and the so-called value-added network, a concept now under consideration by the Federal Communications Commission. In either type, a user (subscriber) will not have to install his own communications equipment and lease circuits, but will still reap some of the savings available from a large, distributed network.

The joint-user shared network operates under the "joint-user" provision of present interstate tariffs. Subscribers share the communications network, but they cannot use it to exchange information (say, a data base) or to transmit messages between themselves. Under the joint-user provision, the common carrier bills the sharing organizations on the basis of their prorated usage (either time or bandwidth) of the circuits. The prorating of the related communications equipment (for example, multiplexers) would have to be agreed on by the sharing companies. Generally, one subscriber acts as the manager of the joint-user network.

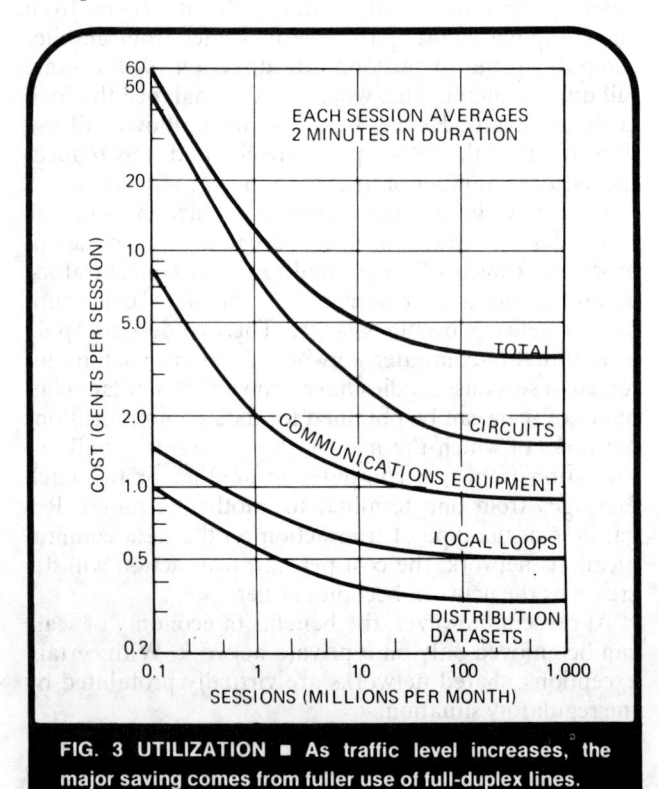

Joint-user shared networks have not proven popular. Although they can yield a lower cost under ideal conditions, there is too much uncertainty about billing and other administrative matters. A joint-use shared network, to operate effectively, would require compatible subscribers—ones that need access from the same geographical centers, that have similar applications (for example, all interactive), and have similar computer equipment to simplify the interface requirements. (Even so, some joint-use networks are in operation. See "Operational joint-use networks," next page. **Ed.**) Although the administrative procedures involved in a joint-use network may be relaxed in the future, the commercial multi-user network may prove more attractive, when and if it becomes available.

Value-added networks

The commercial multi-user network, nicknamed VAN for value-added network, is a recent concept that could offer significant savings in communications costs to both large and small users. In a commercial multi-user network, a separate company would lease circuits from the common carrier, install such communications equipment as multiplexers, concentrators, and datasets, perform such services as error correction and speed conversion, factor in overhead and profit, and then resell custom end-to-end data communications links to subscribers.

Since a VAN implies a very large network, its owner (vendor) would enjoy the benefits of the economy of scale. And even when the vendor adds overhead and profit factors, the user's cost would still be less than if the user built his own private, but smaller, network.

Conceptually, at least, VANs would provide terminal-to-computer and computer-to-computer links, presumably at a high level of data security. The subscriber would have to consider such factors as available geographic coverage, available transmission speeds, maximum message delay times, ease of use, and the vendor's economic viability and quality of service.

VAN service will probably have a charge for connecting each computer and another for connecting each terminal. For each computer, the computer interface charge would be either a fixed fee per month if the computer were connected full time, or a per-unit-time charge if the computer dialed up the network. In either case, the computer interface charge would probably be based on the maximum number of terminals that would simultaneously access the computer, or on the transmission speed of the computer interface, or both.

The terminal connect charges would include such items as a fixed connected charge per session for each dial-up terminal accessing the network, or a flat monthly charge for each hard-wired terminal; a per-unit-time charge for the duration of a terminal connection; a charge based on the maximum rated speed of the line between the terminal and the computer; and a charge based on the number of characters, or messages, transmitted.

Related services to be provided by a VAN are accurate usage and accounting information, network-status information, message-storing facilities, and automatic

FIG. 3 UTILIZATION ■ As traffic level increases, the major saving comes from fuller use of full-duplex lines.

Operational joint-use networks

Although joint-use networks have not proven too popular, some do exist. The latest entry, sponsored by the RCA Corp. and managed by RCA Global Communications Inc., went on line this summer. It extends from coast to coast, uses series 3000 voice-grade lines, and serves those cities in which RCA Corp. and its subsidiaries now have enough traffic to warrant establishing the links—New York, Detroit, Chicago, San Francisco, and Los Angeles.

In addition, RCA has signed up at least 12 joint users, including Detroit Bank and Trust Co., Electronic Memories and Magnetics, Hitachi America, and Mitsubishi Heavy Industries. These users, according to RCA Globcom's Kenneth E. Ryan, will save about 35% compared with present common-carrier tariffs for equivalent service.

The network employs 4,800-bits-per-second transmission, with time-division multiplexing to derive up to 75 separate 75-baud channels. As traffic increases, line speed will be increased to 9,600 b/s and more channels will be added.

Other joint-use networks include: one managed by Dow Jones & Co., serving about five or six other subscribers, besides several hundred locations of its own; Shearson Hammill & Co.'s net, which serves subscribers in Los Angeles, Chicago, and New York; and Computer Dimensions Inc.'s system, which has its hub in Dallas and reaches about 25 subscriber-firms along the West Coast up to Seattle. **Ed.**

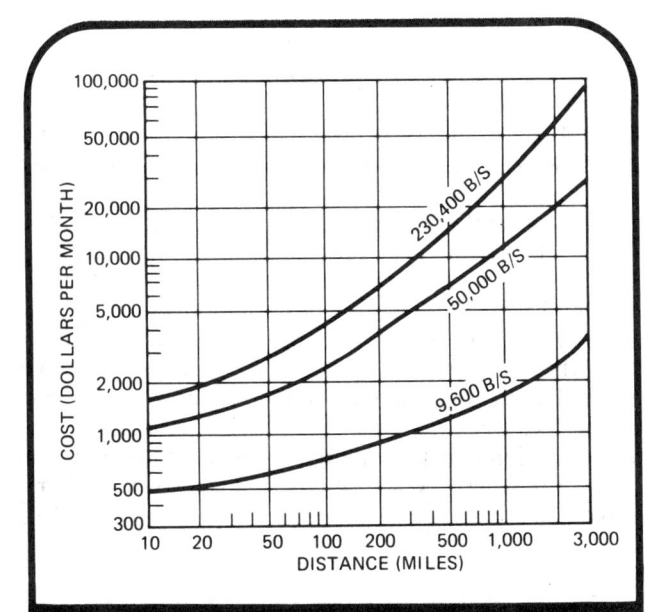

FIG. 4. LEASED LINES ■The monthly cost for lines used for data communications depends on mileage and speed.

dial-out from a computer to a terminal.

A major feature of a VAN service is that a subscriber would deal with one vendor for its data communications network, from the terminal access point to the computer interface, to insure a high-quality service. If a VAN has the geographic coverage, then a user would be able to increase, almost at will, the number of cities to which he desires access.

In many ways, the VAN concept is to data communications as time sharing is to data processing. In either case, a vendor sells a piece of his resources. And just as time sharing did not eliminate the use of in-house data processors, neither is VAN likely to wipe out private data communications networks. VAN will become another alternative for small, medium, and large users of data communications. And, as with time-shared computers, the subscribers likely to obtain the largest incremental benefits will be the small users. However, even with marked savings, a large user could well frown on the implications of placing the total data communications requirements of its business enterprise completely in the hands of an outside VAN organization.

Factors in multi-user networks

Two other factors that will affect the future course of commercial multi-user networks are the clarification of the uncertainties about vendor-owned hybrid teleprocessing services for sale to subscribers, and the effect of pending and future changes in interstate and intrastate tariffs.

The FCC has ruled that, if a service is primarily a data processing function, such as interactive time-sharing and remote-batch activities in which the data communications is incidental, then the service is not regulated. But if the service is essentially communications with data processing incidental, as in computer-switched TWX or Telex services, then the service is regulated. The FCC has not defined the term incidental, so regulatory uncertainty exists about the two kinds of services. Vendors who may want to offer the combination of services called hybrid teleprocessing, using the same data communications network and perhaps even the same host computer, are reluctant to do so for fear of having their circuits disconnected, of being subjected to regulation, or of becoming involved in lengthy and expensive legal and regulatory proceedings.

Thus, at present, if a user organization needs both data-processing and a message-switching service, but cannot justify installing its own private network, then it must buy each service from different vendors. Certain advantages would accrue to the user, in addition to the economy of scale offered by the larger network, if these services could be obtained from a single vendor. Among these are the use of the same terminal for both data processing and message activities and the ability of the data processing computer to access incoming messages and to initiate outgoing messages.

Currently, commercial hybrid teleprocessing is not available because of the regulatory uncertainty. The problem involves where to draw the line between the traditionally regulated message-switching services and the unregulated remote-access data processing services.

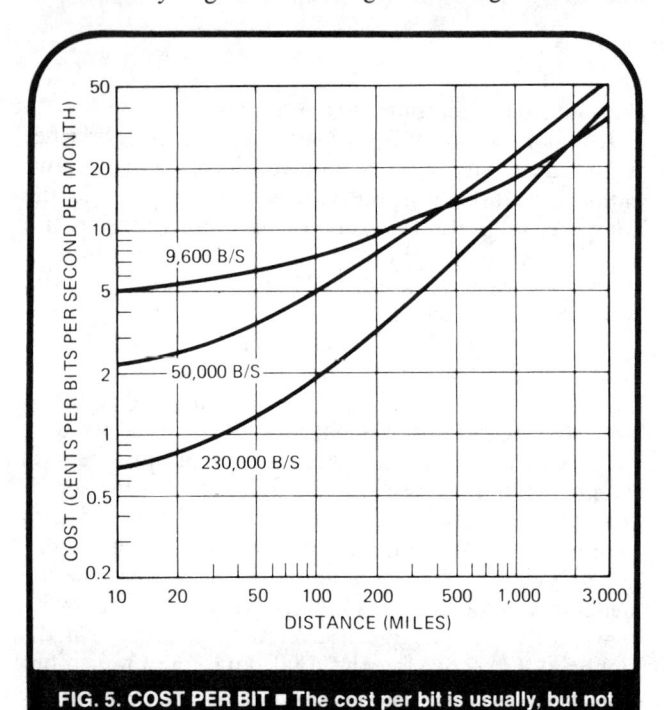

In the area of changes in tariff structures, the specialized common carriers promise improvements that could affect the cost and reliability of large, as well as small, networks. Among these promises are a greater variety of transmission speeds, a shorter minimum billing time on switched circuits, a lower error rate, reduced cost, and a pricing structure that will be less sensitive to distance. However, the specialized carriers proposing nationwide service are still in a regional, start-up situation.

Impact of tariff revisions

In 1973, AT&T submitted an application to the FCC for permission to revise its leased line voice-grade tariff structure. Using present technology and appropriate datasets, these circuits can carry up to 9,600 b/s and are the backbone of most computer-communications networks in present service. Basically, these revisions would divide a route into three classes: short haul, for use on points 25 miles or less apart; high density; and low density. Over 300 locations in the United States, including all large cities, have been designated high-density centers, between which the telephone companies have high-capacity equipment. All other locations are low-density points.

The proposed charges for a full-duplex point-to-point circuit are:

CLASS	COST/MILE
Short haul	$3.00 up to 25 miles
High density	$0.85 over 25 miles
Low density	$2.50 over 25 miles

Figure 6 compares the new circuit costs (including datasets and other fixed costs) with the present costs on a mileage basis. The new tariff would permit the user—not the telephone company—to route his lines in the least-cost manner, so that the user would pick a combination of low-density and high-density routes to meet his needs. Optimizing the line routing under this new tariff may prove quite difficult, particularly for large networks. On balance, though, it appears that for large, geographically distributed networks, the new tariffs would afford the user considerable savings.

Building a data communications network may take several years, and even then the network keeps changing and growing. Therefore, since circuits are the largest single cost in a data communications network, the planner must keep track of new offerings and changes in tariffs. In the future, it is almost certain that circuit costs between major cities will decrease significantly, that tariffs will become less sensitive to distance (as when satellite circuits provide an alternative to terrestrial circuits), that the cost of local loops will increase somewhat through the elimination of the flat rate now available in many locations, and that the specialized carriers will provide high-density circuits between major cities in competition with AT&T and perhaps even with themselves. [EOT]

FIG. 5. COST PER BIT ■ The cost per bit is usually, but not necessarily, less for higher-speed than lower-speed lines.

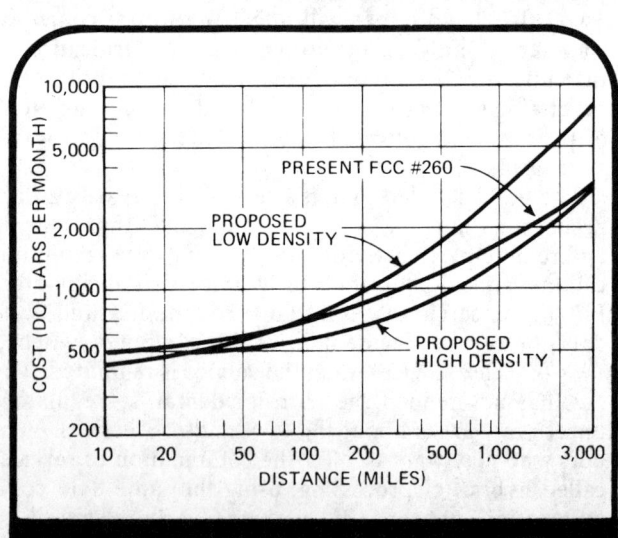

FIG. 6. NEW TARIFFS ■ If the suggested Hi-Lo tariff goes into effect, it will reduce costs for large data networks.

Breaking the logjam in dynamic on-line data base systems

An advanced data communications network, a hierarchy of memories, memory mapping, and a unique operating system combine into a retrieval and update system

Charles F. Pyne
AutEx Inc.
Wellesley, Mass.

Performance

The rapidly improving state-of-the art of transmitting data from remote locations, combined with decreasing costs, has greatly affected the historic role of the central processing unit (CPU) in computer-communications systems. In early installations, data communications traffic was light, and the CPU was dedicated mainly to performing batch data processing activities. Gradually, the data communications throughput has been increased in many systems, with a consequent decrease in data processing volume. In fact, this trend is carried to the extreme in new computer-communications systems, where no data processing is involved at all. These systems are dedicated to supplying an on-line, real-time information update and retrieval service. Here, then, the CPU's main task is to route user traffic from many hundreds of remote inquiry/response terminals to a massive data base, access the associated files, and return messages to the terminals.

When the data communications network which ties the remote terminals to the CPU in a dynamic information system uses today's efficient, highspeed transmission equipment, then the CPU itself can become the bottleneck that limits system response and the number of on-line terminals. The main reason for this choking effect is the batch-oriented operating system—the master software package that orchestrates input/output equipment, including the data communications lines, and the myriad of applications programs undergoing simultaneous processing.

Traditional batch operating systems, combined with a communications access method, have been pressed into service in remote-batch and time-sharing applications. They can perform quite well because the service is basically that of providing remote data processing. But the limitations of a batch operating system on overall performance become more serious as the allowable response time goes down, as the number of terminals to be supported goes up, and as data-processing activities disappear.

The on-line, real-time, transaction-oriented information storage and retrieval system that forms the base of many computer-communication's system's operation is communications and file-management intensive, not data-processing intensive. AutEx' trading information systems are a good example of this. For AutEx, a batch operating system

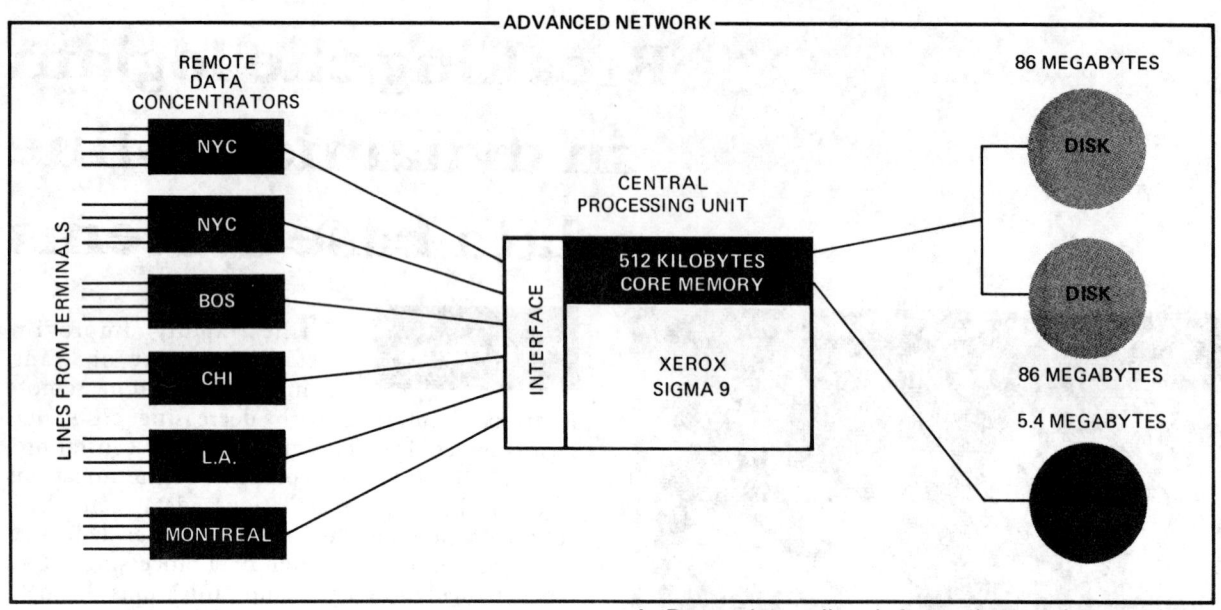

1. Dynamic on-line information update and retrieval data base network employs data concentrators in major cities to speed inquiries to and messages from a Sigma 9 computer.

will not do. The AutEx system is designed to service inquiries from up to several thousand remote keyboard/display terminals and respond to them with assembled messages in less than 10 seconds. No data processing is involved. To reach this goal, the company is installing extensive mass memory, an advanced data communications network, and—most important from the viewpoint of system performance—AutEx is developing its own operating system, called TEX for telecommunications executive, for its Xerox Sigma 9 host computer in Wellesley, Mass. (Fig. 1).

To accomplish fast response and high throughput, the AutEx system includes an unusual virtual-memory technique, and a dynamic file-management method which causes the most recently accessed files to automatically flow to the fast-access core memory and the least-accessed files to flow to the slow-access disk packs. Unlike most computer-communications systems in which the computer and the communications network are considered—at first glance—as separate entities, TEX integrates the data communications network and the central processing unit.

The essential philosophy behind the unique design of TEX is that the performance of a telecommunications system must go beyond the conventional communications requirements for the access and control of lines and terminals. The operating system must also satisfy the special requirements for on-line, real-time, adaptive management of data-base files. Simulation studies have verified the efficiency of the approach, and AutEx' new system is slated to go on line early next year.

AutEx supplies an on line trading information services to traders of stocks and bonds and another to traders of lumber products. AutEx subscribers focus their attention on the central data base. They enter new data in real time and also retrieve data of particular interest in real time. That is, both update and retrieval are dynamic. In fact, the entire AutEx data base is constructed solely from subscriber entries. For example, different on-line subscribers may enter, then change, the price at which they are willing to buy or sell a block of stock. Thousands of changes can be made each hour to the composite data base, so there is a continuous interplay of updates and retrievals from many subscribers. The company expects to add new services, so that eventually the network will contain several thousand terminals that will access several large but independent data bases through the Sigma 9.

Concentrators mean more terminals

The telecommunications executive is designed to efficiently exploit the file handling capability of the central computer and its memories, with the goal of maximizing the possible number of on-line users within the restraints of the configuration. The data communications network must be as efficient and cost effective as possible, avoiding delay of messages or limits to throughput.

To accomplish this, six Interdata Model 50 remote data concentrators (RDCs) will be located in major cities and linked to the Sigma 9 in Wellesley over 9,600 bits-per-second links (Fig. 1). Each RDC can service about 200 keyboard/display terminals. Here, the RDCs, rather than the Sigma 9, perform such routine communications functions as line control, message formatting, intermediate queuing, error control, and network diagnosis. For one thing, these functions do not require the sophisticated processing capability, large core memory, and disk files of the Sigma 9. Nor is the Sigma 9's interrupt structure well suited for such operations. For another thing, the economics favor the use of RDCs. The Sigma 9 is worth about $1 million. AutEx operations statistics show that each RDC relieves the Sigma 9 of 15% of its processing load, worth $150,000, at the cost of about $20,000 for the RDC. The economic details are contained in Table 1.

Even when the bulk of the communications functions is shifted from the host computer to an RDC, the host computer's operating system requires a communications access method whose main job is to provide the host computer with efficient and timely access to the remote terminals. But more important to the AutEx system is the ability to access the data-base files in a reliable, efficient, and dynamic manner—so TEX integrates both communications access and file access in its structure.

Dynamic information, not time sharing

File accessing is commonplace in batch-processing and time-sharing systems, but the requirements for the AutEx dynamic information update and retrieval system are quite different from either, so that it seems worthwhile here to compare each type of system.

In traditional batch-processing environments, programs drive the computer system to produce a desired end result for the customer. Minimizing the job-processing time on the host computer is important, but the overall response time in getting the job from and to the user is usually not critical. In batch operation, any efficient file manipulation technique can be employed to change the files, so long as the final state of the file is correct and consistent. Because file updates in batch systems take place occasionally, say once a day or once a shift, file consistency can be obtained by temporarily taking the file out of the system, constructing intermediate files that can be made internally consistent, and then returning the new updated files to long-term storage.

Time-sharing systems closely resemble dynamic information systems in many ways. The dominant similarities are that time-sharing systems are data communications intensive. They are driven by remote terminals rather than programs, and they usually provide fast response to input messages. The important distinction, however, is that most time-sharing systems emphasize user separation. Therefore, memory space for data bases may be preallocated. Each user will have access to his own data base that only he may or may not change on line, and those data bases used in common cannot be changed at will by any user.

The distinctive characteristics, then, of a dynamic information update and retrieval system, such as the one being developed by AutEx, are access to and update of common data bases by any user at will, control by system users rather than by programs, use of a common memory pool for all tasks rather than preallocated memory for each task, and reduced operating system overhead for swapping files from disks to core so that the physical system can support hundreds and even thousands of terminals.

In particular, the design of TEX has these goals: the system must be responsive to operations in a file-oriented environment when driven by remote terminals; it must use parallel-accessed disk storage efficiently, so that the system can support a high degree of multiprogramming; it must minimize core memory size, by keeping in core only those files of immediate interest; it must maintain files in a consistent state at all times; it must have a hierarchical structure so that changes and additions to files can be easily made at the lowest (appli-

TABLE 1
REMOTE DATA CONCENTRATOR PAYS OFF

	WITHOUT RDC	WITH RDC	SAVINGS
SIGMA 9 PROCESSING REQUIRED FOR LINE CONTROL FUNCTIONS TO HANDLE 200 TERMINALS	20%	5%	15%
RDC COST (AS % OF Σ9 COST)	—	2%	−2%
TOTAL	20% = $200,000	7% = $70,000	13% = $130,000
SIGMA 9 WORTH $1 MILLION			

cations) level without affecting other parts of the system; and it must separate independent processes (that is, services) to promote security control and error isolation.

Here, core memory size is minimized by using a virtual memory technique to map data onto the disk. And file efficiency is obtained by employing a dynamic buffering scheme that allows frequently used data to be retrieved faster than less used data. Thus, dynamic buffering allows the system to rapidly adapt itself to an ever changing user environment, including peak-load conditions.

The dynamic system at AutEx

Figure 1 shows the major elements of the AutEx dynamic information system and their interconnections. Remote terminals in subscriber offices connect to an appropriate RDC through 300 b/s dial-up lines or 1,800 b/s dedicated multidrop lines. Here, transmission is asynchronous. Each of the six RDCs then connects to the Xerox Sigma 9 through 9,600 b/s synchronous private lines.

The Sigma 9 employs a 32-bit word, has a

MEMORY HIERARCHY

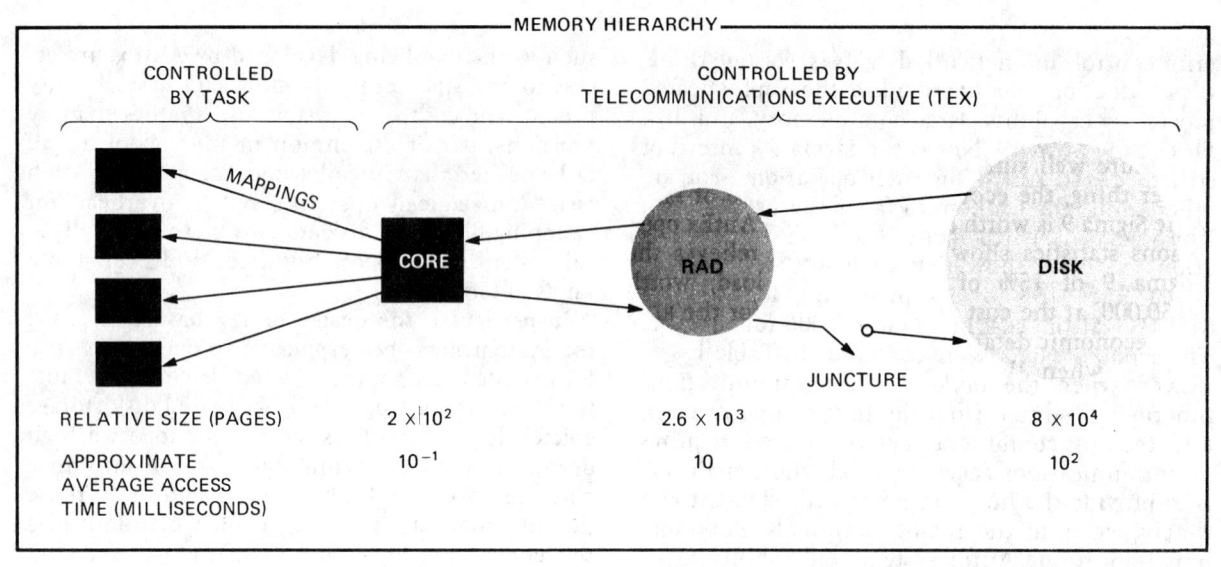

	CONTROLLED BY TASK	CONTROLLED BY TELECOMMUNICATIONS EXECUTIVE (TEX)	
	CORE	RAD	DISK
RELATIVE SIZE (PAGES)	2×10^2	2.6×10^3	8×10^4
APPROXIMATE AVERAGE ACCESS TIME (MILLISECONDS)	10^{-1}	10	10^2

2. Three levels of memory—core, RAD, and disk—provide extensive file capacity with dynamic memory allocation relocating files so that the most popular files stay in core.

memory cycle time of 750 nanoseconds, and has 512 kilobytes of core memory to store the most-frequently accessed data. A fixed-head random-access disk memory, called a RAD, for random-access device, stores 5.4 megabytes of less-frequently accessed data. For bulk storage, the system contains two moving-head disk drives, each with disk packs that can store as many as 86 megabytes of the least-frequently accessed data.

Although the RAD has a relatively slow access time, it has a large number of sectors on its surface. With head-position optimization it can sustain a very high throughput rate as may be needed during peak load conditions.

The Sigma 9 has a hardware memory map (or virtual-memory) feature that divides all core memory into pages, each page containing 2,048 bytes. Every memory reference—that is, access to a file—is considered to fall within a particular 'virtual page' whose address in core depends on the higher order bits of the referenced address. The virtual-page address is converted by hardware registers to a real memory address which may be any place in core memory. Furthermore, the hardware registers can be set to indicate that certain virtual pages are not in core at all. If so, TEX can then retrieve these pages from the RAD or the disk. Also, by appropriate setting of the hardware memory map registers, TEX establishes a different virtual address memory environment for each independent parallel task being processed.

The relationship between individual application tasks, TEX, and the three tiers of memory is shown in Figure 2. The applications tasks, each with its own virtual memory map and other associated context, are the programs that do the actual processing of messages through the system. For example, several application tasks control the traffic to and from the RDCs.

The basic unit of task memory is the 2,048-byte page. When a task is first created, it has no assigned virtual pages. The task is set up by making a MAP supervisor call to TEX requesting a mapping between a certain virtual page in core memory—known as a window—and a particular page in a file on the disk pack. Through a series of such calls to TEX, a complete task sets up a group of windows for both programs and data mapped to pages on the disk. Once done, the task can consider the actual disk pages to be virtually residing within its addressable core memory space. TEX, not the tasks, takes care of such matters as reading and writing, allocating the buffers, and structuring the data to optimize input and output.

Throughput up, access time down

The telecommunications executive maximizes system throughput through efficient disk utilization because of the system's ability to process tasks in parallel and to provide dynamic page buffering. When a given task references a page which is not in core, TEX first places the associated disk request in a queue, then invokes a priority scheduler which suspends the immediate task and starts up another task which is ready to run. Thus, the many parallel tasks build up a set of requests which are placed in queue in a manner related to their disk positions, so that an optimum number of disk requests can be serviced during each rotation of the disk packs.

An even greater advantage in disk utilization, compared with disk parallelism, is gained by adaptive page buffering that responds to the great variance in importance of the numerous file pages being interrogated over the data communications network. Such pages as those containing code, tables, and up-to-the-minute data are used very frequently. But others may not be accessed during an entire day. Many systems preallocate certain task-level data to core, drum, or other high-speed storage media; the data stays put whether accessed frequently or infrequently. This scheme not only complicates the task coding but the files cannot adjust to the changing use of the system from minute to minute and day to day.

Three-level memory

Instead, TEX employs a three-level memory system for holding file pages by using disk, RAD, and core pools (Fig. 2). Initially, all pages reside only on the disk. Then when requested by tasks, the pages are brought into core and mapped (and also saved on the RAD). While the system is running, TEX continuously counts how often each page is accessed during a given period to determine the most frequently and recently used pages. A priority level is then computed for each page.

Thus, at any given moment the highest-priority pages are held in the core pool of about 200 pages, the next highest level of pages are kept in the RAD pool of 2,600 pages, and the 80,000 lowest-level pages are left on the disk. Therefore, even though the programing addresses each page to disk, when a task requests a page the system first checks to see if the page is in core; if not the system checks the RAD and if the page is there it transfers to core; but if not in the RAD the system retrieves the page from disk, transfers it to the RAD, and then to core. In this way the most commonly used pages—in core— are the most rapidly retrieved.

Table 2 shows the average access times, for the three levels of storage, as a function of load as characterized by the number of messages in the queue. In general, the RAD is about 10 times faster that the disk, and the core is about 100 times faster than the RAD. More significantly, though, is the average access time for a number of requests in queue, and this average depends on how many of the pages being accessed reside in core, in the RAD, and in the disk.

For example, suppose that out of a queue length of 12 pages, three pages are in core, four in RAD, and five in disk. Then the average access time for this mix of 12 requests is, from Table II:

$[(3 \times 0.2) + (4 \times 19.8) + (5 \times 162)]/12$
= 75 milliseconds

The 75 milliseconds is a definite improvement compared with the 291 seconds of average access time that would have been taken if all pages had resided only on disk. Even so, in this example the distribution probability of pages being in core, RAD, or disk has purposely been distorted compared with the distribution that would occur in a real operational system in which TEX forces the most frequently accessed files to be in core—thus dramatically reducing the average access time.

To get a better idea of improved average access time offered by the three-level memory system, with dynamic buffering implemented by TEX, a simulation test was run using 200,000 records (16 megabytes) of indexed sequential files. Each file involved three levels of indexing—a master block, a secondary-index block, and a data block—so that each file access required access to three different pages. In this run, the probability distribution showed that core, RAD, and disk each contained about one-third of the indexed pages. The result: the average access time per three-page file was 133 milliseconds compared with the 873 milliseconds it would have taken if all three pages had been on disk. For this case, the dynamic buffering method turns out to be six times faster than when using only disk memory.

This improvement is still conservative in that

TABLE 2
AVERAGE ACCESS TIME DEPENDS ON QUEUE LENGTH

LOAD*	ACCESS TIME (MILLISECONDS/PAGE)			THROUGHPUT (PAGES/SEC)		
	CORE	RAD	DISK	CORE	RAD	DISK
1	0.2	17.3	48	5,000	58	21
2	0.2	18.1	80	5,000	110	25
3	0.2	19.0	109	5,000	158	28
4	0.2	19.8	136	5,000	202	29
5	0.2	20.6	162	5,000	243	31
6	0.2	21.4	187	5,000	280	32
7	0.2	22.2	214	5,000	315	33
8	0.2	23.1	240	5,000	346	33
9	0.2	23.9	265	5,000	377	34
10	0.2	24.7	291	5,000	405	34

*NUMBER OF REQUESTS IN QUEUE

the indexed sequential files were accessed randomly, when in actual practice the users—en masse—concentrate their interest on certain portions of a data base. This concentration drives most files of current interest into the core and RAD memories, which therefore substantially reduces average access time. Studies of actual data-base references by AutEx subscribers indicate that after two hours of operation about 90% of individual files accessed thereafter reside in the RAD or core memories. In a sense, then, the disk becomes an archival storage that can still be accessed automatically and rapidly.

Note, too, in Table 2 how the relative throughput of the RAD increases more rapidly than does the disk as the load goes up. Thus, the buffering scheme is most effective at peak load times, just when the need for efficiency is the greatest.

During operation, the telecommunications

executive itself uses virtual memory and dynamic buffering to good advantage and thus reduces the amount of core memory allocated to its own support. Here, less than 40 kilobytes reside in core memory to drive the hardware inputs and outputs; the rest of the operating system, known as a pseudo-system, required to implement such functions as peripheral conversions, operator communications, file access, and debug software, resides in the disk. The pseudo-system is invoked just as though it is an applications task. Furthermore, conventional operating systems frequently dedicate valuable core memory to such off-line tasks as compiling, job-control language interpreting, and file converting. But TEX can free this "off-line" core without any special programing by assigning virtual pages to these tasks.

The telecommunications executive contributes a significant savings in core memory because of the programing discipline involved in the naming of

Software. Staff develops telecommunications executive on off-line computer.

the virtual pages. As mentioned, TEX assumes that no data resides in core, but is all on disk and that the file pages are mapped by windows. In those conventional operating systems whose programs address all data to core first, if several tasks read the identical data into various addresses in the core (virtual) memory, then the system will have no way of identifying the fact that all the pages are the same. Hence, the conventional approach to virtual memory duplicates pages and wastes core space.

But with TEX, pages always retain their name according to the places where they belong on the disk. If one applications task maps a page, that page moves from disk (or RAD) to core. When other applications tasks refer to (name) the same page of data, TEX immediately directs those references to the real physical address for that page in core. Thus, only a single copy of a page of data need be in core to service many different applications tasks.

As an illustration of the advantages of naming pages, consider the program that will control data communications traffic flow from and to the RDCs. Each RDC consists of the same basic computer. But each will differ in the number and speed of lines between it and the remote terminals, will be independently driven by the host computer and, thus, will require a unique applications task to be executed by the CPU.

Reusable code

Fortunately, though, each concentrator performs the same basic communication's functions and all RDCs use the same code. This code is reentrant, or usable by many tasks, which can be accomplished very easily in the virtual memory system by separating the code from the data. Hence, all tasks—that is, RDCs—map their code pages from the same read-only file; and the necessary data pages such as local variables and scratch storage are mapped from files unique to each task.

In short, the six concentrators share much of the applications-tasks files, which yields a significant savings in core memory and switching time compared with concentrator interface programs implemented on batch operating systems that cannot share files. Table 3 shows that 90 pages of core memory would be permanently allocated for the traditional nonsharing approach that names pages in core to service six remote data concentrators, compared with only 20 pages of core for the TEX sharing system that names pages on disk.

In the shared system, the 10 pages of code are used over and over again but the nonsharing system requires 10 pages for each of the six RDCs. Two data pages are required for either approach for each RDC. Inter-task communications take one page for the sharing system. But this function probably cannot be done in the nonsharing system.

The big savings is in the file pages: For the shared system, no core is allocated since these pages are mapped from the RAD or disk as dictated by the task; but for the nonshared system, 18 pages of core memory must be preallocated for the six RDCs.

If the system is expanded to more than six concentrators, then only two more pages are needed, for data, for each RDC in the TEX system, compared with the 15 pages required by the traditional method for each added RDC.

Since the unique context for each concentrator task is only two pages, and since these pages are accessed very frequently, it is very likely that the TEX dynamic buffering scheme will keep these pages in core, so that the time taken to switch from one RDC task to another in sequence will be less than 0.5 milliseconds (Table 2). Such a rapid response is essential for real-time operation.

By comparison, the nonshared approach is faced with an unpleasant dilemma. Permanently storing 90 pages in core, almost 200 kilobytes of memory, is an expensive proposition. But if the pages are stored on disk, a less costly approach, then the switching time for each file in the task rises to about 50 milliseconds, which is intolerable for a real-time system. One way to reduce the use of core memory and switching time between pages in a nonshared system is to write one task encompassing all concentrators. But doing this involves complex programing, including multithreading, and the software is not amenable to changes when the network is expanded or otherwise modified later.

Telecommunications executive page-sharing allows several applications tasks to retrieve and update the same file at the same time. Frequently, though, these updates involve changing several pages making up one file. But even with on-line operation all pages cannot be changed immediately. Therefore, it is necessary to guard an information retrieving task against seeing a partially updated, or inconsistent, file.

Traditionally, interlocks provide the protection against inconsistency. But interlocking a file means that while an update is taking place, tasks wanting to simultaneously access these pages are forced to wait. When update traffic is heavy and the files are large, the result is a drastic reduction in throughput. And if more complicated interlocks are used on portions of files, the system can even become deadlocked.

Under TEX, however, updating tasks freeze the file pages they are changing, so that all parallel (simultaneous) tasks see only the last consistent file, not partially updated files. But because the pages will be in core, updating is extremely fast. When the task has completed its updating, TEX invokes a function called foist which causes all updated and consistent pages to be simultaneously released for access by other tasks. Should two tasks attempt to update the same page at the same time, the system rejects the foist for one of the tasks. This task simply goes back and repeats its update which is then accepted; the delay is inconsequential.

Quick start up

Computer-communications systems must be able to recover rapidly and reliably from failures of equipment and software making up the total installation. Traditional methods of checkpoint/restart and of reprocessing the transactions have severe drawbacks in large communications-intensive systems. For example, the larger the system, the greater the amount of transactions accumulated between checkpoint intervals and therefore the greater the amount of buffer required to store these transactions.

The checkpoint method periodically saves the data in core memory and the files by dumping them onto disks or tapes. If a failure occurs, the system is started using the consistent files saved during the previous dump operation and all transactions subsequent to the dump must be sent again. But dumping interrupts normal on-line service for a relatively long time, perhaps several minutes during each checkpoint interval. Maintaining a file of all uncompleted transactions that will have to be reprocessed when the system is restarted imposes no interruptions during normal system operation. However, restarting a failed system from transaction-journal tapes takes perhaps 60 minutes.

The telecommunications executive, however, employs a modified checkpoint method, known as the juncture system, which allows operations to continue during a checkpoint dump (Fig. 2). The key to continuous operation is the system's RAD buffer. During normal operation, page changes are written—through the core—on the RAD. But since the buffering is invisible, the applications tasks will

TABLE 3
PERMANENTLY ALLOCATED CORE PAGES TO HANDLE SIX CONCENTRATORS

FUNCTION	AUTEX SHARING SYSTEM	CONVENTIONAL SYSTEM WITHOUT SHARING
CODES	10 × 1 = 10 PAGES	10 × 6 = 60 PAGES
DATA	2 × 6 = 12	2 × 6 = 12
INTERTASK COMMUNICATION	1 × 1 = 1	CAN'T BE DONE AT ALL
FILES	0 (CAN BE MAPPED AS REQUIRED)	3 × 6 = 18 (MEMORY SPACE MUST BE PREALLOCATED FOR READS FROM FILES)
TOTAL PAGES	23	90

see all changes as though they had actually been written on the disk. Normally, then, the disk serves as a read-only memory.

At periodic intervals, say every 15 minutes, all changed pages on the RAD are identified, the juncture switch is closed, and the changed pages are copied on the disk. The juncture interval for writing may take one minute. Furthermore, those updates made during a juncture are segregated on the RAD without affecting the copying of the old RAD files onto the disk. And the core itself need not be dumped because the core serves merely as a place for mapping the files—all of which reside in and are dumped from the RAD.

The telecommunications executive, the three levels of dynamic buffering, page sharing, and the juncture technique all combine to provide a system that is efficient, highly responsive to large volumes of traffic from remote terminals. ■

Packet switching with assorted computers in a private network

A layered approach toward network design permits data transfer between dissimilar operating systems by packet switching

Stuart Wecker
Digital Equipment Corp.
Maynard, Mass.

Networks

A packet switching network can now be built at a price that private organizations can afford. This network design allows many host computers with different architectures and different operating systems to share the same remote peripherals, data files, programs, and even exchange data developed by programs. Moreover, Decnet, as Digital Equipment Corp. calls this concept, can be retrofitted to many existing systems.

Besides providing rapid data communications among scattered locations, Decnet allows its user to pool many varied computing resources and to eliminate redundant programing efforts. Obviously the design owes much to the declining cost of communications processors and the growing refinement of communications software, not to mention all the experience with Arpanet. For instance, control of the network is decentralized, as with Arpanet, with the network facilities being allocated among the different nodes on a contention basis.

Key to linking dissimilar computers in Decnet is the development of standard interfaces between the different layers (or levels) of the hierarchy of communications functions. For example, a computer or data file or terminal in the user layer has to present the same interface to hardware or software in the layer that manages the network.

The DEC design provides for standard interfaces by distributing the communications functions of a data network among three management layers at each node—a user layer, a generalized network control layer called the logical link layer, and a physical-link layer—with a collection of software functions called the Network Services Protocol (NSP). This protocol enables any two systems to serve each other as if they were being operated only in the confines of their own regimes.

NSP may be integrated into existing DEC operating systems. It works equally well with all computers, enabling the 12-bit PDP-8 minicomputer series, for example, with its conventional busing structure, to communicate with the 16-bit, Unibus-based PDP-11s and the 36-bit large-scale Decsystem-10 computers. What's more, the Decnet principle can be applied to mainframes built by other computer makers by using communications processors compatible with DEC's Digital Data Communications Message Protocol (DDCMP).

The layered structure of Decnet allows replacement of DEC-designed protocols with functionally equivalent protocols of other computer manufacturers. Thus, DDCMP at the physical link could be replaced by IBM's Synchronous Data Link Control (SDLC) for communication with IBM-compatible systems (see "Decnet and its relatives").

The size of the implementation effort and amount of code needed to support NSP depends heavily on the local operating system and the interaction with the user hardware and software. Because of its greater involvement with the operating system, Decnet may span a greater range than the DDCMP of the communications processor. Sometimes most effort will go into creating messages, while at other times the stress will be on the interface between the user layer and the DDCMP module. A typical implementation on a DEC PDP-11 would take 1,000 to 3,000 16-bit words of code.

Interchangeability

The software contained in each layer of the hierarchy of communications functions must be completely insensitive to the data contained in the messages. This data is considered transparent to the layer. To this data, preceding layers have appended synchronization bits, control flags, and other fields (groups of bits) that also must be transparent (Fig. 1). The layer receiving all this information then adds or subtracts its own protocol fields and passes the message on to the next layer.

The layers at each node are assigned in a sym-

Decnet and its relatives

Decnet resembles several well-known network systems, including: Arpanet, the forerunner of many of the new networks; Telenet, a commercial network; Canada's Datapac packet switching network which will employ a new protocol called SNAP (Standard Network Access Protocol); and Cyclades, a French network.

All four of these distributed networks give their users the same basic capabilities—they all allow connections to be made and data to be exchanged with input-output devices and programs running in the system. Some have more flexibility than others, some have simpler interfaces, but all use the layered approach.

At the higher levels of network hierarchy, the functions are divided somewhat differently in these systems. Arpanet, for instance, distinguishes between a host computer and an interface message processor (IMP), but in Decnet no such distinction is made.

Because many private networks may consist of only four or five minicomputers, hosts in Decnet act as their own switches. To add switching computers to a small net unnecessarily increases its complexity and cost. For large networks, on the other hand, switching computers can be added to remove the routing and line handling functions from the host system. With such a design, there is no need for a special protocol between host and switch as in the Arpanet with its IMP-to-host protocol, Telenet (virtual connection protocol), and SNAP (Datagram control procedure). These protocols include many of the functions performed by NSP. In these networks, a number of protocols are usually combined to perform functions equivalent to NSP. For example, in SNAP, the Datagram Service and Virtual Call Procedure perform the NSP functions, while in the Arpanet, the IMP-to-host protocol and the host-to-host protocol do the same.

A number of computer manufacturers, including IBM, Hewlett-Packard, and Modular Computer also have designed network hierarchies. Unlike the communications networks previously described, the details of most of these systems have not been made public.

metrical fashion so that the software of the last layer at the transmitter corresponds to (and thereby interfaces with) programs based on the same protocol and forming the first layer at the receiver. Likewise the next to last layer at the transmitter matches the second layer at the receiver, and so forth.

Each node of the network, whether terminal, host computer, file, or whatever, must be designed so that like layers process only the protocols of like layers. If this principle can be maintained, then for example, DEC's DDCMP—the outermost layer in a Decnet design, which transmits the message between network nodes—can be readily replaced by IBM's SDLC at each node, with few interfacing problems or changes to other layers at the node.

The layered operation of Decnet is illustrated in Figure 1. First the user generates the message at a local device such as a teletypewriter, or else data is read out of a file or produced by the user program. This is called the user or conversation layer. The user message might include groups of parity and synchronization bits for device control.

The next layer, called the logical link, manages the routing of the message, and controls traffic. The message is broken down into packets—that is, segments of manageable lengths—and the segments are routed through the network to the selected destination by a variety of paths, as dictated by a routing algorithm, with the segments treated as separate messages. Because of the differing paths, the segments can arrive out of sequence, but they are reassembled in the proper order at the endpoint by NSP, according to message and segment numbers they contain.

Dividing the message into segments or packets and switching these packets into different paths, depending on which are the best available at the time of transmission, prevents long messages from causing bottlenecks. The Network Services Protocol (NSP) provides Decnet with the means for supporting packet switching.

The NSP software segments the message, adds its own control fields, and selects the next node of the

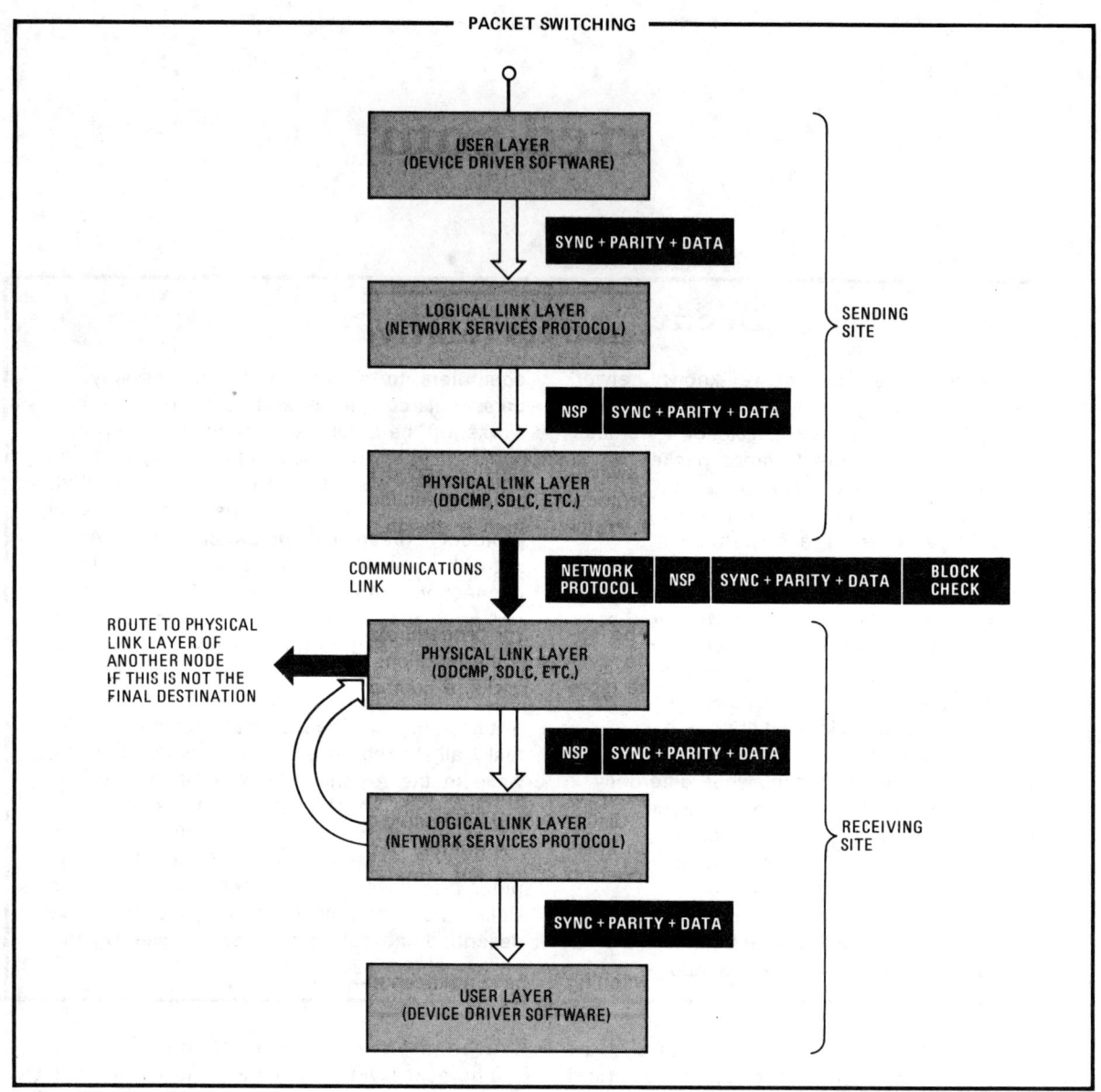

1. Interchangeable layers with clean interfaces permit different computers to be located at the various nodes in the network.

path to initiate packet switching. The control fields appended by NSP consist of a routing header to indicate the source and final addresses, the message and segment numbers, and supplementary information describing the message characteristics, its priority, and the means by which it should be processed. Also produced by NSP are "handshaking" messages for network control such as: creation of a conversation path, requesting the source node to send a message, message acknowledgment, link status information, error detection (but not retransmission), exchange of routing information, interrupting links, and breaking a link.

The data and control messages are then processed by the physical-link layer, which also is locally resident. This layer establishes a communications link to the first node in the path if the message is to make intermediate stops or else connection to the destination if the routing is direct. The physical-link layer also handles error control and recovery in DDCMP. The physical link adds a network-protocol field to the front of the message it receives and a block-check field to the end.

At the receiving node, the message is checked for errors by the physical link, and then the NSP routing header is checked by the logical link to see if this is the final destination. If the message is slated to go elsewhere, a routing algorithm in the logical-link layer passes it to another location, through the physical-link layer. If this is the final location, then the network protocol is stripped away and the message is passed to the user layer for

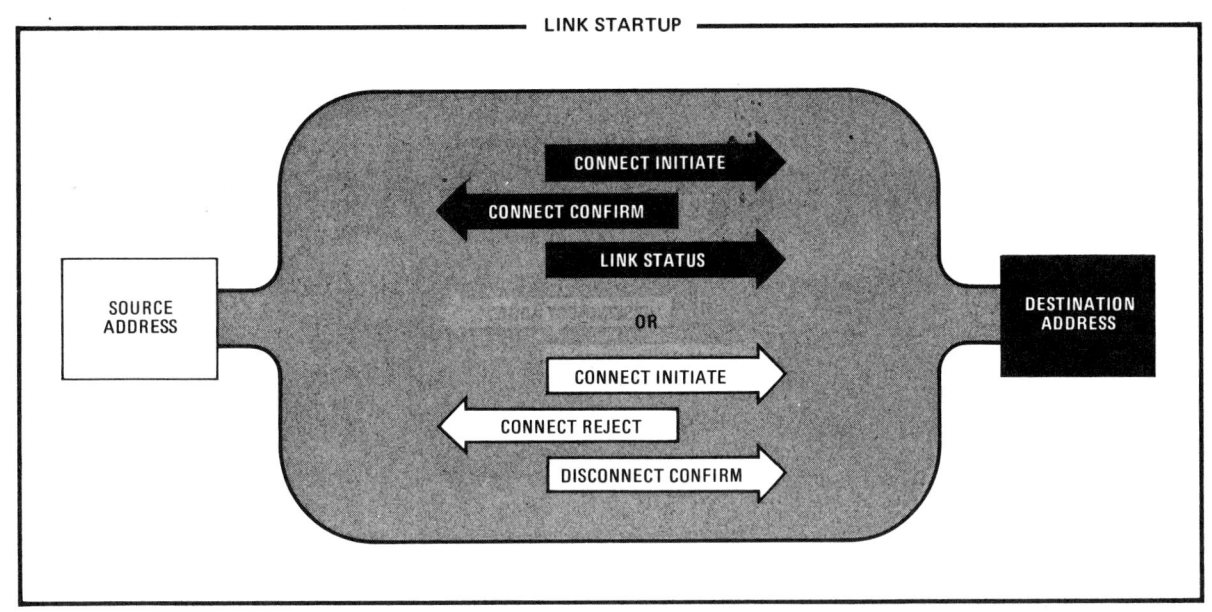

2. A logical link may be established by the message sequence indicated by the upper group of arrows or else the link may be rejected by the lower group.

readout, storage, or other forms of processing.

The various layers may reside partly or wholly in the host computer, and/or communications processor, or terminals. The specific arrangement for a given node depends on its hardware and software configurations. However, to assure clean interfaces, each layer consists of specific, standardized functions, which must interact with each other in a prescribed manner regardless of where they are, which operating system they are part of, and which executive software is used to invoke them.

The only significant disadvantage to the layered approach is some loss of efficiency, since some functions such as error detection and message counting occur at both the NSP and DDCMP levels. However, the additional overhead must be balanced against the complexity and expense of modifying a monolithic system to accommodate different computers and operating systems.

The operation of DDCMP has already been covered in DATA COMMUNICATIONS, Sept./Oct. 1974, pp. 36-46. But the NSP software needs explanation.

Getting from here to there

NSP is mainly concerned with establishing paths and managing the network in such a way that the data packets arrive at the right destinations. The data paths that NSP establishes are called logical links, because it defines them simply in terms of their endpoints at particular network nodes. NSP is not at all concerned with managing the actual physical connections, the job of DDCMP.

The network user has five commands with which to control NSP: connect, transmit, transmit interrupt, receive, and disconnect. The connect command directs NSP to create a logical link between the source and destination devices—say, a teletypewriter and a PDP-8 host; the transmit command initiates the transmittal of data; the receive command authorizes the addressed node—here, the PDP-8—to accept incoming data; and the transmit-interrupt and disconnect commands terminate the message exchange and disable the logical link. Callup of all but the transmit command causes a control message to be developed; the transmit command causes a data message to be sent.

The control messages serve to create the link and to exchange operating data and status messages between the NSPs of the logical-link layers of the nodes involved—here, the teletypewriter and the PDP-8. Along the communication path, the combined control and data message may pass through several intermediate nodes.

Establishing a logical link

When a connect command is issued at a hardware device or by a software process at a network node, the operating system at the node accesses NSP, which produces a connect-initiate message to be transmitted to the destination address by the physical-link protocol (Fig. 2). The message contains the names of the user and/or software process and/or hardware device originating the message exchange, plus any additional pertinent information such as device characteristics and passwords (to identify authorized network users).

The destination address receives the connect-initiate message, and the user or system there decides whether or not to answer. If he or it decides to proceed, a connect-confirm message is returned to the source. If not, then the destination address initiates a connect-reject command, which results in a

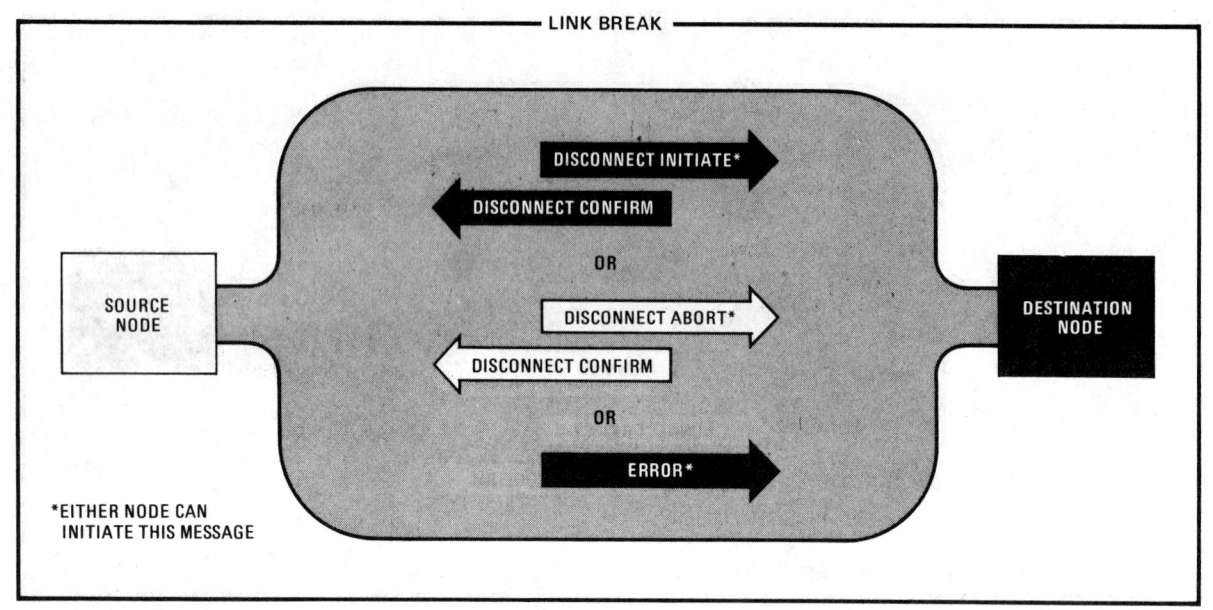

3. The normal sequence for ending a message exchange is shown in the top pair of arrows. The others are used for errors or interrupts.

reject message being returned to the source and the logical link being broken. Reasons for rejecting a message exchange, besides those of security, are detection of an error in the connect-initiate message, a queue at the node, or an incorrect address.

The final step in establishing a connection is for the source to return a link-status message to the destination node in acknowledgment of the connect-confirm message. The acknowledgment is needed to inform the destination node that the return link is established.

Message exchange

The purpose of traffic control is merely to minimize buffer occupancy throughout the network by assuring that the receiving device has a buffer available to store the message. To assure efficiency, the link-status message (when used for traffic control) does not have to request one message at a time but instead can request a number of messages in the request-count field. Another field acknowledges messages received earlier.

In multinode networks the receiver waits a prescribed period of time for a response after sending a link-status message. If nothing arrives, the receiver repeats its link-status message.

When data transfers are only intermittent, an interrupt message may be used to alert an intended receiver that data is forthcoming. This message replaces the transmit message and the link-status return. Since data transfer is intermittent, it is assumed that receiver's buffer is not filled. However, the link-status message is still used for acknowledgment of received messages.

In all types of networks an interrupt message also is used to interrupt normal link operation to inform the receiver of an unusual condition. Because of its immediacy, the interrupt message does not have to be preceded by a request in the form of a link-status message, even for links in continuous operation. The message is received outside the normal buffering process, usually through a local interrupt operating mode.

Messages are segmented if they exceed 576 bytes or some other preassigned value. Each data message may be divided in up to 16 parts, for a total of 9,216 bytes per message. Message acknowledgment and retransmission are performed only after all the segments comprising a message have been assembled at the destination node.

Receipt of a defective message causes the return of an error message. There are three categories of errors. The error message indicates the appropriate category and then lists the specific type of error. Appended to the error message is all of the erroneous message or else its header. The categories of error conditions and specific error types are:
- Message flag error—routing or segmentation is not supported or message count is erroneous.
- Message routing error—destination node does not exist, path is out of service, message queues are full, routing header is invalid, or sequential delivery is impossible.
- Errors in other parts of the message—control message type is not supported, numbering is in error, request count is negative, address invalid, type of connection specified in the message is not supported, or interrupts are too long.

Routing packets

When the routing header of each packet enters a node, it is examined by NSP to determine whether

NSP MESSAGE FORMATS

MESSAGE PART	FIELD DESIGNATION	MEANINGS OF BIT POSITIONS
MESSAGE FLAG	SAME	0 – ROUTE HEADER PRESENT 1 – DESIGNATES CONTROL OR DATA MESSAGE 2 – 3 – MESSAGE HANDLING REQUIREMENTS 4 – (SEE NOTES 1 AND 2) 5 – 6 – RESERVED FOR FUTURE USE 7 – RESERVED FOR EXTENSION
ROUTING HEADER	ROUTING FLAG	0 – MESSAGE PRIORITY LEVEL – 01 FOR DATA MESSAGE, 1 – 10 FOR INTERRUPT, AND 11 FOR CONTROL 2 – DELIVER IN PROPER SEQUENCE 3 – DELIVERY OR RETURN NOT MANDATORY 4 – TRACE THIS MESSAGE 5 – THIS A TRANSMISSION 6 – RESERVED FOR FUTURE USE 7 – RESERVED FOR EXTENSION
	DESTINATION NODE	1 OR MORE BYTES DESIGNATING THE DESIGNATION NODE
	SOURCE NODE	1 OR MORE BYTES DESIGNATING THE SOURCE NODE
DATA MESSAGE (INFORMATION)	DESTINATION ADDRESS	1 OR MORE BYTES DESIGNATING A SPECIFIC DEVICE OR PROGRAM AT THE DESTINATION NODE
	SOURCE ADDRESS	1 OR MORE BYTES DESIGNATING A SPECIFIC DEVICE OR PROGRAM AT THE SOURCE NODE
	MESSAGE NUM	2 OR MORE BYTES INDICATING THE MESSAGE NUMBER AND THE NUMBER OF THE SEGMENT CONTAINED IN THE MESSAGE
	USER DATA	1 OR MORE BYTES OF THE INFORMATION BEING TRANSMITTED
CONTROL MESSAGE (INFORMATION)	COUNT	1 BYTE INDICATING THE NUMBER OF BYTES IN THE CONTROL MESSAGE
	TYPE	1 BYTE DESCRIBING THE TYPE OF CONTROL MESSAGE
	DATA	1 OR MORE BYTES COMPRISING THE CONTROL MESSAGE

NOTES:
1. DATA MESSAGES
   - BIT 2 – MESSAGE NUMBER PRESENT
   - 3 – DESIGNATES INTERRUPT MESSAGE
   - 4 – ACKNOWLEDGMENT REQUIRED
   - 5 – LAST MESSAGE SEGMENT

2. CONTROL MESSAGES
   - BIT 2 – DESIGNATES NUMBERED OR UNNUMBERED LINK
   - 3 – ACKNOWLEDGES INTERRUPT IF SET IN LINK STATUS
   - 4 – COUNT FIELD PRESENT

DATA MESSAGE FORMAT

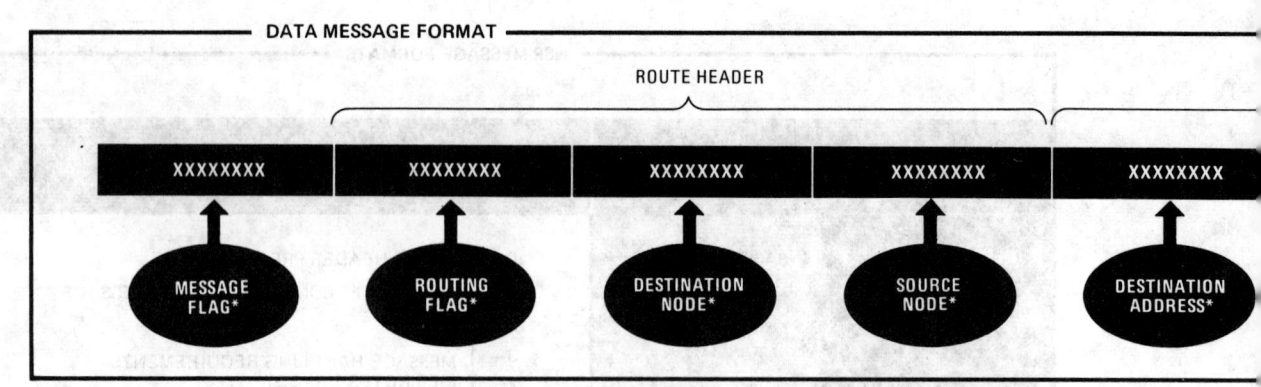

```
         ┌─────────────── ROUTE HEADER ───────────────┐
 XXXXXXX │ XXXXXXX │ XXXXXXX │ XXXXXXX │ XXXXXXX │ XXXXXXX
    ↑         ↑         ↑         ↑         ↑         ↑
 MESSAGE   ROUTING  DESTINATION  SOURCE            DESTINATION
  FLAG*     FLAG*      NODE*     NODE*              ADDRESS*
```

4. The user data field contains the information being transmitted, along with any protocol or synchronization bits added by preceding layers.

or not the node is the final destination. If not, NSP executes a routing algorithm, which associates the destination address with a physical link to the next node along the route, one step closer to the final destination. A properly designed group of algorithms provides for the packets to move over reasonably direct routes, avoiding bottlenecks in the system. For reasons of efficiency, a routing header is not used when the source and destination nodes are directly connected via a physical link.

The routing algorithms range in complexity from simple table lookups to calculations for finding the optimum path for each packet. Therefore, the difficulty lies in deciding which node to send a packet to. The other aspects of routing—receiving a message, checking the header, and sending the message on to another node—are rather straightforward. The specific routing technique used is decided by the user, and it is in no way inherent in the NSP design.

The simplest form of routing is done where there is only a single physical path to a destination node, as in simple hierarchies. In such a case the routing information may be stored in a table that relates the destination address to the physical link. Backup paths also can be listed in case the usual path is busy or out of service. The routing table is usually entered along with the other system software and changed by a programer or operator when the need arises.

For networks that are prone to change, an optional message type, the routing message, permits the routing tables to be automatically updated from a single neighboring node. Other routing algorithms may require additional message types for proper network updating. Good network management requires that routing messages be promptly exchanged between neighboring nodes whenever there is a change.

When the traffic between two nodes exceeds the capacity of the lines between them, several paths may be taken simultaneously. Multiple logical paths can be implemented with NSP by assigning multiple addresses to nodes and different routing paths for each address. Also, multiple physical links can be developed with DDCMP, using a single NSP logical group address.

Destruction of logical links

Logical links may be terminated in three ways—upon request from either the source or destination address, by failure of a communications link, or by failure of an intermediate node. Three types of message sequences are used to terminate a link (Figure 3). One, the normal ending to a message exchange, occurs when either node sends a disconnect after the last message has been processed and its partner replies with a disconnect confirm. The disconnect message contains a field for sending any supplementary information relating to network operation, such as the next time a connection will be established between the two nodes.

Another type of termination occurs when one node sends a disconnect abort. This message is not part of the normal message exchange, but instead is called up for some external reason such as a suspected breach of security or a failure in a connected device or process. No supplementary information is included. The response from the opposite node is a disconnect confirm.

Finally, if there is a breakdown in the path, the link will be disconnected by the return of error messages from the node where the physical link was broken.

Optional messages

All of the messages described so far, except routing path, are necessary for operation of NSP. However, there are other message types (including routing path) that can be included at will in NSP to transmit information on the characteristics of nodes, status information for network management, and network diagnostics. These messages are:

■ Request configuration—updates local files that store information on the characteristics and addresses of other nodes.

■ Configuration—is returned in response to a request-configuration message. It lists node name,

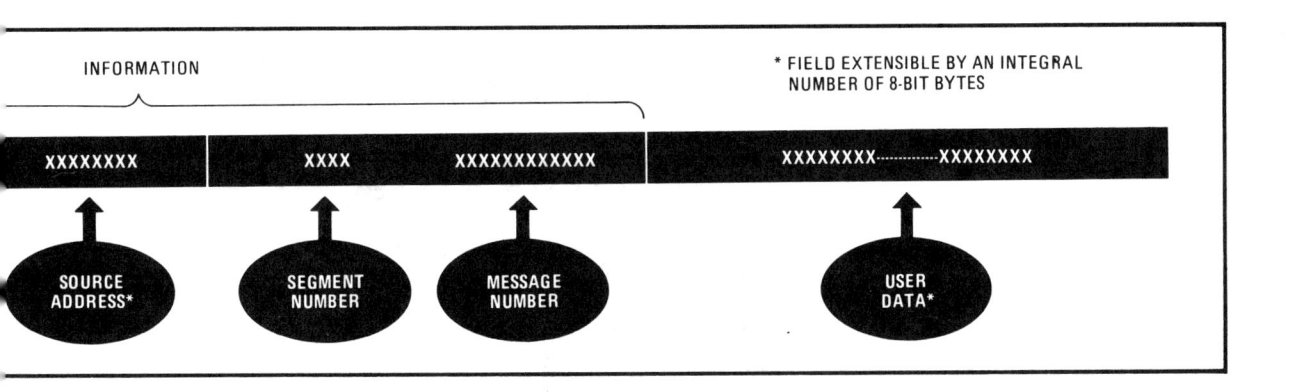

```
    INFORMATION                                    * FIELD EXTENSIBLE BY AN INTEGRAL
                                                     NUMBER OF 8-BIT BYTES

  XXXXXXX      XXXX     XXXXXXXXXXX        XXXXXXX----------XXXXXXX
     ↑          ↑            ↑                       ↑
  SOURCE     SEGMENT      MESSAGE                  USER
  ADDRESS*   NUMBER       NUMBER                   DATA*
```

address, and the NSP features supported at the node, along with the particular versions of the operating system in use, the NSP version, and the DDCMP or other message protocol. A field for optional information also is provided to permit any other significant details to be noted.

■ Request link status—is used on logical links to request the return of a link-status message for the purpose of updating acknowledgment and request information.

■ No operation—tests the physical-link protocol and device drivers at another node without affecting the NSP layer at the site. The message is routed as specified in the routing header and is discarded at the destination node. If it is successfully received, then the device drivers and physical link protocol are judged to be operating properly.

■ Echo—is used for a loopback return from the destination node.

■ Echo reply—is returned by the destination node to the source node sending an echo message. The reply message repeats the data field of the echo message, and the returned data is checked back at the source to determine whether it has been altered in transmission.

■ Trace—records the time at which a message is transmitted from one node to another called a trace node. Software at the trace node performs analysis of the network performance and the routing method.

Message formats

All messages, whether optional or not, consist of three basic parts: the message flag, which identifies the message type and its characteristics; routing header, which provides routing information; and the message itself—either information or control commands from the users. As indicated in Figure 4, each part of the message is composed of fields. (The meanings of individual bits are listed in the table, "NSP message formats," page 61).

The message-flag field lists the characteristics of the message: whether the message is a control, interrupt, or data message, and details of how the message is to be handled at the destination.

The routing-header portion of the message contains a group of fields used only when the nodes are not directly connected by a physical link, so that there is more than one possible path that the message can take. The routing-flag field indicates the message priority, whether it is part of a sequence, whether this message should be traced for network testing, and whether this message is a retransmission of an earlier message. The source- and destination-node fields indicate the respective local sites in the network.

In the information section of a data message, the destination and source address fields call out particular software, files, or I-O devices at the nodes designated in the routing header. They are therefore more specific than the similarly named fields in the routing header.

The next two fields in the data message contain the number of the complete message to be sent by the source and the segment represented by this particular data message. The message-number field can hold 4,096 message numbers, and the segment field can identify 16 segment numbers. The final portion of a data message contains the user data and the control bits generated at the terminal, computer, or data file, in 8-bit groups.

A control message differs in format from a data message in that the information section contains three basic fields: a count field indicating the number of bytes in the message, a type field that specifies the type of control message, and a data field that presents the details of the control message. The data field terminates a control message.

An important feature of NSP is that it allows for expansion of many of the fields to provide for future, more complex networks and network-management software. For example, the fields containing the sources and destinations can be extended beyond one byte, and therefore they can reference more than 128 addresses. To implement this feature, the last bit of the extensible address field, ordinarily a zero, is set to change the next field into an extension of the present field. ■

The user's role in connecting to a value added network

Once a user opts to buy services from a value added network, it's necessary for him to do some initial work, mostly some reprograming

Richard B. Hovey
Telenet Communications Corp.
Washington, D.C.

Technology

Early in 1975, data started flowing at 56,000 bits a second over the circuits of a unique nationwide common carrier service called a value added network. This new approach to data communications combines old and new transmission facilities and adds to them a form of intelligence to improve the performance.

The value added network (VAN) is different from present data transmission services and from private data networks in both the enhanced and extensive offerings to users and the sophisticated technology it employs. The technology, called packet switching, makes it possible for the value added carrier—the implementer and operator of the VAN—to provide any user, large or small, with the kind of fast-response, error-free, low-cost-per-transaction data transmissions now available only to companies that have invested in their own large private networks.

In essence, the value added carrier (VAC) takes advantage of the substantial economies of scale resulting from one very large network— fully utilizing such expensive resources as transmission lines and concentration equipment by sharing the network among the VAN's subscribers. The VAC passes on a portion of the consequent savings to the individual user-subscribers through a tariff charge based mainly on traffic volume.

Beyond the simple economics, leasing existing communications facilities allows the carrier to obtain just as much transmission capacity for each location as is required by the traffic load. This provides the flexibility to adapt quickly to subscriber traffic and geographical demands, and permits the incorporation of new transmission offerings—such as satellites and AT&T's Dataphone Digital Service— as they become available.

Conceptually and technologically, VAN's have their origin in the Arpanet, a nationwide consortium of computers at numerous research centers tied together over a packet switching network. However, the Arpanet is operated in behalf of the government to support research activities of various Federal agencies, not as a common carrier facility.

This lack of availability of VANs to individual user-companies in commerce and industry was remedied in 1973 when the Federal Communications Commission approved the concept of value added networks, determined that VACs should be

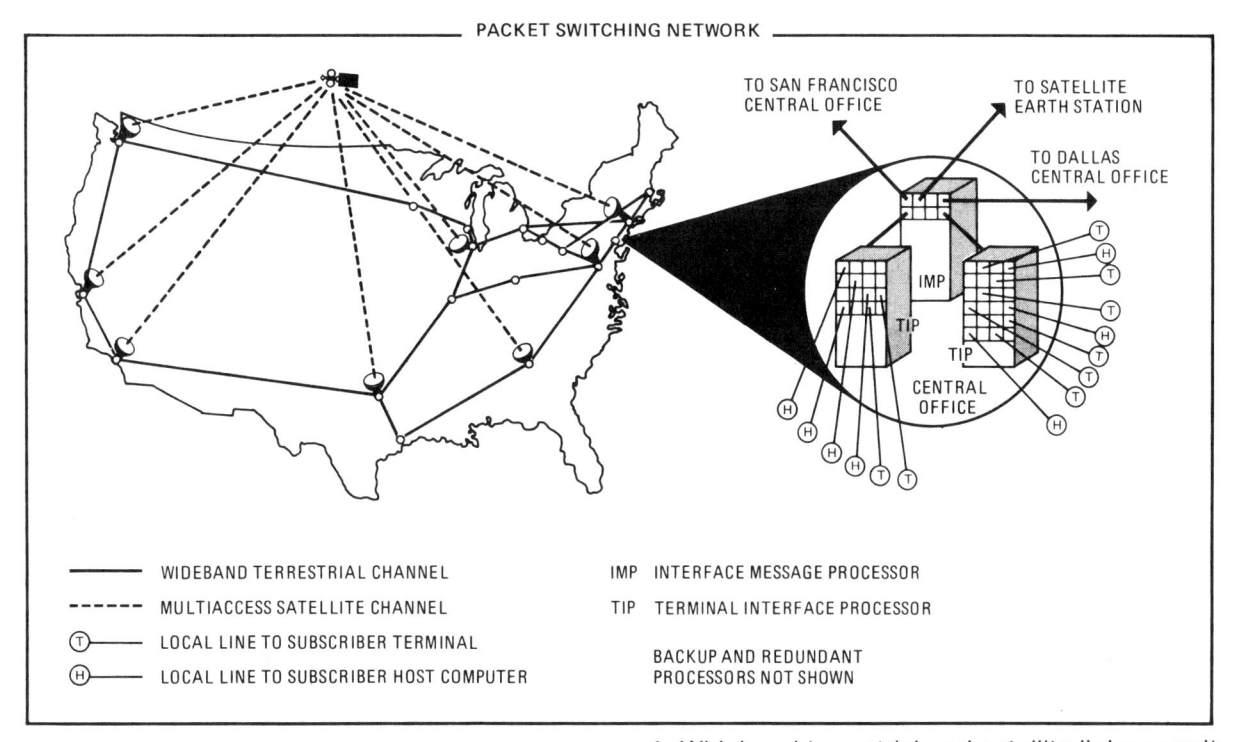

1. Wideband terrestrial and satellite links permit different packets of bits in a subscriber's message to take alternative paths from source central office to destination central office.

regulated as common carriers, and declared an open entry (non-monopoly) policy permitting potential public network operators to propose a value added network so long as they applied for FCC approval. Thus, VACs will be regulated, but will operate in a somewhat entrepreneurial and competitive environment. Telenet Communications Corp. received FCC authorization as a VAC on April 16, 1974.

In addition to FCC's advocacy, the viability of the VAC concept was enhanced when AT&T amended its FCC tariff 260, for private voice-grade and wideband lines, to permit VAC's to "resell" the transmission channels they leased from AT&T to VAN user-subscribers. More recently, AT&T specifically included provision for lease of dedicated lines to VACs for resale in its new FCC tariff 267 for Dataphone Digital Service. Furthermore, specialized and satellite carriers have agreed to provide wideband transmission facilities to VAC's.

The FCC has authorized a VAC to provide the data communications network required to connect a user's terminals and computers. Specifically, connections can be made from terminal to computer, computer to terminal, and computer to computer. Transmission will be available to users at all common speeds, ranging from that for the slowest teleprinters up to 56,000 bits per second (b/s) for high-speed data transfer between two computers.

A value added carrier such as Telenet, however, will not supply data processing services nor will it provide hybrid data processing involving both data processing and data communications services. What a VAC will do is simply permit data transfer between any two or more dynamically selected user stations, where a station is defined as a data terminal or host computer. To accomplish this, the VAC leases long-haul wideband lines and invests in computerized interface and switching equipment, high-speed modems, diagnostic facilities, and the like to construct a network. The VAC programs the computerized interfaces to provide such services for customers as code conversion, speed conversion, and error detection and control. In addition, the computers at the VAC's monitoring centers gather data for user traffic statistics and billings and provide other network related services.

A prospective user of a VAN service must consider two areas: the technical and operational tasks and responsibilities which are the factors discussed here; and the possible impact of new communications services on the user's operating and organization structure. For although immediate cost savings may be achieved, of greater long-run significance is the new flexibility users will have for accommodating growth in existing systems and for implementing new remote and multi-computing applications based on overall communications needs and not on present technical and economic constraints.

As mentioned, the user must make a certain technical and operational effort to be able to match up with a VAN. The amount of effort is quite small if the user just wants to connect a popular terminal; more effort, mainly some reprograming, may be required to connect a host computer. That is, while the VAC operates the network, for the subscriber to employ the network properly is, at least during startup, a joint project involving both the VAC and the user. These points will be detailed following this description of how the packet-switched VAN itself goes about providing data communications transmission services to a multitude of diverse users.

In time, several VANs may be in operation, so users will have some choices in selecting competitive and alternative services. It appears likely that all VANs will use packet-switching technology and all will be configured in substantially the same way—but each, certainly, with some technical and tariff differences. The Telenet network should serve, then, as representative of what will be available in the next few years (Fig. 1). An 18-city network is expected to be on-line by the end of 1975. Telenet has leased-medium- and high-speed lines from transmission carriers to carry data traffic among host computers and terminals. In addition, transmission capacity will be leased from domestic satellite carriers to serve both as primary wideband channels and secondary back-up circuits to the terrestrial lines.

One or more central offices are to be located in each of the cities functioning as nodes in the network. Note that each node city has two or more (full duplex) lines connected to it.

This redundant access to a central office serves several useful purposes. For one thing, multiple lines permit traffic to be simultaneously routed over parallel channels between source and destination central offices. For another, multiple access lines permit traffic to immediately reach its destination via an alternate route in the network should one of the lines become degraded with excessive noise or go out altogether.

Packet switching nodes

But most intriguing from a technical and operational viewpoint is that different packets comprising one message may be delivered along different routes. For example, suppose a user-subscriber located in a suburb of Seattle wants to transmit an 8,000-bit message to Houston. The message is delivered to the VAC's central office in Seattle over a dial-up or leased-line, and is formatted into eight 1,000-bit packets in high-speed core memory by a special processor there.

The first packet is released from buffer and, for example, travels to and through the central offices in San Francisco, Los Angeles, Dallas, and on to Houston—being error-checked over each hop of the journey and buffered at Houston to await the arrival of the other seven packets.

The second packet, however, might have to take a different route if the line to San Francisco is busy at that moment. In such a case, the packet could take the route from Seattle to Houston, via Minneapolis, Milwaukee, Chicago, St. Louis, and Dallas. And so on for the other packets in the message.

Once a message is accepted by the network, the VAC assumes responsibility for error-free transmission. Therefore, at the sending office the buffer for a packet is not released until the next receiving office acknowledges correct receipt. If received in error, the buffered packet can be retransmitted. When all packets have arrived without error at the buffer in Dallas, the message is released from the central office to the recipient terminal or computer. The recipient gets the packets in the same order they were sent.

Although the original message may have started out, say, at a 4,800 b/s message rate, once it enters the network it is transmitted at 50,000 or 56,000 b/s. Even including electrical propagation, queuing, and acknowledgement delays, a packet will proceed from any source office to any destination office in, on average, one-third of a second.

Central office interfaces

The inset in Figure 1 shows the two major pieces of equipment in a Telenet central office, the terminal interface processor (TIP) and the interface message processor (IMP). Each TIP has hardware/software ports to accept data from a user's terminal or host computer. Terminal data can be a character, a block of characters, or some segment of a long, con-

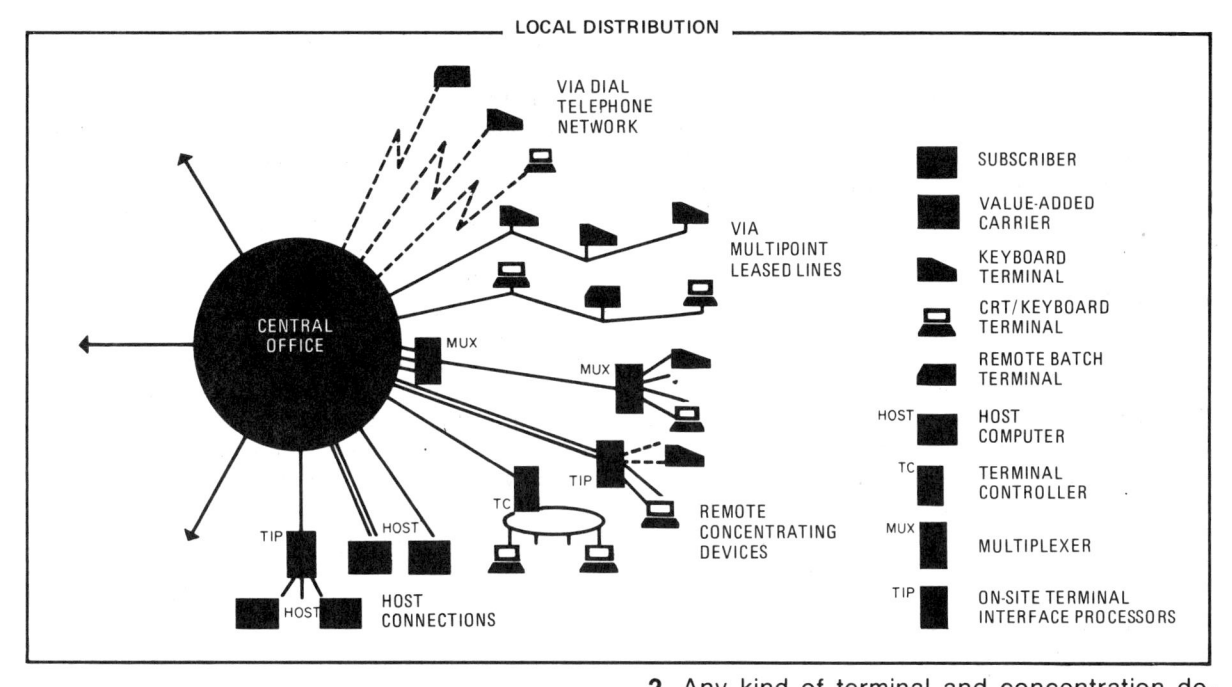

2. Any kind of terminal and concentration device can have access to a value added network's central office over dial-up lines and point-to-point and multipoint private lines.

tinuous message. Even though data from the several terminals may be in different lengths, codes, and speeds, the TIP formats the output data stream into the standard packets used for internal network transmission.

At the user's option, all data entered from a character-oriented terminal attached to the Telenet network can be translated into a single, well-defined 'virtual terminal' format by the TIP. Conversely, data sent from the host computer in 'virtual terminal' format is translated back into a form compatible with the terminal. This mechanism can remove much of the burden of code conversion and terminal support from the host computer, which supervises the single 'virtual terminal' type to communicate with the diverse terminal models connected to the network.

The number of TIPs actually located in each central office depends on the expected number of terminals and host computers being serviced. However, each central office will have at least two TIPs, each handling a share of subscribers even though neither one may be fully loaded under normal flow conditions. If a TIP should go out of service, either on a planned or emergency basis, all its connections to users can be instantly switched to back-up interface equipment.

The terminal interface processor are multiply connected to IMPs which route the standardized packets of data over the optimal long-haul links. The technique of dynamically routing packets individually along one of several alternate paths is used to minimize end-to-end transmission delay, spread traffic evenly throughout the network, and increase reliability. At each TIP and IMP along a route, packets are checked for errors and, if necessary, retransmitted. Flow-control techniques within the network (invisible to users, as are the packets themselves) ensure that the fast-access core memories of the processors do not become overloaded while maintaining high channel utilization. Therefore a packet cannot get lost for lack of buffer capacity at any node.

Each interface message processor and terminal interface processor periodically reports observations about itself and its environment to network monitoring center processors. These computers watch the instantaneous state of the network—warning of network components whose capacity may need to be increased and initiating remote diagnostics and repair activity when necessary. The system is designed such that the failure of an individual component can be immediately detected and its tasks simultaneously absorbed by one of its functionally redundant counterparts without interrupting service.

Interface message processors and terminal interface processors are stored program processors. Because they employ software, not hardwired logic, new programs can be readily added to permit the processors to interface with new types of terminals and host computers and provide new terminal support functions. In fact, new or modified programs can be sent over the network itself from a central location to all, or selected, IMPs and TIPs to make a new service immediately available to all subscribers without having to halt the network.

User terminals can reach the central office in a number of ways, including dial-up lines, multipoint, and point-to-point leased-lines, and using such concentrating devices as multiplexers, remote data concentrators, and terminal controllers. In some instances, VAC-owned TIPs may be installed at the user site (Fig. 2). The lines and devices in color represent items supplied by the subscriber and those in black designate those supplied and/or under control of the VAC.

Terminal access

The criteria for choosing either a dial-up line or a dedicated line to access the central office are substantially the same as those employed in configuring private data communications networks. In short, such factors as transmission speed, acceptable busy-signal incidence, response time, volume of traffic, length of individual transmissions or transactions, and whether line use is substantially continuous or mostly occasional, must be considered. Because these lines extend only to a local central office, cost is less of a determining factor than in an extensive private network.

In addition to establishing a physical connection over a line to the VAN's nearest central office, the terminal must also establish a logical connection through the network to some destination computer. For data terminals, the logical connection can be established in either of two ways depending whether the terminal has a dedicated or dialed connection to the central office.

When a terminal on a dedicated line is merely required to converse with one particular computer on the network, that terminal's TIP port may be pre-initialized to the desired transmission parameters and host computer address. When the terminal is switched on-line it will be automatically connected to the preselected computer in less than one second and the computer may respond by printing out its own name, such as MH COMP, on the terminal's display. Then the terminal-computer-terminal dialog can proceed just as if there were a direct physical circuit between the two station sites.

For a dial-up terminal, the process is similar, except that the operator may first have to provide a short command identifying his terminal model and a second command specifying the host computer with which it is to be connected.

Simple commands

In the majority of cases, the terminal operator will be concerned with, at most, the command that defines his terminal model and the commands required to establish and break the connections to host computers. From time-to-time, however, some users may wish to communicate other instructions directly to the TIP. This is accomplished by issuing additional commands, most of which set a transmission parameter or mode. For example, it is feasible for the operator to change the speed at which an asynchronous terminal sends and receives data, assuming, of course, that the terminal and the associated modem can support the selected transmission speed.

In addition, a VAN quite efficiently provides a number of simple functions beyond pure data transfer. For example, local editing is available for unbuffered typewriter-oriented terminals and displays. Data entered on these terminals is normally accumulated by the TIP prior to being forwarded to the destination host computer for processing. Before the buffered data leaves the TIP, it may be edited by the operator on a character or line basis, typically to correct keyboard input errors. Again, this editing facility can be preset or the user can issue short commands that define characters which when typed will delete the last preceding character from the data just entered or all the data in the current line of input.

These and other functions are strictly optional; their ultimate utility may depend on the nature of the user's particular application. In short, some terminal users will want to become familiar with a limited repertory of commands and facilities. For others, the network interface can be pre-initialized to exactly those parameters and facilities required, eliminating the need for any commands during routine usage. In either case, the user can adapt the network to his needs with ease and little training.

Normally, the VAC will build a software module into each TIP's program for each class of terminal. For popular terminals this software is part of the VAN service. Therefore, at the outset about the only concern a user may have is to make sure the VAN can handle his type of terminals. But, if the terminal is very special, or brand new on the market, the user may have to help defray the cost of unique software development.

Computer access

Since a host computer converses simultaneously with many terminals or other hosts through the network the physical connection to a VAN's central office will usually be via a dedicated local point-to-point access line operating at up to 56,000 b/s. (A second access line, perhaps of a lower speed, may be installed for reliability or increased throughput capacity.) There is little a subscriber has to do to obtain a physical connection to the network. But the user must get involved in planning for logical connections of host computers as the VAN takes on more and more of a subscriber's load.

Format, signaling, and error control protocols enable a subscriber's computer to send and receive data over the carrier's system. These protocols must be incorporated in the user's system in the form of software routines. The protocol software may reside in either the host computer, when the

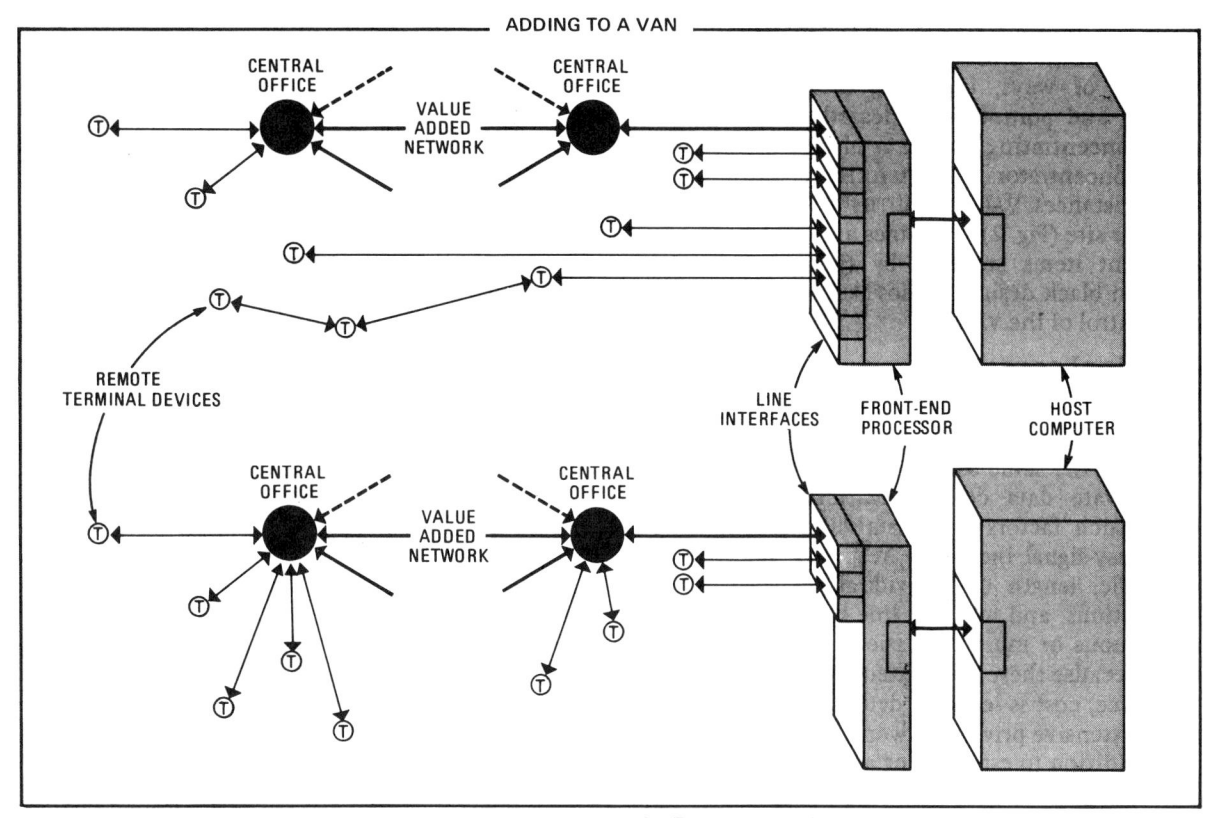

3. By connecting to the nearest central office, users can convert point-to-point and multipoint links into less costly local access links.

interface between the communications lines and the host computer is a hardwired transmission control unit, or in a programable processor that front-ends the host computer. To minimize the effort required to connect a host computer to a VAN, the interfaces have been designed to emulate standard communications devices on a conventional data control links—something the host "already knows about." Thus, the changes required because of the use of VAN links may involve no more than reparameterizing the communications macros in present vendor-supplied packages.

Consider the system configuration at the top of Figure 3. Here, several point-to-point lines and one multipoint line go to a hardwired transmission control unit. The top left most terminals were previously accessed via an expensive, transcontinental line. As a means of initially testing the VAN service, the user has selected to access these two terminals through the network. Therefore, the TIP, in the central office closest to the host computer, is programed to emulate a multipoint, binary synchronous line. (The terminals might actually be on a separate point-to-point line connected to their local central offices.) Accordingly, few changes in the software macros have to be made by the user in the computer's communications software.

System use of the VAN could stop with the assignment of just two terminals to the network. However, VAN utilization can grow to the stage shown at the bottom of Figure 3. Here, all terminals connected to the multipoint line and all dedicated line terminals, except the two closest to the host computer, are connected to nearest central offices. Thus, using the value added network has eliminated several point-to-point lines and one multipoint line with minor impact on user programing. The subscriber also needs fewer line terminations at his central office.

If the subscriber also opts to convert from hardwire transmission control unit to a true programable front-end processor he can remove considerable inefficiencies from the host computer, but at the initial cost of programing the front-end processor. (Such programing costs are not attributable to going on a VAN.) If this change is made at the same time the subscriber integrates his system onto a VAN, he may want to incorporate in it a software interface with more efficiency and flexibility than the emulation interface provides. If a customer wants to connect several computers to one location to the network, the carrier may determine that it is to the user's advantage to install a TIP on the subscriber's premises. With regard to software modifications, the carrier may either contract to develop the software for the user's computer or assist the subscriber in having its in-house personnel develop the software. ∎

Where to spend money to improve system availability

Methodical analysis and a few calculations locate optimum places to invest in redundancy and better maintenance

Dixon R. Doll
DMW Telecommunications Corp.
Ann Arbor, Mich.

Tutorial

A lot of money can be wasted in maintaining a high level of over-all up-time, or availability, in a data communications system. Often it's not clear where to install redundant equipment for backup, or whether to provide for preventive maintenance and fast repair services, or both. Whatever the route, it's necessary to have an orderly method for pinpointing the critical trouble areas if availability-enhancement dollars are to be well spent.

All parties to a system acquisition should agree on the definition and method to be used to compute availability. And because different suppliers are usually assigned responsibility for different equipment in the over-all network, availability contributions must be determined separately for each subsystem, such as terminals, lines, and communications processors, as well as for the network as a whole.

Of course, keeping adequate records of the frequency of failure for each piece of equipment, and how long each failure knocks one or more terminals out of service, is the starting point for calculating availability and for focusing on those areas of the network that will benefit most from availability-enhancement dollars.

The prime operational criterion for a data communications network is that all terminals should always be able to provide service to the using company's remote locations. Therefore, a common way of measuring system (network) availability, a, is:

$$a = (TOH - TNOH)/TOH$$

where TOH is total possible hours of terminal operation per month and TNOH is total hours the terminals cannot operate during the month.

Consider a typical multipoint (multidrop) network in which five terminals are connected over one private line to a central processing unit (CPU) by way of a programable front-end processor, as shown in color in Figure 1. Here the number of terminals that go out of service depends on the nature and location of the failure.

A modem failure, say lack of power, at a remote terminal location takes only that terminal out of service. But the loss of modem power at the central site (or failures of the front-end processor or the CPU) will end traffic flow to all five terminals. In another instance, if the transmitter of the malfunctioning remote modem remains on, then no other terminal on the line can send data. In this case, a single remote-modem malfunction takes out all five terminals.

Multipoint networks spanning long distances usually go through several telephone exchanges connected by interexchange channels (IXC). If a break occurs in an interexchange channel, one or more terminals will go out of service. The average number of terminals that go out per failure, NTPF, on a multipoint line depends on the probability of failure in a given segment of the circuit (P_i) and the number of terminals (NT_i) affected per failure in segment i.

For the multipoint line shown in Figure 1:

$$NTPF = P_1(NT_1) + P_2(NT_2) + P_3(NT_3) + P_4(NT_4) + P_5(NT_5)$$

Here, NT_3 means, for example, that a break in segment 3 will take out three terminals.

If probability of failure is assumed to be equal in each of the five segments, then the failure probability for each segment is 0.2 (The sum of all such probabilities must always equal 1).

Calculating availability

Thus, for the multipoint line in Figure 1, the average number of terminals going out of service due to a random failure in any segment is:

$$NTPF = 0.2(1) + 0.2(2) + 0.2(3) + 0.2(4) + 0.2(5)$$

for an average of 3 terminals per failure.

To calculate the total number of hours that terminals are out of service during a month, the multiple multipoint network in Figure 1 is assumed to have 10 multipoint lines and each multipoint line handles five terminals. If either the CPU or the front end goes down, the failure will knock out all 50 terminals in the network. But failure at a central modem will impact only five terminals.

Suppose there are five outages of the long-haul lines each month, each outage lasting an average of four hours; three outages of remote terminals, each outage lasting an average of two hours; four outages a month for remote terminals, each outage lasting an average of two and a half hours; one outage per month for the front-end processor and the CPU, each lasting four hours; and five outages per month, lasting two hours, on average, for the central modems.

Such information can be consolidated, as in Table I, to arrive at the total amount of terminal outage, TNOH, in the complete network. Here, TNOH is 526 hours. And if each terminal is to be in operation 10 hours a day, 20 days a month, the TOH is 50 x 10 x 20 = 10,000 hours a month. Therefore, from the system availability equation:

$$a = (10,000 - 526)/10,000 = 0.9474$$

Partial availability

Knowing the availability of different subsystems can be important because responsibility for the performance of different portions of the network may be assigned to different parties. These partial availabilities can be determined in a manner similar to

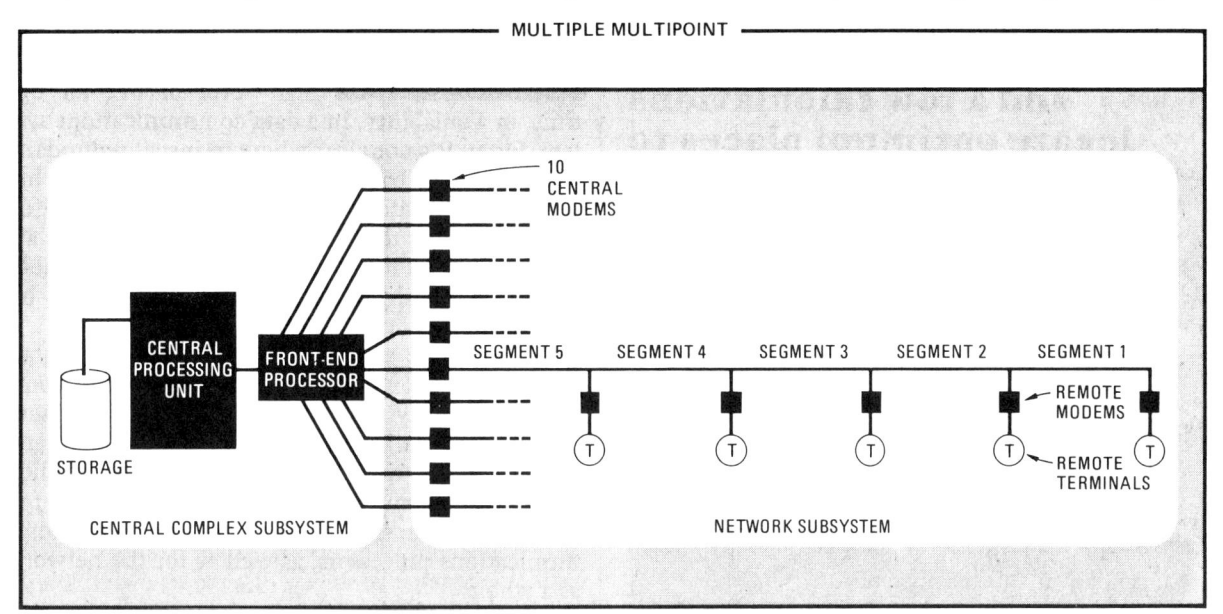

1. Failure of the front-end processor will take five terminals out of service in single multipoint line (shown in color) and all 50 terminals when the other nine multipoint lines (shown in black) are on the multiple multipoint network.

that for over-all system availability.

Figure 1 also shows how the 50-terminal network can be divided into two logical subsystems. Then, a_n is network subsystem availability and a_c is the central complex subsystem availability. Using the failure data in Table I:

$$a_n = (10{,}000-76)/10{,}000 = 0.9924$$
$$a_c = (10{,}000-450)/10{,}000 = 0.9550$$

Thus, in this example most terminal outages come from central complex subsystem failures.

What probability means

Another way of gaining insight to system availability is probability of failure, P, the ratio of the number of failed-terminal-hours to the possible number of operational-terminal-hours. Thus, the concept of this probability invokes the concepts of failure frequency and duration of failures.

In general:

$$P = 1 - a$$

Using the numerical data for the example network, the probability of failure for the total system:

$$P = 1 - 0.9474 = 0.0526$$

which means that, on average, 5.26% of the time during the operating day one terminal in the system is not available for operation.

In the same way, the probability of downtime within a subsystem can be determined by:

$$P_c = 1 - a_c \quad \text{and} \quad P_n = 1 - a_n$$

where P_n is network subsystem failure probability and P_c is central complex subsystem failure probability. Thus, again using the numerical data in Table I, the probability of terminal outage caused by a failure within the network subsystem is 0.0076, and for the central complex it's 0.0450.

In general, then, the probability of failure in the total network, P_t, is:

$$P_t = P_n + P_c = (1 - a_n) + (1 - a_c)$$

Similarly, the impact of terminal downtime, the converse of availability, can be determined for a particular device. For example, using the data in Table I for the CPU:

$$a_{cpu} = (10{,}000-200)/10{,}000 = 0.9800$$
$$P_{cpu} = 1 - a_{cpu} = 1 - 0.9800 = 0.0200$$

On average, therefore, 2% of the terminals in the total network are not available for operation because of a failure within the CPU

Spending money effectively

Now, to locate the most cost-effective places for spending availability-enhancement dollars, refer again to the 50-terminal network described in Figure 1. Here, each additional hour of terminal uptime during a 10,000-terminal-hour month raises the system availability by 1/10,000, or 0.0001. Similarly, each additional of hour of CPU uptime raises system availability by 50/10,000, or 0.0005. That is, each device or subsystem will have its own relative return (depending on how many terminals go down because of its failure) for each additional hour of its operation.

Table II shows eight alternatives for improving uptime in the example system, with the relative return for each contained in the second column. The

---

TABLE I

FAILURE LOCATION	MEAN TIME TO REPAIR (HOURS)	NUMBER OF TERMINALS AFFECTED	NUMBER OF FAILURES PER MONTH	NUMBER OF DOWNTIME HOURS PER MONTH	
LONG-HAUL LINES	4 (AVG)	3 (AVG)	5	60	NETWORK
REMOTE MODEMS	2 (AVG)	1	3	6	
REMOTE TERMINALS	2.5 (AVG)	1	4	10	
FRONT-END	4	50	1	200	CENTRAL COMPLEX
CPU	4	50	1	200	
CENTRAL MODEMS	2 (AVG)	5	5	50	
TOTAL NUMBER OF DOWNTIME HOURS ⟶ 526					

TABLE II

	ONE HOUR OF ⋯	RELATIVE RETURN	COST ($/MO.)	RELATIVE RETURN/$	PRIORITY FACTOR	OPPORTUNITY RANK
1	⋯ REMOTE TERMINAL UPTIME =	0.0001	50	2.000×10^{-6}	1	4
2	⋯ REMOTE MODEM UPTIME =	0.0001	60	1.667×10^{-6}	1	5
3	⋯ REMOTE LOCAL LOOP UPTIME =	0.0001	100	1.000×10^{-6}	1	6
4	⋯ IXC UPTIME =	0.0003	800	0.375×10^{-6}	1	7
5	⋯ CENTRAL MODEM UPTIME =	0.0005	60	8.333×10^{-6}	1	1
6	⋯ CENTRAL LOCAL LOOP UPTIME =	0.0005	250	2.000×10^{-6}	1	4
7	⋯ FRONT-END UPTIME =	0.0050	1,000	5.000×10^{-6}	1	2
8	⋯ CPU UPTIME =	0.0050	2,000	2.500×10^{-6}	1	3

third column lists the monthly cost of obtaining the relative return for each alternative. This cost would result from installing redundant equipment, improving maintenance service, installing diagnostic equipment, or any combination.

For instance, $800 a month is the cost for redundant interexchange channels for one multipoint link. Sixty dollars a month for each remote modem, on the other hand, might include a premium for better-than-normal maintenance from the vendor or its service organization.

Column four contains the relative return per dollar. The larger this number, the better the opportunity for spending availability-enhancement dollars. Column six shows the ranking of each alternative. Thus, for this example, the best place to spend uptime dollars is at the 10 central modems (alternative five), which is therefore given a rank of 1. Similarly, in this example, the least effective place to spend money is for a redundant multipoint line (alternative four).

Establishing priority

The ranking in the sixth column presupposes that each terminal is equally important to company operations. Thus, the same priority (1) is assigned to each alternative as shown in the fifth column.

Cost may not be the only determinant in setting priority factors, however. Suppose the data communications network supports both inquiry/response and remote job entry (RJE) applications, and that it's more important to company operations to favor the inquiry/response application. In this case the inquiry/response terminals could be given a priority factor of 5 and the RJE terminals a factor of 2; all other subsystems and devices would be assigned their appropriate priority factors. Then the opportunity rank can be found from the product of the

COMPARING REPAIR COSTS

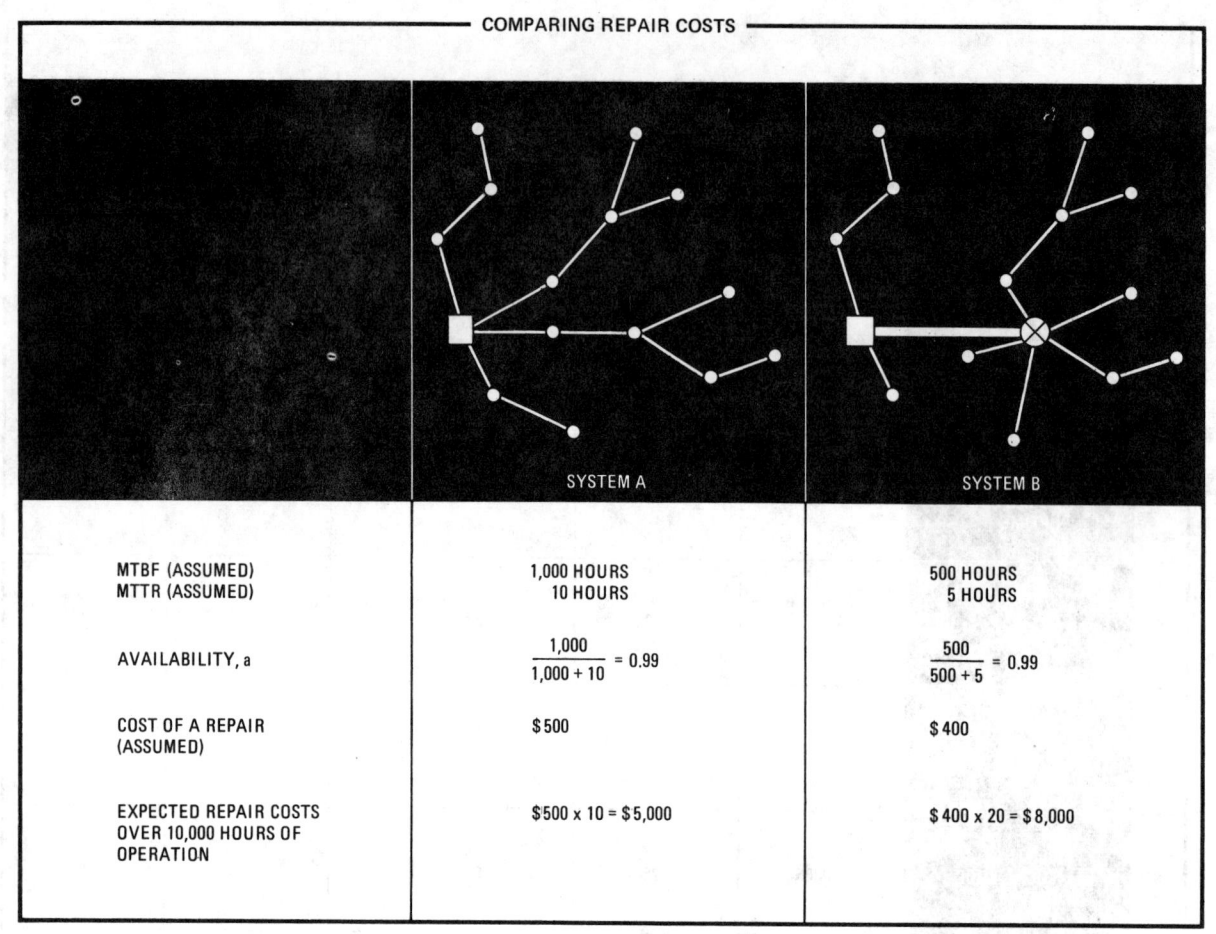

	SYSTEM A	SYSTEM B
MTBF (ASSUMED)	1,000 HOURS	500 HOURS
MTTR (ASSUMED)	10 HOURS	5 HOURS
AVAILABILITY, a	$\dfrac{1{,}000}{1{,}000 + 10} = 0.99$	$\dfrac{500}{500 + 5} = 0.99$
COST OF A REPAIR (ASSUMED)	$500	$400
EXPECTED REPAIR COSTS OVER 10,000 HOURS OF OPERATION	$500 × 10 = $5,000	$400 × 20 = $8,000

2. Even with the same availability resulting from equipment failures and cumulative downtime, different configurations can have different repair costs.

relative return per dollar and the priority factor for that alternative, the largest number again indicating the best opportunity.

The important point here is that the method for targeting expenditures remains the same for any network configuration and its applications. But actual rankings also depend on the relative importance of individual elements in the network.

Statistical data often throws considerable light on preconceived notions about where to place redundant equipment. Studies by DMW Telecommunications Corp. have shown system availabilities ranging from 0.97 to 0.99. By far the most significant source of outages has been the local loops between customer terminal locations and common carrier central offices. Such outages can be 100 times more significant (in terms of out-of-service terminals) than the loss of a CPU or communications controller. And installing a redundant CPU or controller might only increase availability by a factor of 10. Several organizations are therefore taking another look at the wisdom of installing redundant CPUs and communications controllers, looking instead at ways to improve local-loop uptime.

The foregoing discussion of availability and the use of historical data may permit network managers and designers to relate various device and subsystem failures to the behavior of the entire system over an extended time period. But there is another perspective on availability, one that considers equipment *reliability* and the cost of repairing a failure. Here:

$$a = (MTBF)/(MTBF + MTTR)$$

where MTBF is mean time between failures and MTTR is mean time to repair.

Consider two different network configurations, each having the same system availability (Fig. 2). Note, however, that one system has 10,000/1,000 = 10 failures, while the other system has 10,000/500 = 20 failures during the same 10,000-hour period. Also note that the average cost of repair for each failure differs for each system. Thus, even though their availability is the same, one system costs $5,000 for repairs during a 10,000-hour period, while the other system costs $8,000.

How to calculate network reliability

Determining the reliability of devices and their use in series, parallel, and series/parallel links helps to anticipate system failures

Dixon R. Doll
DMW Telecommunications Corp.
Ann Arbor, Mich.

Tutorial

Because data communications equipment will fail from time to time, users need to know how failed equipment will affect network operations. Sometimes a failure in the network merely causes one or a few terminals to go out of service for several hours, but sometimes network performance is reduced to the point of a complete system shutdown. Fortunately, the equations and examples given here provide a simple way to calculate reliability and to obtain worthwhile quantitative answers to such questions as:

- How can I estimate system availability, in terms of total number of hours per month that terminals can communicate, so that I can judge the over-all performance and adequacy of a network's design and of product construction and maintenance?
- What are the chances that a terminal I want to use will not be available when I want to run a job?
- What is the probability that I can complete a lengthy run, such as a remote-job-entry task, without being interrupted by a failure anywhere in the terminal-to-host-computer link?
- How much will adding redundant equipment at strategic places ameliorate unreliability and improve system availability? Is the improvement worth the cost of redundancy?

At stake, then, in the search for high reliability are satisfactory network performance, minimum capital investment, improved network configuration, selection of reliable equipment, and reduced maintenance costs.

Within the context of systems in general, including data networks, reliability is defined as the probability that no failure will occur in a given time period. Conversely, unreliability is the probability of failure within a given time period. One measure of the reliability of a device or system can be evaluated in terms of the mean time between failures, or MTBF.

Implicit in this view of reliability are the assumptions that:

- Equipment "burn-in" and software debugging have been completed before the operating time period (for measuring MTBF) begins.
- The operating time period of interest never extends beyond the useful life of the equipment or system.
- Failures occur at random.
- The number of system failures in any given time

RELIABILITY FINDER

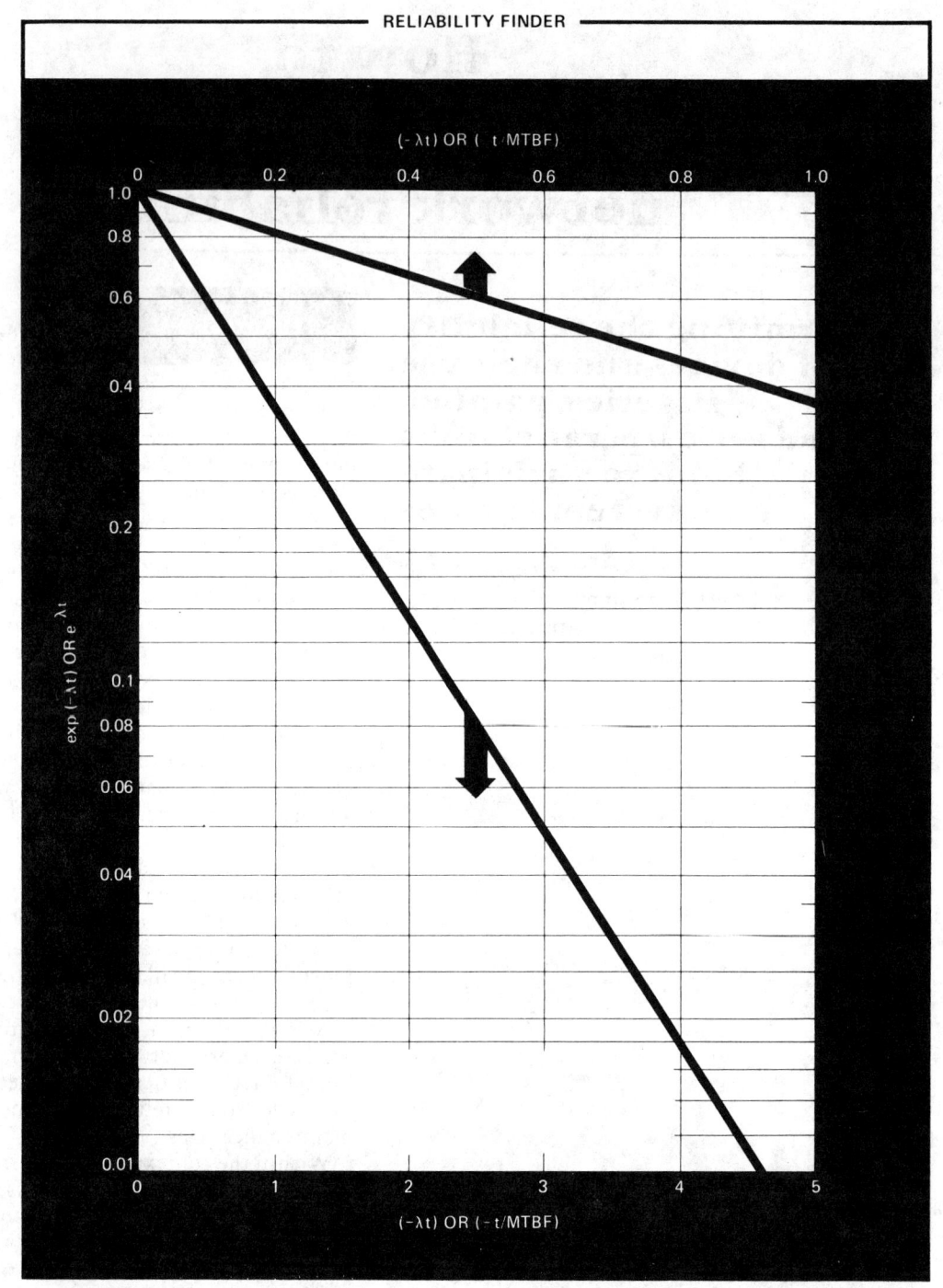

1. Graphs give values of probability that a device or subsystem will not fail as determined from mean time between failures, or failure rate λ.

period is the same for all equally long periods.
• Equipment operates in a reasonable, specified environment and in a specified manner.

Under these assumptions, it is reasonable to mathematically approximate the reliability of a device (or system) as:

$$R(t) = e^{-\lambda t}$$

Here, e denotes the base of the natural logarithm (2.7183), λ is a constant called the average failure rate, and t is the time instant for which the device reliability is desired.

A more convenient form of that equation is:

$$R(t) = exp(-\lambda t) \qquad (1)$$

Here exp means exponential. The constant for the average failure rate, λ, is:

$$\lambda = 1/(MTBF) \qquad (2)$$

Thus Equation 1 states that as MTBF increases, the probability of failure decreases and the average duration of failure-free operation increases.

What is the probability that a device will not fail in 500 hours when its MTBF, as determined from

BASIC LINK TYPES

(a) SERIES CONNECTION

(b) PARALLEL CONNECTION

(c) SERIES/PARALLEL CONNECTION

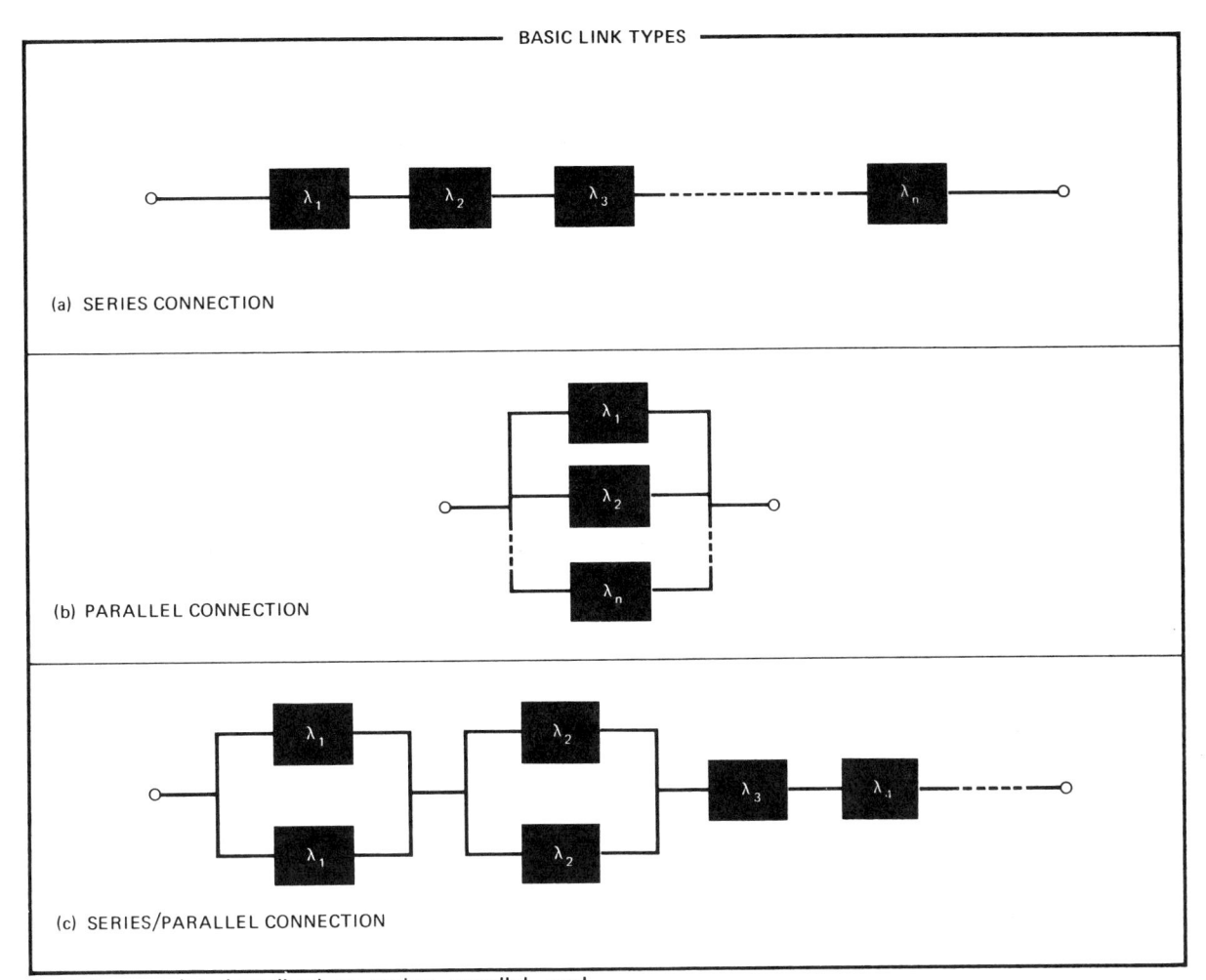

2. Links can be described as series, parallel, and series/parallel connections, each having its own equation for over-all reliability.

operating experience, is 1,000 hours?

Because the device fails on average once every 1,000 hours, λ is 1/1,000. Using Equation 1 with $t = 500$, then:

$$R(t) = exp(-500/1,000) = 0.607$$

Values of the exponential expressions can be readily determined from exponential tables (Reference 1) or on a scientific calculator, or (without too much precision) from the graphs in Figure 1.

The interpretation of the value 0.607 is that there is a slightly better than 60% chance that the device with an MTBF of 1,000 hours will run for 500 consecutive hours without a failure, with the 500 hours starting at any arbitrary instant. Conversely, there is a 40% probability the device will suffer a failure during any arbitrary time period of 500 consecutive hours.

Connection types

In order to ascertain the reliability of an end-to-end link in a data communications network—between, say, a remote terminal and a host computer—it is necessary to define the link as a series, parallel or series/parallel connection (Fig. 2).

A typical series link is shown in color in Figure 3. It includes every device and line from the remote-job-entry (RJE) terminal to the central processing unit (host computer). To find the link's reliability, the equivalent average failure rate of the complete link, λ_l, must be computed and inserted into Equation 1.

In a series connection, a failure in any device in the link will put the entire connection out of action. That is:

$$R_l = R_1 \times R_2 \times R_3 \ldots R_n \qquad (3)$$

where R is link reliability. Furthermore, the equivalent average failure rate of the link is the sum of the failure rates of the individual device:

$$\lambda_l = \lambda_1 + \lambda_2 + \lambda_3 + \ldots + \lambda_n \qquad (4)$$

Therefore:

$$MTBF = 1/(\lambda_1 + \lambda_2 + \lambda_3 + \ldots \lambda_n) \qquad (5)$$

The series link shown in color in Figure 3 consists of six devices and elements: the RJE terminal

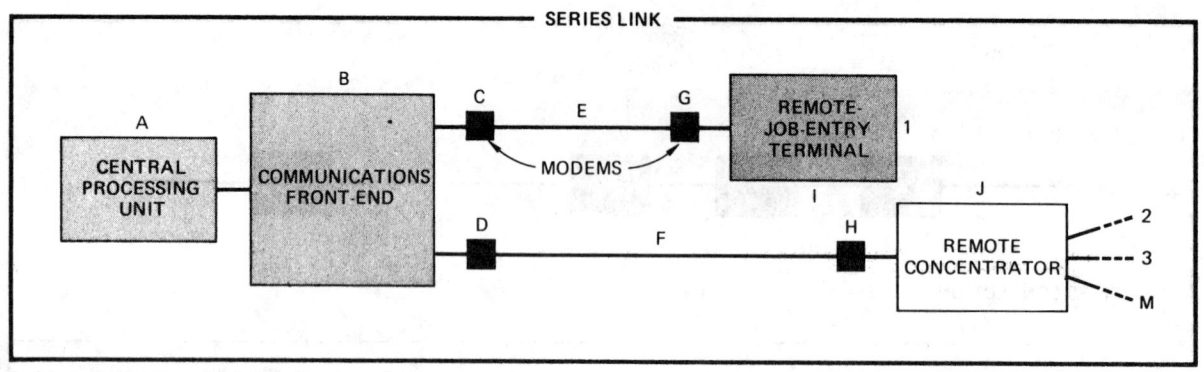

3. Link from remote-job-entry terminal serves as example for determining how long the link can operate without a failure.

(I), the terminal modem (G), the line (E), the central-site modem (C), the communications front end (B), and the central processing unit (A). A failure in any of these elements affects the RJE-terminal user (1).

Equation 5 yields the mean time between failures as seen by user 1. That is:

$$MTBF_1 = 1/(\lambda_I + \lambda_G + \lambda_E + \lambda_C + \lambda_B + \lambda_A) \quad (6)$$

where
- λ_I = failure rate of the RJE terminal
- λ_G = failure rate of the RJE terminal's modem
- λ_E = failure rate of the transmission line from the RJE terminal
- λ_C = failure rate of the central-site modem serving the RJE terminal
- λ_B = failure rate of the communications front end processor
- λ_A = failure rate of the central processing unit

When the hypothetical failure-rate data contained in Table I is inserted, the equivalent average failure rate as seen by the RJE-terminal user works out at:

$$MTBF_1 = 1/[(1 + 0.2 + 2 + 0.2 + 5 + 10) \times 10^{-3}]$$
$$= 54.3 \text{ hours}$$

Therefore, the reliability for this link is:

$$R(t) = exp(-t/MTBF_1) = exp(-t/54.3) \quad (7)$$

The equations and calculations developed so far can produce the important answers to two practical questions:
- What is the probability that the RJE-terminal user can transmit a two-hour job to the host computer without a link failure during that period?
- What is the availability of the remote-job-entry terminal to the user?

The answer to the first question can be obtained from Equation 7, using $t = 2$ hours:

$$R(t) = exp(-2/54.3) = 0.9625$$

The interpretation here is that 96 out of every 100 two-hour job attempts will be processed without link failure. But four times out of a hundred, the job will be aborted by an individual failure.

Mean time to repair

The answer to the question on terminal availability requires the introduction of the concept of mean time to repair (MTTR), or more specifically an average MTTR embracing all the devices in the link. When a device fails, some time will elapse before it can be repaired and restored to service. The longer the MTTR, the lower the availability of the terminal to the user. The MTTR is obtained from operating experience, and each device in a series link will have its own MTTR value. Average MTTR, then, is one value for the link that takes into account all the individual devices' MTTRs.

The average value of the MTTR is the sum of the individual devices' MTTRs, with each MTTR multiplied by its own failure probability. That is:

$$AVG\ MTTR = \sum_{i=1}^{n}(MTTR_i)(P_i)$$

The sum of the probabilities of failure of the devices in the link must add up to unity. To find the individual failure probabilities requires the mathematical step called normalization. That is:

$$P_i = \lambda_i / \sum_{i=1}^{n} \lambda_i$$

Therefore:

$$AVG\ MTTR = \frac{(\lambda_1 MTTR_1) + (\lambda_2 MTTR_2) + \ldots (\lambda_n MTTR_n)}{(\lambda_1 + \lambda_2 + \ldots \lambda_n)}$$

Using the failure-rate and MTTR data contained in Table I for the link shown in color in Figure 3, then average MTTR as seen by the user of the RJE link is:

$$\frac{[(1)(3) + (0.2)(2.5) + (2)(4) + (0.2)(2.5) + (5)(5) + (10)(2)]10^{-3}}{18.4 \times 10^{-3}}$$

which equals 3.1 hours.

Reference 2 and other standard reliability handbooks define system availability, a, as:

$$a = (MTBF)/(MTBF + MTTR)$$

so that, for the situation in the series-connection

TABLE I

DEVICE OR SUBSYSTEM	FAILURE RATE (PER 1,000 HOURS OF OPERATION)	MEAN TIME TO REPAIR (HOURS)
REMOTE JOB ENTRY TERMINAL (I)	$\lambda_I = 1.0$	3.0
MODEM AT RJE SITE (G)	$\lambda_G = 0.2$	2.5
LINE TO RJE SITE (E)	$\lambda_E = 2.0$	4.0
MODEM AT CENTRAL SITE (C)	$\lambda_C = 0.2$	2.5
COMMUNICATIONS FRONT-END (B)	$\lambda_B = 5.0$	5.0
CPU (HARDWARE, SOFTWARE) (A)	$\lambda_A = 10.0$	2.0

numerical example:

$$a = (54.3)/(54.3 + 3.1) = 0.946$$

Consequently, the user of the RJE terminal can be sure that, on average, the terminal will be available for communications with the host computer 946 out of every 1,000 operating hours, and that once the operator starts a two-hour job the run will continue to completion 96% of the attempts.

An alternative and perhaps more direct way of calculating the reliability of a series connection is to use an equivalent relationship derived from Equations 1, 2, 3, and 4A, namely:

$$R(t) = exp[-(\lambda_1 + \lambda_2 + \lambda_3 + \ldots + \lambda_n)t] \quad (4B)$$

Here, the reliability of the entire series connection is obtained simply by summing the individual failure rates of each device.

The use of redundant, or hot standby, devices and lines is particularly common in computer-based data communications systems. Two devices—perhaps two communications front ends—are placed in parallel with each other, but only one device has to be on line for the network to be operational. If the operating unit fails, then the standby unit is promptly placed on line. A diagram of a generalized parallel connection is contained in Figure 2b.

Finding redundant reliability

Defining R_c as the probability of the parallel connection not failing and P_c as the probability of the parallel connection failing, then:

$$R_c = 1 - P_c = 1 - [(1-R_1)(1-R_2)\ldots(1-R_n)] \quad (8)$$
$$= 1 - (P_1 \times P_2 \times P_3 \ldots P_n) \quad (9)$$

Suppose a computer-based communications system uses redundant central processing units, each with an MTBF of 500 hours.
- What is the probability that the parallel CPU combination will operate (not fail) for 500 hours?
- How does this performance compare with the reliability when using just one CPU?
- What is the net mean time between failures of the parallel combination?

Using Equation 1:

TABLE II

NUMBER OF PARALLEL STAGES	RELIABILITY OF EQUIVALENT SUBSYSTEM	MTBF OF EQUIVALENT SUBSYSTEM
1	$e^{-\lambda t}$	$\dfrac{1}{\lambda}$
2 EQUAL*	$2e^{-\lambda_1 t} - e^{-2\lambda_1 t}$	$\dfrac{3}{2\lambda_1}$
2 UNEQUAL	$e^{-\lambda_1 t} + e^{-\lambda_2 t} - e^{-(\lambda_1 + \lambda_2)t}$	$\dfrac{1}{\lambda_1} + \dfrac{1}{\lambda_2} - \dfrac{1}{\lambda_1 + \lambda_2}$
n EQUAL	$1 - (1 - e^{-\lambda_1 t})^n$	$\dfrac{1}{\lambda_1} + \dfrac{1}{2\lambda_1} + \dfrac{1}{3\lambda_1} + \text{---------} \dfrac{1}{n\lambda_1}$

* λ_1 = FAILURE RATE FOR EACH DEVICE

TABLE III

	DEVICE NUMBER	MTBF (HOURS) (ASSUMED)	FAILURE RATE, λ (FAILS PER 1,000 HOURS)
TERMINAL	1	1,000	1
LINE	2	500	2
COMMUNICATIONS CONTROLLER	3	250	4
CPU	4	250	4
STORAGE SYSTEM	5	200	5

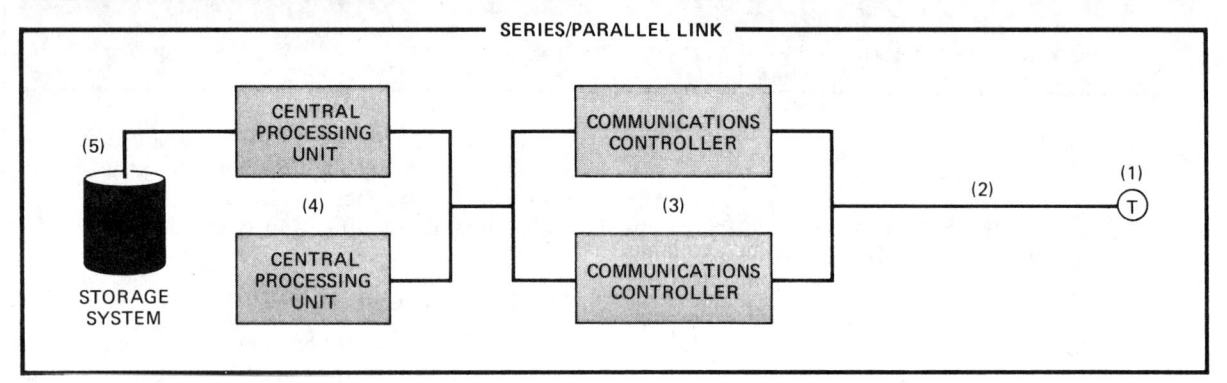

4. A combination series and parallel (redundant) connection serves as another example for determining probability of no failure, or reliability.

$R_1 = exp(-t/500)$ and $R_2 = exp(-t/500)$

Therefore:

$P_1 = 1 - [exp(-t/500)]$ and $P_2 = 1 - [exp(-t/500)]$

Hence, using Equation 8:

$R_c = -[1-exp(-t/500)][(1-exp(-t/500)]$
$R_c = 2\,exp(-t/500) - exp(-2t/500)$

For $t = 500$ hours, then:

$R_c = 2\,exp(-1) - exp(-2) = 0.601$

It is important to note that since the function R_c is not purely an exponential, the comparison of reliability is only valid for the first 500 hours—not an arbitrary 500 hours.

By comparison, one CPU having an MTBF of 500 hours yields a reliability of:

$R = exp(-500/500) = 0.368$

while one CPU having an MTBF of 1,000 hours has a reliability of:

$R = exp(-500/1,000) = exp(-0.5) = 0.606$

For the situations discussed here, the redundant-CPU connection definitely improves reliability, but at substantially double the cost. However, taking steps to improve the MTBF of a single CPU from 500 hours to 1,000 hours will probably be less costly and yet will provide the same reliability as a redundant configuration.

Of course, the key practical problem for data communications users is their lack of ability to control these MTBF variables for individual devices.

Table II contains the equations for the equivalent reliability of several types of parallel configuration. Here, the equivalent subsystem can be treated mathematically as being one element in a series connection, even though the actual equipment is linked in parallel. This table also contains the MTBF of the equivalent subsystem. Thus, for the preceding case, the appropriate equation is $3/2\lambda_1$ (both communications front ends have the same 500-hour MTBF). Therefore, the net MTBF is:

$3 \times 500/2 = 750$ hours

That is, even if one communications front end fails at the end of 500 hours and is replaced by the hot standby unit, statistically the combination will last another 250 hours—during which the network remains operational while the failed front end is being repaired.

Series/parallel connections

In practice, if redundant equipment is used at all, the actual configuration will be a series/parallel combination. The equations for series connections can be used to determine the reliability and equivalent MTBF for a series/parallel connection (Fig. 2c).

Using the hypothetical data contained in Table III for the configuration in Figure 4, compute the equivalent MTBF and reliability (R_e) as seen by the user of the terminal, T, for an interval of $t = 200$

5. Centralized diagnostic and stand by equipment shorten downtime of failed links at Manufacturers Hanover Trust Co. in New York City.

hours. Here, treating the configuration as if it were a series connection only and using Equation 3:

$$R_e = R_1 \times R_2 \times R_3 \times R_4 \times R_5$$

where R_1 is terminal reliability, R_2 is line reliability, R_3 the reliability of the parallel controllers, R_4 the reliability of the parallel CPUs, and R_5 the storage-system reliability. Thus, at $t = 200$:

$R_1 = exp(-200/1,000) = 0.819$
$R_2 = exp(-200/500) = 0.670$
$R_3 = 2exp(-200/250) - exp(-400/250) = 0.697$
$R_4 = 2exp(-200/250) - exp(-400/250) = 0.697$
$R_5 = exp(-200/200) = 0.368$

Therefore the equivalent reliability is:
$(0.819)(0.670)(0.697)(0.697)(0.368) = 0.0981$

Thus, there is a slightly less than 10% chance that the connection will be sustained without failure in the first 200 hours of operation.

Here $MTBF_1 = 1,000$, $MTBF_2 = 500$, and $MTBF_5 = 200$ and, from Table II:

$$MTBF_3 = 3/2\lambda_3 = 3,000/8 = 375$$

and

$$MTBF_4 = 3/2\lambda_4 = 3,000/8 = 375$$

Therefore, from Equation 5:

$$MTBF = \frac{1}{\frac{1}{1,000} + \frac{1}{500} + \frac{1}{375} + \frac{1}{375} + \frac{1}{200}} = 75 \text{ hours}$$

Diagnostics help

The only real options open to designers who want to increase system availability are to select reliable equipment in the first place and to install redundant equipment and lines wherever the improvement in network availability outweighs the penalty of extra cost. Reliability is up to the vendor. That is, only in rare instances will the user work with the vendor to upgrade the reliability of the equipment. But the user can look into competitive equipment and talk with people who have installed the equipment in which he's interested.

Since system availability, in terms of usable terminals, depends on both the reliability, or MTBF, and the downtime as measured by mean time to repair, or MTTR, users can overcome the consequences of marginal reliability by speeding up fault isolation and diagnosis. This is the reason for the current strong trend toward installing a full range of diagnostic features as an integral part of the data communications network. These diagnostic capabilities have already appeared in hardware form (Fig. 5).

In the future greater emphasis will be placed on diagnostic routines driven by software which will minimize the need for human involvement in the tedious tasks of network troubleshooting. Included in these diagnostics will be those for analog line impairments, bit-error-rate tests including bit-pattern generators, and protocol tests. ∎

References

1. "Mathematical Tables From Handbook of Chemistry and Physics", Chemical Rubber Publishing Co., 1957.
2. Dixon R. Doll, "Where to Spend Money to Improve System Availability", Data Communications, Vol. 3, No. 4, November/December 1974.

Design workshop: configuring an actual data system

Start with the specifications for a realistic law-enforcement network, then go through a step-by-step selection of equipment and lines, aided by an analysis of available options

Elliot Nestle
*Interdata Inc.
Oceanport, N.J.*

The reality of data-communications-system design will be demonstrated here by first specifying the parameters of a fairly complex and widespread network and then examining the major decisions involved in selecting suitable data-communications equipment and transmission services.

The goal is to design a nationwide law-enforcement network that provides immediate retrieval and update of data-base files relative to motor vehicles. Thus, the system must allow a police officer to have access through a terminal to a central data base in the investigation of possible stolen cars, revoked drivers' licenses, drivers' addresses, and vehicle descriptions. Furthermore, the central computer should be able to send reports on issued summons and other specific or general information related to motor-vehicle control, either to a given terminal, to all terminals in a region, or to all terminals on the system.

About 1,500 terminals will be needed, some in remote geographical areas and others in metropolitan areas. Most terminals will be in police stations, others in homes of officers operating in low-population-density areas, and some in police cars. The terminals must operate under extreme conditions of temperature, humidity, vibration, and dust—and every terminal must be the same throughout the country. They must cost less than $1,000 each.

Because of population distribution, and thus police distribution, the terminals tend to form clusters in which terminals are within about 50 miles of each other. But clusters may be hundreds and perhaps thousands of miles from each other. And the furthest terminal may be 3,000–4,000 line miles from the host data-base computer at a central site.

The traffic from each terminal is low, about one message every 10 minutes, or no more than 10% usage a day. The police officer will manually key in an inquiry message of perhaps 40 to 60 characters, and the data-base computer will generate a message about 300 to 1,000 characters long in response to his inquiry. Thus, on the average, the ratio of computer output to terminal input, in characters, is about 5:1.

Some other essential factors are that the officer must receive a response in no more than 6 seconds, and that the terminal must produce hard copy since the officer may have to physically carry the response message during an investigation and because the response is an official record of police activity.

From the viewpoint of the total traffic reaching the data-base computer from the 1,500 terminals, the arrival rate averages about 40 messages a second. However, about 25% of these messages will be rejected by the host computer because of errors in input keying or errors occurring on the line during transmission. Further, a traffic analysis shows that due to 24-hour operation and a 3-hour time differential across the nation, the ratio of peak traffic to minimum traffic will not be greater than 5:1. (Some localized on-line information-retrieval systems have a ratio of 100:1 or more.)

Since the whole nationwide motor-vehicle law-enforcement network depends on data stored in the central host computer, the computer system must be 99.8% reliable and have an availability of 99.4%. That is, the system may be operating reliably, but as far as the police officer is concerned, it may be temporarily unavailable. Momentary delays or short busy signals can be tolerated. However, during a delay, the computer system must let the officer know he is still connected, perhaps by periodically printing BUSY until the computer can process his inquiry.

While the data base will not contain information vital to national security, both the data base and the transmission network connecting the terminals to the data base must be secure enough to prevent unauthorized persons from obtaining information to which they are not entitled by law, regulation, or social responsibility. Furthermore, the transmission network must be low-cost and either be readily available from the common carrier or capable of being developed rapidly and operating in time for system start-up.

CHOOSING THE SYSTEM'S COMPONENTS

Using the stated system specifications and conditions for the nationwide law-enforcement network, the next task is to make a series of wise decisions when selecting the procedure by which a terminal will have access to the host computer and when choosing the type of terminal, the network configuration, the line concentration equipment, the carrier facilities, and the data-base configuration.

In the description that follows, the reader is given a set of alternatives at each major decision step, and he chooses the alternatives he thinks most appropriate. Then the rationale for the author's choice will be discussed. At each step in the design, some kinds of equipment are excluded. Obviously, for a different type of data-communications system, one operating under a different set of application conditions, these alternatives will be perfectly valid.

Selecting the access procedure

A) Polling

B) Contention

The choice is B, contention. The specifications permit only a 6-second response time, about half of which, it turns out, is used up at the host computer during the retrieval of information from disk files. The sequential polling of even 20 to 30 terminals on each of 50 to 75 multidropped lines by software would use up more than the remaining 3-second communications response time. Furthermore, the traffic analysis indicated that message flow was fairly smooth with a small peak-to-minimum usage over a one-day period. That is, using contention, the host computer is available to any police officer any time he decides to send an inquiry, but with polling, he would first have to signal his desire to transmit and then wait his turn to get access to the computer.

Selecting a terminal

A) Buffered batch terminal.

B) CRT terminal with cluster of closely situated terminals, controller, and hard-copy printer.

C) CRT terminal without cluster controller but with hard-copy printer.

D) Specially designed terminal.

E) Electronic keyboard with non-mechanical (thermal, ink spray, etc.) hard-copy printer.

F) Mechanical keyboard with mechanical hard-copy printer.

The choice is E, the electronic keyboard with non-mechanical thermal hard-copy printer. The buffered batch terminal is out of the question. Such terminals provide superior features (message buffering, high speed, extensive error control) not needed for this application. Further, the cost well exceeds the $1,000 limit. The CRT terminal, with or without a cluster controller, is inappropriate in this application, since terminals are many miles apart and the required primary readout is hard copy, making the temporary CRT display superfluous. In either case, cost exceeds the limit.

A specially designed terminal would probably be the best choice, resulting in the lowest cost in view of the quantity needed. However, it was not selected because of the excessive development and field-trial time. The mechanical keyboard terminal has the lowest cost of commercial terminals, but it was not chosen because of anticipated difficulty in obtaining routine maintenance and spare replacements in remote geographical areas, and because its bulkiness would make it unsuited for vehicular operation.

The electronic keyboard printer selected for this application sells for less than $1,000, is readily available, and prints at 30 characters (equal to 300 bits) a second—fast enough for on-line inquiry response from a computer to a human operator. It is quiet, an important factor when installed in homes and vehicles. A 15-character-per-second terminal would also be an acceptable match with the keying and reading speed of the officer. However, the 30-character-per-second terminal does reduce transmission time per message from the computer and thus increases utilization of the line in conjunction with the multiplexers and concentrators that may also be employed in the overall system.

Selecting a network configuration

A) Each of the 1,500 terminals connected point-to-point to the central data-base computer.

B) A few point-to-point lines, optimally routed to the computer, with many terminals multidropped from these lines.

C) Terminals connected point-to-point to a time division multiplexer (TDM), then point-to-point from several TDMs to a remote data concentrator (RDC), and then point-to-point from several RDCs to the central data-base computer.

D) Terminals connected point-to-point to TDMs and RDCs as in C, but with concentrators multidropped to the host computer.

The choice is C, point-to-point from terminals to multiplexers to concentrators to the host computer. Choice A is unacceptable because it would mean paying for private-line charges for well over a million miles, due to the large number of terminals and their wide geographical dispersion. However, choice A would provide the shortest response time and offer the highest reliability because there would be multiple access to the data base. The loss of a modem or line would affect only one terminal. In addition, using fewer devices, such as multiplexers and concentrators, offers increased system reliability. Under other circumstances, choice B would be a valid decision, but it was not selected for this system because doing so would mean using a polling procedure to access the computer—which would increase response time and also raise the cost of the terminal to support a polling operation.

Choice C was selected because it reduces line costs through the line concentration techniques provided by the TDMs and the RDCs, while still yielding acceptable delays and response times. For this system, the short response time permitted will not allow storing and queuing of messages, so no mass storage will be utilized. Furthermore, this configuration has the advantage over others in that it will permit orderly, low-cost network expansion in the future.

Choice D would cost somewhat less than C, but the amount saved would not warrant the resulting degradation in response time due to polling or the added cost for pollable terminals. This configuration is quite valid, though, for many data-communications systems having a different set of operational specifications.

Selecting the best carrier facility

A) High-speed (voice-grade) private lines for the entire network.

B) Public switched (dial-up) voice-grade telephone network.

C) Low-speed private lines for terminal communications and high-speed private lines for multiplexer/concentrator communications to the computer.

D) Public switched telegraph network.

The choice is C, low-speed private lines from terminals and high-speed lines for the other communications links. Using the switched telephone or telegraph networks cannot be tolerated because they are available to the public and thus would make it too easy for an unauthorized person to gain access to a data base. Furthermore, the public telegraph network is not normally fast enough to carry the high-speed data traveling between the computer and the concentration equipment. Choice A would mean using high-speed lines for low-speed terminal links, and the cost would be excessive.

In choice C, the low-speed links can carry the 300-b/s data streams to and from the terminals, while the voice-grade lines can carry the 2,400-b/s data streams between the computer and the concentrators and multiplexers. (As more terminals are added and traffic increases, the high-speed links can be upgraded to 4,800-b/s simply by installing faster modems.) The common carrier can condition the normal low-speed telegraph line to provide the bandwidth needed to carry 300 b/s. Since the lines do not carry traffic related to national security, there is no need to impose any special transmission security (as distinct from data security) on thee lines. Thus, the common carrier is not presented with any unusual engineering or construction demands in providing the communications links on a timely basis.

Selecting line-cost-reducing equipment

A) Programable remote data concentrators (RDCs) only throughout network.

B) Hard-wired time-division multiplexers in conjunction with programable remote data concentrators.

C) Frequency-division multiplexers only.

D) Hard-wired time-division and frequency-division multiplexers, no remote data concentrators.

The choice is B, hard-wired TDMs in conjunction with programable RDCs. Using only remote data concentrators (choice A) would be a case of overkill: RDCs are expensive and, although they can perform more than simple line concentration of low-speed terminals, there are many places in the network where RDC performance is unnecessary. Using FDMs only (choice C) may involve low-cost-per-channel equipment, but each channel is dedicated to a given terminal whether or not data is on the line, so that over-all voice-grade-line utilization is poor for such a system handling so many terminals. Furthermore, each of the 1,500 FDM channels would require a channel set at the computer, since the host computer cannot perform demultiplexing when using FDM.

Using hard-wired TDMs and FDMs (choice D) still keeps the cost low and is often the best choice for line concentration. But, in this law-enforcement network using only a combination of multiplexers will require installing the number of high-speed lines dictated by the peak—not average—traffic from the terminals. This is so because in FDM a given bandwidth is dedicated to a given channel and in TDM a given time slot is dedicated to a given channel. In neither case can the data in each band or time slot be averaged to reduce the number of high-speed lines. Furthermore, the handling of character-interlaced data at the host computer would be rather inefficient and present severe loading problems.

In choice B, though, each TDM concentrates the lines from many terminals at a low cost per channel. The remote data concentrator first demultiplexes the data stream from each TDM. Then the RDC concentrates the high-speed outputs from several TDMs and smooths or averages the over-all flow of data to provide more efficient line utilization and permit the use of fewer long lines. The RDC can perform error control on data coming from terminals before the concentrated data blocks are sent to the computer.

Selecting a data-base computer configuration

A) One large host computer using a hard-wired transmission controller to interface to the communications lines.

B) One smaller host computer with a programable front-end processor (PFEP) to interface the lines.

C) Two redundant smaller host computers with redundant PFEPs.

D) Redundant large host computers with redundant hard-wired transmission controllers.

E) Several geographically distributed smaller computer systems.

The choice is C, two redundant smaller host computers interfaced with two redundant programable front-end processors. Choice E is not acceptable because security and political considerations dictate that all data be held at one central location. The systems specifications set a high value on reliability, enough to justify the installation of dual-data-base systems. Thus choices A and B are not acceptable.

For new—and even old—installations, the use of a hard-wired transmission controller may now be moot. Even IBM, which formerly sold the popular 270X series of hard-wired transmission controllers, has discontinued them in favor of the programable front-end processor, which IBM calls the 3705 communications controller. It is unlikely that any new, fairly complex data-communications system would employ a hard-wired transmission controller. Thus, choice D is out.

The important word in the final choice—using redundant smaller host computers with redundant programable front-end processors—is smaller. A PFEP relieves the host computer of much of its overhead in performing data-communications tasks, so a smaller host computer can be chosen.

There is another side to this price-performance tradeoff. If the host computer is fully loaded—that is, operating at an average of 70% of its CPU time—then adding data communications without a PFEP may require an expensive upgrading of the host computer. For example, if the law-enforcement system used a host computer and a hard-wired transmission control unit, the CPU might spend about 30% of its time merely doing housekeeping chores on the inputted data. This would leave only 40% of the CPU time for accessing the data-base files and other data-processing tasks. But with a PFEP screening incoming messages for errors and formating the messages for the host CPU, the load on the host CPU could drop to 10–15%. This 15–20% increase in available CPU time—and related reduction in memory—is significant enough to obviate the need to upgrade the host computer. The cost of a PFEP installation is usually less than that for a host-computer upgrade.

In summary, the design of a data-communications system involves a sequence of steps, with the decision at each step intertwined with those taken earlier and later. Although design often must invoke formal, mathematically based procedures, the planner's experience, intuition, and his willingness to make a decision can bring him a long way towards final design.

part 6
data data data data data data data data data data
channel performance

Error control: keeping data messages clean

To avoid serious errors in data streams at the receiving terminal, data-communications systems may use one of several error-control techniques, but each reduces net data throughput

Karl I. Nordling
*Paradyne Corp.
Clearwater, Fla.*

Because of the noisy environment surrounding telephone lines, errors are introduced into the messages transmitted between terminals and the host computer. The higher the quality of the lines, the fewer the errors will be. Electrically induced errors—in contrast to keying errors caused by an operator—are caused by crosstalk between adjacent lines, signal distortion on the lines, and impulse noise from lightning, switches, and other electrical sources. These disturbances change digital 1s to 0s—and vice versa—in unpredictable patterns.

A number of techniques have been devised to control transmission errors. The type of control to be chosen depends on many factors and constraints imposed by available terminals, computers, software, line-control discipline, inherent delays, and network configuration.

When the dial-up telephone system provides the network, connections between two points will be routed over various lines at different times. The bit-error rate for each connection depends on the quality of lines involved in each connection. For example, a remote line printer will run at 200 lines per minute on one connection, but may be slowed down to 100 lines per minute on another connection because of excessive retransmission resulting from transmitted data blocks containing errors when received.

And despite the efforts of telephone companies to keep the error rates on leased private-line links below specified values, the user should—in his own self-interest—monitor the bit-error rate to make sure that it does not rise above the prescribed standard.

An error-control method must fall into one of two categories, each with appropriate error-coding techniques:
- Error detection with retransmission (ARQ), using either the stop-and-wait or the go-back-two method of block-transmission discipline.
- Forward error correction (FEC) using—for example—the Hamming code, as explained later, to provide error correction at the receiving terminal.

In ARQ—for automatic request for repeat—the sending terminal divides the data stream into blocks of suitable length, each block usually containing from one to seven records, wherein the data contained on one punched card or in one line of print usually constitutes a record. Error-detection coding is then added to each block being transmitted. The receiving device—the host computer, a programable front-end processor, or often a terminal—checks each block for errors. It requests a repeat transmission of a block if an error is detected; otherwise the next block is transmitted. In forward-error

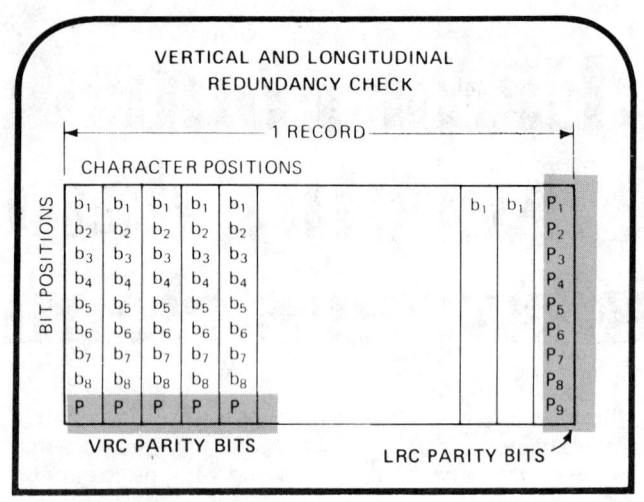

1. Adding parity. Using vertical and horizontal parity costs little in redundant bits, but it sharply increases message accuracy.

correction (FEC), checking-code bits added to the data permits the receiving equipment not only to detect, but also to correct, errors introduced during transmission. Thus, with FEC, no retransmission is involved. However, while error detection and correction proceeds, FEC does require extensive redundancy, intricate logic procedure, and large buffers to store transmitted and received blocks.

Redundancy checks

For error detection and retransmission, the two common error detection codes added to the block are:
- Vertical and longitudinal redundancy check (VRC/LRC) and
- Cyclic redundancy check (CRC)

Fig. 1 shows how these redundant bits are added to the transmitted block for VRC and LRC. For VRC, a 0 or 1 parity bit, P, is added to each character in the record as shown at the lower left. A parity bit, P_n, is added to each bit position across all characters in the record for LRC, as shown at the right of Fig. 1. When a bit error occurs during transmission, the character's parity bit is inverted, and thus it is the reverse of the parity convention established within the system. That is, if the 0 or 1 parity bit is added to the character to make the binary field an even number of 1s (called even parity), and an odd number of 1s is received, then the receiver knows an error has occurred. Frequently, as in tape-to-tape transmission, VRC is used without LRC. Parity check for LRC works similarly to that for VRC.

As Fig. 2 shows, cyclic redundancy check (CRC), consists of adding bits—about 12 to 25—to the entire record or entire block. In CRC checking, the data block can be thought of as one long binary polynomial, P. Before transmission, equipment in the terminal divides P by a fixed binary polynomial, G, resulting in a whole polynomial Q, and a remainder, R/G. That is,

$$\frac{P}{G} = Q + \frac{R}{G}$$

The remainder, R, is added to the block before transmission, as a check sequence k bits long. The receiving device divides the received data block by the same G, generates an R, and checks to ascertain if the received R agrees with the locally generated R. If it does not, the data block is assumed to be in error and retransmission is requested.

CRC long-division analogy

The error-detecting properties of CRC are illustrated in the table, which contains a selection of blocks, P, and two values of G to demonstrate the importance of the proper selection of the divisor. Although the data is actually a stream of binary bits, decimal numbers are used here by way of analogy to clarify the CRC technique. For G = 100, there are 100 possible remainders, ranging from 00 to 99.

If the error occurs in these two lowest bit positions, the received remainder will not coincide with the transmitted remainder, and the error will be detected. For this value of G, if the error affects only one more-significant digit—say the error changes the 8 in 4893497 to a 7, the remainder—97—is unchanged, and the error goes undetected. Thus 100 proves to be a very poor choice for G.

If, instead, the value of G is 103, then, should the 4 in 5040201 be changed by an error during transmission to a 3, the remainder changes from 102 to 93. The receiving terminal tries to match the received R = 102 with the calculated R = 93, spots a mismatch, declares an error, and requests retransmission.

What actually happens in CRC is that the n-bit data block is treated as a binary polynomial of the form:

$$P(x) = b_n x^n + b_{n-1} x^{n-1} + b_{n-2} x^{n-2} + \ldots + b_3 x^3 + b_2 x^2 + b_1 x^1 + b_0$$

where the b_i are 0 and 1, according to the state of the corresponding bit position in the block. The divisor, G(x), is a polynomial of the same form as P(x). The operation is polynomial division, which can be readily implemented by software or special-purpose integrated circuits.

The effectiveness of CRC can be summarized for a check sequence k bits long as follows:
- All data blocks with an even (or odd) number of errors are detected if k is even (or odd).
- All data blocks with burst errors less than k bits long are detected.
- All data blocks with a total number of error bits less than M, where M is about k/4, are detected.

Of all the remaining error patterns, one in 2^k are undetected, so that CRC error-detection can be made increasingly effective at little cost for additional redundancy.

In ARQ systems, a 25-bit CRC code added to a 1,000-bit block would allow only three bits in 100 million to go undetected. That is, for a 2.5% redundancy, the error rate is 3×10^{-8}.

With CRC, the whole block must be sent, even if the error occurs early in the transmission, in order for the error-detection technique to work. If, instead, the 1,000-bit block is formatted into 125 eight-bit characters and a VRC parity bit added to each character block, the resulting redundancy is one out of nine bits, or 11%, yet the undetected error rate is as much as 5×10^{-3} for typical bit-error rates occurring on telephone lines. Even so, VRC is chosen in such applications as remote transfer of

CYCLIC REDUNDANCY CHECK

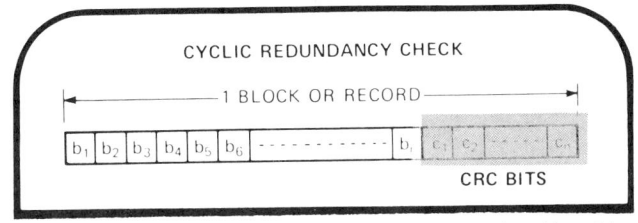

————— 1 BLOCK OR RECORD —————

| b_1 | b_2 | b_3 | b_4 | b_5 | b_6 | - - - - - - - | b_i | c_1 | c_2 | - - - | c_n |

CRC BITS

2. Block check. Several cyclic redundancy-checking bits added to the data block greatly reduce undetected errors in the message.

HOW CYCLIC REDUNDANCY CHECK DEVELOPS ERROR DETECTION CODE
$P/G = Q + R/G$

G = 100		
P	Q	R
→ 4893497	48934	97 ←
4893498	48934	98
4893499	48934	99
4893500	48935	00
4893501	48935	01
4893502	48935	02
→ 4783497	47834	97

G = 103		
P	Q	R
5040200	48933	101
→ 5040201	48933	102 ←
5040202	48934	000
5040203	48934	001
5040204	48934	002
5040205	48934	003
→ 5030201	48836	093

information from tape to tape, since the character parity is inherent in the source data. Also, VRC allows the receiver to terminate as soon as an error has been detected, which will improve channel efficiency somewhat.

Besides a choice of error-control codes, the other major way to classify ARQ systems is by their retransmission procedures. Two common methods are stop-and-wait ARQ and go-back-two ARQ.

Stop-and-wait ARQ

In stop-and-wait ARQ (Fig. 3), the transmitting terminal stops at the end of each block and waits for a reply from the receiving terminal as to the block's accuracy (ACK) or error (NAK) before transmitting the next block. The dead time between blocks reduces the effective data rate of the channel. When two-wire systems are used with common half-duplex terminals, the channel must be turned around once to receive the reply and then again to resume sending. This dead time is a significant portion of total connection time.

Depending on error-rate probability and block length, the effective throughput can go as low as 50% of the rated channel speed. Using a four-wire system with stop-and-wait ARQ eliminates line turnaround, so throughput can be increased, although the efficiency still depends on block length.

In go-back-two ARQ systems, the dead time is eliminated by use of a simultaneous reverse channel for ACK/NAK signaling. After each block, the reverse channel carries back to the sending terminal either the ACK or NAK response. A typical sequence is shown in time profile in Fig. 4. On completion of transmission of block i, the transmitter immediately starts sending block (i + 1). During the transmission interval of this block, the ACK-i reply to block i is returned over the reverse channel so that the transmitter knows before starting to send block (i + 2) whether or not block i was correctly received. If an error was detected, as shown for the case of block (i + 1), a NAK comes back over the reverse channel, the terminal goes back two blocks, and starts retransmitting block (i + 1) and then (i + 2). In short, an ACK control signal—which will be transmitted most of the time—permits blocks to be transmitted continuously, and that a NAK causes a two-block back-up.

Continuous go-back-two ARQ enables efficient operation, at about 90% of the line/modem rated speed, provided the block length is selected properly and used consistently—usually about 1,000 bits per block. It has a sharp optimum peak of net data throughput versus block length, independent of turnaround time or bit-error rates.

In the forward-error correction operating procedure, errors are detected and corrected at the receiving terminal. Numerous FEC codes have been developed—one of the earliest and simplest being the Hamming code (Fig. 5). The Hamming code can detect and correct only a single-bit error; others can correct multiple-bit errors.

How FEC works

The Hamming code is based on a seven-bit block—four data bits and three check bits. The three check bits are used in three parity checking operations on three different subsets of the bits in the block. As shown in Fig. 5, the rectangle labeled C1 encloses the bits checked by parity-bit C1, rectangle C2 encloses the bits checked by parity-bit C2, and rectangle C3 encloses the bits checked by parity-bit C3. The circled bits are the parity bits, and the others are the data bits.

The receiving equipment performs the three parity checks and notes the combination of passes and fails. The receiving equipment's logic then determines, according to the table in Fig. 5, which bit, if any, is in error. Then, the logic corrects the error by inverting the erroneous bit.

Note, however, that if two bits are in error, say b_2 and b_3, the results C1 = fail, C2 = fail, and C3 = pass would be interpreted as an indication that only b_2 had changed. This could be corrected, but—in this simple configuration—b_3 could neither be detected nor corrected.

Larger codes can be constructed to detect and correct multiple errors in the block, but at the expense of high redundancy. For example, assuming a block length of 1,000 bits and a maximum error burst length of 25 bits, the FEC procedure would require a minimum of 165 bits, compared with 25 bits for a cyclic redundancy check. Furthermore, if the burst length exceeds 25 bits, then the probability of an undetected error is certain with FEC, and three in 100 million with CRC. FEC does not require a return channel, and, since data flow is continuous, it does not suffer from reduced throughput resulting from turnaround delays.

From this discussion, it would appear that the optimum error-control system for a network with telephone-line transmission links would be go-back-two ARQ with a cyclic error-detection code. However, the choice of an actual error-control procedure depends on constraints imposed by available terminals, computers, software, and line discipline. Available choices are ex-

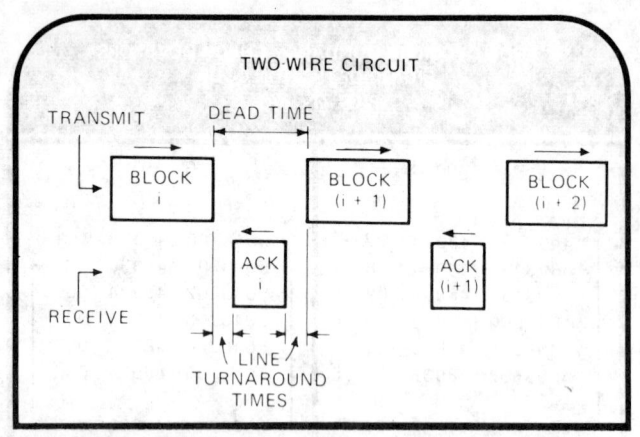

3. Stop-and-wait. In this common ARQ scheme, the receiving terminal sends back an acknowledgment after each block.

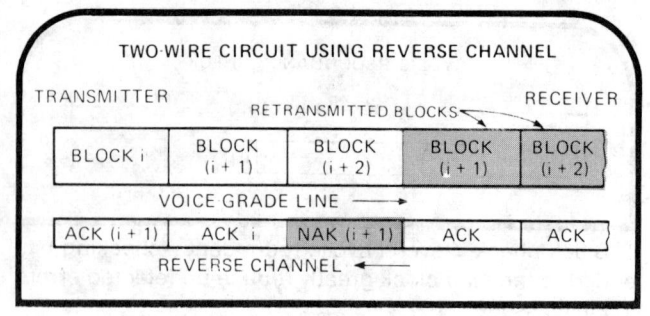

4. Go-back-two. Transmitter continuously sends blocks until the receiver senses an error, then it retransmits two blocks.

plored for two common applications: low-speed interactive and high-speed batch systems.

Low-speed terminals limit choices

Since low-speed interactive systems are usually time-shared, many remote users have access to a computer independently. To serve their requirements, a fairly wide range of error-control approaches has evolved through development of many commercial and company-owned time-sharing systems. No error control at all is quite common when the transmission is an administrative message, in which errors can be spotted and corrected in context by the reader. Error control schemes include echoplexing, FEC, or some variation of ARQ.

The main constraint on the designer of a low-speed interactive system is that most low-speed terminals are not well equipped for error control. The low-speed local links and the high-speed time-division-multiplexed link are subject to errors.

On low-speed links, the terminal's characteristics determine the approach to error control. Because these terminals are usually asynchronous, they deliver data at random, and they can't be stopped under control of the modem interface. Further, they are usually unbuffered, so that if the receiver is busy or shut down and can't accept a character when the terminal delivers it, the character is lost.

This combination of characteristics limits error control of local links to some kind of FEC technique. FEC requires a large number of redundant bits, which mandates a modem bit-rate substantially higher than that for the basic terminal speed, and it also requires the addition of FEC electronics at both the terminal and the computer-related multiplexer or concentrator. These extra costs effectively discourage introduction of error control into low-speed links because the overriding economic incentive is to make the system accessible to all types of terminals, without special attachments, and at lowest cost.

In high-speed multiplexed links, the situation isn't quite as rigid. The designer may elect to use FEC and thereby trade a significant loss in effective bit rate for an improvement in error rate. For instance, a system having a modem/line basic bit rate of 7,200 b/s and using a rate ⅔ FEC code—the block contains ⅔ data bits and ⅓ redundant bits—provides a maximum effective bit rate of 4,800 b/s and error reduction of about 1,000 to 1.

The designer must decide if the error-reduction improvement offered by FEC offsets the higher raw error rate from the increased bit rate and is worth the extra cost of higher-speed modems and the cost of conditioned lines to handle the higher signaling speed. The only effective way to get an answer is to measure actual error rate and burst characteristics on the private voice-grade line.

Another option for the high-speed link is ARQ error control between the time-division multiplexers. Because in the ARQ mode the high-speed link is periodically stopped for retransmissions, extensive buffer memory must be added to each TDM to store continuous low-speed inputs for later transmission when the high-speed line is available for sending blocks. ARQ with store-and-forward of temporarily held messages achieves a vast improvement in error rate, but the buffers add to the cost. In addition, a round-trip delay, which depends on line loading and the severity of the basic error rate, is introduced.

High-speed batch error control

High-speed batch systems involve the continuous transmission of a mass of data—a batch—in a short time. Among such systems are computer-to-computer, terminal-to-terminal, and tape-to-tape transmissions. In these systems, the user has the least degree of control, since they make use of line protocols that are thoroughly entrenched and fairly rigid in their configurations. (The line protocol is the set of rules that defines the message and response sequences that permit two devices to communicate with each other). These protocols almost universally employ stop-and-wait type error control procedures, which, as already pointed out, seriously reduce the effective throughput on the line. The approaches available to the user to alleviate this efficiency problem are limited to the use of four-wire instead of two-wire lines, the use of long blocks instead of short ones, and the use of multileaving instead of one-way transmission. Some of the considerations involved in evaluating these approaches are discussed below.

SYSTEM DELAYS. In communications links using four-wire circuits, system delays may cause turnaround losses, even though the line presents no serious turnaround delay. This type of delay commonly occurs in computer-to-computer and tape-to-tape systems where double-buffering is not employed. For instance, many tape-to-tape systems operate as follows: A block of data is received from the transmitter, stored in a buffer, and checked for errors. Before returning a reply to that block, the receiver records the block on tape, then backspaces the tape and rereads the block to insure that it has been recorded correctly.

If an error is detected when the block is reread, the receiving terminal replies with a NAK, causing the sending terminal to retransmit. This type of system has turnaround delays of about 150 to 300 milliseconds, in addition to line turnarounds. Thus, systems using high-speed modems, 4,800 b/s and up, have communications-link efficiency that is quite low—typically about 50%, especially since they usually use one record per block. The problem can easily be resolved by double buffering, i.e., the tape is written from one buffer while the transmission line fills a second buffer, thus eliminating the wait for the tape write and operation check.

The same considerations apply in computer-to-computer and computer-to-terminal systems, although here the problem is not so prevalent because these systems commonly employ double buffering. However, when a user writes his own program for a computer-to-computer application, he should take care to write the program using a double-buffering technique so as to avoid the inefficiency described above.

DIAL-UP LINKS. In many cases, it is impossible to use four-wire full-duplex lines to avoid line-turnaround delay. This is true when the dial-up network is the basic data-communications medium. There are many reasons—reliability, accessibility, economy—why the dial-up network may be the most desirable choice. Because the dial-up network is two-wire in nature, the stop-and-wait type of error control causes a severe efficiency problem, especially at high bit rates. One attempt to alleviate this problem, provided by the IBM binary synchronous communications (BSC) protocol, is known as multileaving. Multileaving reduces the effect of turnaround losses by interleaving output data with input data as follows:

When a block of data has been output to a terminal, the terminal replies with a block of data to the computer instead of a simple ACK/NAK message. Thus, for each pair of line turnarounds, one block of data is transmitted in each direction, which works out to one line turnaround per block, instead of two necessary for normal one-way transmission.

In practice, the effect of multileaving on line efficiency is not overwhelming. For typical remote job entry systems, both card reader and line printer operate simultaneously only about 20% of the time. Therefore, multileaving reduces the inefficiency of the stop-and-wait method by 50% for 20% of the time—yielding a 10% improvement. However, even that is better than nothing, and when combined with the simultaneous operation of the peripherals, it has achieved popular use on the types of terminals and operating systems that can support simultaneous operation.

REVERSE-CHANNEL SYSTEMS. In tape-to-tape systems, the user has one further degree of freedom in that he may elect to use a line protocol based on the use of a simultaneous reverse channel. These systems use the reverse channel for ACK/NAK signaling. This type of protocol can be more efficient than the conventional stop-and-wait protocol because of shorter delay times between blocks; however, in many cases, the improvement is vitiated by the use of a single buffer.

ERROR-CONTROL MODEMS. An approach that can eliminate the inefficiencies of the BSC line protocol for high-speed half-duplex batch terminals operating on two-wire circuits is to use an error-control modem, such as the Paradyne Bisync-48. This modem employs a continuous-type ARQ algorithm that performs the error control for the link without turnaround losses, and it contains the necessary hardware adaptors to make the modem suitable for use in IBM Bisync links without need for software or hardware changes.

Speed dictates new techniques

Perhaps even more efficient methods of error control will result when computer and terminal makers take more cognizance of communications requirements and of developments now going on in higher-quality, higher-speed, transmission links at lower costs. For example, when domestic satellites become available in the near future, they will carry voice-grade communications. If these circuits are used for data communications in much the same way as terrestrial circuits, the round-trip electrical propagation delay—from land-based transmitting station to land-based receiving station—will add about 500 milliseconds to the dead time. Thus, if the conventional stop-and-wait error-control method is used, as practically mandated by present line protocols, then channel efficiency will drop to 25%-45% utilization.

Since it is unlikely that users will scrap existing computers and data-communications equipment with a well-entrenched line-control protocol, the user may have to instruct the common-carrier not to transmit data messages via satellite circuits.

And demand by users to take advantage of satellite communications may force computer makers at last to improve line protocols and modify equipment to provide both economical data processing functions and economical communications links capable of operating at a high net data throughput for the complete data-communications data processing system.

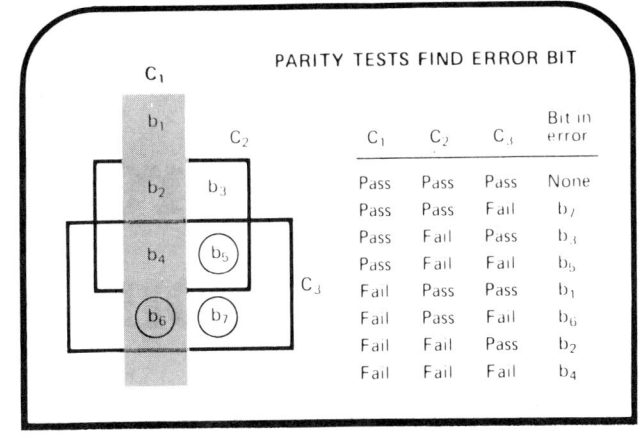

5. Hamming code. Used in forward-error correction, logic tests on received data and check bits correct a one-bit error.

	C_1	C_2	C_3	Bit in error
	Pass	Pass	Pass	None
	Pass	Pass	Fail	b_7
	Pass	Fail	Pass	b_3
	Pass	Fail	Fail	b_5
	Fail	Pass	Pass	b_1
	Fail	Pass	Fail	b_6
	Fail	Fail	Pass	b_2
	Fail	Fail	Fail	b_4

Hit analysis leads to reduced data errors from a microwave link

Continuous observation of hits that produce data errors can locate sources of troubles on long-haul microwave-carrier circuits and thus improve the performance of a user's data communications system

Edward J. Henley
MCI Communications Corp.
Washington, D.C.

PERFORMANCE

During February 1973, an extensive 664-hour test on MCI's microwave link between Chicago and St. Louis demonstrated quite vividly that a long-haul communications link designed, installed, and operated specifically for carrying data can, in fact, perform substantially free of hits—a variety of electrical disturbances that cause data errors.

For the whole month, less eight hours for special tests, only 250 hits occurred. Each hit lasted no longer than a few milliseconds. This performance is much better than the phone company's published objective of no more than "15 hits in 15 minutes," or 39,840 hits during the same 664-hour period. Furthermore, MCI's microwave link operated completely hit-free for a continuous period of 64 hours. The month's hit performance is plotted in bar-chart form as a function of the hour of the day (Fig. 1).

When a carrier can supply minimum-hit lines, the ultimate beneficiary is the data communications user because it allows him to free up communications resources and time. Designers of computer-based data communications systems are accustomed to a high incidence of data-error-producing hits on communications lines.

They have learned to live with the situation by installing error-control procedures—usually calling for retransmission of blocks or messages—to prevent false data from propagating through the communications network and eventually into the organization's operations. Although necessary, error-checking and retransmission must be regarded as overhead in that it wastes such communications-network resources as lines, datasets, multiplexers, remote data concentrators, front-end processors, and even host computers.

The primary responsibility for reducing the error-control overhead rests with the common carriers because, unless a hit occurs, there is no need for error correction. Thus, carriers have the obligation to provide lines that are as free of hits as is technically and economically feasible. When this is accomplished, the user can then relax error-control procedures to a point consistent with actual hit incidence. This, presumably, will reduce error-control overhead by, for example, permitting use of longer data blocks.

To design a data communications network, keep it running well, and execute follow-on performance audits (see "Auditing brings insight," next page), the user must be aware of the kinds of lines employed throughout his network, the communications technology on which they

> **Auditing brings insight**
>
> Minimizing the impact of hits on data communications systems requires a continuing and independent technical audit of line performance by the common carrier and of over-all network performance by the subscriber. Since it is unlikely that data communications people can properly audit their own performance, the demand for such an audit must stem from carrier management and user management.
>
> Consider, for example, one benefit of an technical audit performed by an independent group reporting directly to MCI's top management. This group was charged, among other things, with running the tests whose results are depicted in Fig. 1 and analyzing the results to independently arrive at ways to improve operation. Note in Fig. 1 that more hits occurred during daylight hours than at night, contrary to the usual performance of a microwave link which should exhibit more fading—hence contribute more hits—at night. Further, hit incidence was much greater on weekdays than on Saturdays and Sundays.
>
> What was going on along the system during weekdays, but not at night, to induce more hits? The answer: routine maintenance, during which equipment would be manually switched from one channel to another. This switching would occasionally register as a hit. Recognizing the problem, the equipment vendor modified the hardware, and operations management evolved a new routine so that maintenance can be accomplished during the day, and as necessary, without creating inadvertent hits. Therefore, line performance is expected to be even better than indicated in Fig. 1.

are based, and the time and magnitude profiles of the hit-incidence performance that can be reasonably expected from each type of link. To this end, the balance of this article describes the use of microwave technology for long-haul communications, sources of hits, and the instrumentation for gathering the hit data.

Sources of errors

Barring catastrophic failure, errors occur from two major sources: deviations from prescribed steady-state parameters, which result in easily recognized, more or less continuous streams of erroneous data, and hits that occur at random, inducing short-lived error situations.

Among the more important steady-state line parameters are frequency response, envelope delay, and distortion. Frequency response and envelope delay are related to each other electrically and, in general, the longer the line, the less satisfactory these parameters could become. The common carriers improve the line to obtain relatively flat response within a voice-grade band by using a technique called conditioning. Good-quality datasets (modems) can improve frequency response and envelope delay even further by using equalization to maintain the flat response.

Distortion occurs because of nonlinearities introduced by poor synchronization of carrier facilities, by equipment for gain expansion and gain compression, and by interchannel modulation. Again, a quality dataset can compensate for the effects of line distortion.

Once a leased line and dataset are adjusted for such steady-state parameters, the link should perform in a rather stable manner and merely require slight adjustments, perhaps once every three months. Therefore, for purposes of discussion, steady-state line parameters will be excluded as error sources.

The data-error-producing sources—hits—of concern here can be classified as amplitude hits, phase hits, impulse noise, and dropouts. Such hits can occur on any type of data communications circuit, but they will be discussed with particular reference to a long-haul microwave system, reviewed next.

Figure 2 shows one hop in a microwave link which uses the space-diversity technique to circumvent the atmospheric and terrain-configuration effects on the quality of the signal going on a line of sight from one tower to another tower.

Space diversity

In space diversity, usually the upper antenna of a pair on a tower is used for receiving, as well as transmitting. The lower antenna is used for receiving only.

Under certain fading conditions, ground reflections may be so strong that the amplitude of reflected signals R_U and R_L may be equal to the desired line-of-sight signals U and L. Suppose, for example, that the direct and reflected signals have the same amplitude, but the path length of the upper reflection is longer, so that R_U arrives at the antenna 180° out of phase with the direct path U. Then, the signals will cancel, and the upper antenna at tower B cannot deliver a useful signal for amplification and retransmission to the next tower. A similar out-of-phase situation could occur with the relationship of L to R_L. Or, perhaps, reflected signals could be in phase and reinforce the direct signals.

Generally, when atmospheric and terrain conditions cause fades, the signal at one antenna will be better than the signal at the other antenna. To profit from this situation, in space diversity, each antenna has its own receiver, including automatic gain control. A diversity switch monitors the output of both receivers, selects the better output, and feeds it to the antenna transmitting to the next tower along the path.

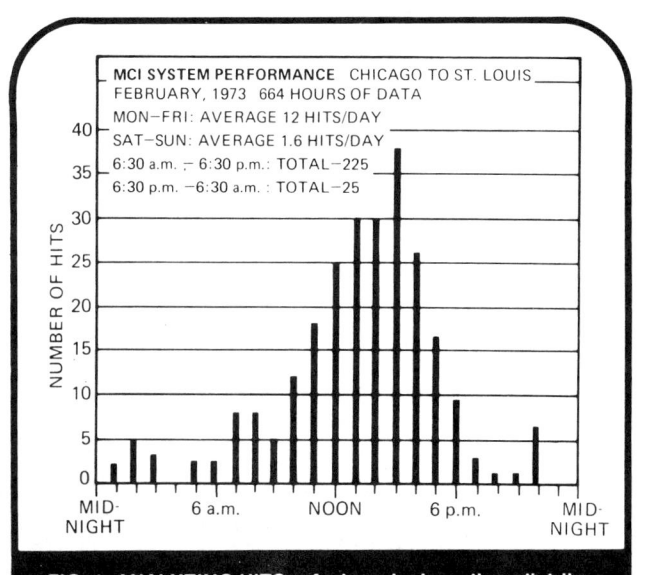

FIG. 1. ANALYZING HITS ■ An hour-by-hour time distribution of hits caused by electrical disturbances yields diagnostic information about microwave carrier performance.

Figure 3 contains actual simultaneous traces of signals received at a top antenna and a bottom antenna, as well as at the diversity switch which chooses, at any instant, which antenna has the better signal. Two "windows" are used here to focus on when the switching has occurred from one antenna to the other. A diversity switch in good working order will not itself cause a hit.

More than likely, the line-of-sight electrical path lengths from the transmitter to the two receiving antennas will differ, so that when switching occurs, there will be a difference in electrical phase angle—and if it is large enough, it will be classed as a phase hit. The cure for this is the insertion of a length of cable in the shorter-path-length facility so that both have the same electrical length. The procedure for correcting phase difference is called DADE, for differential absolute delay equalization.

During fading conditions, and particularly during deep fades that occur at night in the late spring and summer, the reflected signals may dominate either or both of the line-of-sight signals. Therefore, and despite good DADEing, when the space-diversity switch chooses the better signal, the electrical lengths are again unequal, thus creating a phase hit.

Frequency diversity

The effects of fading can also be reduced by another microwave technique called frequency diversity, which relies on the diversity switch to choose the better signal at two different frequencies rather than two different path lengths. MCI's link between Chicago and St. Louis uses frequency diversity. It is DADEd to 6°. To conserve the microwave spectrum, the Federal Communications Commission now allows the construction of frequency-diversity routes only under special circumstances. As a result, space diversity is used in initial construction of many long-haul microwave systems. It can be shown, also, that space diversity has certain technical advantages over frequency diversity, provided that the DADEing at each hop is satisfactory. The objective on the MCI space-diversity system under construction between Chicago and New York is a phase shift of 3 electrical degrees or less. Further, to raise performance by reducing transmission loss, and thus dynamic range of required amplification, the towers will be spaced an average of 23 miles apart, compared with industry custom of 27 miles.

In the tests run by MCI, a phase difference of 10° or more is classed as a phase hit. This threshold level is consistent with the present state of the art of modulation/demodulation techniques in high-speed datasets. That is, with a phase hit of 10° or less, a quality dataset should be able to remain in synchronism. But for larger phase hits, the dataset may temporarily go out of sync, and during the resynchronizing interval, its data stream is erroneous.

When a phase change occurs at a baseband line frequency (up to 8.3 megahertz), the same phase change appears on the voice-frequency (voice-grade) channel that is subsequently demultiplexed from that frequency.

Both equipment and propagation effects may cause differences in amplitude of the demodulated signals at the output of the regular and the diversity radio paths. When traffic switches from one facility to another, this difference introduces a step change in signal level which, if large enough, will cause the dataset to generate an erroneous bit stream while its automatic gain control acts. Even when switching takes place at the radio frequencies of a microwave system, an amplitude hit may result from distortions in gain-frequency response characteristics of the demodulated signals.

An amplitude hit is defined here as a 6-dB-step change in level. This assumes that level changes smaller than this value can be compensated for by the dataset's circuits. Depending on the time relationship of the amplitude hit with respect to the data signal, an amplitude hit may appear simultaneously with a phase hit.

Impulse noise

Impulse noise derives from such sources as lightning strikes, power-load switching, motor startups, and even the opening and closing of small relays and switches. Generally, impulse-noise bursts can be as short as a few microseconds and as long as several milliseconds. The short-haul, or local, wire and cable links in urban or industrialized areas are thus more susceptible to impulse noise—both in amplitude and frequency of occurrence—because of the large amount of electrical equipment operating in these areas. Long-haul microwave routes, however, usually traverse rural areas, which have a sparse amount of electrical equipment and thus are relatively "quiet" (Fig. 4).

Even so, microwave links do suffer from occasional impulse hits. Here, an impulse is considered a hit when it equals or exceeds 71 dBrn0 (VB), where dBrn0 means decibels above a zero reference-noise level of 1×10^{-12}

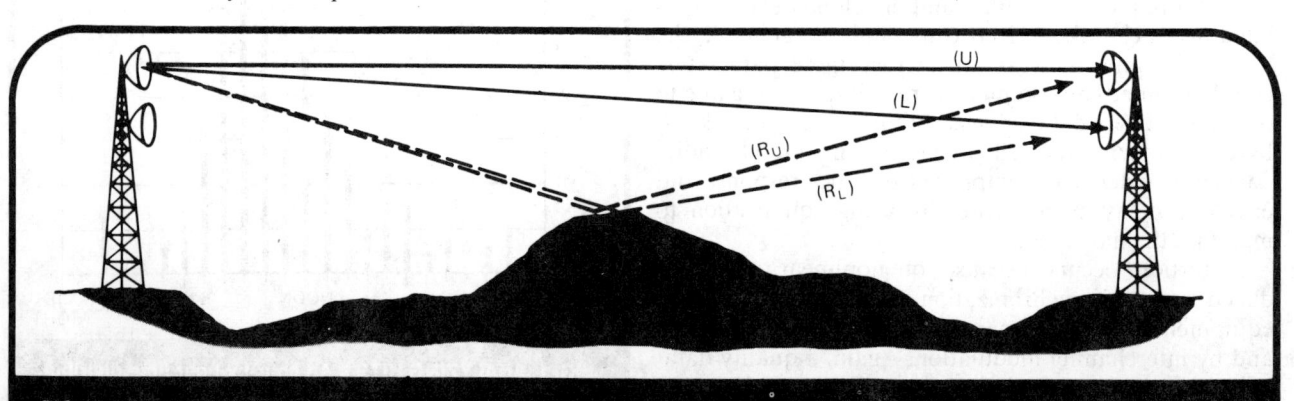

FIG. 2. SPACE DIVERSITY ■ Using two channels, the space-diversity microwave link can choose the stronger of the two signals to overcome the effects of fading and forward the stronger signal on to the system's next microwave tower.

watts. VB, for voice band, means the frequency-response characteristics of a filter used in taking noise measurements. Thus, 71 dBrn0 (about 10 microwatts) is slightly more conservative than the more customary 72 used by the phone companies. Because datasets are generally designed to work well at noise inputs of less than 72 dBrn0, only impulses of 72 dBrn0 or greater may cause the dataset to generate errors.

Although impulses less than 71 dBrn0 are not considered hits, and thus will not create data errors, it was decided to study the distributions of impulses at three lower power levels and correlate them with specific causes. Figure 5 shows the results of tests run at 54, 60, and 66 dBrn0. At such high sensitivities, the number of measured impulse spikes per hour is quite high, and their distribution depends on the source. For example, deep fading creates a phenomenon which, while not truly impulse noise, behaves similarly to impulse noise. The deeper the fade, the more noise spikes are created at a given threshold level. This is shown by the colored bars in Fig. 5.

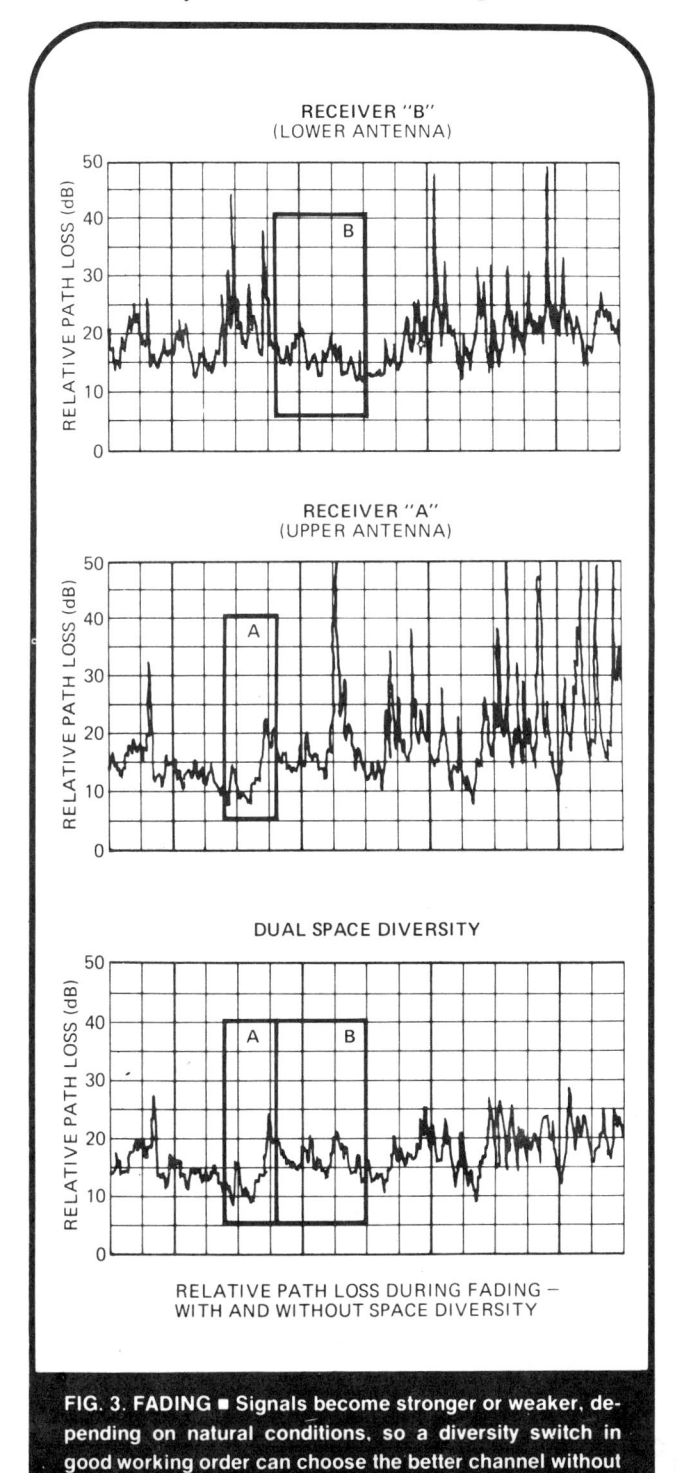

Noise created by electrical-equipment operation—which would go unnoticed at 71 dBrn0 by the dataset—also shows up in the distribution at high-sensitivity thresholds, as shown by the white bars in Fig. 5. Each time a nearby farmer, for example, starts up a water pump, this could register as a spike at one of these threshold levels. Continuous monitoring at such high-sensitivity noise thresholds leads to identification of the noise source and its rapid correction.

Dropouts are amplitude hits that last longer than 300 milliseconds. They can be caused, as examples, by a defective diversity switch or by a transient voltage spike (impulse) of such magnitude as to saturate an amplifier and clip the signal until saturation disappears. Although no dropouts occurred during MCI's test period, the impact of dropouts cannot be ignored. This point will be discussed later.

Test instrumentation

The performance data presented in Figs. 1 and 5 was obtained from a comprehensive instrumental setup. Test equipment to measure threshold levels of impulse noise, phase hits, and amplitude hits were connected to counters. In addition, logic provided a count of a coincident amplitude and phase hits. A signal was also provided to count every time the diversity switch at St. Louis operated. Besides this, all these signals were recorded on a multi-event strip chart recorder. Figure 6 shows a portion of a test run, which has been redrawn

FIG. 3. FADING ■ Signals become stronger or weaker, depending on natural conditions, so a diversity switch in good working order can choose the better channel without causing a phase shift, or phase hit, which causes errors.

FIG. 4. BUCOLIC ■ Microwave links traveling across country are not too susceptible to electrical noise.

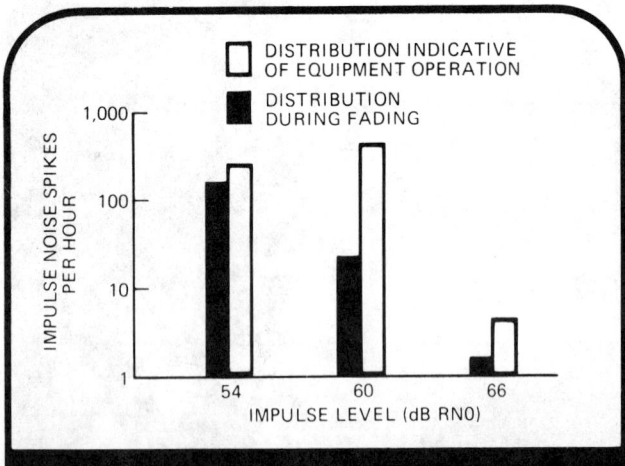

FIG. 5. SENSITIVE ■ Even though noise impulses at the levels shown won't cause data errors, it's wise to observe their distribution to anticipate future difficulties.

to provide a readable presentation of hit events.

The counters were read every hour on the hour by an operator. Figure 1 displays the total count of all hits for the month, by hour of the day. This information has proven so useful that the instrumentation will be redesigned for automatic, unattended operation. In a sense, the counters provided a composite characteristic of line performance. The strip chart's main application was to permit the detailed observation and interpretation of the relationship between specific hits and other identifiable activities.

Over a long communications circuit, even a very short hit can stretch into a longer hit burst, which has the same effect as a dropout because of the time taken by automatic gain- and frequency-control circuits to resettle to a steady-state condition. Under these circumstances, a short hit can cause a chain reaction of delays in the radio equipment, the multiplex equipment, and the datasets.

Measuring performance

In the data communications environment, it has been customary to measure the performance of a data channel in terms of bit-error rate (BER), where bit-error rate is the number of bits in error relative to the total number of transmitted bits. That is, 10 error bits out of 20,000 transmitted bits is an error rate of 5×10^{-4}. BER measures effect, not cause. And such a measure of performance may be reasonable when the hit distribution is fairly uniform in time and the user is concerned with performance averaged over a long period.

However, bit-error rate gives no insight to the real impact of hits: whether a hit caused a 1-bit error which, because of error control procedures, in turn required the retransmission of a 1,000-bit block, or whether a stretched hit caused a longer error burst which overlapped two blocks and caused the retransmission of 2,000 bits.

From the user's operational viewpoint, performance is measured more accurately by the time distribution of hitless intervals and of the hit-burst lengths. Such information can then be more meaningful in specifying acceptable system performance and developing an efficient error-control system. For example, assume that data is transmitted over a leased line for display on a video terminal. If hits occur frequently, then the system may have to be designed to request retransmission of the line (of display) in which the error occurred. This procedure would involve substantial error-control overhead. And it could be disconcerting to the operator if several lines are retransmitted for each page displayed by the terminal.

However, if a hit occurs infrequently, and, even if it stretches, then it may be more efficient to transmit a page at a time, in a continuous manner, and retransmit the whole page on the rare occasion of a hit.

The exact approach taken by the user to monitor his communications depends on many system factors, one of which is the hitless quality of his lines. But the quality of all lines will be different, and thus it may be beneficial for the user to install his own monitoring equipment, similar to that used in testing MCI's microwave link. He can then observe actual line performance, develop an operational plan suited to actual hit distribution, and anticipate any trend toward degradation of the line, and isolate faulty links for rapid restoration of service by the common carrier. [EOT]

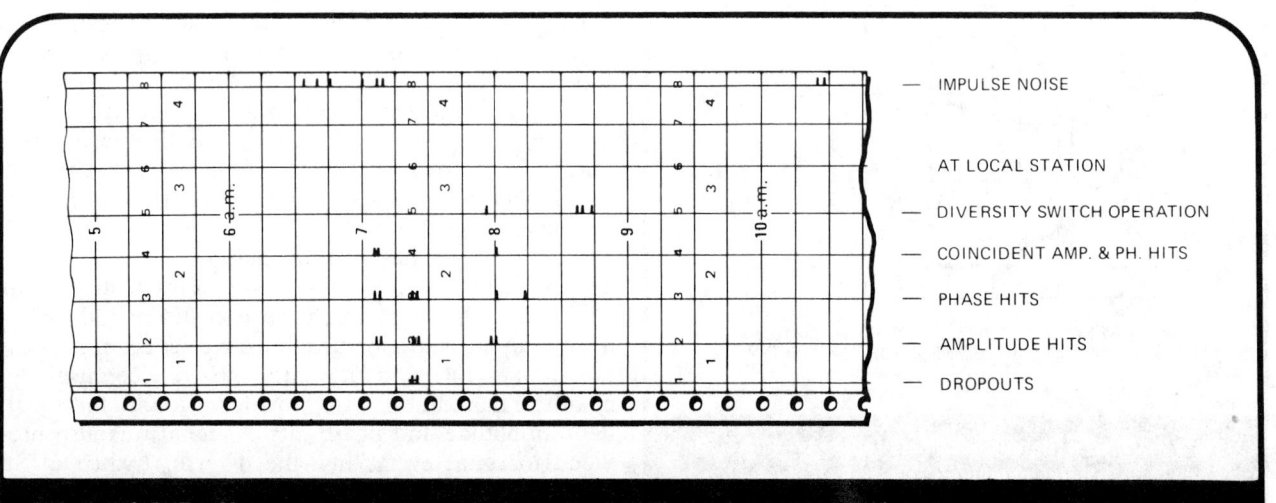

FIG. 6. INSIGHT ■ Multipen strip-chart recorder keeps track of time and coincidence of hits to permit detailed analysis and troubleshooting; meanwhile the hits are counted to produce the hour-by-hour time distribution of hits as in Fig. 1.

Getting peak performance on a data channel

How much usable data can be pumped down a channel depends on eight operating factors, ranging from modem speed to line protocol

Carl N. Boustead and Kirit Mehta
International Communications Corp.
Miami, Fla.

Throughput

Characterizing a data channel in terms of its transmission speed is a convenient way of specifying the maximum instantaneous data rate. But this rating usually masks a more essential specification—the number of usable characters per second that can be transmitted. The raw transmission speed of the equipment and transmission lines making up the data channel is what the user pays for, and, in general, the faster the equipment, the more it costs.

Net data throughput, the rate that usable characters are received, depends not only on the equipment, but also on how efficiently the user integrates the various elements—and there are many elements—that make up the data channel. Unless the channel's elements work in concert to attain the peak usable character rate, the extra cost of higher speed equipment and lines may be wasted.

Consider the data channel consisting of a transmission line, two modems, two terminals, and a particular way of implementing error control. For each channel, with a given type of system arrangement, there is a set of operating conditions that will provide maximum net data throughput (NDT). NDT, a measure of information-transfer efficiency, distinguishes between total number of text, overhead, and retransmitted characters sent by a terminal and the number of correct text characters that actually are received by a terminal or computer in a specified length of time.

One of the more important parameters in establishing NDT is the modem's speed. But, contrary to popular belief, raising the modem's speed in an attempt to increase NDT will not work in every case. Usually a faster bit rate from a more expensive modem does significantly increase the net data throughput, but a faster modem will in a relatively few isolated cases actually reduce the usable throughput. In a few other cases the marginal increase in NDT won't be worth the extra cost for the faster bit-rate modem. Besides the bit speed, the over-all systems operating parameters cannot be ignored when optimizing throughput. The other important parameters that can affect data-channel performance include such things as how rapidly the modem can resynchronize itself, the total time between consecutive blocks in a message, the error environment, the method of error control used in a particular link, the number of characters in a text

Calculating data channel throughput

For half-duplex operation, the net data throughput, in characters per second, can be obtained from:

$$NDT = \frac{K_1(M - C)(1 - K_2K_3E)^M}{\frac{M}{R} + \Delta T} \qquad (1)$$

NDT is net data throughput in bits per second
K_1 is the number of information bits per character
K_2 is the total number of bits per character
E is the bit error rate
C is the average number of noninformation characters per block
K_3 is the multiple-bit-error discount
M is total block length
R is the modem's speed in bits per second
ΔT is the time interval between blocks

Holding every factor constant at a given set of conditions and varying block length M results in the curve in Case I, shown later. But other factors can also be varied to provide specific insight to channel performance, as shown in other cases.

The curve of optimum block length (M*) as a function of bit error rate, shown in Case V, is obtained from this equation:

$$M^* = \frac{C - R\Delta T}{2} + \left[\frac{R^2\Delta T^2}{4} + \frac{CR\Delta T}{2} + \frac{C^2}{4} - \frac{R\Delta T + C}{\log_e(1 - K_2K_3E)}\right]^{1/2} \qquad (2)$$

Equations derived by ICC Institute. NDT is also called TRIB, for transfer rate of information bits.

---

block, and the number of overhead characters required to transmit the text block.

In fact, getting peak performance requires knowing everything about how the data channel's line, modems, and terminals interact with each other.

Factors in throughput

Moreover, at least eight factors determine net data throughput. Because the interactions between them are so complex, obtaining a qualitative understanding—much less a quantitative result—of the relationships becomes difficult. The complexity is demonstrated by an equation that combines all pertinent factors to yield NDT. This equation, together with a companion equation, is presented in "Calculating data channel throughput."

Fortunately, understanding data channel behavior can be simplified by using graphic plots derived from the equations. The plots, which show interaction of varying values of the channel factors being considered, demonstrate not only how a data channel operates under a given set of circumstances, but also how to improve the channel's performance. In other words, the eight plots presented later as case studies indicate ways to obtain the most NDT for the lowest system cost.

Before discussing these curves, it is necessary first to define the relevant factors and put them in operational perspective.

NET DATA THROUGHPUT is the rate of the number of information characters accepted by the receiving station (terminal or computer) compared with the total transmission time required to get these characters accepted. In the simplest data communications system operating in one direction with no error control, a 4,800 b/s modem can send 480 10-bit characters each second. But suppose errors on the channel force the rejection of 80 characters during the second. Then the number of information characters accepted by the receiver is 400 during that second. Therefore, the NDT is 400 accepted characters a second—well below the 480 characters transmitted each second.

INFORMATION BITS PER CHARACTER (K_1) depends on the code being used on the data channel. For example, for EBCDIC code K_1 is 8, for ASCII code it's 7, and for Baudot code it's 5. (EBCDIC means extended binary coded decimal interchange code; ASCII means American standard code for information interchange.)

TOTAL NUMBER OF BITS PER CHARACTER (K_2) includes, in addition to K_1, bits for parity check and for the serial start-stop bits to delimit a character when using asynchronous transmission. Thus, for example a 7 bit ASCII character transmitted with 1 parity bit, 1 start bits, and 2 stop bits would have a K_2 of 11. The difference between K_2 and K_1 is considered to be the number of overhead bits per character.

BIT ERROR RATE (E) is the number of bits in error, relative to some total number of bits transmitted. Thus, 5 bits in error out of 100,000 transmitted is a bit error rate of 5×10^{-5}. The bit error rate (BER) is a function of transmission line characteristics, noise, and the ability of the modem to perform under these conditions with a given coding scheme. Unless otherwise stated, the assumption is made that errors are randomly distributed in time.

MULTIPLE-BIT-ERROR DISCOUNT (K_3) takes into account the fact that, even when the external cause of bit errors (impulse noise, for example) is randomly distributed, the bits in error coming from the modem tend to occur in groups, or bursts. One reason is that in a modem with an adaptive equalizer, the cause of the error is propagated through the modem's tapped-delay filter. During this propagation interval, the modem will usually produce closely grouped fault bits. Necessary coding methods in higher speed modems also contribute to this bursty bit error condition.

Because of this burst characteristic, several errors usually impact the same character or data block. And no matter how many bits are in a character or block, a single bit in error is all it takes to require a retransmission. Therefore, K_3 discounts the effect of burst errors on retransmission. K_3 is defined as the occurrence of errors divided by the number of bits in error as measured over a long time with a particular type of modem on a particular line. It is always less than 1. The range of K_3 typically varies between 0.2 and 0.5. For example, network tests show that K_3 would be 0.3 for most adaptively equalized modems.

TOTAL BLOCK LENGTH (M) is the total number of characters in a block, including, as shown in the figure, the start-of-header characters, the start-of-text character, the message text, the end-of-text character, and the block-check characters. Only the message-text characters are considered as information characters. They are shown in color.

NONINFORMATION CHARACTERS PER BLOCK (C) are all the characters in a transmitted block that are not message characters. These are shown in white in the figure. These characters, and such other control characters as PAD and SYN required to keep a channel properly synchronized for bits, characters, and blocks are considered to be overhead characters. As can be seen from the numerator of Equation 1 in the panel, the larger the number of overhead characters (C), relative to the number of message characters (M) in a block, the lower the net data throughput.

RATED MODEM SPEED (R) is the nominal continuous bit-transmission rate in bits per second.

TIME INTERVAL BETWEEN BLOCKS (ΔT) is the dead time in seconds that occurs, for any reason, between the end of one block and the beginning of the next block in a given message transmitted in one direction. Included in this time interval are delays caused by such operating factors as:

- Line turnaround time when operating over a two wire, half duplex, link which can become particularly burdensome because two line turnarounds are involved in transmitting a block and then an acknowledgment. Trying to reduce the ratio of line turnaround time to total transmission time is one of the main reasons for increasing the number of bits transmitted in each block.

- Waiting for an acknowledgment signal (ACK/NAK/WACK) over a low speed reverse channel when operating on a two wire link, or over a full speed return channel when operating on a four wire link.

- The time allotted to a modem to retrain its equalizer (if any) and to regain synchronization. Called the CLEAR TO SEND (CTS) time, this interval can be reduced to a negligible amount by operating the modem in a continuous carrier (carrier up) mode over a four wire link.

- Electrical propagation interval required for a signal to traverse a link from the sending end to the receiving end—and back again for the error-control acknowledgment signal. Any buffering and queuing delays can be included here, particularly if the data block passes through one or more network nodes containing communications processors such as programable front-end processors and remote data concentrators.

Error controls modify throughput

In the eight graphic cases that follow to demonstrate the impact of various operating variables, the error-control method plays a very important role, since each method makes a different contribution to the time interval between blocks and thus to the reduction of the channel's net data throughput. The three methods considered here are:

- HALF DUPLEX OPERATION using a two wire link and employing the Binary Synchronous Communications (BSC) data link control. The major contribution to the time interval between blocks is the 300 milliseconds per block required to turn around the echo suppressor in a dial-up line, 150 milliseconds for each direction. Furthermore,

(continued on page 187)

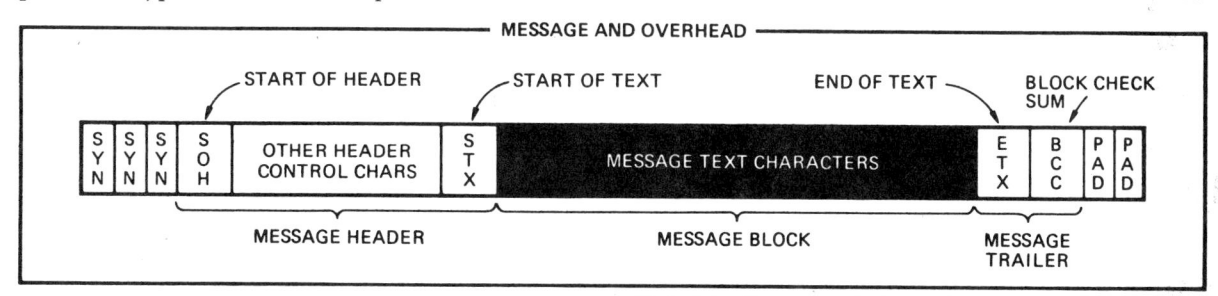

MESSAGE AND OVERHEAD

| SYN | SYN | SYN | SOH | OTHER HEADER CONTROL CHARS | STX | MESSAGE TEXT CHARACTERS | ETX | BCC | PAD | PAD |

START OF HEADER — START OF TEXT — END OF TEXT — BLOCK CHECK SUM

MESSAGE HEADER | MESSAGE BLOCK | MESSAGE TRAILER

Net data throughput depends, among other things, on the relationship of the number of message characters to the total number of characters in the block, including the overhead control characters.

Eight examples of data channel performance

CASE I. HOW BLOCK LENGTH AFFECTS NET DATA THROUGHPUT

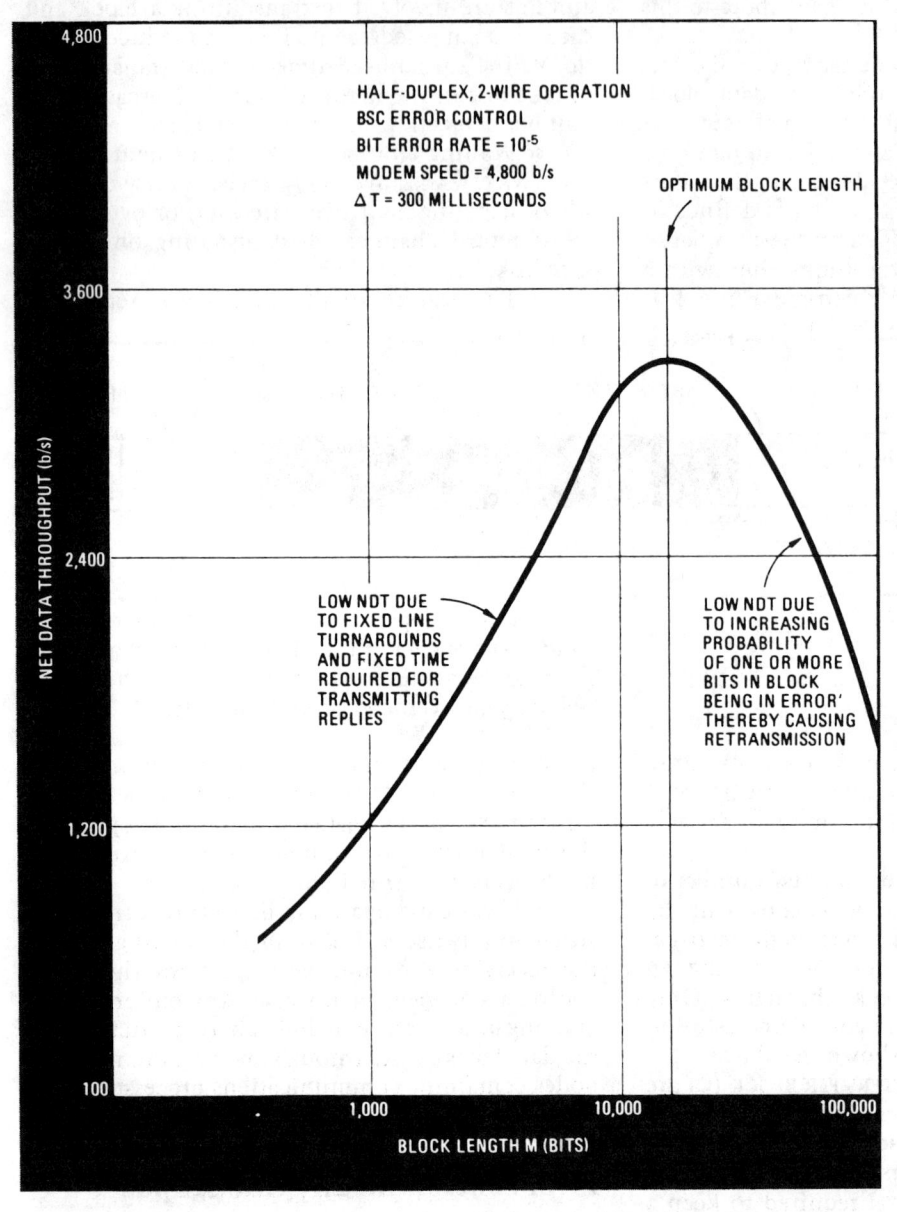

[Graph: Net Data Throughput (b/s) vs Block Length M (bits), log scale from 1,000 to 100,000. Y-axis from 100 to 4,800. Curve peaks around block length of ~15,000 bits at approximately 3,400 b/s, labeled "OPTIMUM BLOCK LENGTH."

Annotations on graph:
- HALF-DUPLEX, 2-WIRE OPERATION
- BSC ERROR CONTROL
- BIT ERROR RATE = 10^{-5}
- MODEM SPEED = 4,800 b/s
- ΔT = 300 MILLISECONDS
- LOW NDT DUE TO FIXED LINE TURNAROUNDS AND FIXED TIME REQUIRED FOR TRANSMITTING REPLIES
- LOW NDT DUE TO INCREASING PROBABILITY OF ONE OR MORE BITS IN BLOCK BEING IN ERROR' THEREBY CAUSING RETRANSMISSION]

Using the equation for net data throughput (Equation 1) to prepare this curve, all factors are held fixed except block length M. The more significant fixed parameters are listed next to the curve. The shape of the curve shows the particular block length that yields the highest net data throughput (NDT) for the given data channel.

To the left of the peak, the lower NDT for shorter block lengths is caused by the impact of the long turn-around time and reply time, relative to block-transmission time, that characterizes half duplex two-wire operation using conventional error detection and retransmission schemes. Furthermore, to the right of the optimum, net data throughput deteriorates when block lengths become too long, because even one erroneous bit impacts a long block and causes retransmission of the entire block—thus once again reducing the net data throughput.

This curve also shows that, under the stated conditions, the peak NDT is about 30% less than the modem speed. At best, then, using a 4,800 b/s modem this data channel can transfer just about 3,500 usable bits each second, not the 4,800 b/s called for by the modem's speed rating. If a 3,600 b/s modem were used, the transfer rate of information bits would decrease proportionally. And it doesn't take much of a difference from the optimum block length to drastically reduce the net data throughput.

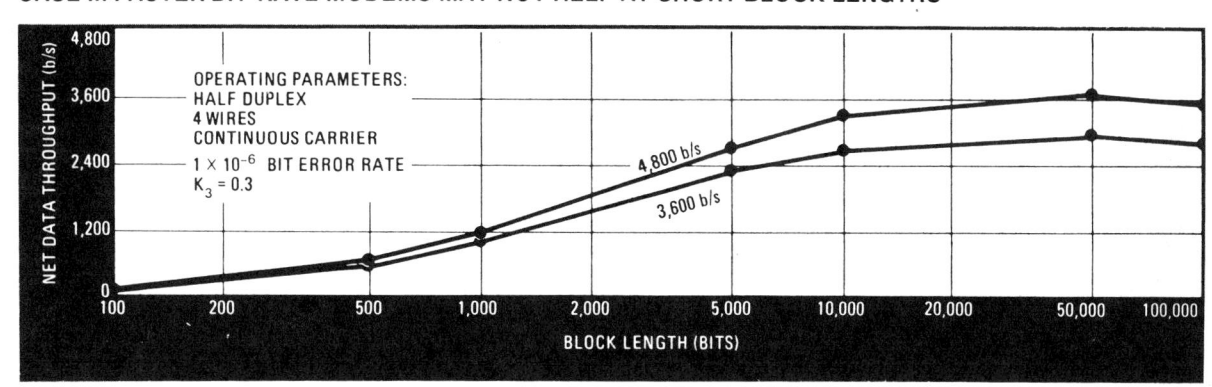

CASE II. FASTER BIT-RATE MODEMS MAY NOT HELP AT SHORT BLOCK LENGTHS

Throughput as a function of block length is compared here for modems of two different speeds, operating in half duplex, four-wire mode. The modems run with continuous carrier, so the CLEAR TO SEND (CTS) time is essentially zero.

For the given error rate of 1×10^{-6}, the optimum block length of the two data channels—one using a 4,800 b/s modem and the other a 3,600 b/s modem— is about 50,000 bits. At the optimum block length, the NDT for the 4,800 b/s channel is substantially better than that for the 3,600 b/s channel. But when block lengths are short, say about 1,500 bits or less, there isn't any significant improvement in throughput by operating at the higher transmission rate.

In this range of up to 1,500 bits per block, NDT is largely determined by the relatively long turnaround times (due to modem and line propagation and terminal reaction time) and the time interval required to transmit block acknowledgments or response to idle polls. For the medium block lengths, more information is transferred in a given time, relative to the turnaround dead time, so throughput goes up for both channels.

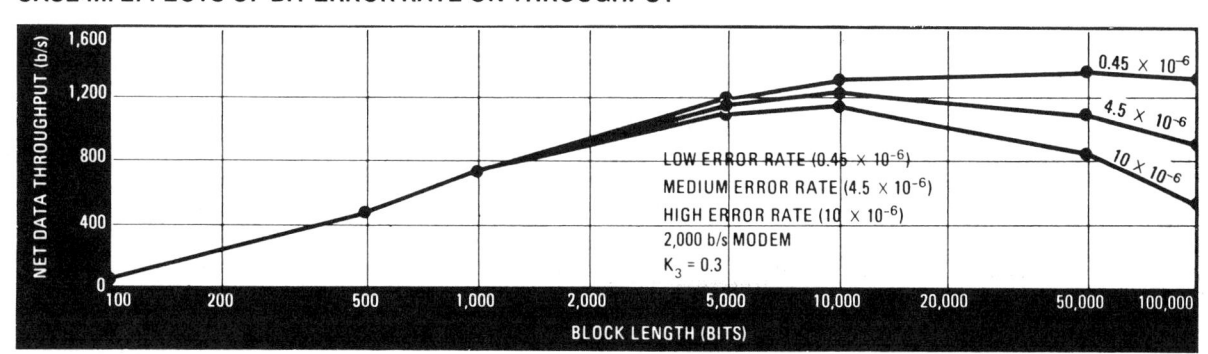

CASE III. EFFECTS OF BIT ERROR RATE ON THROUGHPUT

For the 2,000 b/s modem and channel used in this case, the bit error rate can vary by as much as 20 to 1, and there will be little noticeable difference in net data throughput for blocks that may contain as many as 5,000 bits. Again, the long dead time introduced by line turnarounds between blocks dominates NDT more than does the error environment.

But difference in throughput due to error becomes apparent when blocks become longer than 5,000 bits. Above this value, obtainable NDT starts to decrease rapidly when BER goes up (10×10^{-6}). But when BER is extremely good (0.45×10^{-6}), the throughput continues to increase with block length up to about 100,000 bits per block.

When everything else remains steady, it would seem that if one could be certain that the error would not exceed 0.45×10^{-6}, then a block length of 100,000 bits would be a wise and efficient choice. But, as the next case shows, this choice can be risky.

Eight examples of data channel performance

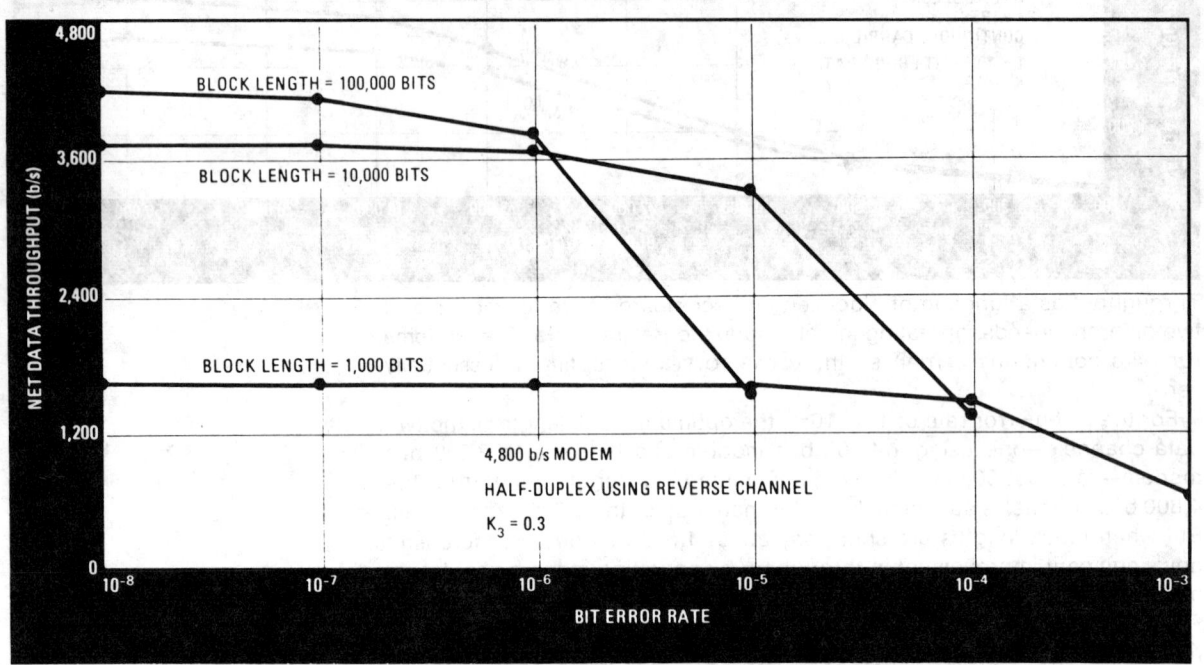

CASE IV. SENSITIVITY OF BLOCK LENGTH TO BIT ERROR RATE

[Graph: Net data throughput (b/s) vs Bit error rate, showing three curves for block lengths of 100,000 bits, 10,000 bits, and 1,000 bits. 4,800 b/s MODEM, HALF-DUPLEX USING REVERSE CHANNEL, $K_3 = 0.3$]

For a data channel of a given speed, here 4,800 b/s, the net data throughput is a function of the selected block length. But the ability of the channel to sustain the NDT at the given block length is itself dependent on the error environment to which the channel is exposed. As shown, a 1,000-bit block length (that is, the shorter block lengths) is relatively insensitive to the bit error rate. In this case, the designer can be assured of an NDT close to 1,600 b/s, even though the error rate becomes as bad as 1×10^{-4}. Thus, if the transmission line has poor or variable error quality, a 1,000-bit block seems a wise choice.

However, a 100,000-bit block can be used between satellite links, where the error rate is expected to be low (less than 1 erroneous bit in a million bits) and the nature of the applications make long block transmissions worthwhile. Note, however, that with longer block lengths, NDT deteriorates so rapidly as to be worthless, once the bit error rate exceeds the nominal design value.

CASE V. HOW OPTIMUM BLOCK LENGTH DEPENDS ON BIT ERROR RATE

In Case I, the optimum block length is shown for a particular bit error rate. For any set of conditions, there is an optimum block length for each bit error rate. Graphically, this means that location of the optimum block length and its corresponding NDT shifts as the BER is varied. Another curve, connecting the resulting peak values is then the plot of optimum block length as a function of BER. Such a plot can be obtained from the second equation in the panel.

Using this equation, the curves shown for Case V were calculated for the conditions indicated. These curves reveal that at zero bit error rate, to the left, the optimum block length approaches infinity. Zero BER is unrealistic. However, in the BER ranges found on most data channels, the optimum block length is still very sensitive to BER; the worse the error rate, the shorter is the optimum block length. Note also how the number of bits per character (K_2) affects the optimum block length, measured in characters and not bits, at any given error rate: the fewer the total bits per character, the longer the optimum block length in number of characters.

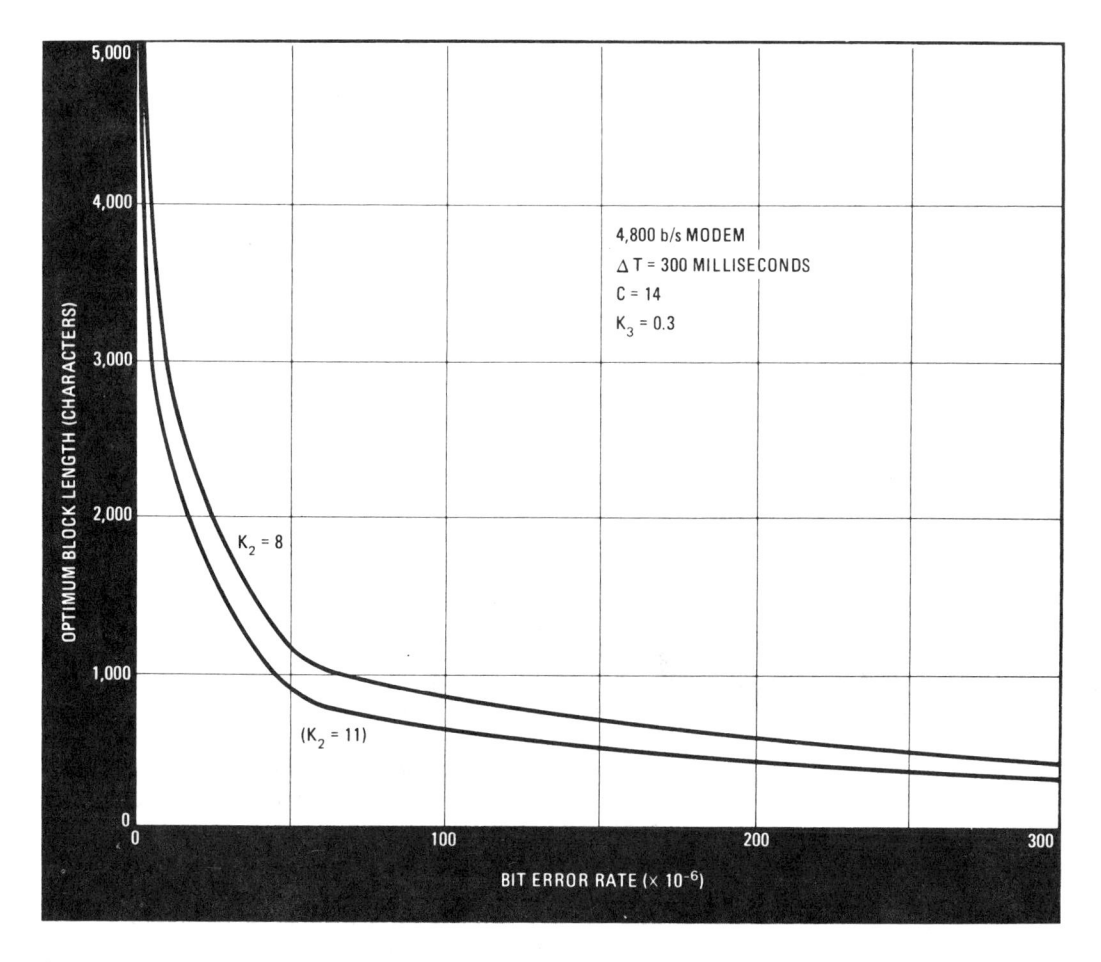

[Figure: Optimum block length (characters) vs. Bit error rate (× 10⁻⁶), for 4,800 b/s MODEM, ΔT = 300 milliseconds, C = 14, K₃ = 0.3, with curves for K₂ = 8 and (K₂ = 11).]

CASE VI. IMPACT OF CLEAR TO SEND TIME ON THROUGHPUT

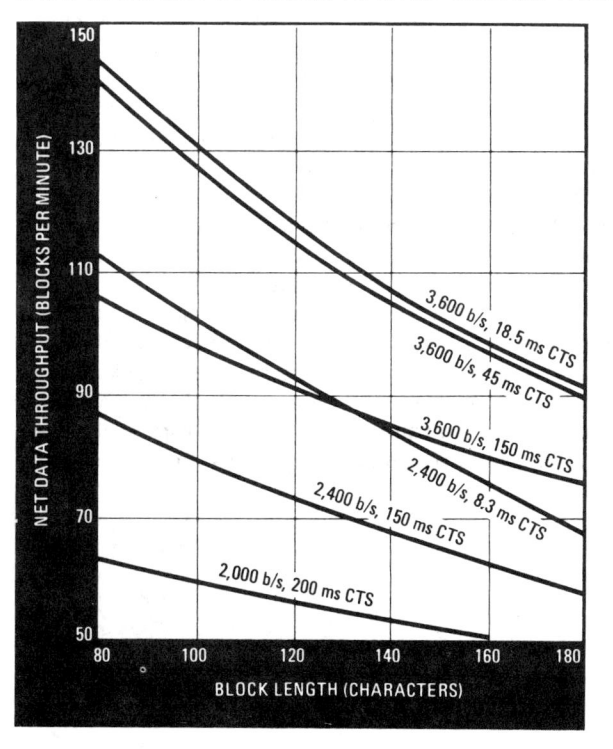

[Figure: Net data throughput (blocks per minute) vs. Block length (characters), with curves for 3,600 b/s, 18.5 ms CTS; 3,600 b/s, 45 ms CTS; 3,600 b/s, 150 ms CTS; 2,400 b/s, 8.3 ms CTS; 2,400 b/s, 150 ms CTS; 2,000 b/s, 200 ms CTS.]

Even though a terminal indicates during the handshaking procedure that it is ready to transmit data, actual transmission can't start until the associated modem gets its receiving circuits ready to receive error-free data. The time for the modem to get into this ready state is called CLEAR TO SEND (CTS) time.

Different types of modems have different CTS times. In general, modems employing automatic equalization have longer CTS times than those that are manually equalized. The reason is that an auto-equalized modem requires "retraining" time, while the equalizer adjusts to changing line parameters.

In the examples of the 3,600 b/s modems, shown in color, the effect of increased CTS time is very apparent. All other things being equal, the shorter the CTS time is, the better the throughput. The two middle curves show dramatically that faster bit-rate modems are not always more efficient than slower bit-rate ones. At a block length of 135 characters or less, the 2,400 b/s modem has a better throughput than does the 3,600 b/s modem. The reason is that the faster bit-rate modem has a longer CTS time than the slower bit-rate modem, which defeats the advantage of faster transmission rate.

Eight examples of data channel performance

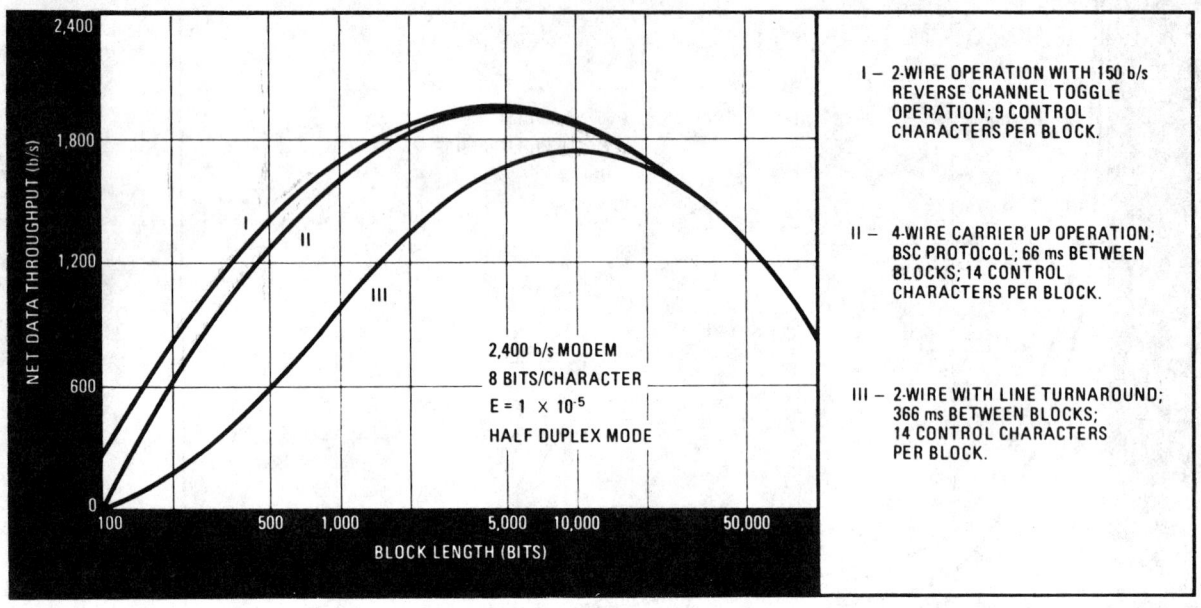

CASE VII. HOW ERROR CONTROL METHOD AFFECTS THROUGHPUT

2,400 b/s MODEM
8 BITS/CHARACTER
$E = 1 \times 10^{-5}$
HALF DUPLEX MODE

I — 2-WIRE OPERATION WITH 150 b/s REVERSE CHANNEL TOGGLE OPERATION; 9 CONTROL CHARACTERS PER BLOCK.

II — 4-WIRE CARRIER UP OPERATION; BSC PROTOCOL; 66 ms BETWEEN BLOCKS; 14 CONTROL CHARACTERS PER BLOCK.

III — 2-WIRE WITH LINE TURNAROUND; 366 ms BETWEEN BLOCKS; 14 CONTROL CHARACTERS PER BLOCK.

The most common mode of data channel operation is half duplex. In this mode, two terminals communicate with each other over a pair of primary wires, but only one terminal can transmit at a time. And when operating half duplex, three ways of implementing the popular stop-and-wait error control technique are intrinsic to IBM's Binary Synchronous Control (BSC) data link control.

Every time a block is sent, the transmitter must wait for a signal from the receiver as to whether or not the block was received without error; if an error occurs, the block must be retransmitted. The three ways of acknowledging are: using a reverse channel that sends back a 0 or 1 toggle to indicate correct reception; using another pair of wires (hence called a four-wire system) to return the ACK (acknowledge) or NAK (negative acknowledge) replies; and using a two-wire system in which the line is turned around to send back the ACKs and NAKs. Each method introduces a different amount of dead time (ΔT), which therefore affects the available net data throughput. Surprisingly, the four-wire method is less efficient in NDT than the reverse channel method. The reverse channel method requires 56 milliseconds between blocks, compared with 66 milliseconds between blocks for the four-wire method. Not so surprisingly, the two-wire method requiring longer turnaround times shows the poorest performance.

CASE VIII. EFFECT OF ROUND TRIP PROPAGATION DELAYS

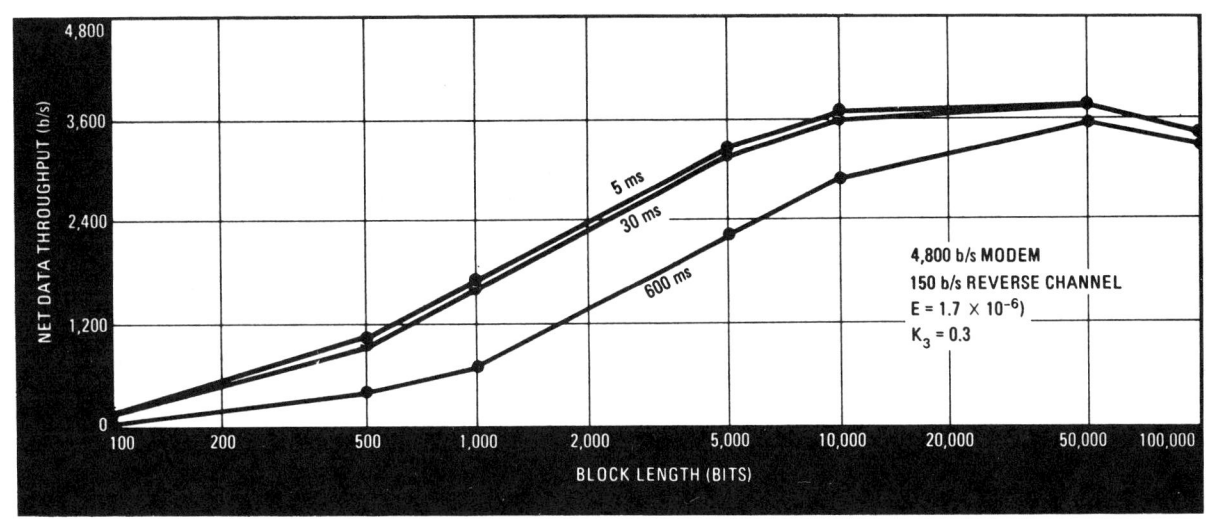

[Graph: Net Data Throughput (b/s) vs Block Length (bits), showing three curves labeled 5 ms, 30 ms, and 600 ms. Conditions: 4,800 b/s MODEM, 150 b/s REVERSE CHANNEL, $E = 1.7 \times 10^{-6}$, $K_3 = 0.3$]

Among the several components of channel dead time (ΔT), the propagation delay of the data traveling over the transmission link can also have some negative impact on the net data throughput. For terrestrial telephone lines, a rule of thumb for the delay is 10 milliseconds for each 1,000 miles of line, one way. Thus, transmissions between nearby locations can add 5 milliseconds to dead time, and transmissions 1,500 miles long will add 30 milliseconds. For satellites, the total distance is about 50,000 miles, and this round-trip delay totals about 600 milliseconds.

The difference in throughput for these three delay times is shown here as a function of block length. For delays of 5 and 30 milliseconds, the throughputs are substantially the same because the more important factor in reducing throughput at short block lengths is the stop-and-wait time inherent in the automatic request for the repeat error control used here.

Throughputs on a satellite link are shown here, in color. However, it is unlikely that any satellite channels requiring high net data throughput would operate with a half duplex stop-and-wait block repeat error control. Instead, satellites will probably employ full duplex operation and multiple frame numbering, as in IBM's new Synchronous Data Link Control (SDLC)—with multiple acknowledgment. This procedure on satellite links will yield higher net data throughput.

(continued from page 181)

the time equal to the transmission of forward control characters and reply control characters, to permit the sending and receiving modems to synchronize, must be included in the total time interval.

■ HALF DUPLEX, LOW SPEED TOGGLE OPERATION also employs a two wire link, but a low speed reverse channel option on the modem provides a narrow-band channel for transmitting error-control acknowledgments. The reverse channel usually works at 150 b/s, and one way of acknowledging is to toggle the reverse channel from a 1 to a 0 condition, or send some similar signal, to indicate the acceptance of a correct forward block. Thus, the toggle delay time is equal to that for one bit, or 6.7 milliseconds. Furthermore, the carrier for the reverse channel is always on, or "up," so that no line turnaround time is involved. But a delay similar to the transmission time for forward control characters is required for imparting the information on the low speed channel.

■ HALF DUPLEX, HIGH SPEED REPLY OPERATION employs a BSC protocol and four wire link, with each pair of wires carrying data at the modem's rated speed in both the forward and reverse directions. Here, as in the first case, the time interval between blocks includes that for forward and reverse control characters and electrical propagation delays. Carriers in both directions are always up. Although the reply-channel speed here is much faster than the toggle change on the reverse channel operation, six control characters are required in sending the acknowledgment reply.

The lengthy captions for the graphs provide further insight to the relationship between modem speed, and hence cost, and the range of performances that can be obtained from a data channel. ■

Taking a fresh look at voice-grade line conditioning

AT&T's new D-conditioning for private voice-grade lines helps high-speed modems operate at improved error rates

Karl I. Nordling
Paradyne Corp.
Largo, Fla.

Performance

Private lines leased from the telephone company are not consistently perfect for high-speed data transmission. Still, today's fast, 4,800- and 9,600-bit-per-second modems with their automatic equalizers do a fine job of keeping error rates low—and these error rates can be further reduced by D-conditioning, AT&T's newest way of minimizing line impairments.

D-conditioning is not an alternative to the well-established C-conditioning. Granted, each improves the data characteristics of private voice-grade lines, but they do so in different ways and can be used either separately or in conjunction, depending on a user's needs.

Furthermore, it's not hard to grasp the difference between the two types of line conditioning and when they will help (or hinder) a modem's operation. But first it's necessary to understand the kinds of line impairments that induce data errors in telephone lines.

The 3002 voice-grade private line is the type of channel most commonly used for high-speed data communications. It is subject to many line impairments, 11 of which have been identified so far. Four of them may be controlled by the telephone company to the limits expressed in the FCC tariff no. 260, and four more are controlled by AT&T to limits specified by its internal practices (Table I).

The first four are the main subject of this article. To this group belong attenuation distortion and envelope-delay distortion—controlled through C-conditioning—and signal-to-noise ratio and harmonic distortion—now controlled through D-conditioning. Whenever the telephone company exacts a monthly charge for these two forms of conditioning, it is required by the tariff to keep these line impairments from exceeding the limits set out in the tariff.

The second group consists of impulse noise, frequency shift, phase jitter, and echo. Traditionally, because the telephone company knows that customers intend to transmit data over certain lines, it tries to maintain the lines within certain limits which it sets itself, even though not obligated by tariff to do so.

The three remaining kinds of line impairments—phase hits, gain hits, and dropouts—are not controlled at present. They are transient electrical events, which are triggered by such disturbances as

TABLE I

IMPAIRMENTS OF 3002 VOICE-GRADE PRIVATE LINE	BASIC 3002 LINE CONTROLLED TO LIMITS SPECIFIED BY AT&T INTERNAL PRACTICES		3002 LINE CONTROLLED TO TARIFFED CONDITIONING SPECIFICATIONS
	BASIC	C	D
ATTENUATION DISTORTION	X	X	SAME AS BASIC
ENVELOPE-DELAY DISTORTION	X	X	
SIGNAL-TO-NOISE RATIO	X	SAME AS BASIC	X
HARMONIC DISTORTION	X		X
IMPULSE NOISE	X		
FREQUENCY SHIFT	X	SAME AS BASIC LINE	
PHASE JITTER	X		
ECHO	X		
PHASE HITS	NOT CONTROLLED		
GAIN HITS			
DROPOUTS			

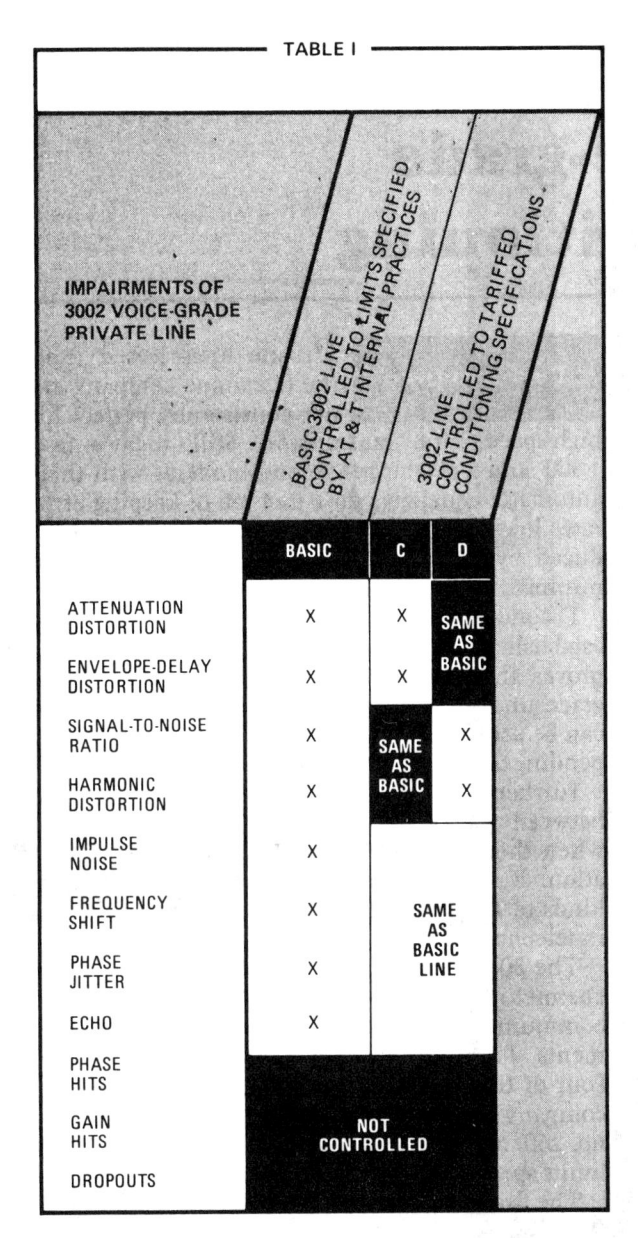

switching transients, atmospheric electricity, and signal fading.

Attenuation distortion

The limits for attenuation distortion and envelope-delay distortion of a basic unconditioned 3002 line are shown in Figure 1. (The limits of the basic line are controlled by AT&T practices, not by tariff.)

Attenuation distortion comes about because a signal going through a channel is attenuated by different amounts as the frequency of the signal changes. The telephone company's facilities are basically designed to transmit speech. And in speech, most of the information is contained in frequencies between 300 and 3,000 hertz. Even severe attenuation of the signal above 2,000 Hz does not interfere too much with comfortable listening.

For data transmission, however, the emphasis changes for economic reasons. The price per unit time for a given telephone channel is the same, regardless of how much data per unit time is sent over it, so it usually pays to send as much data as possible. Transmitting information can loosely be defined as reporting the change of state of some variable. Thus to send a lot of information per unit time, it's necessary to transmit many state-changes per unit time.

For binary digital data, the state changes involved are between 0 and 1, and these can translate fairly directly into signal excursions between plus full-scale and minus full-scale. Thus, the more bits per second sent, the more plus-to-minus excursions per second the channel must handle. The ability of a channel to transmit rapid plus-to-minus excursions is precisely what is meant by its high-frequency response. In other words, high-speed data transmission requires good high-frequency response, which is the same as requiring low attenuation distortion. Attenuation distortion at a given frequency is measured relative to that at 1,700 Hz, and is stated in decibels (dB). C-conditioning improves the high-frequency response of the basic 3002 channel.

Envelope delay distortion

Technically, envelope delay is defined as the rate of change with respect to frequency of the phase difference between the output signal and the input signal of a channel. Envelope-delay distortion occurs because the amount of delay is not constant with frequency (Figure 1). That is, high-frequency signals are delayed by an amount different from low-frequency signals. But because the human ear is nearly insensitive to it, envelope-delay distortion has not been a tightly controlled parameter in traditional telephone (voice) channel design.

For data transmission, the situation again is different. Regardless of the modulation scheme used, digital signals are always transmitted as pulses of one form or another, with the digital information contained in some aspect of the pulse shape. Therefore, if the information is to be reliably detected at the receiving end, the channel must preserve the critical aspects of the pulse shape.

But delay distortion interferes with accurate detection. For example, at the leading or falling edge of the pulse, which contains high-frequency components, the signal is subjected to the delay which the channel imposes at high frequencies. Where the signal changes slowly, as at the top of a pulse, the signal is subjected to a different delay, one which the channel imposes at low frequencies. Where the difference between these two delays is great, various parts of the pulse arrive at varying times and cause the reception of a distorted version of the original signal.

Furthermore, if successive pulses are spaced close together in time, as they will be at fast transmission rates, distortion from one pulse spills over

Limits of bandwidth-related C-conditioning

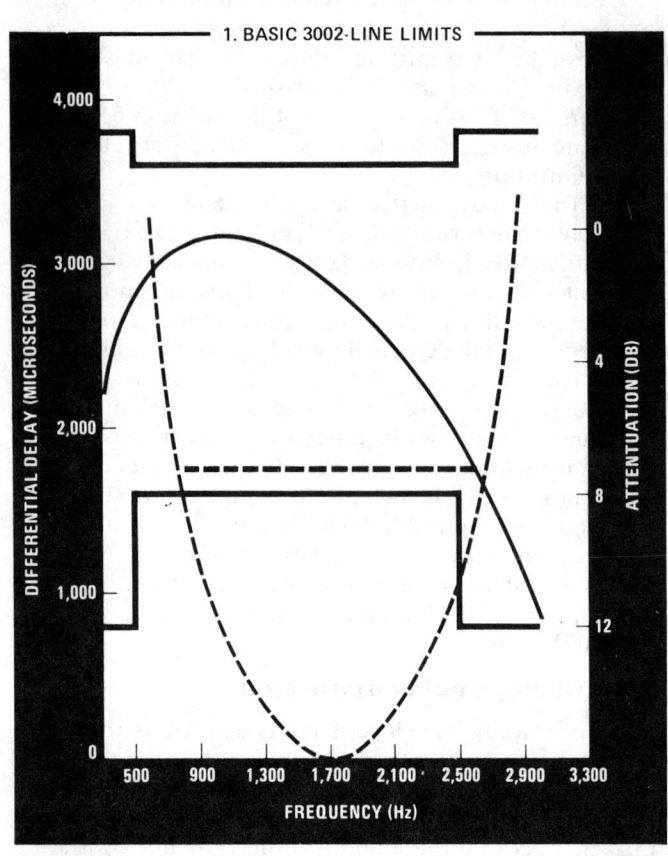

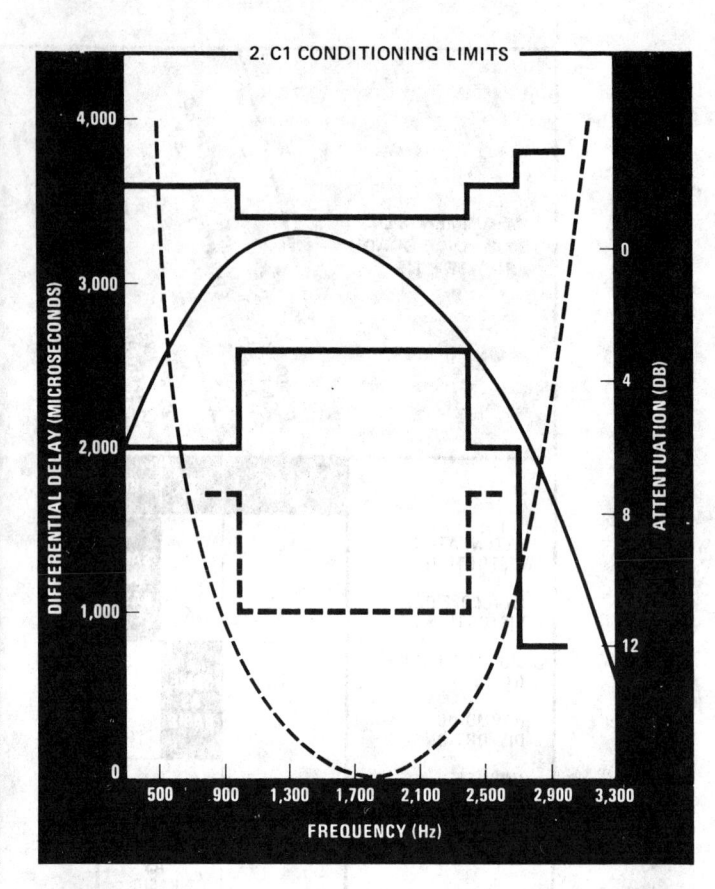

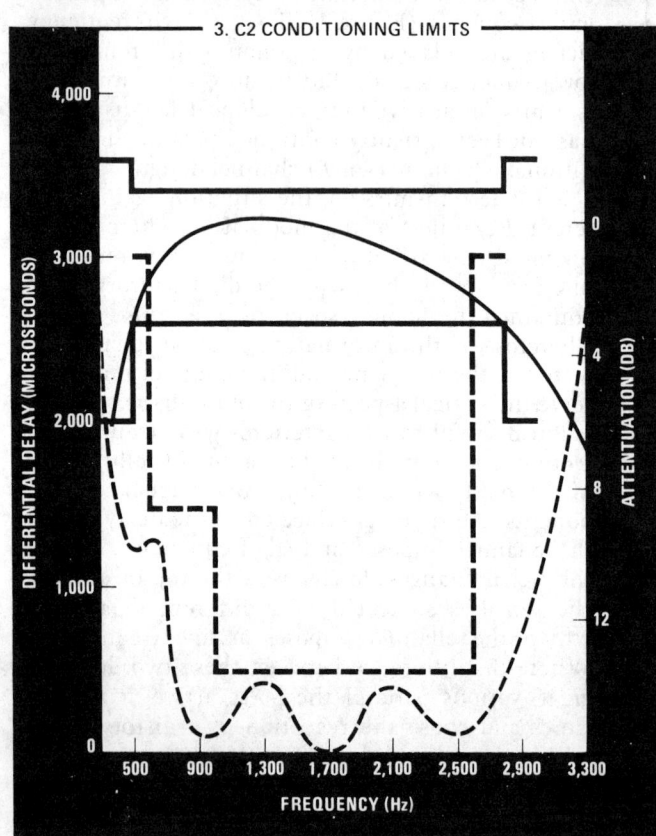

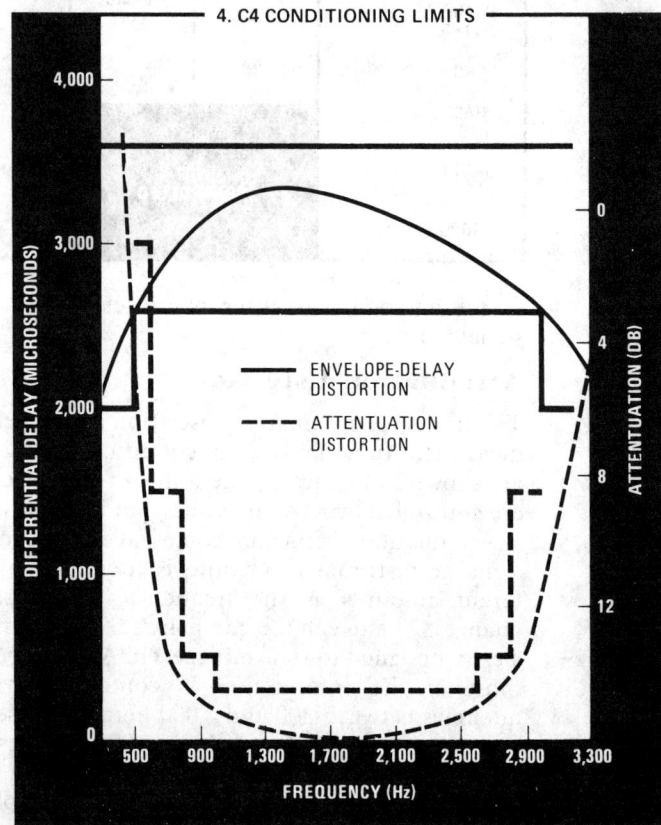

1. BASIC 3002-LINE LIMITS

2. C1 CONDITIONING LIMITS

3. C2 CONDITIONING LIMITS

4. C4 CONDITIONING LIMITS

—— ENVELOPE-DELAY DISTORTION
---- ATTENUATION DISTORTION

FIGS. 1-4. The attenuation distortion and envelope-delay distortion for a basic (unconditioned) type 3002 private voice-grade line are shown in Figure 1. The limits for these parameters for the three types of C-conditioning are shown in Figures **2, 3,** and **4**. Note that as the degree of conditioning increases, the more the attenuation and envelope-delay distortion decreases at the ends of the frequency band.

into the adjacent pulses, causing all pulses to be further misshaped. (This phenomenon is technically known as intersymbol interference.) Envelope-delay distortion is normally expressed as the differential delay in microseconds, at a given frequency, relative to differential delay at 1,700 Hz.

What C-conditioning does

C-conditioning is the term given to the special provisions furnished by the telephone company to make voice-grade lines meet tariff specifications for attenuation distortion and envelope-delay distortion. Included in these special provisions is the attachment of loading coils and other filter arrangements to the line.

C-conditioning is available in five varieties—C1 through C5. C1, C2, and C4 conditioning apply to voice-grade private lines between customer sites, while C3 and C5 apply to lines which are part of some large network, such as CCSA (common control switching arrangement) and international links.

The heavy colored lines in Figures 2, 3, and 4 show the attenuation-distortion and envelope-delay limits for C1, C2, and C4 conditioning. The black lines represent a typical conditioned line within these limits. The higher the degree of conditioning, the more the attenuation distortion and delay distortion are reduced at both ends of the frequency band.

C1, C2, and C4 conditioning can be provided for any point-to-point 3002 line. C1 and C2 conditioning can be provided on any multipoint 3002 line without restriction on the number of points. C4 conditioning, however, is not provided on multipoint channels with more than one central and three remote points. The telephone company charges for line conditioning by the month, but there is no one-time installation charge for C-conditioning. The monthly interstate rates for C-conditioning are contained in Table II.

The required degree of conditioning depends on the bit rate that the channel must support and the type of modem used. Older modems, with crude automatic equalizars or with manual equalizers, cannot fully compensate for extensive attenuation and delay distortion, and they usually require a degree of C-conditioning that depends on bit rate. The modem manufacturer will usually tell the user which grade of conditioning is needed.

Sometimes a channel may be within C2 specifications without conditioning and allow a pair of modems to operate satisfactorily without their user needing to specify and pay for C2. Relying on this to happen is risky, however, since the phone company is under no obligation to maintain that channel at that level of performance. If the phone company reroutes or otherwise changes a line, the result might be a channel that meets only basic 3002 specifications. Then transmission quality degrades. Thus, to assure satisfactory performance, the user must specify the degree of conditioning the modem requires.

An old trick has been to order a channel with C2 or C4 conditioning and then order a change to an unconditioned line after a couple of months of operation. In the past, the telephone company might have left the conditioning in anyway. But, with the growing demand for data channels, the phone company will now probably remove it.

New high-speed modems, such as the Paradyne M-96, are equipped with powerful automatic equalizers which do much of the same job as C-conditioning. With such modems, therefore, C-conditioning generally does not help performance significantly and in some instances may slightly increase the error rate. The reason for this is that an unconditioned line usually has a smooth delay characteristic (Figure 1). The electrical equipment added to the line to bring it within C2 or C4 specifications, however, may create a delay curve with steep slopes and ripple as shown in Figure 3, which may be more difficult for an automatic equalizer to deal with than the smooth curve of the unconditioned line.

D-conditioning is an option recently introduced by the phone company specifically for 9,600-b/s operation. It specifies limits for noise and harmonic distortion. The basic 3002 channel is specified by internal AT&T practice to have a signal-to-noise ra-

TABLE II

COSTS OF LINE CONDITIONING (INTERSTATE)

C1 (POINT-TO-POINT)	$ 5.25/POINT/MONTH
C1 (MULTIPOINT)	10.50
C2 (POINT-TO-POINT)	19.95
C2 (MULTIPOINT)	29.45
C4 (POINT-TO-POINT)	31.55
C4 (MULTIPOINT)	37.85
D1 (POINT-TO-POINT)	$ 14.20/MONTH PLUS $170 INSTALLATION CHARGE
D2 (TWO-OR THREE-POINT)	47.50/MONTH PLUS $162 INSTALLATION CHARGE

tio of not less than 24 dB, a second-harmonic distortion of not more than –25 dB, and a third-harmonic distortion of not more than –30 dB. Here, the signal-to-noise ratio of 24 dB means that the noise power is approximately 6% of the received signal, which is an unpleasantly high noise level for the ear. However, most modems can operate satisfactorily at this noise level if no other impairments are present.

D-conditioning

Harmonic, or nonlinear, distortion occurs because attenuation varies with signal amplitude. As an example, a 1-volt signal excursion around zero volts (+0.5 to –0.5 V) may be attenuated by a factor of one half while a 1-V excursion around +4 V (3.5 to 4.5 V) may be attenuated by a factor of two thirds. A sine wave sent through such a channel will be flattened at the peaks. This is the same effect as would result from adding harmonics of low amplitude to the signal in the first place. Therefore, this form of distortion is called harmonic distortion and is measured in terms of the amount of second- and third-harmonic content that would cause the same amount of flattening.

Noise and nonlinear distortion are critical to data transmission because, again, they interfere with the faithful reproduction of the transmitted pulse shapes at the receiver. Furthermore, in contrast with the intersymbol interference caused by delay distortion, their effect is irreversible. No amount of sophistication in the receiver modem can re-create the original signal shape once it has been changed by these impairments.

Therefore, the only protection against harmonic distortion is to make the signal inherently resistant to it. The way to do that is to maximize the difference between unique pulse shapes that have to be distinguished by the receiver. In other words, a signal that uses four voltage levels to define two bits per pulse is more resistant to noise and nonlinear distortion than if it uses eight levels to define three bits per pulse. However, to attain a data rate of 9,600 b/s on voice-grade lines, modems require many modulation levels (and some modems more than others).

For those modems, D-conditioning provides the answer. Such conditioned channels meet the following specifications:

Signal to C-notched noise:	28 dB
Signal to second-harmonic distortion:	35 dB
Signal to third-harmonic distortion:	40 dB

D1 conditioning is offered for point-to-point channels and D2 for two-or-three point channels. The monthly charges for D-conditioning are contained in Table II. These charges are for the channel as a whole rather than for each termination point as for C-conditioning.

D-conditioning provides a comfortable margin for 9,600-b/s performance for most modems. In fact, some 9,600-b/s modems will operate satisfactorily on lines which meet only the basic 3002 specifications for noise and harmonic distortion. Still, because a minimum of these impairments gives modems greater tolerance for the other, less tightly controlled, impairments, it is usually worth paying for D-conditioning.

Being relatively new, D-conditioning is not yet available throughout the country. Consequently, both the user and telephone company staff may be unfamiliar with it, so that it is often believed that D-conditioning is an alternative to C-conditioning. Actually, both types of conditioning can be provided on the same 3002 channel to limit the four line impairments controlled by tariff.

Other impairments

The second group of line impairments will not be discussed here, except to note that, when a line falls outside the limits set by AT&T engineering practices, the telephone company will usually attempt to correct whatever's wrong—impulse noise, frequency shift, phase jitter, or echo. The user's leverage lies in the fact that the impairment limits are published in AT&T documents and that the effects of any excessive line impairments on data-transmission error rates are common engineering knowledge.

As for the third group of line impairments, phase hits, gain hits, and dropouts are currently not controlled to any limits either under tariffs or otherwise. Still, if a user can demonstrate that their presence in a channel renders it unsuitable for data transmission, he has a strong claim for corrective action since data transmission capability is what he is paying for. The people in the telephone company are likely to respond to requests for correction since they are aware of the impairments and their effects.

The problem, of course, is in demonstrating that one or more of them are present in the channel. The user who has access to specialized test equipment to measure the impairments affecting data transmissions can tell the telephone company exactly what problems the line has. The repair men and maintenance engineers are more likely to appreciate such information than to argue. Specific details help them do their job.

More often than not, though, when some exotic impairment makes a line useless for data communications, the main obstacle to its rapid correction is the fact that neither the user nor the telephone company employees have the test equipment required to identify it. As for the tariffed parameters of C- and D-conditioning, however, the local telephone company must be able not only to measure them but to demonstrate to the user that they are as specified. ∎

part 7
data data data data data data data data data

data LINK CONTROLS

Taking a fresh look at data link controls

If the potential of computer-based data communications systems is to be fully exploited, then data link control, also called line discipline and line protocol, must be made simpler and more efficient

J. Ray Kersey
International Business Machines Corp.
Research Triangle Park, N.C.

PERSPECTIVE

Data link control today is undergoing a thorough reappraisal. It came into its own in the early 1960s when remote terminals and other computer input/output devices began, in earnest, to be connected to distant computers over communications lines. Basically, data link controls have two functions: to establish and terminate a connection, and to assure message integrity during the passing of data between two stations. A data link control uses control characters that differ in bit pattern from data characters in a given code set.

A succession of improved data link controls (DLCs) has been developed, each providing an expanded repertoire of control functions in recognition of the expanding application of remote data processing and message switching over data communications networks. The details of a specific data link control, also known as line discipline and line protocol, are described later.

In retrospect, though, the availability of additional control characters resulting from new, expanded code sets brought about an embarrassment of riches. Instead of a few DLC characters simply being assigned the basic task of establishing a connection between two stations—say a communications processor and a remote data terminal—and assuring the integrity of messages passing between these stations, the additional control characters in the DLC were given the responsibility for such other things as peripheral-device control, network control, and formatting of messages.

Stated another way, evolving DLCs provided more control functions, but they did not contemplate the future growth of data communications and the ever increasing requirement for more efficient and economical use of computers, terminals, and communications lines. They did not permit the economic utilization of full-duplex links. They place a burden on equipment makers and data communications users in terms of both hardware and software for their implementation. They also lack flexibility and extendability for meeting requirements of system expansion.

In short, the potential for data communications systems using computers requires improvement of data link controls. Within the standards community, this endeavor is being undertaken by the American National Standards Institute which, on Feb. 15, 1973, issued its sixth draft of ANSI X3S3.4/475 "Proposed American National Standard for Advanced Data Communications Control Procedures (ADCCP)."

To help put the subject in perspective, it is worth-

Evolution of data link controls

How DLCs have evolved can be illustrated from the progression of IBM DLCs through the years. Since 1960, IBM has developed three data link controls, each improving on its predecessor. The first was GPD, for general purpose discipline, and the second was STR, for synchronous transmit/receive. Since many systems using these DLCs are yet in use, they are reviewed to show some factors that have influenced DLC development. The third DLC is the present-day binary synchronous communications (BSC). BSC has been employed for virtually all synchronous data communications products related to the IBM 360 and 370 families of computers.

GPD was an early IBM DLC for start-stop transmission. It used the binary-coded-decimal code which provided two unique DLC control characters and had three additional multi-use characters for a total of five control characters. This was an advance over the use of Baudot code which, having no intrinsic control characters, had to rely totally on data-character sequences for link control. GPD permitted terminal-to-terminal operation, based on a design for implementing central control from one station.

However, GPD allowed options and was at times implemented in ways that impaired compatibility between products designed to take advantage of its options. Further, line control, device control, and device addressing were intermixed. Finally, its scope was narrow, in that GPD applied only to start-stop, half-duplex operation.

STR was more tightly defined, more efficient, and more reliable than GPD. Designed specifically for synchronous transmission, it provided a continuous hunt for the unique synchronizing character. Its 4-out-of-8 code set provided six uniquely defined control characters. Probably the most distinctive features of STR were that it separated data link control from device control, it permitted either half-duplex or full-duplex operation, and it was structurally extensible for conversational operation.

However, STR, too, had some limitations. Among them were that it had no addressing structure, and thus was restricted to point-to-point operation. It also assumed that all stations have equal capability and that the error recovery procedure be initiated by the transmitting station.

BSC, like STR, was designed specifically for synchronous transmission. It is capable of operation on leased or switched point-to-point communications facilities and more importantly, added, above STR, an addressing capability for use on multipoint facilities. By use of EBCDIC (or ASCII), the code-handling ability was substantially improved over STR's 4-out-of-8 code (70 valid combinations). Further, it added the ability to divide messages into short blocks (via use of ETB) for checking purposes.

But, as new requirements continue to evolve, particularly with the trend to more interactive terminals, BSC, too, can be expected to have its sphere of application exceeded. Among the characteristics of BSC are that it is code-dependent, character-oriented, operates in half-duplex mode, and requires the terminal to initiate certain error recovery procedures. Moreover, because it was designed primarily for use with unattended batch terminals handling lengthy transmissions, its efficiency in an interactive environment is not ideal.

Other properties of BSC are that line control is intermixed with device addressing and control and that error detection is applied to data characters but not to control and address characters. Finally, like STR and GPD, BSC does not permit the programing for a particular peripheral device to be done in a common manner for both those devices located in a computer room and those served remotely via communications facilities.

---

while at this time to review the history and present situation of data link controls, by way of a discussion that concentrates mainly on IBM's family of data link controls (see "Evolution of data link control," above).

What control characters do

Data link control is a discipline: by invoking defined control characters, it provides an orderly and efficient way of assuring that, among other things, a remote terminal is in a ready condition, and that the terminal will send data when instructed, will receive data when instructed, and will advise the sending terminal when it receives erroneous data. Since the same physical link carries both data (text) and control characters, the DLC must distinguish between the data and control characters available within a code set.

With respect to control of remote devices, there are many analogies to the control of an input/output (I/O) device when it is physically close to the computer:

In the computer room, the control of data transfer to an I/O device is handled by programs. Under DLC, the station designated as the control station (which might be a front-end processor) establishes the data-link connection and maintains order during a transmission.

In the computer room, the starting of an input device occurs when an address and a read command are sent to the device. Under DLC the starting of a remote sending terminal is accomplished by the control station sending a polling address to the remote station.

In the computer room, the awareness of the end of an I/O operation occurs by sensing of the status (electrical level) at the device's electrical-interface leads. Under DLC, though, sensing the end of a remote I/O operation requires the sending of a specific control character.

In the computer room, the formatting of data to an I/O device is implicit in a line of print, an 80-column card, or the like. But under DLC, the formatting of data requires the transmission of header control characters and/or block-structure control characters.

In contrast, there are a number of I/O-device control functions that are not analogous to computer-room operation, but rather are prompted by the topology of the data communications system: the remoteness of the I/O device; the sharing by multiple stations of common communications facilities by means of polling, selection, and contention; and the fact that messages may proceed through several nodes made up of communications processors along the route. Among the major nonanalogous functions are:

- The transmission of a station address code (ID) to assure proper connection through the network.
- The use of header characters, such as job number, which must be machine-sensible, provides the information necessary for the handling, including routing, of a message through a multinodal switching system.

- The use of variations in DLC procedures to permit the full code set, including those characters normally used for controls, to be transmitted as text information. This "transparent" mode is necessary when it is desired to send packed decimal, binary bits of data, or other uniquely coded information that might otherwise present code patterns that could be confused by machines as being control characters.
- The requirement for dividing messages into block segments of lengths determined by and suitable to the error environment of a given physical data link.

Influences on DLC design

Many factors have influenced the design of data link controls, not the least of which are historical precedents. Other factors are the expansion of code sets from 5 to 6, 7, and then 8 bits, including escape capability; the wider acceptance of synchronous transmission; the nonseparation of the controls for link, devices, and networks; and the use of more extensive methods of error control. Those who have little experience with or need a refresher on data link controls may first want to read "How a data link control works"

The introduction of the character-oriented teleprinter in 1915 paved the way for machines capable of receiving data transmissions without human intervention. This was accomplished through the use of a limited set of device-control characters available within the 32 characters obtainable from the five-level Baudot code that the teleprinter's electromechanical transmitter/distributor generates. Although such control characters for teleprinters were not data-link-control characters under today's meaning, the approach taken then affects present line disciplines. One influence is that today's DLCs are still character-oriented; that is, the control functions are represented by characters in the code being transmitted. A second influence is that each teleprinter on the network was essentially the same and each had limited, but equal, ability to generate or respond to various control signals.

Still today, in practically every case, each of the terminals on a network has equal capability for initial error recovery procedures. Specifically under IBM's BSC (binary synchronous communications) DLC, each remote terminal must be equipped to send an ENQ (ENQUIRY) control character when no valid response is received to a transmitted block. In other words, the responsibility for error recovery is still distributed throughout the network, at the remote terminals, rather than handled totally by a central control station in the form of a front-end processor.

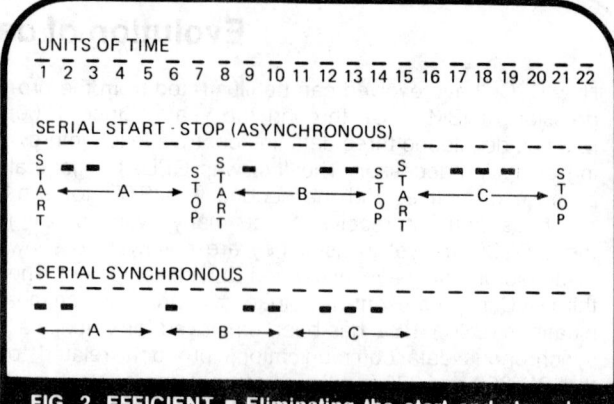

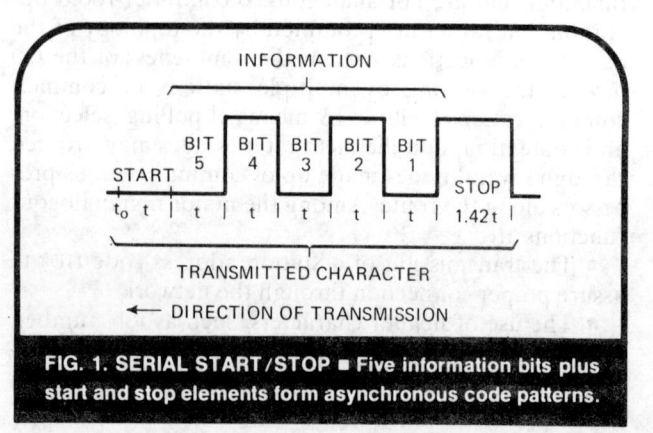

Synchronization and information coding directly influence how a particular DLC must be designed and then implemented.

Consider first the distinctions between asynchronous and synchronous transmission. Figure 1 contains the five-level Baudot code used in many teleprinters. The five information bits representing a character are preceded by a start bit (0) whose duration is one unit of time and a stop bit (1) whose duration is, in one version, 1.42 units of time. These start and stop bits separate one character from another and bit-synchronize the receiving station with the transmitting station. Sending the information and start-stop bits in sequence is called serial start-stop, or asynchronous, transmission (Fig. 2).

Synchronous transmission does not use start-stop bits. Instead, the line is sampled at regular intervals, called the clock rate, to receive and record information (Fig. 2). Synchronous transmission permits more information to be passed over a circuit during a given time interval because no transmission time is wasted on the start and stop bits. Figure 2 shows that three characters, when transmitted at the same speed, require 21 units of time (assuming a one-time-unit stop bit) when asynchronous and only 15 units of time when synchronous.

A distinction must be made between bit synchronization, character synchronization, block synchronization, and message synchronization. In asynchronous transmission, bit and character synchronization is handled by the start-stop bits.

Synchronous transmission, though, is more complicated. Bit synchronization can be achieved by a master-clock signal generated by a modem. Other modems or business machines (say a terminal) derive their clock signals from the 0 and 1 transitions occurring in the received serial data stream. Deriving the bit-synchronizing signal from the data stream itself, called self-clocking, overcomes the effect of propagation delay between distant stations and the tendency of electronic circuits within the modems to drift. During each line turnaround, transitions will not occur, and bit synchronization will be lost. To assist the initial establishment of bit synchronization following each line turnaround, IBM's BSC DLC causes a control character sequence called PAD SYN SYN to preface each transmission. For

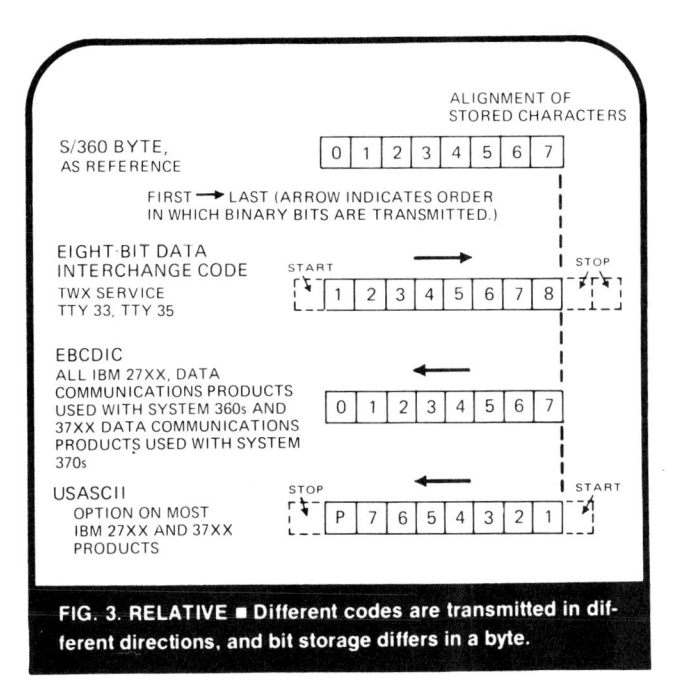

```
                              ALIGNMENT OF
                            STORED CHARACTERS
  S/360 BYTE,              ┌─┬─┬─┬─┬─┬─┬─┬─┐
  AS REFERENCE             │0│1│2│3│4│5│6│7│
                           └─┴─┴─┴─┴─┴─┴─┴─┘
          FIRST → LAST (ARROW INDICATES ORDER
              IN WHICH BINARY BITS ARE TRANSMITTED.)

  EIGHT-BIT DATA
  INTERCHANGE CODE  START          ────→            STOP
  TWX SERVICE        ┌╌┐ ┌─┬─┬─┬─┬─┬─┬─┬─┐ ┌╌┐
  TTY 33, TTY 35    └╌┘  │1│2│3│4│5│6│7│8│ └╌┘
                         └─┴─┴─┴─┴─┴─┴─┴─┘

  EBCDIC
  ALL IBM 27XX, DATA              ←────
  COMMUNICATIONS PRODUCTS  ┌─┬─┬─┬─┬─┬─┬─┬─┐
  USED WITH SYSTEM 360s AND│0│1│2│3│4│5│6│7│
  37XX DATA COMMUNICATIONS └─┴─┴─┴─┴─┴─┴─┴─┘
  PRODUCTS USED WITH SYSTEM
  370s
  USASCII            STOP          ←────            START
  OPTION ON MOST     ┌╌┐ ┌─┬─┬─┬─┬─┬─┬─┬─┐ ┌╌┐
  IBM 27XX AND 37XX  └╌┘ │P│7│6│5│4│3│2│1│ └╌┘
  PRODUCTS               └─┴─┴─┴─┴─┴─┴─┴─┘
```

FIG. 3. RELATIVE ■ Different codes are transmitted in different directions, and bit storage differs in a byte.

this DLC, the realization of PAD is a set of alternating 0s and 1s, a pattern which has a strong impact on forcing a receiving clock circuit to track the transmitted bit rate.

Character synchronism in synchronous transmission is (in this DLC) accomplished by recognizing the control characters, called SYN, whose code pattern is 00110010. This pattern sets the electronic equipment to recognize, by bit count, the beginning and end of each successive control or data character.

Staying in sync

On some occasions during ordinary data transmission the combination of data characters may not provide sufficient transitions to allow the self-clocking mechanisms to stay in sync. To avoid this, the DLC causes SYN characters to be inserted into the data stream at periodic intervals.

Block and message synchronism is more a matter of framing, or recognizing, the beginning and end of a block or message than it is of exact time dependency. Framing in synchronous transmission is accomplished by such control characters as STX for START OF TEXT and ETB for END OF TRANSMISSION BLOCK.

Information coding has also been subject to evolution, with some aspects influenced by telegraph transmission and others influenced by computer design. Figure 3 contains major codes now in wide use for computer-based communications. The 8-bit Data Interchange Code associated with TWX service and many teleprinters is compatible with the character size of most computers. Note that this code uses eight bits in a synchronous transmission mode and 11 bits when in an asynchronous transmission mode. In the latter case, the data stream is actually bit-synchronous but character-asynchronous, a situation which has come to be called isochronous transmission.

Most DLCs are designed for a particular code. Therefore, it is important for users and equipment makers to have clear understanding of the nature of that code and, particularly, the number of data bits per character, whether or not the code uses a parity bit for error control, the alignment of a transmitted character relative to its storage in memory or temporary buffer, the order in which the bits are transmitted, and whether start and stop bits will be present.

The EBCDIC code (for extended binary coded decimal interchange code) contains eight bits, and thus allows up to 256 (2^8) characters. Among its uses, EBCDIC has been employed in all IBM data communications products for System 360 and System 370 computers.

The USASCII document X 3.4 (for United States of America Standard Code for Information Interchange, and often simply called ASCII) defines a seven-level code, resulting in 128 characters. The use of an eighth bit for parity check, the assignment of extra bits for start-stop transmission, and bit sequencing are defined by collateral ANSI documents.

The top of Fig. 3 depicts the bit positions of an eight-bit byte stored in the IBM 360. The alignment of the bits of an EBCDIC character corresponds exactly, as would be expected, with bit alignment of the IBM 360 byte. Note that in EBCDIC bit 7 is transmitted first and bit 0 last. In ASCII, however, bit 1 is transmitted first although bit 1 fits into bit 7 in the IBM 360 byte. The eight-bit data interchange code, moreover, transmits its bit 1 first, but bit 1 fits into the bit-0 position of the IBM 360 storage byte.

The number of DLC control characters provided by a code significantly influences the design of the DLC for use with the code. The Baudot code used with early teleprinters provides no true data-link-control characters. The 4-out-of-8 code employed in IBM's synchronous transmit/receive (STR) DLC has six control characters. One of IBM's earlier DLCs, the GPD (for general purpose discipline), used the binary-coded-decimal code which provided two unique control characters. More recently, IBM's BSC DLC utilizes nine single control characters as well as many two-character sequences to extend the DLC's repertoire substantially.

As pointed out earlier, DLCs in wide use today are character-oriented. Thus, even though the control discipline of a given DLC may be identical for two different codes, the code realization of the DLC will be different. Such is the situation with BSC, which can be implemented in either EBCDIC or ASCII. For example, EOT in EBCDIC is 00110111, but EOT in ASCII (with odd parity for synchronous transmission) is 00000100. But despite the fact that all present DLCs are code dependent, there is no intrinsic necessity for this state of affairs. Under the proposed ANSI standard, the architecture of future DLCs would be independent of code structure.

For any code, the whole question of how many control characters are needed and how they should be used in a DLC requires reexamination. Present approaches invite a conflict. On the one hand, the need to provide more functions and more flexibility encourages the development of DLCs having more defined control characters or control-character sequences. On the other hand, such a richness of control characters or sequences complicates the hardware and programing situation. For example, suppose only one terminal on a link requires, because of its particular application, a rather complex control repertoire made up of numerous two-character sequences to define the extended functions. This com-

How a data link control works

IBM's binary synchronous communications data link control (BSC DLC) can be regarded as having two distinct functions: an establishment-and-termination procedure, and a message-transfer procedure. Each takes a special form, depending on the application.

Thus there is one establishment-and-termination procedure which applies to leased point-to-point links, another to switched point-to-point links, and yet another to nonswitched multipoint centralized operation. Similarly, one message-transfer procedure applies to data communications using transparent text capable of transmitting, in the text-data stream, all 256 characters obtainable within the eight-level EBCDIC code. Another message-transfer procedure, called nontransparent text, applies to those systems not using, in text, the 10 characters defined in EBCDIC for data link control.

A third message-transfer variation is that BSC DLC with nontransparent text can also be implemented using the seven-level ASCII code. Character code-pattern realization may or may not differ from EBCDIC, depending on the particular character.

Thus the BSC DLC is actually a family of alternatives, each member of which satisfies the operational requirements of a particular type of data communications system. An example affording insight into how BSC DLC works for nonswitched multipoint centralized operations with nontransparent text is given in Figs. A and B.

In order to discuss Figs. A and B, it is necessary to define the functional roles of communications processors, remote data terminals, and remote peripherals collectively called stations. On a multipoint data link with centralized operation, the central station—usually a processor—is permanently designated the control station, and all other stations are designated tributary stations.

In its polling mode, the control station determines, at any moment, which tributary station is to become the master station and have the opportunity to transmit information to the control station acting as a slave station. Alternatively, in its selection mode the control station may command a particular tributary station to assume slave-station status and receive information from the control station acting as a master station. The control station is also responsible for re-establishing order on the link, via an error recovery procedure, should control be lost for any reason during an exchange with a tributary station.

The control characters for BSC using EBCDIC are defined in the table. The basic control-character set is implemented with one 8-bit byte. The set is extended by using 2-byte sequences for certain characters—EVEN ACKNOWLEDGE (ACKO), for example, and all others below it in the table.

It will be recognized that there is a reply phase involved in both the establishment-and-termination (control) and message-transfer (text) modes. Control characters employed in the reply phase may have one meaning when the link is in its control mode and another when in its text mode. For example, in the control mode an ACK response to a selection means the selected station is ready to receive. But in the text mode an ACK means that the last block that was transmitted has been received correctly at the remote location.

The polling-and-selection establishment procedures, along with a termination procedure, are shown in Fig. A. Establishment is initiated by the control station sending an EOT control character (path 1) to reset all tributary stations from data mode to control mode so they can "look" for their address to be sent by the control station.

The control station may now poll or select. The prefix or address designates a polling sequence (path 3) and which of the tributary stations may now transmit information. The polling sequence ends with an ENQ. When polled, a tributary having a message to send enters the message-transfer procedure (path 8) and becomes a master station. Or, if the tributary has no message to transmit (6), it may decline master-station status by sending an EOT (15) to terminate the logical connection. Then, master-station status reverts to the control station through path 16 back to point A in Fig. A.

If an invalid or no reply is received in response to a polling sequence (7), the control station again resets all tributary stations by sending EOT (15) to terminate before resuming polling or selection.

The prefix may alternatively designate a selection sequence (4) and which of the tributary stations is asked to be a slave station and receive information from the control station acting as a master station. The selection sequence also ends with ENQ. If a selected tributary is ready to assume slave-station status, it responds positively with an ACKO (5) and enters the message-transfer procedure (8). However, a WACK (11) may be sent in lieu of an ACKO to cause a delay (15, 16, A, 4, 5) prior to entering message transfer.

If, however, the selected tributary is not ready to receive information, it responds with a NAK (12) and does not assume slave-station status. When the control station receives the NAK, it may select the same or another station (15, 16, A, 4) or revert to polling (15, 16, 3) depending on the order prescribed by a poll-and-select algorithm.

If the selected tributary station cannot receive because it has previously entered a transmit mode and a polling sequence is required first, it sends an RVI (10). If an invalid or no reply (9) is received by the control station to a selection, it may reselect the same tributary (15, 16, A, 4)

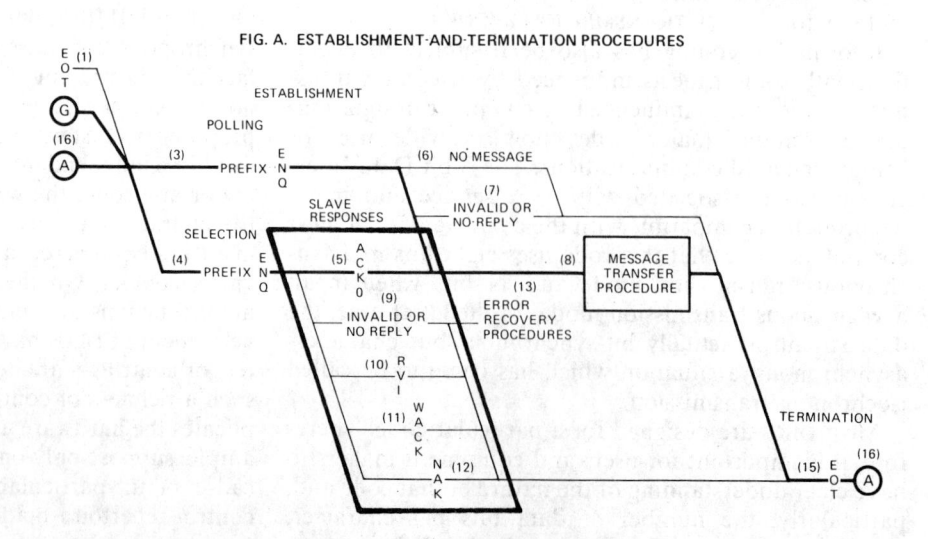

FIG. A. ESTABLISHMENT-AND-TERMINATION PROCEDURES

or a different tributary (15, 16, 2, 4), or enter error recovery procedures (13).

The establishment of a valid logical connection is denoted by the heavy line in Fig. A. At (8), BSC enters the message-transfer procedure. Figure B details the message transfer which starts at E. Note that, once the message transfer procedure has been entered, it can be reentered, without requiring the establishment of a new connection, whenever it is necessary to transmit a new or additional block of header/text (A), or to retransmit a block of header or text (B), or to transmit a header or text as a conversational reply (C).

FIG. B. NONTRANSPARENT MESSAGE-TRANSFER PROCEDURES

When the message from a master station starts with a heading, the transmission begins with an SOH control character (2). Also, an intermediate block which continues a heading starts with SOH. However, if the message has no heading (3), the message starts with STX. Further, an intermediate block that either begins or continues a text starts with STX.

Independent of the blocking required for transmission purposes, segments of a message may be checked by inserting an ITB character followed by a block check (5). In this sequence, the STX is optional. And no line turnaround occurs as a result of an ITB. Note that the F, 5, F loop permits the checking of multiple segments within a message block.

Blocks that complete a text—not a transmission—end with ETX (6). And intermediate blocks of header or text end with ETB (7). Either of these control characters is immediately followed by a block check (BCC). (The block check is two characters and they are not to be interpreted or considered as control characters.) In addition to ending blocks with an ETX or ETB, the master station can abort a transmission at will by sending an ENQ.

After one of the three ending control characters has been sent, the master station waits for a reply from the slave station. Because BSC applies only to half-duplex data transfer in which the transmission can go in two directions but only one direction at a time, there is a slight delay for modem-and-line turnaround. The slave then transmits one of several replies to the master station.

If the originally transmitted message block was received correctly by the slave station, the slave's usual reply is an ACKN (9). ACKN alternates between ACK1 and ACK0. Or an RVI (10) may be sent to request that, relative to text, the roles of sending and receiving station be reversed. In response to either an ACKN or an RVI, and following another turnaround, the master station may continue to transmit header or text (16, A). Of course, when the message, possibly including composite texts, is complete, the connection terminates with an EOT (17, G), and BSC reverts to its establishment procedure, Fig. A. Sometimes, when stations are properly equipped, a conversational reply (11, C, 2–7) may be sent in lieu of an ACKN. Also, a WACK (12) may be sent instead of an ACKN to cause a delay (18, 19, 9, 16, A) prior to receiving the next header or text.

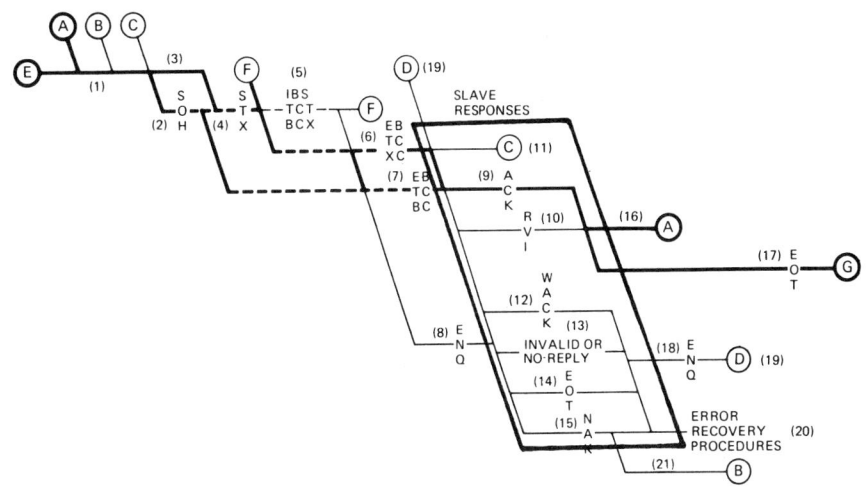

If the block was not received correctly or was aborted via an ENQ (8) from the master station, the slave's usual reply is NAK (15). In response to the NAK, the master station may retransmit the block (21, B, 2 or 3, 4, etc.), or if the number of retries (NAKs) exceeds some predetermined count, the master station may enter error recovery procedures (20).

If the master station detects an invalid or no reply (13), it may send an ENQ (18) to ask the slave station to repeat its reply (19, D, 9–15). Alternatively, the master station may immediately enter error recovery procedures (20).

A slave station may choose to abort the logical connection for any of several reasons (e.g., a hardware malfunction). In this case, an EOT (14) may be sent as a response to the master station. In this context, an EOT response from a slave station indicates a negative acknowledgment (nonacceptance) of the last block received. The master station will then immediately enter error recovery procedures (20).

Data link control characters defined for BSC using EBCDIC

SOH	Start of heading	ACK0	Even positive acknowledgment	XITB	Transparent end of intermediate block
STX	Start of text	ACK1	Odd positive acknowledgment	XETB	Transparent end of transmission block
ETB	End of transmission block	WACK	Wait before transmit, a positive acknowledgment	XSYN	Transparent synchronous idle
ETX	End of text				
EOT	End of transmission				
ENQ	Enquiry	DISC	Mandatory disconnect	XENQ	Transparent block cancel
NAK	Negative acknowledgment	RVI	Reverse interrupt	XTTD	Transparent temporary text delay
SYN	Synchronous idle	TTD	Temporary text delay		
ITB	End of intermediate transmission block	XSTX	Transparent start of text	XDLE	Data DLE in transparent mode
DLE	Data link escape	XETX	Transparent end of text		

plexity means, in a character-oriented DLC, that every terminal on the same link, whether or not it needs the added capability, must recognize the full repertoire.

An alternative to using an extended control character set, in which each character may or may not have a particular context depending on whether or not the DLC is establishing a connection or passing a message, is to rely on positional significance. Positional significance, as now considered in the draft ANSI standard, means that one specific field (say, bits i through j in a sequence following a unique delimiter) will be dedicated to control. The unique delimiter would serve for character synchronization and transmission-block delineation. Since the positional control field can contain any number of bits, then it is possible within this one control field to define up to 2^n unique data-link-control operations associated with establishing, maintaining, and terminating a connection, transferring data, and assuring message integrity. For compatibitility, n, the number of bits in i through j, once chosen, should remain constant.

The use of positional significance can open up new and powerful improvements in data link control. Among these possible improvements are the numbering of successive blocks to detect whether transmitted data has been lost or duplicated, the implementation of new response functions such as HOLD TRANSMISSION, simplification and extension of addressing schemes, and improved error control prodedures.

Strengthening error control

Consider the increased power of error control that could be achieved by using some of the 2^n unique DLC operations to number successive blocks. At present, when an error is detected through the use of a block check, the receiving terminal responds (in BSC) with a NAK, then the sending terminal, which stops sending while waiting for an acknowledgment, retransmits the block.

An alternative to this "stop-and-wait" method is the continuous transmission "go-back-two" method. Two successive blocks are always held in the sender's buffer. Transmission continues until a NAK is received—in which case the sending terminal retransmits the last two blocks.

With the added functional capability of the positionally significant control field, this continuous "go-back" method can be extended to more blocks; that is, the numbering modulus of blocks and acknowledgments can be increased. Suppose, for example, the sender's buffer can store eight blocks, and successive blocks are numbered from 0 through 7, back to 0, and so on. In this method, the transmitting station sends, along with the data block, a send-sequence count and a receive-sequence count. The receiving terminal also maintains corresponding send and receive counts.

In operation, the transmitting station sends a block of data, N, then increments its send count to N + 1. If the receiver's receive count is also N and no error has occurred during the transmission, the send and receive block numbers are the same. The receiver not only notifies the sender, via a channel in the opposite direction, that it has received a block accurately, but actually verifies this by identifying the block number (the receiving station's receive count) under consideration. On receipt of a good block, the receiver's receive count is also incremented by one to N + 1. Under error-free operation, transmission and reception continues without interruption. This dual count mechanism can be used either to selectively retransmit one block or go back N blocks when a block is missing or in error.

For example, suppose errors are detected in block 2. With block numbering and a large buffer, the transmitting terminal, which at a particular instant may be sending block 6, is informed by the receiver that block 2 was received with errors. The transmitting terminal then goes back 4 blocks (6 - 2) to retransmit blocks 2 through 6 before resuming transmission of the message.

Or suppose a transmitter sends block 4 but for some reason the receiver, which expects block 4, never gets it, perhaps because of loss of synchronism. Then the next block received is block 5. The send and receive counts do not match. Recognizing that it has not received the anticipated block 4 in sequence, the receiver tells the sender to retransmit block 4, which it may do by slipping it in between blocks 7 and 0. This is called selective block retransmission and is more complex than the go-back-N-method described previously.

Impact on efficiency

These more complex error-control procedures, which can more readily be implemented with a positionally significant control discipline, could markedly influence the efficiency of future data communications systems.

Such error control procedures can overcome the detrimental impact of buffering and queuing delays that occur at every node in a complex data communications system using intermediate communications processors. For example, suppose that the transmission rate is 4,800 bits per second and that a block consists of 1,000 bits. Therefore, about five blocks are transmitted each second. Further, suppose the end-to-end queuing delay is one second. This means that the transmitter is sending block 5 about the same time that the receiver is accumulating block 1. If the transmitter stores eight blocks then this discrepancy need not slow down the actual data rate. But if the transmitter stores only two blocks, then it will have to wait and retain these blocks in buffers until the transmitted blocks can propagate through the system, and an acknowledgment is returned, before the DLC will permit the transmitter to safely send the next block.

Future directions

Any new approach to the design of data link controls must allow sufficient open-endedness for future requirements. Thus, design must address not only hardware and line economics, but more importantly, include programing support economics along with improved availability, serviceability, security, and ease of modification and expansion. For these reasons, modularity of function is important; that is, there should be functional separation of I/O device and network control from data link control. Also, there is little, if any, reason to treat point-to-point links differently from multipoint links.

Further, DLC for half-duplex communications should be a pure subset of DLC for full-duplex communications. A further aim should be to promote compatibility at the application program interface for the same device, whether it is used in the computer room or at an outlying location as a remote peripheral.

Synchronous data link control

Long awaited and much discussed, SDLC holds out the promise of greater efficiency in the utilization of data communications resources

J. Ray Kersey
International Business Machines Corp.
Research Triangle Park, N.C.

Tutorial

Achieving the potential benefits available in computer-based data communications systems requires a data link control that permits fuller utilization of resources—terminals, computers, communications processors, transmission links—than is feasible with present data link controls. Experience over the past decade has shown that a data link control (DLC), sometimes called a line protocol, should be straightforward and universal. The main tasks for a DLC should simply be to establish and terminate a logical connection between two stations—for example, between a computer and a remote terminal—and to achieve high-throughput, accurate transmissions between stations. For universality, the DLC should be able to service a multiplicity of operational configurations in a simple and unambiguous manner. In retrospect, then, the sphere of application of IBM's binary synchronous communications (BSC) DLC has been exceeded by new requirements (Ref. 1). Therefore, IBM's new synchronous data link control (SDLC) has been developed to meet these new requirements by simple, yet significant advances in the DLC structure.

Some major features of SDLC are that it is independent of code structure; it separates peripheral device control and network control functions from link control functions; it is bit oriented and does not use control characters (which sometimes have had multiple meanings); it is unambiguous as to the data link control functions to be performed and to the direction of transmission; it uses one standard frame format for information transfer as well as for link supervision and control; and its error checking and correction technique tests a complete frame—including address, control, information (text), and error-redundancy sequences—whereas previous DLCs usually only checked the information sequence.

Synchronous data link control applies to serial-by-bit synchronous transmission between buffered stations on a data transmission link using centralized control. SDLC operates on data transmission links that may be customer owned, leased from a common carrier, or part of the public, switched (dial-up) telephone network. Furthermore, SDLC applies to both half duplex and full duplex operation. It can be employed in point-to-point, multipoint, and loop configurations. (The use of SDLC in loop

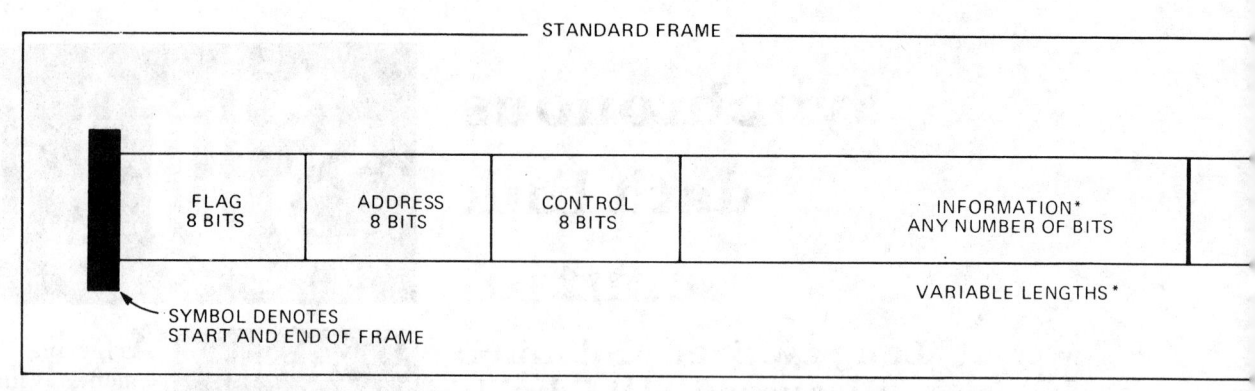

1. Under SDLC all frames conform to a standard format—with multiple functions such as acknowledgment of a previous frame contained in an information transfer.

configurations, is discussed in Reference 2.) Six examples of its usage in half- and full-duplex operation are given later.

Synchronous data link control employs "two sequence" frame numbering. That is, each buffered station sends a number, or count, sequence to identify the frame being transmitted, and the count sequence of the frame it expects to receive next from the other station. The send and receive counts, eight of each, are independent of each other, thus permitting unbalanced-load transmissions in either direction. SDLC uses the go-back-N-frames error correction technique. BSC also uses retransmission for correcting errors but it can only have one unacknowledged block (frame) outstanding at any time, while SDLC can have as many as seven frames outstanding at any instant. Stations buffered for eight frames in combination with the SDLC's go-back-N-frames error correction technique achieve higher net data throughput by reducing the number of line turnarounds in half duplex links and by reducing the effects of unbalanced and variable loads between the two directions in both half duplex and full duplex links. (References 1 and 3 review simple retransmission and go-back-N error control.)

Centralized control

Before describing the simple architecture of SDLC, and advantages resulting from this simplicity, it is first necessary to define centralized control. Here, one station is designated the primary station. As examples, a host computer, or a remote data concentrator, or a front-end processor would be a primary station. It retains control of the link at all times and exercises control through a prescribed set of commands and responses as defined in the DLC. The primary station is also responsible for initiating error recovery procedures when it detects errors or other exception conditions.

All other stations on the link (not the network) that can "connect" to the primary stations are designated as secondary stations. They receive the commands and information sent by the primary station and then react with suitable responses and their own information (if any) to be sent to the primary station.

Under SDLC, all transmitted frames—for information transfer, supervisory control, and miscellaneous control requirements—conform to a standard format (Fig. 1). And, multiple functions are handled within a single frame. For example, an information frame sent by the primary station to a secondary station is used to transfer information. The same information frame can be used to acknowledge to the secondary station, using a receive count, that one or more frames previously sent have been received and accepted as correct. On receiving a valid error-free acknowledgment, the secondary station releases (clears) an appropriate number of frame buffers, readying the buffers for new inputs. Meanwhile, the same information frame from the primary station can poll the secondary station to find out if it has any frames to transmit.

Furthermore, an information frame from the secondary station to the primary station can, along with the information field, be used to acknowledge one or more frames received from the primary station and to indicate whether a frame is the final frame in the transmission. When the primary station senses this "final frame" bit, it can go on to polling other secondary stations.

As mentioned, frames can contain counts of received frames and transmitted frames that are independent of each other. Independent frame sequence numbering compensates for a variety of information flow conditions that can occur during link operation. For example, SDLC can accommodate unbalanced information flow, unequal information frame length, and variability in the amount of information to be transmitted.

Unbalanced information flow means that more information (volume) flows in one direction, say from secondary to primary, than in the other direction in remote job entry (RJE) applications.

| FRAME CHECK SEQUENCE
16 BITS | FLAG
8 BITS |

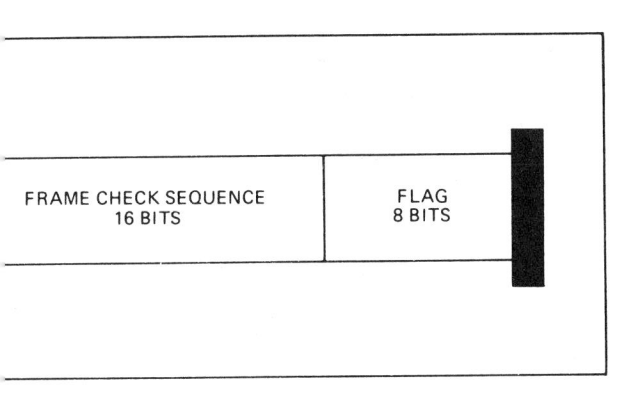

Unequal information frame length means that there will be short frames in one direction and long frames in the other direction. This is characteristic of inquiry/response systems.

Variability in the amount of information to be transmitted can mean, for example, that there is a wide variation in the number of frames in queue at different secondary stations, and that this wide variation can occur at different times at one secondary station. Such variations happen, for instance, at secondary stations that are situated in different time zones.

Just how SDLC handles these flow compensation tasks will be shown later in some examples.

Transmission is by frame

Architecturally, SDLC differs from previous DLCs in the fact that it uses positional significance, not control characters. Positional significance means that the position of a given bit sequence following or preceding a unique delimiter, called a FLAG, will be dedicated to a prescribed function, such as ADDRESS, CONTROL, or FRAME CHECK SEQUENCE. Under SDLC, all transmissions occur in frames and each frame must conform to the structure in Figure 1. The only variation is that the INFORMATION field may or may not be present.

As shown, a frame starts with the 8-bit FLAG sequence, 01111110, followed in position by an ADDRESS sequence, a CONTROL sequence, an INFORMATION sequence (if present), a FRAME CHECK SEQUENCE, and ending with another FLAG sequence. Each station attached to the data link continuously hunts for the FLAG sequence and an ADDRESS sequence. In multi-point operation, for example, a secondary station must detect a FLAG immediately followed by its own ADDRESS to get on line.

When the primary station transmits the 8-bit station ADDRESS sequence, it thus designates which secondary station is to receive the balance of the transmitted frame. When a secondary station transmits, the ADDRESS tells the primary station which secondary station originated the frame. A secondary station must recognize its valid address before it can accept a frame and take any action on the contents of that frame. Also, the primary station will accept a frame only when it contains the address of a secondary station that has been given permission to transmit. For data integrity reasons, the ADDRESS sequence appears within each frame. This also provides flexibility so that, for example, the primary station can interleave receptions from several secondary stations without intermixing individual-station information transfer. Using straight binary coding, the 8-bit ADDRESS sequence can differentiate between (up to) 256 terminals.

Controlling SDLC

The 8-bit CONTROL sequence follows the address field. The CONTROL sequence is the kernel of SDLC and will be described in considerable detail following a short discussion of the INFORMATION, FRAME CHECK, and end FLAG sequences.

The INFORMATION field may vary in length, including different lengths in sequential frames making up a complete transmission. The data may be configured in any code structure, including straight binary, binary coded decimal, packed decimal, EBCDIC, ASCII, and Baudot. That is, the INFORMATION field may be used to convey any kind of code. However, the content of the field must be self-defining by actual or implied means. For example, peripheral device control characters, such as CARRIAGE RETURN, will actually be part of the INFORMATION field, while whether the code being used is ASCII or EBCDIC may be implied in the address of a specific terminal designed for a specific code. Furthermore, whether a frame contains an INFORMATION field at all depends on the particular CONTROL format transmitted in the frame.

Because there is no restriction on the bit patterns that may appear between the end of the start FLAG and the beginning of the end FLAG, the transmitted data stream may contain six or more contiguous 1s and this pattern could be interpreted as a FLAG, and inadvertently terminate an incomplete frame. To circumvent this, once the start FLAG has been completed the transmitting station starts counting the number of contiguous 1s; when five 1s occur, the transmitter automatically inserts a 0 following the fifth 1. The receiver, too, counts the number of contiguous 1s. When the number is five, it inspects the sixth bit; if a 0 the receiving station drops the 0, resets its counter, and continues receiving. But if the sixth bit is a 1, then the receiving station continues to receive and act on the pending FLAG.

The FRAME CHECK SEQUENCE is included in all SDLC frames to detect errors which may occur during transmission. This field is 16 bits long and immediately precedes the beginning of the end-of-frame FLAG. The contents of the FCS field, based on a cyclic redundancy check, is a remainder polynomial numerator derived from a division of the

INFORMATION FORMAT

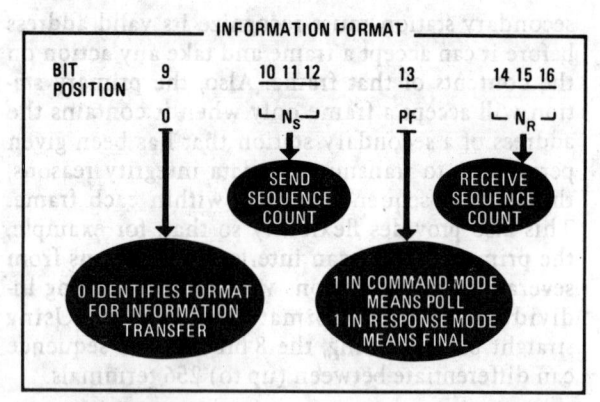

BIT POSITION	9	10 11 12	13	14 15 16
	0	N_S	PF	N_R

- 0 IDENTIFIES FORMAT FOR INFORMATION TRANSFER
- SEND SEQUENCE COUNT (N_S)
- 1 IN COMMAND MODE MEANS POLL / 1 IN RESPONSE MODE MEANS FINAL
- RECEIVE SEQUENCE COUNT (N_R)

2. The control field in its information format is used by primary and secondary stations to transfer an information field.

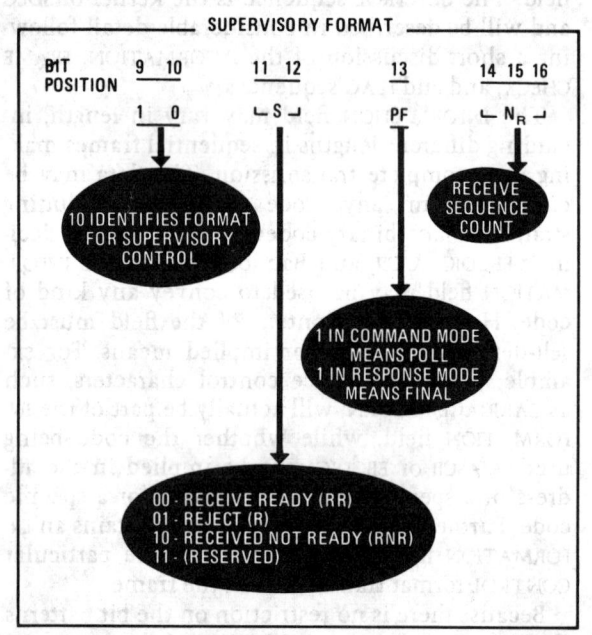

SUPERVISORY FORMAT

BIT POSITION	9 10	11 12	13	14 15 16
	1 0	S	PF	N_R

- 10 IDENTIFIES FORMAT FOR SUPERVISORY CONTROL
- 1 IN COMMAND MODE MEANS POLL / 1 IN RESPONSE MODE MEANS FINAL
- RECEIVE SEQUENCE COUNT
- 00 - RECEIVE READY (RR)
- 01 - REJECT (R)
- 10 - RECEIVED NOT READY (RNR)
- 11 - (RESERVED)

3. The control field in its supervisory format is used for such things as acknowledging frames and requesting retransmission.

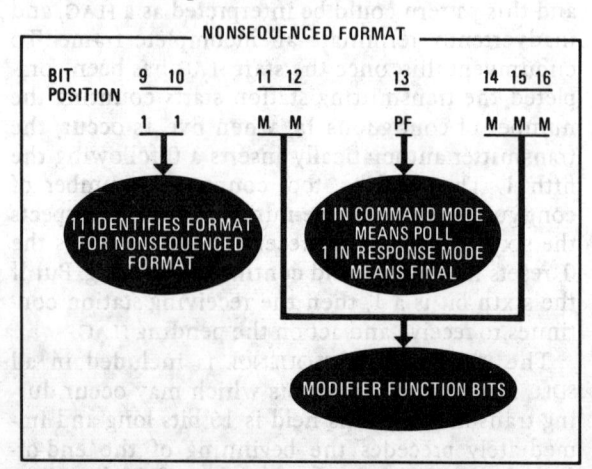

NONSEQUENCED FORMAT

BIT POSITION	9 10	11 12	13	14 15 16
	1 1	M M	PF	M M M

- 11 IDENTIFIES FORMAT FOR NONSEQUENCED FORMAT
- 1 IN COMMAND MODE MEANS POLL / 1 IN RESPONSE MODE MEANS FINAL
- MODIFIER FUNCTION BITS

4. The control field in its nonsequenced format can provide up to 32 functions without changing frame count sequence.

transmitted data by a generator polynomial (Ref. 3). The generator polynomial for SDLC is:

$$G(x) = x^{16} + x^{12} + x^5 + 1$$

All data transmitted between the start FLAG and the end FLAG is included in the checking accumulation, except those 0 bits inserted to prevent unwanted FLAGS.

Implicit in SLDC is that a primary station uses CONTROL to tell (command) the addressed secondary station what operation it is to perform. The secondary station uses CONTROL to react (respond) to the primary station.

The CONTROL field takes on any one of three formats depending on whether the field is to indicate:
- Information transfer (Fig. 2.)
- Supervisory commands/responses (Fig. 3.)
- Nonsequenced commands/responses (Fig. 4.)

The information format is used by the primary station when it wants to transfer an INFORMATION field to a secondary station, and by the secondary station when it wants to transfer an INFORMATION field to the primary station (Fig. 2.). A 0 in the 9th bit after the end of the start FLAG identifies the CONTROL field as being in the information transfer format. (During transmission, the 9th bit is sent first, followed by the 10th bit, and so on.)

A primary station inserts a 1 in the 13th-bit (PF) position to inform (POLL) the secondary station to initiate transmission (Fig. 2). The secondary station inserts a 1 in this position to respond to the primary station to indicate when a frame is the FINAL one in a transmission. Otherwise, the 13th bits stays 0. A transmitting station, whether primary or secondary, uses the N_S subfield to indicate the count sequence, or number, of the frame being sent. Thus, N_S means "I am now sending you frame number N," where N can go from 0 to 7. Similarly, N_R means "I am acknowledging error-free receipt of sequences up to (N_R-1) that you previously sent and I am now looking for frame number N." The use of independent frame-sequence counts provides for the detection (and eventual retransmission) of missed, erroneous, or duplicated frames in either direction. Each of these two subfields, (N_S and N_R), contains three binary positions, so each can count up to eight frames; that is, of course, 0 through 7.

Handshaking

At the completion of a frame, the transmitting station updates its send-sequence count. The station receiving the frame compares the contents of the received N_S with the contents of its receive-sequence-count field (N_R) and, if they are equal, accepts the frame (provided no error is indicated by the FRAME CHECK SEQUENCE mechanism). The receiving station's receive count (N_R) is then updated. Thus, the contents of the N_R field in the received frame acknowledges to the receiving station that frames through sequence number (N_R-1),

204

Six examples of SDLC in action

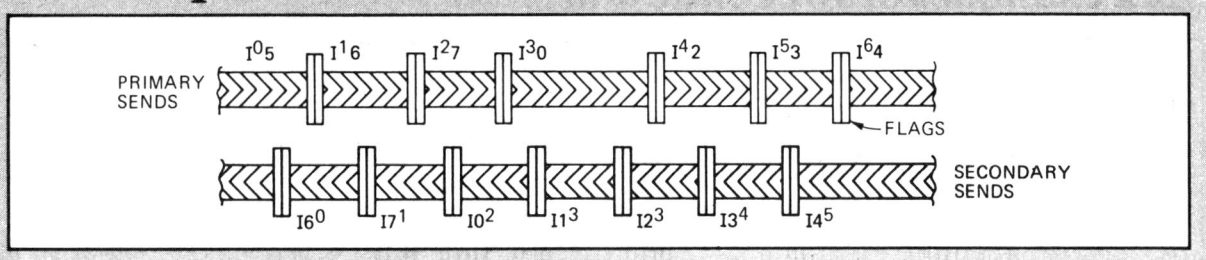

5. Full duplex information flow. In full duplex operation, information flows in both directions at the same time. In SDLC, point-to-point (shown here) and multipoint operation is the same. Time increases from left to right. The legend associated with the start of each frame—a shortened version of the content of the 8-bit CONTROL sequence carried in the frame—is all that is needed to describe a link's operation under SDLC. The I means INFORMATION transfer. The number immediately following the I is a station's send count, N_S, which can range from 0 through 7. And the next number is the station's 0 through 7 receive count, N_R. Thus, for example, the I1,6 related to the primary station means an INFORMATION transfer with a send frame numbered 1 and with the primary station looking for the secondary's frame 6. Controls referring to the primary are shown in color and to the secondary in black.

A station does not acknowledge (accept) a frame in transit until the frame has been completely received and the 16-bit FRAME CHECK SEQUENCE has been tested and found correct. (The primary/secondary dialog shown here represents error-free transmission.)

At the upper left, the primary station has its frame 0 in transit. When that frame is completed, shown by the flags, the primary starts sending frame 1 and tells the secondary it is looking for its frame 6. At that instant, the secondary has its frame 6 in progress. This secondary frame also carries a receive count of 0 to the primary—thus telling the primary at completion of the secondary frame 6 that the secondary acknowledges correct receipt of primary frame 7. This acknowledgment clears frame 7 (and all preceding unacknowledged frames, if any) from the primary's buffer. The secondary's receive count then advances to 1. That is, the secondary is primed to receive the primary's next frame to be completed—the ongoing frame 1.

The primary and secondary frames at the left of the figure are about equal length, so that send and receive counts are completed and acknowledged, and advanced, on a one-for-one basis for both the primary and secondary station. But two secondary frames are completed during the transmission of primary frame I3,0. Note how the independent counts of send and receive sequences are automatically advanced and acknowledged without having to stop transmission in one direction to wait for frame acknowledgment. As it should, the primary's send count advances by 1, from 3 to 4 (as indicated in color), without interruption. But the primary's receive count jumps from 0 to 2, which occurs because two secondary frames, with send count 0 and 1, are completed during the primary I3,0 frame. These two frames therefore are kept in the secondary station's buffer until the completion of primary frame I3,0.

If an error or other exception condition occurs, buffered 0 and 1 frames at the secondary remain available for retransmission. Here, though, the primary acknowledges the acceptance of secondary send counts numbered 0 and 1, and the secondary station releases buffers for these two frames. That is, the next primary frame, I4,2, means the primary is sending the next frame (4) in sequence, and the primary is now looking for the successful transmission of secondary frame 2—which is, at that instant, in transit and will be acknowledged by the next secondary frame following the completion of the ongoing primary frame.

Six examples of SDLC in action

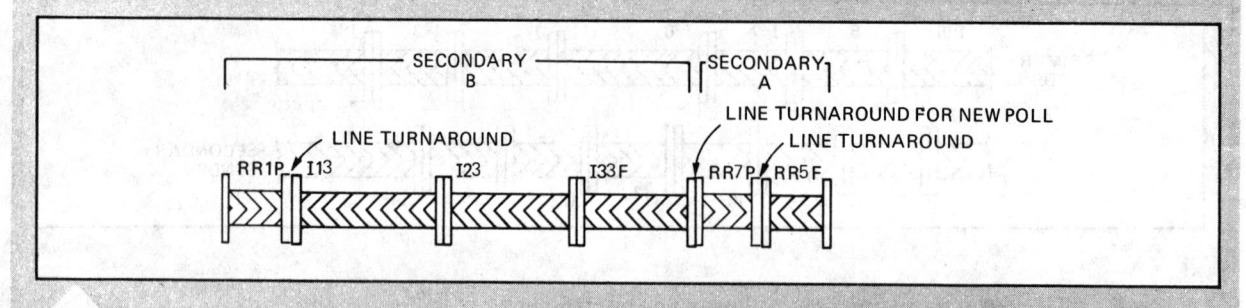

6. Half duplex operation. In half duplex operation, shown here, one pair of wires is used to send frames in either direction, but in one direction at a time. Each time a line must be turned around to reverse transmission direction can consume up to about 300 milliseconds, as required to disable echo suppressors in the link and to permit modems to perform automatic equalization. With binary synchronous communications (BSC) data link control, the line is turned around for each block, with a corresponding consumption of transmission resources. With synchronous data link control, net data throughput can be increased because line turnarounds are reduced by using one supervisory format frame to initiate transmission of up to 7 information frames from the secondary.

Here, a supervisory frame represented, for example by RR1P, means that the primary station is in a RECEIVE READY state, the station has a receive count of 1 (that is, looking for the secondary's frame number 1), and is polling the secondary to find out if it has any frames to transmit. Polling is initiated by a 1 in the 13th bit after the end of the start FLAG (Fig. 3).

After the secondary receives and accepts the primary's RR1P supervisory sequence, the line is turned around and the secondary starts to sequentially transmit the frames stored in its queue. Here, three frames have been stored, waiting for a poll. The secondary sends the three frames without any intervening line turnarounds. The F, for FINAL frame, in I3,3F is a secondary's way of telling the primary it has finished transmitting its buffered frames. To accomplish this, a 1 is inserted in the 13th bit of the INFORMATION frame (Fig. 2).

When the primary station detects the secondary station's F bit, a line turnaround is executed, and the primary can go to the next secondary on its polling list. The two frames at the right show the dialog between the primary station and another secondary. Here, RR7P means that the primary is in a RECEIVE READY state, that it is polling the new secondary, and that the last frame it acknowledges from that secondary is number 6, and it's waiting for frame 7.

However, as shown, the secondary does not have any INFORMATION frames to send at that moment so it responds with RR5F—meaning that the last frame accepted from the primary was number 4, it is waiting for number 5, and doesn't have any frames to send so is signing off with an F for FINAL frame. The primary station can then continue polling.

It should be noted, but not shown here, that on full-duplex multipoint links, an F-response from the secondary tells the primary that the receiving link to the primary station can be made available to some other secondary station. Also not shown, but implicit in the use of SDLC on half-duplex links is that the primary station can send one or more INFORMATION frames to the secondary, followed by one or more INFORMATION frames from the secondary—except that no more than seven unacknowledged frames can be outstanding in any one direction. If the limit is reached then the line must be turned around to permit frames to be acknowledged and the buffers cleared for new text. Note that a one-for-one information frame exchange represents a "conversational" operation with no frames used exclusively for control purposes.

Six examples of SDLC in action

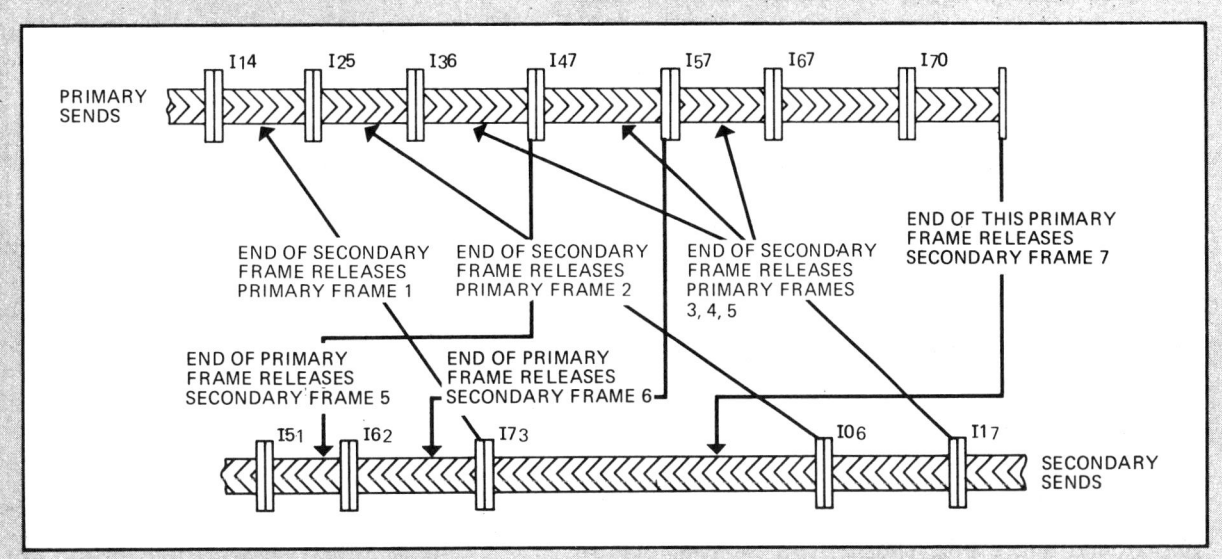

[Diagram: Full-duplex timing diagram showing PRIMARY SENDS stream with frames I14, I25, I36, I47, I57, I67, I70 and SECONDARY SENDS stream with frames I51, I62, I73, I06, I17, with arrows indicating:
- *END OF SECONDARY FRAME RELEASES PRIMARY FRAME 1*
- *END OF SECONDARY FRAME RELEASES PRIMARY FRAME 2*
- *END OF SECONDARY FRAME RELEASES PRIMARY FRAMES 3, 4, 5*
- *END OF THIS PRIMARY FRAME RELEASES SECONDARY FRAME 7*
- *END OF PRIMARY FRAME RELEASES SECONDARY FRAME 5*
- *END OF PRIMARY FRAME RELEASES SECONDARY FRAME 6]*

7. Buffer release. In this example of full-duplex, point-to-point operation, the secondary data stream contains a frame that exceeds the length of three primary frames. Again, multiframe acknowledgment with automatic buffer release permits smooth transmission flow. The operation of this full-duplex link is given by the legends in the illustration. Note here how the completion of secondary frame I6,2—meaning the secondary is sending its frame 6 and is expecting the primary frame with a send count of 2, upper left—acknowledges primary send frame 1 and releases that buffer. Note how the long frame in the secondary data stream causes the N_R count to jump from 3 to 6, because three primary frames are completed during this time interval. That is, the secondary's N_R count indexes from 3, to 4, to 5, to 6, but the secondary frame doesn't release primary buffers until the end of the secondary's send count of 0, which carries with it its N_R count of 6. Therefore, the acknowledged 6 releases primary buffers through frame (6−1), or 3, 4, and 5.

This acknowledgment procedure allows buffer release at the earliest possible time, unless one chooses to use a supervisory frame (with RR) for the purpose of buffer release. For example, a supervisory frame of RR3 inserted between the secondary frame I6,2 and I7,3 would acknowledge and release primary buffer 2 significantly sooner. Similarly, a supervisory frame of RR6 could be inserted between I7,3 and I0,6 to acknowledge and release primary buffers 3, 4, and 5 earlier.

8. Multiframe acknowledgment. In this example of full-duplex, point-to-point operation, the frame length of the traffic being transmitted by the primary station is much longer than the frame length of the traffic from the secondary station. Because each station independently keeps track of its send and receive counts, the number of frames sent in one direction does not depend on the number of frames in the other direction. But no more than seven unacknowledged frames can be outstanding at any instant since the modulo 8 count mechanism can handle just eight frames. The exchange of INFORMATION frames in both directions between the primary station and the secondary station flows smoothly because one receive count can acknowledge up to seven frames.

Here, one primary frame acknowledges three secondary frames. For example, the completion and acceptance (error-free transmission is assumed here) of receive count 4 in the primary's I3,4 frame acknowledges accurate receipt of secondary frames 1, 2, and 3. Note that the secondary's receive count remains unchanged for several transmitted secondary frames, until one of the longer primary frames is completed. An example is secondary frames I2,2, I3,2 and I4,2—all waiting for the completion of primary frame I2,1. Specifically, secondary I4,2 means the secondary is starting to transmit its frame 4 and is looking for and waiting for the completion of primary frame 2. A similar smooth flow of INFORMATION frames will also take place when primary frames are shorter than secondary frames.

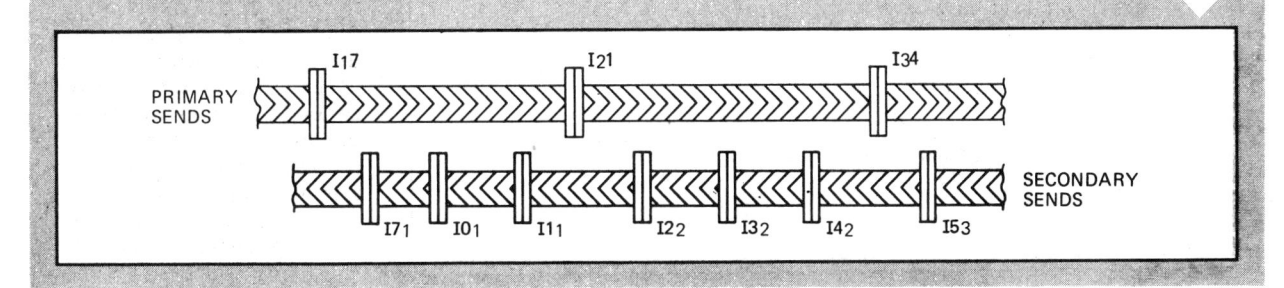

[Diagram: PRIMARY SENDS stream with frames I17, I21, I34 and SECONDARY SENDS stream with frames I71, I01, I11, I22, I32, I42, I53]

Six examples of SDLC in action

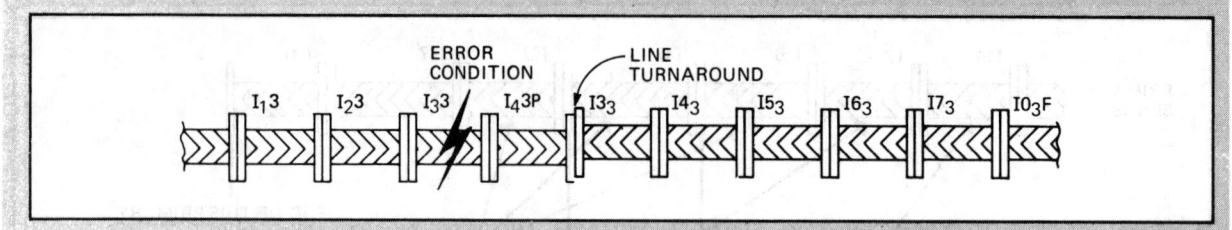

[Diagram: frames I_13, I_23, I_33 (ERROR CONDITION), I_43P, LINE TURNAROUND, $I3_3$, $I4_3$, $I5_3$, $I6_3$, $I7_3$, $I0_3F$]

9. Error correction. Whether used in full-duplex or half-duplex links, SDLC permits traffic to continue in one direction even though an error condition has occurred in the other direction. In the illustration (of half duplex operation), the primary sends frames 1 through 4, during which frame 3 is impacted by an error condition. In SDLC the primary continues sending until a last frame of a transmission is denoted by the poll bit, P. The P bit initiates a line turnaround and the secondary starts sending the frames in its transmission. Here the final secondary frame is 0, as noted in the frame I0,3F. The F-bit causes the line to turnaround again (not shown) and the primary starts transmitting again. That is, the primary backs up and starts retransmitting at frame 3 because the secondary frames carry an N_R count of 3—meaning that the secondary has only given positive acknowledgment to, and cleared the buffers for, up to primary frame 2. The primary then sends frame 4 again, as well as any new frames that may have been deposited in the primary's buffer. This error correcting mechanism works in the same manner when the error occurs in a secondary frame.

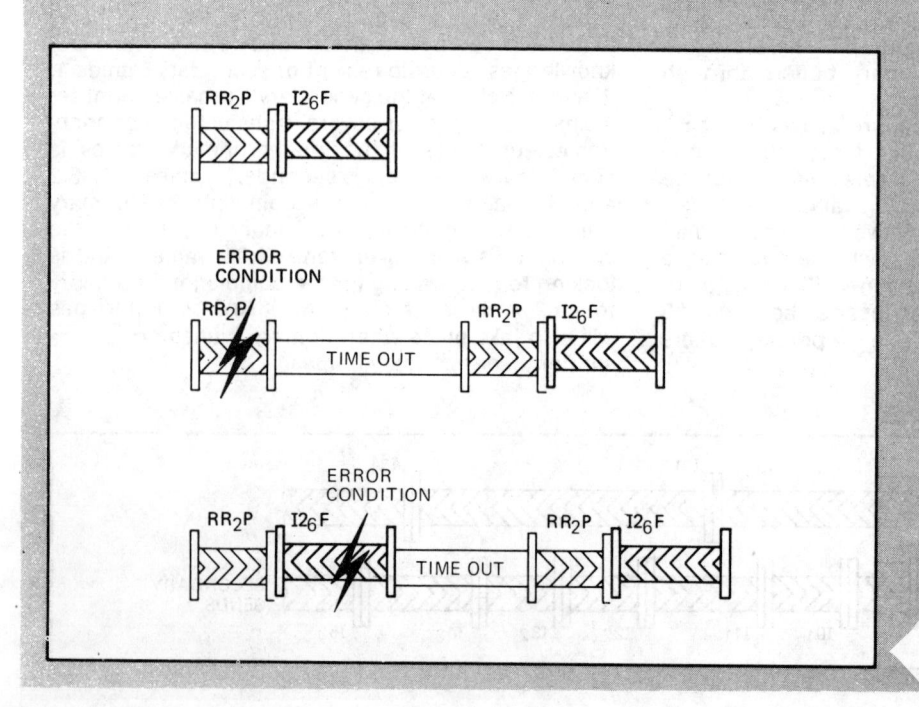

[Diagram: Three scenarios showing RR_2P and $I2_6F$ exchanges with ERROR CONDITION and TIME OUT intervals]

10. Consistent. Synchronous data link control's error recovery procedures provide consistent command/response exchanges whether the transmission is free of error (top), or an error occurs in the primary's polling frame (middle), or an error occurs in the secondary's final INFORMATION frame. When an error takes place in the primary's polling frame (middle), say due to lack of synchronism, the frame may either get lost altogether or be missed by the secondary. For such a situation, SDLC includes a timeout interval: if the primary does not receive a valid response from the secondary in a given time, then the primary will back up and retransmit its last polling frame. In the same manner, an error in the secondary's final information frame may cause the frame to be missed by the primary. Then, following a timeout interval, the primary will repeat its poll and look for a valid secondary frame. System recovery for errors occurring in ongoing information frames (not a final frame) was explained in the previous example.

which the receiving station had sent previously, have been accepted and corresponding buffers may be cleared.

The supervisory format is identified in the CONTROL field by a binary 10 pattern in the 9th and 10th bits after the end of the start FLAG. (Fig 3). The supervisory format is used to acknowledge I frames, to request retransmission of I-format frames, and to inhibit the sending of I frames. These alternative supervisory functions are accomplished by the supervisory subfield.

A frame containing a supervisory format never contains an INFORMATION field. Such a frame still follows the standard structure of Figure 1, with the INFORMATION field interpreted to be of zero length.

Besides providing data integrity by accepting or rejecting information frames, the supervisory control format regulates the flow of information frames by invoking such controls as RECEIVE READY and RECEIVE NOT READY.

As shown in Figure 3, supervisory commands and responses provide three functions: RECEIVE READY (RR) when bits 11 and 12 are binary 00, REJECT (R) when these bits are binary 01, and RECEIVE NOT READY (RNR) when the bits are binary 10. Bit pattern binary 11 is reserved.

A station sends RECEIVE READY (00) when it wants to acknowledge to another station that it has received error free information frames with sequence counts through (N_R-1) and that it is ready

Management. SDLC manages channel linking of IBM's 3790 communications system (along left wall), controlling up to 16 terminals, to 3704 or 3705 communications controller itself linked to distant host computer.

to receive additional frames. N_R is carried in the CONTROL field in the 14th, 15th, and 16th bits after the end of the START flag.

Turning down data

A station sends REJECT (01) when it does not receive or does not accept, due to sequence error, the transmitted frame with an N_S count corresponding to the N_R count contained in bits 14, 15, and 16 of the transmitted frame. Frames with sequence numbers through (N_R-1) are acknowledged, but all frames received after the given N_R count are rejected. REJECT is transmitted just once for each exception condition, and no more than one REJECT may be outstanding at any instant. The exception (REJECT) condition is cleared (reset) when the receiving station receives an INFORMATION frame with an N_S count equal to the existing count in the receiving station N_R subfield.

The RECEIVE NOT READY supervisory format indicates the receiving station has a condition, such as all buffers being full (BUSY), so that it cannot receive any more frames that require buffer space. Again, frames with sequence numbers (N_R -1) are acknowledged but frame N_R and any subsequent frames are not accepted. When a station receives an RNR response, which means the other station is BUSY, it must not send any additional frames containing an INFORMATION frame until the other terminal reports that the condition has been cleared. The busy station signals when it is ready for additional traffic by sending an RR (meaning buffer space available) or an INFORMATION format frame with a zero length information field.

The NONSEQUENCED format in the CONTROL field is identified by a binary 11 pattern in bit positions 9 and 10 after the end of the start FLAG. As Figure 4 shows, five modifier bits are allocated to distinguish up to 32 different NONSEQUENCED functions. NONSEQUENCED frames transfer information without regard to send and receive sequence counts. And they do not alter station sequence counts. Examples of NONSEQUENCED control functions are the polling of secondary stations without affecting sequence numbers, broadcasting, exchanging terminal identification information between primary and secondary stations, and commanding a secondary station to go "ON HOOK" (primarily on dial-up links). Typical NONSEQUENCED response functions are to reject invalid commands and to perform initialization and diagnostic procedures pertaining to data link control.

Bootstrapping

Synchronous data link control has four modes of operation: normal response mode, asynchronous response mode, normal disconnected mode, and asynchronous disconnected mode. The modes are operational rules that define such things as the conditions under which a secondary station may initiate transmission and the conditions for expected response times (timeouts).

In the normal response mode, a secondary station may initiate a transmission only after being directly polled by the primary station. In asynchronous response mode, a secondary station may initiate a transmission on an unsolicated basis at any appropriate respond opportunity. What constitutes an appropriate respond opportunity depends on how the link is configured or being operated. In half duplex operation, a secondary station may initiate an asynchronous response transmission after receiving 15 contiguous 1 bits (defined as an IDLE state) on its receive channel. In full duplex operation, the secondary station may initiate transmission at any time.

Four operating modes

Sometimes secondary stations are not ready or equipped to assume operation in either the normal or asynchronous response modes. The two initialization modes—normal disconnected and asynchronous disconnected—allow the primary to bootstrap the secondary station into operation by modifying or suspending normal protocol requirements. The initialization modes are, therefore, considered to be temporary measures.

The preceding pages contain six examples of the use of synchronous data link control for both half duplex and full duplex operation and under conditions that have and do not have errors during transmission. In these examples, the contents of the CONTROL field tell what is happening on the link, frame by frame.

As a concluding note, it is pointed out that SDLC has been implemented in several recently announced IBM products—the 3790 communications system is an example—and is supported in IBM's new access method called VTAM for virtual telecommunications access method. ■

REFERENCES
1. J. Ray Kersey, "Taking a fresh look at data link controls," Data Communications Systems, Vol. 2, No. 1, September 1973, pp. 65-71.
2. "Synchronous Data Link Control General Information Manual," IBM SRL GA 27-3093.
3. Karl I. Nordling, "Error control: keeping data messages clean," Electronics Deskbook Data Communications Systems, Vol. 1, No. 1, 1972, pp. 91-95.

Advanced link control runs full and half duplex on various types of nets

Offering the capabilities of the latest data link controls, this new protocol can operate on DEC's existing hardware

Stuart Wecker
Digital Equipment Corp.
Maynard, Mass.

Throughput

Present data link controls, especially binary synchronous communications (BSC), are so wasteful of data communications resources that great emphasis has been placed in many quarters to develop more efficient approaches. To fulfill this need, a few computer manufacturers and certain users have designed their own protocols to permit communications among the various elements of data networks.

Now Digital Equipment Corp. (DEC) has also produced its own data link control, which operates on existing hardware and can be implemented on many operating systems. For many communications networks built around DEC computers, the new protocol can be implemented with few, if any, hardware changes and only a moderate amount of software development. For users planning DEC systems, the data-link control may entail no more hardware investment or software development than previous protocols.

The new control, called DDCMP (Digital Data Communications Message Protocol) closely parallels the operational features and architecture of IBM's synchronous data link control (SDLC) and the latest draft standard of American National Standards Institute's ADCCP (Advanced Data Communications Control Procedure).

The control's main functions are to establish communications between two stations and to ensure the integrity and correct sequence of data passing between the stations. Furthermore, DEC's protocol operates in full-duplex and half-duplex modes and in point-to-point and multipoint configurations (Figs. 1, 2, and 3) over serial and parallel links. Just how DDCMP compares with SDLC, ADCCP, and BSC is discussed in the panel, "How DDCMP stacks up," page 217.

DDCMP was designed to operate most efficiently over channels having a high probability of errors at speeds ranging from 400 to 50,000 bits per second. In theory, the protocol will operate at any speed, but it may not be the optimal technique for all transmission methods. There are no provisions for forward error correction. The software for DDCMP is initially being included in an operating system that is an updated version of LIPS (Laboratory Information Processing System), but it will become part of all future DEC network and communications software products.

Implementing the protocol

DDCMP is designed to operate on existing models of DEC's PDP-8, PDP-11, PDP-15, and DECsystem-10 computers, but implementation of the protocol on systems for which the software already is running may require some effort. Several models of Digital's PDP-11 family can handle DDCMP efficiently. For synchronous transmission, DP-11 and DU-11 interface processors provide character interrupt, and DQ-11 provides double-buffered DMA (direct memory access). For asynchronous links DL-11 and DH-11 provide character interrupt.

For present systems, DDCMP can be retrofitted to existing software by writing the code to perform the data-link control functions, according to specifications furnished by DEC, and interfacing the program to the existing operating system. Writing the code to perform DDCMP functions is a fairly straightforward job and requires only 1,500 to 2,000 words of storage in a PDP-11, a 16-bit minicomputer. One customer reports completing the programing in only three weeks. The difficult part of the job may be to interface the new software with the operating system that is presently installed and running.

Another problem may be to apply the protocol to remote "unintelligent" terminals. As in retrofitting other new data-link controls, terminals require some processing capabilities and storage to be able to handle the protocol. This is not a problem if terminals already are connected to a local terminal-control unit, such as a minicomputer, that can take on the DDCMP functions.

But terminals that connect directly to data channels will require additional hardware. This hardware may be in the form of a minicomputer or microprocessor-based control unit at each terminal, or, for economy, several terminals can be clustered around one shared terminal-control unit. The alternative is to replace these terminals with intelligent terminals.

Choosing the hardware

The hardware requirements for all stations are almost identical. From a practical point of view, however, the control station in a multipoint network should be reliable and have adequate processing speed to maximize network efficiency.

All stations must be capable of these functions:
• Processing 8-bit data segments.
• Synchronizing on the DDCMP synchronization character.
• Providing 1,500 16-bit words of storage, or equivalent, to hold the DDCMP control program for all point-to-point stations or a tributary in a multipoint configuration. A control station in a multipoint network requires 2,000 words for protocol storage. (These figures are based on implementing the protocol on Digital's PDP-11 series of minicomputers.)
• Using the count field to find the end of the message.

To handle all of the factors listed above, each station must contain at least a simple processor, usually a minicomputer. For minicomputers with serial links and interfaces providing character-interrupt operation, the entire protocol can be implemented in software.

In an interface minicomputer with direct-memory-access (which is block-oriented), reception of data requires double or chained buffering. Otherwise, the system will have only one bit time to respond to the interface and set up for the data field's count after receiving the header. This may be difficult or impossible on high-speed lines. One way to alleviate this problem is to use fixed-length data messages so that the software can ignore the count field and assign fixed-length buffers. This means lowering the line efficiency or increasing software overhead by blocking and unblocking logical messages into fixed-length DDCMP transmission messages.

In some minicomputers, the overhead for DDCMP may be so great that a special DDCMP hardware controller may be needed. A simple DDCMP controller might perform some hardware functions, and the software would take over the others. For example, the controller could look for a sync character and then check the following character to determine whether it is an SOH (indicating a data message) or ENQ (protocol message).

Receipt of the SOH character would condition the controller to accept the next 14 bits as the count field, and to read the appropriate amount of data. The software would then be responsible for analyzing the remainder of the header and performing protocol functions.

An alternative approach might be to build a complete DDCMP processor to handle all protocol functions. Such a device could be treated by the host computer as a block-oriented peripheral.

Outlining the DDCMP format

Data is transmitted in variable-length messages of a prescribed format, which consists of the data along with the control and error-detection information needed to check and order this data at the receiving end (Fig. 4). Besides data messages, there are protocol messages for acknowledgment, initiating message exchanges, and error recovery.

There are two parts to a data message—the header, which contains the control information and the data itself. DDCMP allows a receiver to accept a maximum of 255 messages before an acknowledgment must be returned to the transmitter.

The protocol is inefficient for short messages, since each message requires a 10-character overhead for the message-header and block checks.

Data messages are numbered so that they may be assembled in the proper sequence at the receiving station. The reason these messages sometimes get out of order is that when an error is detected or a message is lost during transmission, the messages must be retransmitted.

Message numbering begins with an initialization

(continued on page 215)

Sequence of operation

1. POINT-TO-POINT, FULL DUPLEX

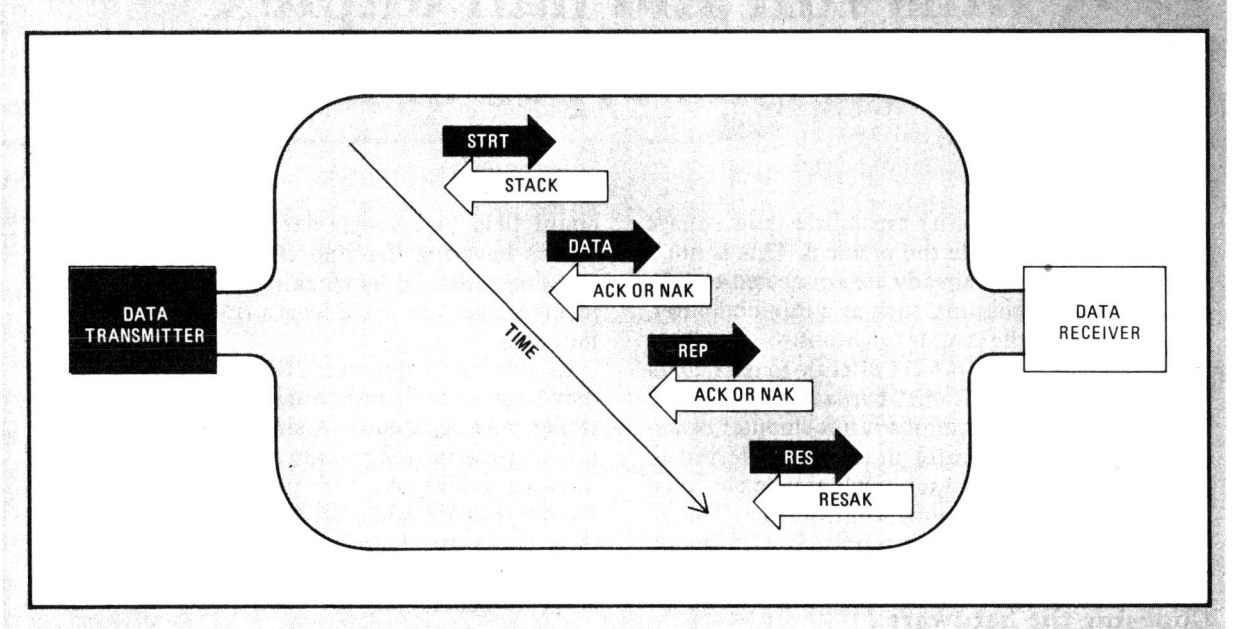

Operation in the point-to-point, full-duplex mode enables both stations to transmit and receive simultaneously. Message flow in one direction is independent of the flow in the other; therefore, this discussion relates only to the message transmitter. The station requiring link control begins by sending a start message (STRT). The proper response is a start acknowledge message (STACK). If no response is received within a predetermined interval (usually twice the round-trip propagation delay) or if the reply is different from a STRT or STACK, then the transmitter sends another STRT.

The STRT message includes the number of the first data message (usually 0) to be sent. Each message received is checked for eligible control characters and a 16-bit cyclic redundancy check is performed to test for errors. If the message is found to be correct, an acknowledgment is returned. The acknowledgment may be sent as a separate ACK message, or may be sent implicitly as a number in a data message going the other way.

If the receiver is busy at the time it is ready to acknowledge a given data message, the acknowledgment will be sent later, along with any others requiring acknowledgment, by sending the number of the latest received message. This implies that all previously unacknowledged messages also are correct.

If the redundancy check indicates an error, the receiver sends a NAK (negative acknowledgment) with the number of the last correct message. Then the transmitter resends the erroneous message, plus all others subsequently transmitted. Other reasons for NAKing are receipt of a message that does not have a legal start character (SOH, ENQ, or DLE, memory overflow, or failure at the receiving station).

When an acknowledgment is received at the transmitter all buffer locations holding acknowledged messages are made available for new messages. If a NAK is returned, the transmitter stops sending new messages, sends new sync bits, and retransmits all messages, starting with the NAKed one.

Whenever the transmitter holds unacknowledged messages and has received no replies for twice the average round-trip times (as determined by a timer), it sends a REP (reply message) containing the number of the last transmitted message. The proper response is an ACK or NAK. The reason for lack of a reply may be a failure at the receiver, a data link breakdown, or simply a traffic backup at the receiving station. The timer is reset either after each ACK or NAK is received, or after transmission of a REP.

On receiving a REP, the receiving station compares the message number in it with that of the last good message received. If the numbers are equal, the receiver sends an explicit or implicit acknowledgment. If the message numbers do not match, a NAK is returned with the number of the last good message received.

Message-header synchronization. If there are unacknowledged messages and an acknowledgment is received in which the message number is not one of those outstanding, or the last one acknowledged, an error has been made in number synchronization. In response, the transmitter sends a reset message (RES) with the lowest message number that it is still holding and the transmitter then waits for a reset acknowledge message RESAK.

The receipt of a RES message at the receiver adjusts the number of the expected message to the indicated value, sends a RESAK, causes the error to be logged, and the receiver to wait for the unacknowledged messages to be retransmitted. If, however, the RESAK is not received, outside intervention is required to restart the message exchange.

Sequence of operation

2. MULTIPOINT, FULL DUPLEX

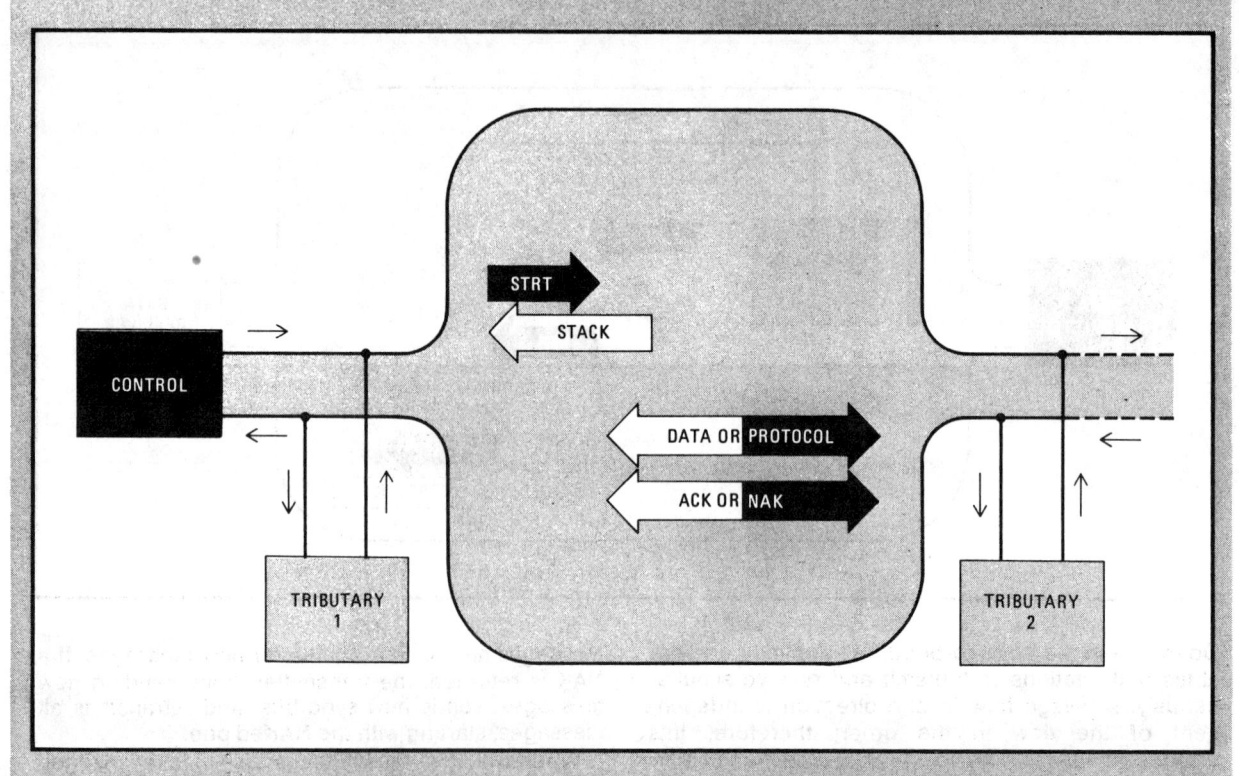

In multipoint operation, message exchange may only be initiated by the control station, which selects tributaries periodically on the frequencies and in the priority selected by the user. Only control and tributary communicate; tributaries cannot ordinarily transmit to each other. In addition, tributaries cannot interrupt the normal sequence for priority exchanges, but, instead, must wait until they are selected.

In selecting, the control sends a STRT message to the appropriate tributary, with the select bit set. If there are no data or protocol messages awaiting transfer, the tributary responds with STACK (start, acknowledge) with the final bit set, to return ownership to the control station. If messages are to be sent, the tributary still returns a STACK, but with the final bit reset, followed by the messages. In the last message, the final bit is set to transfer ownership back to the control.

Error recovery is handled in the same manner as in a point-to-point system. If the cyclic redundancy check indicates an error, or else if there is an illegal start character or a failure at the receiver, then a NAK is returned and all messages beginning with the one NAKed are retransmitted.

Note that for full-duplex operation, the final bit should not be transmitted by the control station, and the select bit should not be sent from the tributary.

Tributaries do not ordinarily require timers unless a particular application warrants it. A timer can be eliminated because tributaries sending responses to the control station must wait until the tributary is selected again, an indefinite time interval.

Sequence of operation

3. HALF-DUPLEX OPERATION

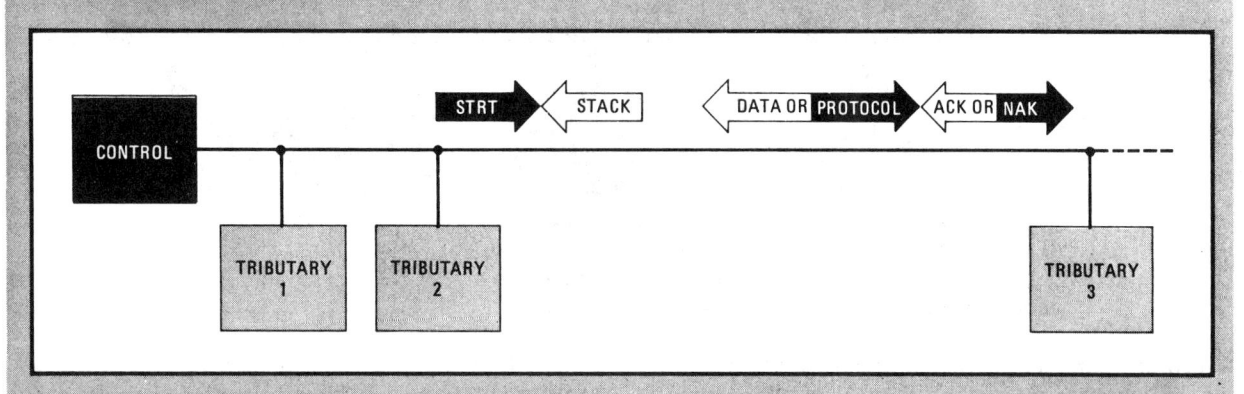

Unlike full-duplex operation, where both transmitters and receivers may be active at all times, the half-duplex mode (both point-to-point and multipoint) requires that only the receiver or transmitter may be active at a given instant, and stations alternate between transmitting and receiving. When the line is inactive, both stations in a point-to-point system must alternate ownership of the link.

Point-to-point. (not illustrated). When a station initiates a transmission, it turns its transmitter on and receiver off. After the transmitting station sends the STRT message, the transmitter shuts off, and the receiver turns on to await the STACK (start, acknowledge) message. Upon receipt of a STACK, the first station transmits a block of messages and then transfers ownership (via the select and final bits) to the other station.

The other station may ACK (acknowledge) or NAK (not acknowledge). If a NAK is received, the select and final bits in the last message transfer ownership back to the first station, and the erroneous message is retransmitted. When the messages are correct and no messages are awaiting transmission from the receiver, then the transmitter responds to the transmission with an ACK message.

If both stations attempt to send a STRT message at the same time, then there will be a line-contention problem. To circumvent this, the response-timing interval of one station is lengthened to permit the other station to take command before the first station can send another STRT message. An additional factor that must be added to the timing interval is the turn-around time of the line, a function of the line and the modem. In all other respects, operation in the half-duplex point-to-point mode is identical to that of full-duplex point-to-point mode.

Multipoint operation (diagram). Half-duplex multipoint operation is treated by the control station as a series of point-to-point half-duplex connections. The control station first selects tributary 1. Any messages to be sent are transmitted to the control station. If no messages are outstanding, the tributary replies with an ACK (with the select and final bits set), whereupon tributary 2 is selected.

(continued from page 212)

number (usually zero) as specified by the start message (one of the protocol messages). The number field in the header of each succeeding data message is incremented by one. The numbers range from zero to 255, and they may wrap around. (Note that protocol messages are not numbered.)

The receiving station acknowledges receipt of correct and in-sequence messages by returning the number of the last correct message received, either in the response field, in data messages transmitted in the opposite direction, or in a separate acknowledgment message.

Messages certified correct at the receiver usually are not separately acknowledged. To provide efficient use of the data link, data messages are acknowledged when the receiving station has a data message to return to the transmitter. When a message is acknowledged, it implies that all preceding messages have passed their CRC checks, contain allowable header characters, and have been cleared from the receiver's buffer.

Whenever an error is detected, a negative acknowledgment (NAK) message returned to the transmitter. The NAK contains the number of the last correctly received message (acknowledging all messages through the number given), and the succeeding messages must be retransmitted.

All control and data characters are 8 bits long, unless otherwise specified. A data message is preceded by a sequence of two or more synchronization characters. The message begins with the header portion, which contains control information for use by the protocol software.

Analyzing the format

In the data message format diagramed in Figure 4, the first header character is the ASCII character SOH (10000001), indicating start of the message. The

DATA MESSAGE FORMAT

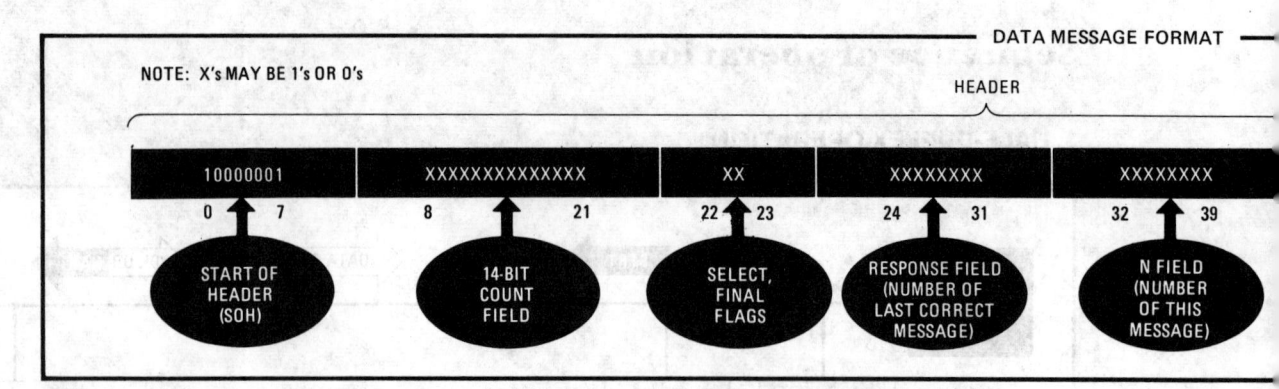

NOTE: X's MAY BE 1's OR 0's

10000001	XXXXXXXXXXXXXX	XX	XXXXXXXX	XXXXXXXX
0 — 7	8 — 21	22 — 23	24 — 31	32 — 39
START OF HEADER (SOH)	14-BIT COUNT FIELD	SELECT, FINAL FLAGS	RESPONSE FIELD (NUMBER OF LAST CORRECT MESSAGE)	N FIELD (NUMBER OF THIS MESSAGE)

HEADER

4. A data message always begins with an SOH character, which causes the receiving station to treat the various fields according to the message's format.

second character in the header is a 14-bit count field, which indicates the number of characters in the data portion of the message.

The count field is used by the receiver to locate the end of the data field. This eliminates the need to scan for an end of data marker, as required by SDLC. Since the data field does not have to be scanned, there is no way data characters can be confused with control characters, and the data, therefore, is fully transparent to the protocol.

Following the count field are two flag bits—select and final—that designate control of the link. Next is the response field. This indicates the number of the last good data message received at the station transmitting the outgoing message, which is the usual acknowledgment method for correctly received data messages.

The NUM field contains the number of this message. It is followed by the ADDR field, which contains the address of the station to which this message is being sent. In a multipoint system as many as 255 destinations may be addressed. The final segment of the message header is a 16-bit field used by the receiver for the header CRC.

Since the contents of the data field have no effect on the protocol, any coding or control characters may be embedded in the data. The only exception is that if the bit count of this field is not a multiple of eight, additional zeros must be added to fill out the last character. The last field is the cyclic redundancy check. Note that, unlike SDLC, DDCMP has two CRC fields—one for the header and one for the data portion of the message.

Separate checks of the control and data portions of the message tend to improve system response time because the receiving station can make use of the header information as soon as the header block check is found to be correct. However, on the negative side, this approach reduces the efficiency of line utilization because it adds 16 bits to the communications overhead.

The CRC field is the remainder left after division of the data stream by a polynomial. Called CRC-16, this algorithm produces a unique bit sequence for the combination of bits being checked. On the receiving end, the calculation is repeated and the block check is subtracted from the remainder, yielding a zero result.

Protocol messages

There are seven types of protocol messages. The purpose of each message is described below, and format details are presented in "Protocol message formats." Selection of transmitting and receiving

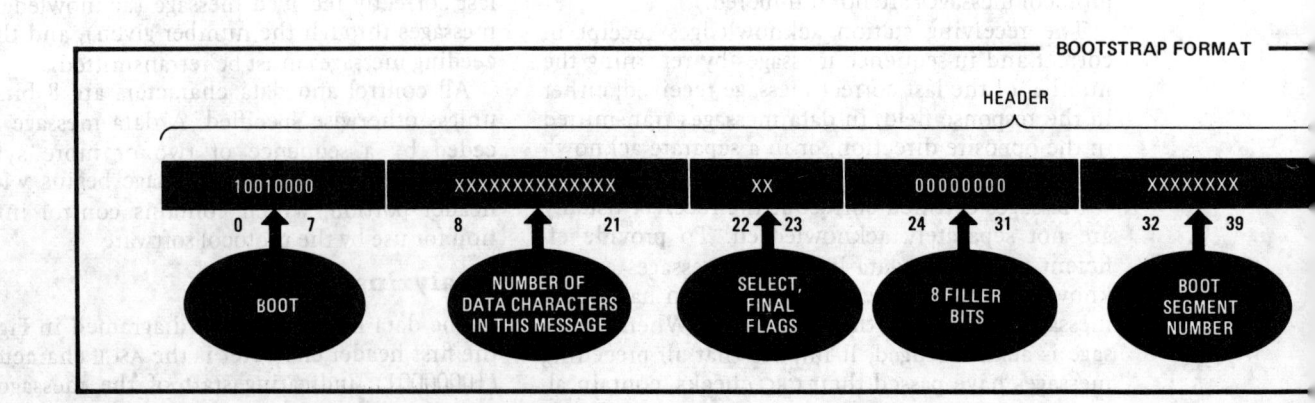

BOOTSTRAP FORMAT

10010000	XXXXXXXXXXXXXX	XX	00000000	XXXXXXXX
0 — 7	8 — 21	22 — 23	24 — 31	32 — 39
BOOT	NUMBER OF DATA CHARACTERS IN THIS MESSAGE	SELECT, FINAL FLAGS	8 FILLER BITS	BOOT SEGMENT NUMBER

HEADER

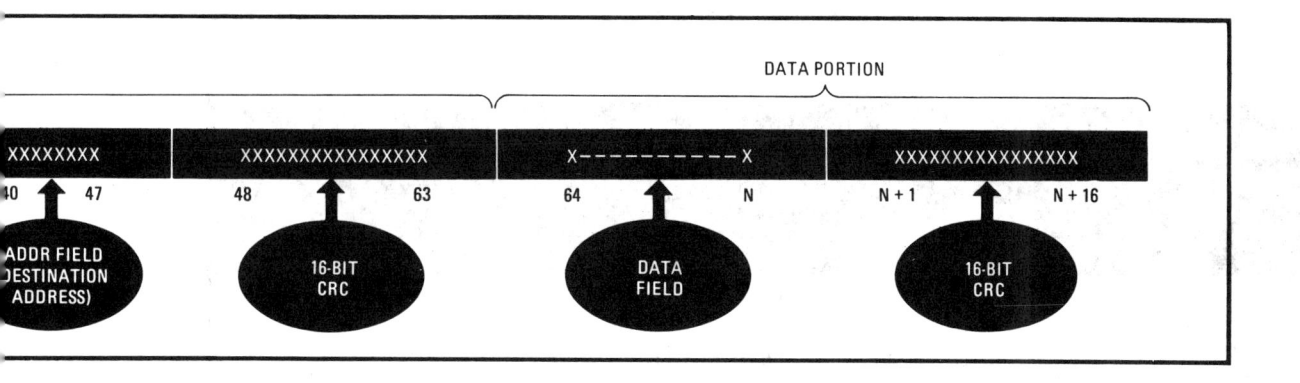

stations is based on the conditions of the select and final bits in these messages. The select bit polls another station, whereby the on state of the select bit invites the station specified by the message's address field to transmit. The final bit relinquishes ownership of the line after completion of the current message.

Each protocol message is preceded by at least two synchronization characters. The message begins with character ENQ (00000101). The next two fields designate the type and subtype of the protocol message. The RESP field contains message-numbering information.

The NUM field provides numerical information to supplement the RESP field, as shown in the notes for the table. The meanings of the RESP and NUM fields vary, depending on the type of protocol message. ADDR designates the destination of the message, which is followed by the last field, the CRC. Detailed descriptions of the protocol messages are given below.

Acknowledge message (ACK). Whenever acknowledgment is required for a protocol message, or if a data message is not forthcoming to acknowledge earlier data messages, an ACK is sent. The SUBTYPE and NUM fields, which are not used, contain all 0s. The RESP field holds the number of the last correct message received at this station.

Negative acknowledgment (NAK). If a data or protocol message is rejected for any reason, a NAK is returned to the transmitter from the receiver. The message number in the RESP field of the NAK is the number of the last received message that was correct. The reason for NAKing is indicated in the SUBTYPE field, as follows: header-block-check error, 000001; data-block-check error, 000010; REP message response, 000011; buffer temporarily unavailable, 001000; receiver overrun, 001001; message too long, 010000; and header-format error, 010001.

Reply to message number (REP). A reply message is sent from the transmitter to the receiver if the transmitter has not heard from the receiver within a predetermined period and unacknowledged messages are outstanding. The NUM field contains the number of the last message sent to the receiver. The response to a REP message is either an ACK to acknowledge the last received message or a NAK to indicate the last message verified by the receiver as correct.

Reset message number (RES). This message is sent by a transmitter to a receiver to cause the receiver to reset its receiving message number to that of the NUM field. This message is used when

5. The data field of a bootstrap message contains the program to be loaded into a remote computer or an intelligent terminal.

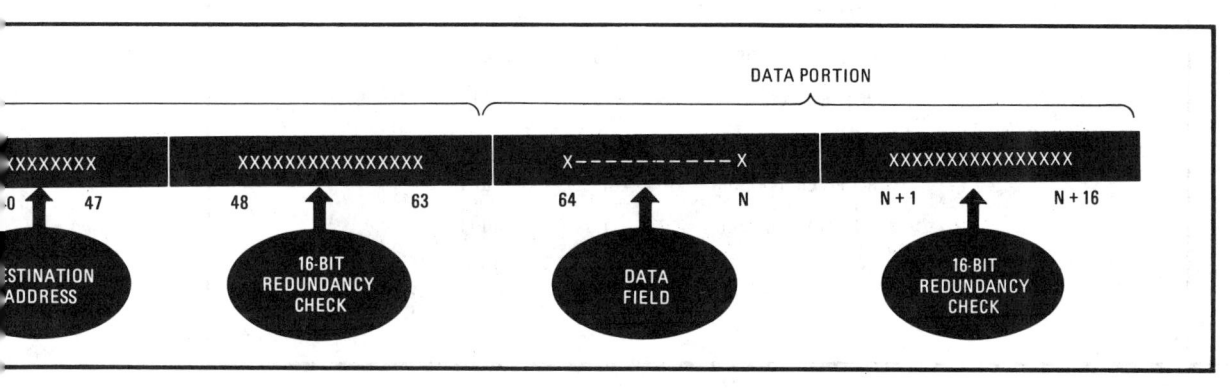

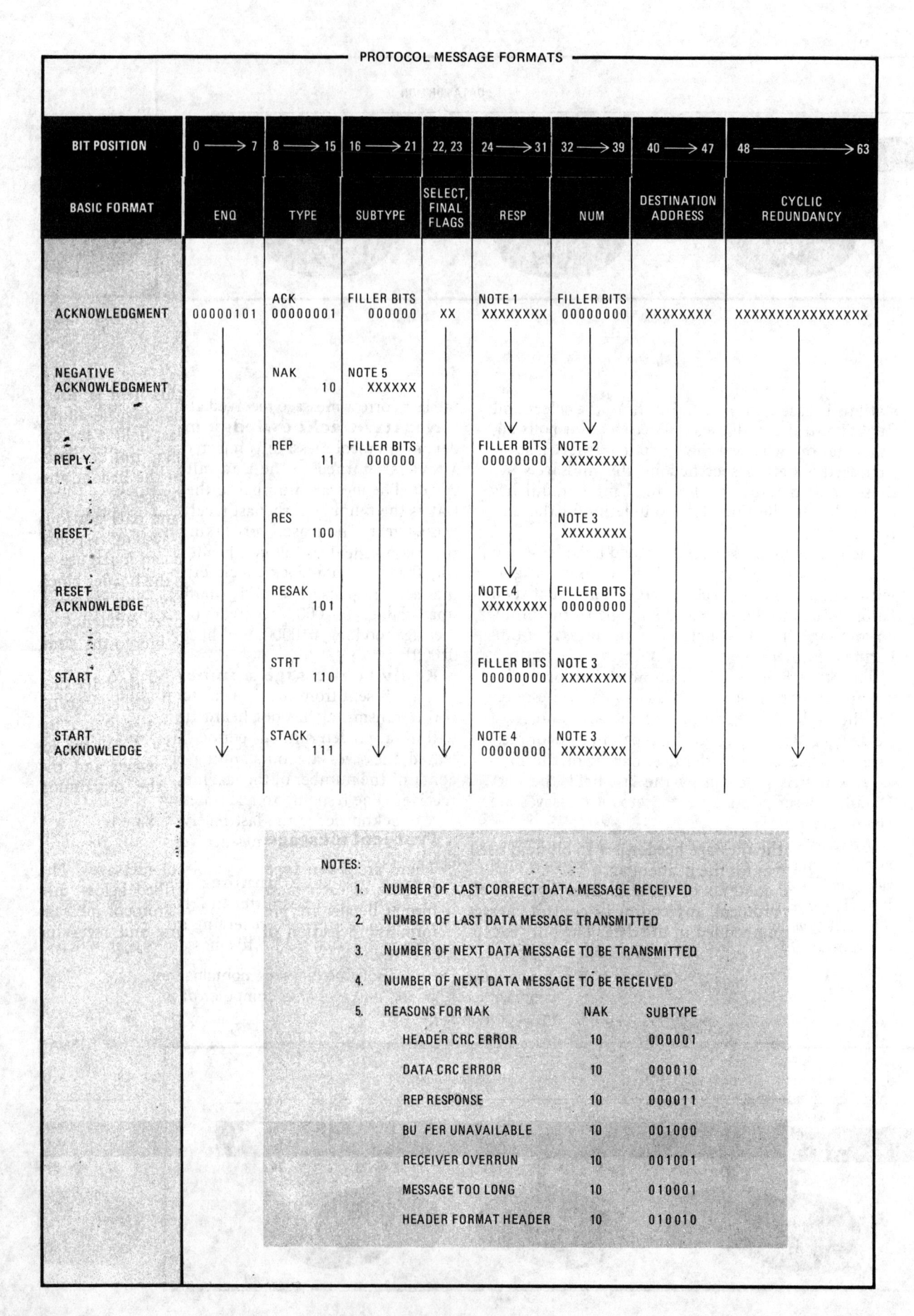

PROTOCOL MESSAGE FORMATS

BIT POSITION	0 → 7	8 → 15	16 → 21	22, 23	24 → 31	32 → 39	40 → 47	48 → 63
BASIC FORMAT	ENQ	TYPE	SUBTYPE	SELECT, FINAL FLAGS	RESP	NUM	DESTINATION ADDRESS	CYCLIC REDUNDANCY
ACKNOWLEDGMENT	00000101	ACK 00000001	FILLER BITS 000000	XX	NOTE 1 XXXXXXXX	FILLER BITS 00000000	XXXXXXXX	XXXXXXXXXXXXXXXX
NEGATIVE ACKNOWLEDGMENT		NAK 10	NOTE 5 XXXXXX					
REPLY		REP 11	FILLER BITS 000000		FILLER BITS 00000000	NOTE 2 XXXXXXXX		
RESET		RES 100				NOTE 3 XXXXXXXX		
RESET ACKNOWLEDGE		RESAK 101			NOTE 4 XXXXXXXX	FILLER BITS 00000000		
START		STRT 110			FILLER BITS 00000000	NOTE 3 XXXXXXXX		
START ACKNOWLEDGE		STACK 111			NOTE 4 00000000	NOTE 3 XXXXXXXX		

NOTES:

1. NUMBER OF LAST CORRECT DATA MESSAGE RECEIVED
2. NUMBER OF LAST DATA MESSAGE TRANSMITTED
3. NUMBER OF NEXT DATA MESSAGE TO BE TRANSMITTED
4. NUMBER OF NEXT DATA MESSAGE TO BE RECEIVED
5. REASONS FOR NAK

REASONS FOR NAK	NAK	SUBTYPE
HEADER CRC ERROR	10	000001
DATA CRC ERROR	10	000010
REP RESPONSE	10	000011
BUFFER UNAVAILABLE	10	001000
RECEIVER OVERRUN	10	001001
MESSAGE TOO LONG	10	010001
HEADER FORMAT HEADER	10	010010

How DDCMP stacks up

DDCMP is designed to operate over clocked (synchronized) full- or half-duplex channels, switched or direct links, point-to-point or multipoint networks, and serial or parallel transmission facilities. Further, it will accommodate both synchronous and serial start-stop (asynchronous) modes. The SDLC and ADCCP protocols share all of these capabilities, with the important exceptions being that they cannot operate on bit-parallel or asynchronous lines. Another characteristic shared by all three line protocols is that they are designed primarily for communications between intelligent stations, rather than communications with simple terminals. The nature of the transmission medium (e.g. dial-up or leased line, cable or microwave transmission) does not affect operation for any of the three protocols.

In comparison with IBM's binary synchronous communications (BSC), the most common protocol today, both DDCMP and SDLC have the advantage of providing full-duplex, as well as half-duplex operation. BSC requires different formats for leased point-to-point links, switched point-to-point and nonswitched, multipoint operation. But DDCMP and SDLC can accommodate all types of configurations with the same message formats. An important difference between DDCMP and SDLC is that SDLC has no count field. Instead, the end of the data field is marked by a unique flag character (01111110). To avoid the problem of mistakenly interpreting a string of six 1s in the data stream as an end flag, special circuits at the transmitting station must count the number of consecutive 1s. When five in a row are detected, a 0 is automatically inserted if this sequence is not the end of the data field. This technique is called bit-stuffing.

The receiving station constantly searches for five consecutive 1s. If the sixth bit also is a 1 and the seventh is a 0, the sequence is interpreted as an end flag. But if the sixth bit is a 0, it is dropped, and the search for the end flag is resumed.

In DDCMP, the data is never scanned for special characters. The only requirement for the data field is that it contains a multiple of 8-bit characters. If it does not, 0s must be added to fill it out. This feature allows DDCMP to operate with many existing interfaces and controllers.

The number of unacknowledged messages permitted to be outstanding for the three protocols also differs. BSC allows only one unacknowledged block of some defined length, SDLC allows seven frames with no restrictions on length, and DDCMP permits up to 255 unacknowledged messages with a maximum of 16,000 characters per message, to be received at any station. An advantage to operating with large numbers of unacknowledged blocks is efficient use of satellite links, where there are long delays. Both SDLC and DDCMP include cyclic redundancy checks of control messages, while BSC does not.

A worthwhile feature of DEC's protocol is that it permits bootstrap startup of a remote intelligent terminal of a communications processor so that all stations on the network can have their operating software, including the DDCMP control program loaded from one control site.

---

message-number synchronization is lost or if the numbering sequence must be changed.

Reset acknowledge (RESAK). Acknowledges receipt of a reset message. The RESP field contains the next expected data message number, which is the same as the number in the NUM field of the corresponding reset message.

Start (STRT). Used to begin message exchange, the start message is sent by a station seeking to transmit to another station, to cause the latter to prepare to receive. The NUM field informs the receiver of the number of the first data message to be transmitted.

Start acknowledge (STACK) is transmitted by the receiver to acknowledge receipt of a start message. As in the RESAK message, the RESP field contains the next data message number expected to arrive at the receiver (same as the number in the NUM field of the start message). This serves as an acknowledgment of the message number. The NUM field of a STACK contains the number that will be assigned to the first data message transmitted by this station.

A special type of protocol message is the bootstrap (BOOT), which is used to load a program into a remote computer and then transfer control to this program (Fig. 5). BOOT messages are like data messages with the following exceptions:

- No acknowledgment is required.
- BOOT segment numbers always start at zero, eliminating the need for a start sequence.
- The header of a BOOT message starts with the ASCII character DLE (10010000).

Following a synchronization sequence, the BOOT message begins with a header character that identifies this type of message. The second field is a 14-bit character count indicating the number of characters to be sent. Next are the link control flags and an 8-bit filler field.

The next field contains the number of this segment of the bootstrap program. Segmentation of programs into several messages is required when the number of characters exceeds reliable capacity of the data link or 16,000. The last two fields of the header are the destination address and CRC for the header. The data portion of the message contains the program to be loaded. Following the data is the second CRC. ∎

BDLC—a link-control method that can handle up to 127 unacknowledged messages

Burroughs' bit-oriented protocol operates either half or full duplex for high throughput

Michael J. Bedford
Burroughs Corp.,
Detroit, Mich.

Protocols

The new generation of bit-oriented protocols is much more efficient than earlier character-oriented protocols because the newer methods have the inherent ability to operate on full-duplex lines and they are insensitive to formally structured information codes. In addition to these efficiencies, Burroughs data link control protocol (BDLC) can operate with a maximum of seven to 127 unacknowledged transmissions, and it is fully compatible with high level data link control (HDLC), the worldwide standard soon to be finalized by the International Standards Organization (ISO). Moreover, a version of BDLC that can be interfaced to IBM's bit-oriented synchronous data link control (SDLC) will be available.

Until now Burroughs users had only the character-oriented Basic protocol (not to be confused with the high-level language of the same name), which is more limited than BDLC. Among the advantages of BDLC over Basic is that transparency is integral to BDLC (transmission control is unaffected by the data or the information fields in the frame); primary and secondary relationships are always assigned to stations on a data link; information frames are independently numbered for retransmission of an individual frame or series of frames received with errors; the frame structure is standardized; and the procedure itself is modular. The significance of modularity is that specific combinations of commands and responses can be configured to best suit a given application.

Unlike BDLC, older protocols such as Basic have rigid, character-based message structures, and they reserve subsets of the character code for transmission control. This has inhibited standards development and reduced throughput efficiency. Data treated in this manner often requires code conversion, expansion (packed numeric digits to characters), or fragmentation (nonformatted data broken into fixed-length bytes) prior to transmission at the data source, and then reconversion at the destination.

Moreover, several line turnarounds may be required to accomplish transmission of a single message. However, with its transparency, code independence, and bit streaming, BDLC eliminates these conversions, and with improved error detection and error recovery it reduces the overhead required for error control and produces more accu-

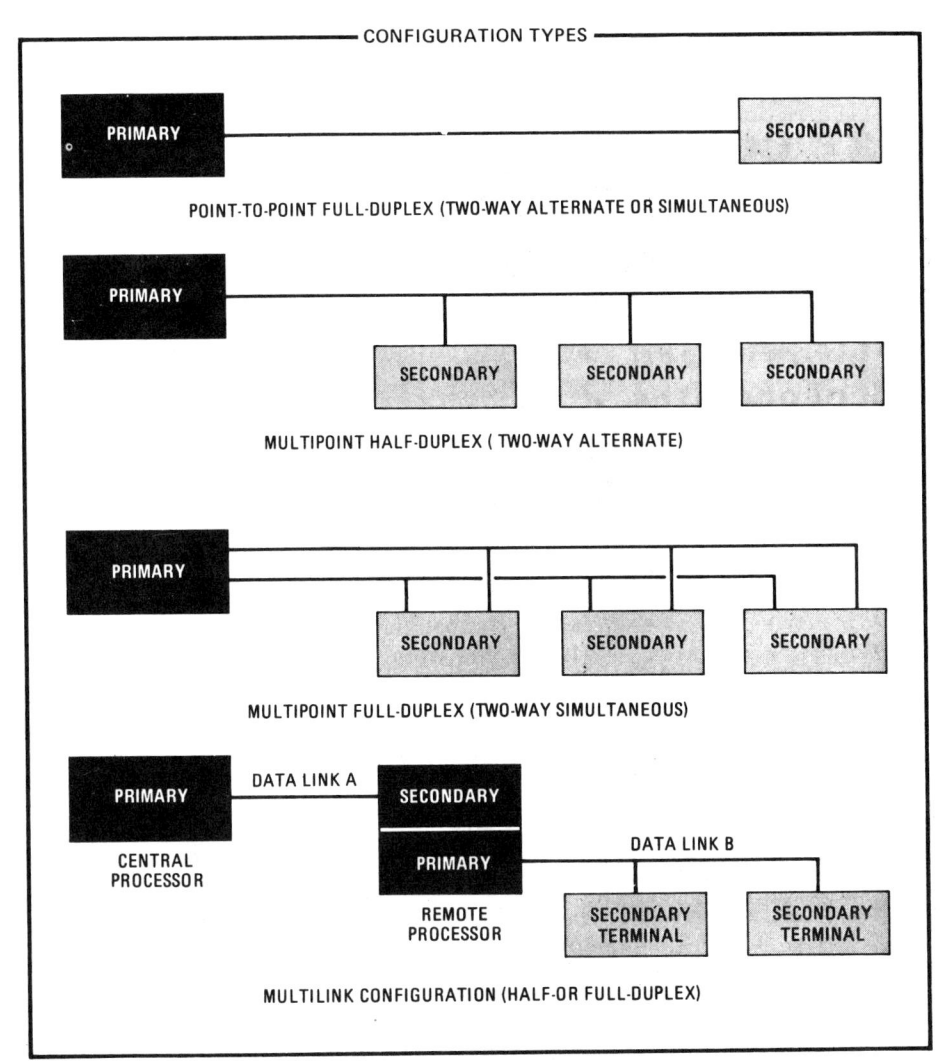

1. BDLC can operate over point-to-point and multipoint full- or half-duplex links, or in multilink configurations that include both full- and half-duplex communications lines.

rate, flexible, and economical data transmission.

The protocol provides for two-way alternate or two-way simultaneous synchronous transmission in switched or nonswitched networks and point-to-point or multipoint data links, as illustrated by the configuration types in Figure 1.

Secondary stations can operate in either the normal response or the asynchronous response mode. In the normal response mode, a secondary station may initiate transmission only as a result of receiving explicit permission to do so from the primary. In the asynchronous response mode, a secondary station may initiate transmission without explicit permission from the primary.

During normal operation, a station may transmit up to seven consecutively numbered frames (or 127 frames with an extended control field) without requiring an intervening acknowledgment. A poll bit is set within the control field of a frame to request a response or acknowledgment. The transmitting station includes a send sequence number to identify the frame being sent. It also inserts a receive sequence number to denote the next expected frame and to acknowledge receipt of previously received frames.

The new protocol can be retrofitted to present Burroughs equipment capable of synchronous operation. For example, only a single board change and reloading of firmware is required to convert the following Burroughs products: TC 3500 intelligent terminals, DC 140 intelligent communications controllers, and TC 1700 bank-teller terminals.

The new TC 5100 series of intelligent, interactive, and batch terminals also offers BDLC, and a newly announced communications controller, B776, provides for BDLC devices to be used in Basic networks and Basic devices in BDLC networks. The memory requirements of BDLC are compared with those of Basic for a typical TC 3500 intelligent terminal in "How much firmware?"

BDLC vs Basic

As mentioned earlier, BDLC is inherently more efficient that Basic. For example, BDLC acknowledgments are included in data messages, whereas in Basic, a separate acknowledgment message must be returned for each data message. Also, unlike the Basic protocol, BDLC is designed to operate two ways simultaneously, using full-duplex lines. Some other features of BDLC that are not available in Basic are the ability to send several messages to several different remote stations in one contiguous transmission and the ability to include a polling command in an output message.

All of this adds up to significant differences in the ways in which Basic and BDLC perform in such applications as inquiry systems, one-way traffic, and distributed networks.

Inquiry system. Possibly the most com-

(continued on page 223)

BDLC versus Basic

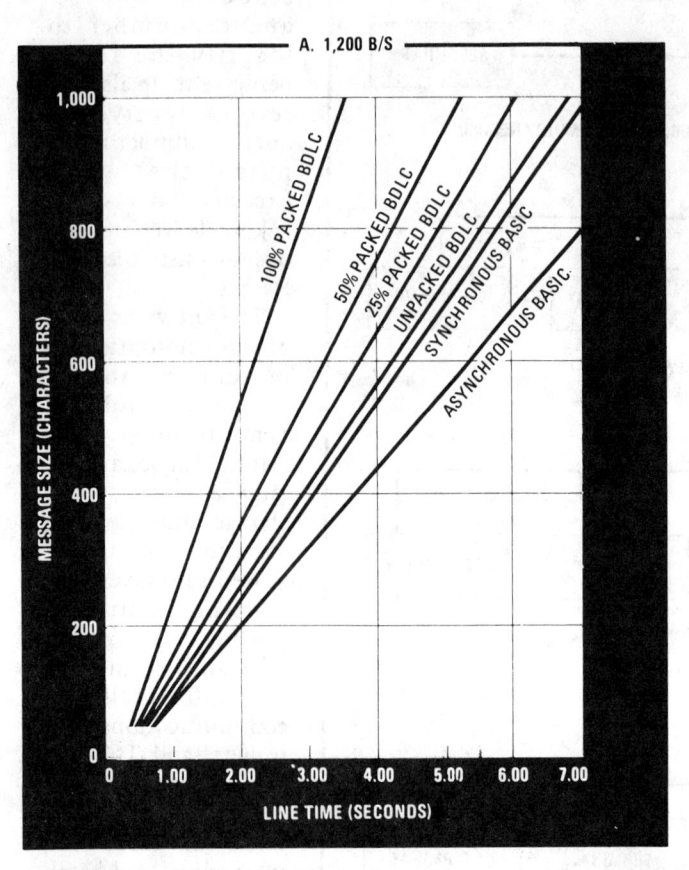

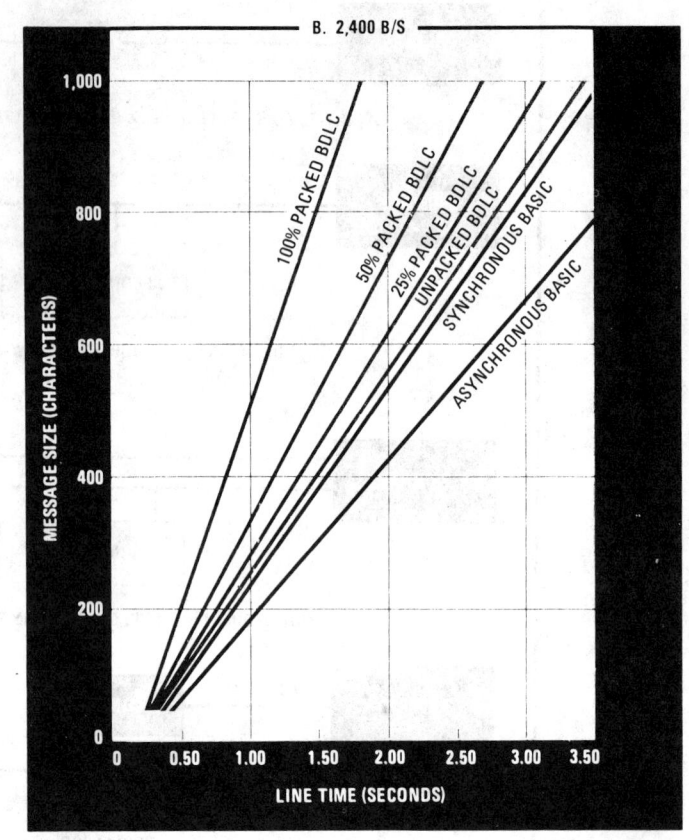

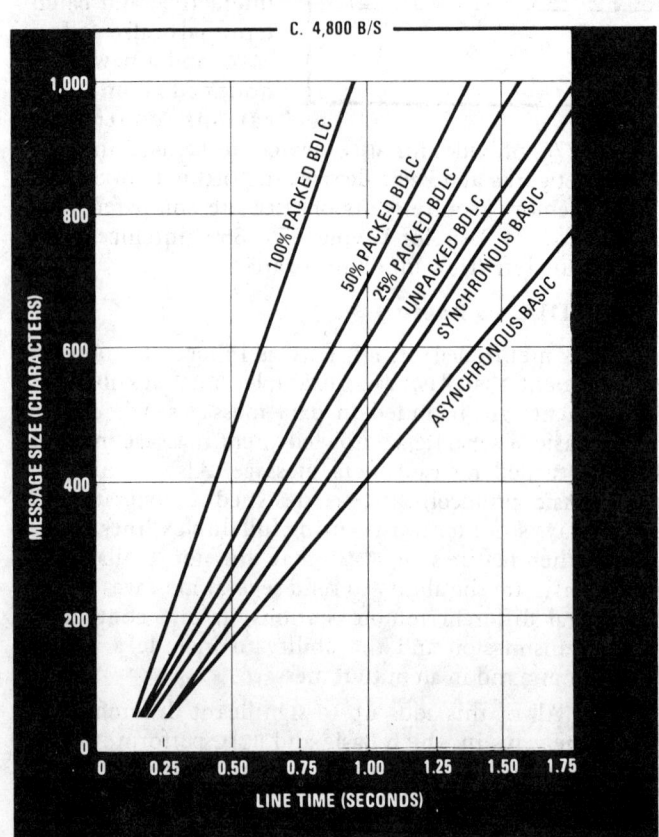

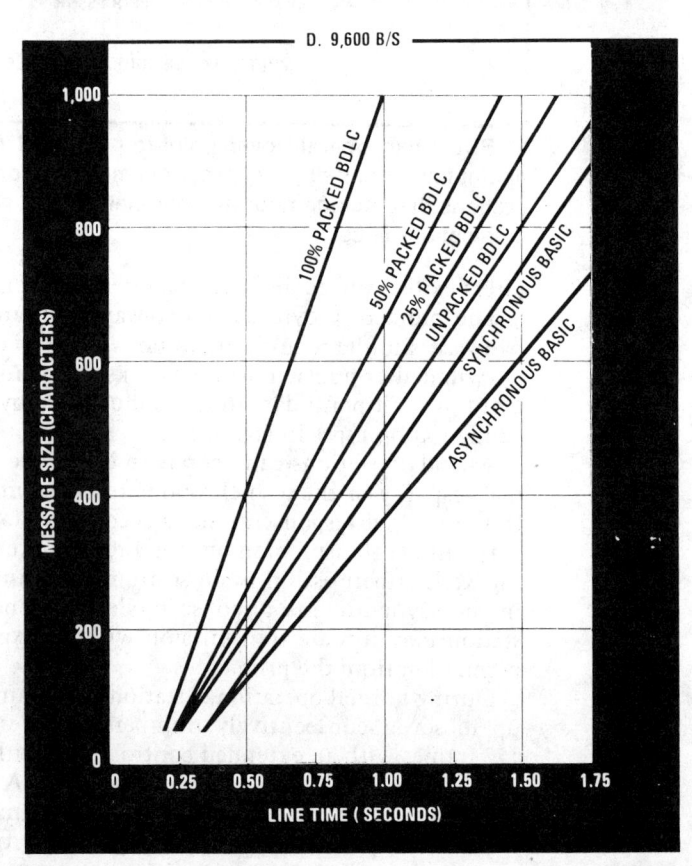

(continued from page 221)

2. The transmission times for various forms of BDLC and Basic are plotted as functions of message sizes for line speeds of 1,200 bits-per-second, 2,400 b/s, 4,800 b/s, and 9,600 b/s.

monly implemented data communications system is the one-message-in/one-reply-out type of multipoint inquiry system. In this application BDLC saves three to five line reversals by eliminating separate acknowledgments and by using the message sent in response to an inquiry to acknowledge the inquiry. Another advantage is that any character size can be handled by BDLC, whereas Basic has a rigid character structure. Because of this, BDLC allows the packing of data into bit streams for further improvement in line utilization.

A detailed comparison can be made of the line efficiency of an 8-bit inquiry system operating in BDLC and one operating in Basic. The set of graphs in Figure 2 shows relative line-time usage for asynchronous Basic, synchronous Basic, unpacked BDLC, 25% packed BDLC, 50% packed BDLC, and 100% packed BDLC as functions of total message sizes. Note that the term packed means that numeric data can be represented in 4-bit increments rather than 8 bits, as in Basic.

The figures suggest that BDLC is always the more powerful discipline. But that's not always true. When used on 1,200-b/s lines to send 100-byte messages, asynchronous Basic provides about 20% shorter response times than either synchronous Basic or unpacked BDLC. The explanation is that asynchronous Basic spends relatively little time on unsuccessful polling attempts.

Therefore, one conclusion to be drawn is that with slower multidrop lines and message sizes up to approximately 100 bytes, asynchronous Basic is still more efficient than BDLC, ignoring all other system costs.

For the purpose of illustration, 1,200-b/s asynchronous operation was compared with 1,200-b/s synchronous. But this is not a practical comparison, since the synchronous hardware is made for higher line speeds than asynchronous.

Another conclusion is that as message sizes increase, the performances of synchronous Basic and unpacked BDLC approach each other, saving about 20% line time compared to asynchronous Basic. (This is to be expected—8 bits against 10 bits, synchronous against asynchronous.) At all speeds, for the larger message sizes, and in half-duplex inquiry applications, BDLC significantly outperforms Basic when the message data can be packed.

One-way traffic Consider a system in which there is only output to one or more terminals. Examples are end-of-day report writing or inputs of batched data. In either case, BDLC can improve throughput since acknowledgment need only be transmitted after seven or more messages, so that fewer line turnarounds are required.

For example, in sending seven 132-byte messages to a remote job entry printer, BDLC requires 24% to 28% less line time than asynchronous Basic between 1,200 and 9,600 b/s, and 10% to 13% less than synchronous Basic. Clearly, as message sizes decrease or line speeds increase, or when hardware delays increase, the efficiency of BDLC grows and becomes still more of an advantage.

Distributed network system. Powerful concentrators that control their own small networks yet are linked to central computers are found in more and more systems. In such distributed networks, Basic may continue to be used on the local network, but BDLC would have advantages in the concentrator-to-central-system connection, either with point-to-point or multidrop links.

For point-to-point half-duplex lines, the advantages of BDLC over Basic are similar to those discussed in the one-way-only application, one advantage being the ability to group several messages into one transmission. Similar efficiencies can be found in BDLC conversational point-to-point communication involving fairly heavy traffic, whose throughput would be superior to conversational Basic in the half-duplex mode.

BDLC becomes advantageous when data is packed in the point-to-point environment. For example, in the one-way only application where the seven 132-byte messages are sent at 9,600 b/s, the advantage of BDLC with seven unacknowledged frames is 23%. If 50% of the data can be packed, the line-time saving of BDLC over Basic could increase to 40%.

In point-to-point half-duplex systems BDLC's advantages over Basic are even more impressive when the hardware delays are significant. This will certainly be the case for satellite connections, where the long distances cause long delays.

Note that in actual practice, transmission sizes in a distributed network are generally much larger than in an inquiry system, since several logical messages might comprise a single transmission to

---

More about protocols

Other line protocols are described in the following DATA COMMUNICATIONS articles:
- "Taking a fresh look at data link controls," September, 1973, p. 65-71—a look at character-oriented protocols such as IBM's binary synchronous control.
- "Synchronous data link control—SDLC," May/June, 1974, p. 49-60—a discussion of IBM's bit-oriented protocol.
- "Advanced link control runs full and half duplex," Sept./Oct., 1974, p. 36-46—a description of DDCMP, Digital Equipment Corp.'s bit-oriented protocol.

STANDARD BDLC FORMAT

| 01111110 | XXXXXXXX | XXXXXXXX | XXXX _____ XXXX | XXXXXXXXXXXXXXXX | 01111110 |

- 8-BIT FLAG
- 8-BIT ADDRESS
- 8-BIT CONTROL FIELD (EXPANDABLE TO 16 BITS)
- VARIABLE LENGTH INFORMATION FIELD
- 16-BIT FRAME CHECK FIELD
- 8-BIT FLAG

NOTE: X's MAY BE 1's OR 0's

3. Three fields have variable lengths—the address field, expandable in 8-bit increments; control field, which can accommodate up to 127 unacknowledged frames; and the information field.

and from the concentrator or controller, regardless of the protocol used.

Surprisingly, BDLC's greatest efficiency advantage over synchronous Basic is with small messages, and this advantage increases slightly with line speed. In the theoretical best case for BDLC at half-duplex, where no individual polls or acknowledgments are necessary, where every frame is 100% packed, and with the same hardware delays as before, a 500-byte message transmitted at 9,600 b/s would experience a 61% saving in line time in BDLC over asynchronous Basic and 52% in the case of the synchronous Basic mode.

The limiting values are also quite interesting. They show that in a half-duplex distributed processing system, one can expect average line-time savings of up to a third, using unpacked BDLC in networks where the hardware delays are highly significant. However, in networks where the hardware delays are not significant compared to actual transmission time, the savings of unpacked BDLC over synchronous Basic are negligible and approximately 20% better than asynchronous Basic.

Frame formats

In BDLC and other bit-oriented control procedures conforming to the ISO and the proposed U.S. standards, the fundamental unit of transmission is the frame. In all of the proposed standards, the frame formats conform to a common structure of defined fields, as shown in Figure 3.

■ **Flag bit sequence.** All frames start and end with fixed 8-bit sequence called a flag. Frame synchronization is achieved by detection of the leading flag, and the end of the frame is signaled by an end flag. A single flag sequence may be be transmitted to signify the end of one frame and the beginning of the next frame.

To assure that other elements in the frame are not misinterpreted as a flag bit sequence, a process known as zero insertion is applied at the time of transmission to all fields bounded by delimiting flags. This process inserts (stuffs) a 0 bit into the bit stream each time five consecutive 1s are transmitted. The receiver deletes these inserted 0s to reconstruct the original bit stream. These functions may be performed by firmware or hardware at the transmitting station and at the receiver. The method of implementation depends on the equipment involved.

■ **Address field.** This field contains the address of a secondary station. The field begins immediately after the leading flag of transmitted frames. It is normally 8 bits long, providing 256 addresses. An address extension capability permits expansion of the address field in 8-bit increments to accommodate additional secondaries.

■ **Control field.** The control field conveys commands from the primary station, responses from a secondary, and sequence numbers of transmissions in a data link.

The protocol allows up to seven unacknowledged frames to be outstanding at a given time. Each station maintains independent sequence numbers for transmitted and received frames. The field may be extended from 8 to 16 bits, extending the command/response capabilities and expanding the sequence numbers from 3 bits to 7 bits, which in turn extends the potential number of unacknowledged frames to 127. In fact, a deliberately unbalanced or asymmetrical system can be operated, in which many short frames are transmitted in one direction and fewer but longer frames are transmitted in the other direction, without the requirement for frequent line turnarounds solely for acknowledgments.

A poll bit solicits a response or sequence of responses from a secondary station. In the normal response mode, the final bit is set by a secondary station to indicate the final frame transmitted in response to a poll command. In the asynchronous response mode, the final bit is included in the first response frame following receipt of a poll bit to indicate the final frame of a sequence.

The control field includes a means to identify the type of frame sent to indicate to the receiver

CONTROL FIELD FORMATS

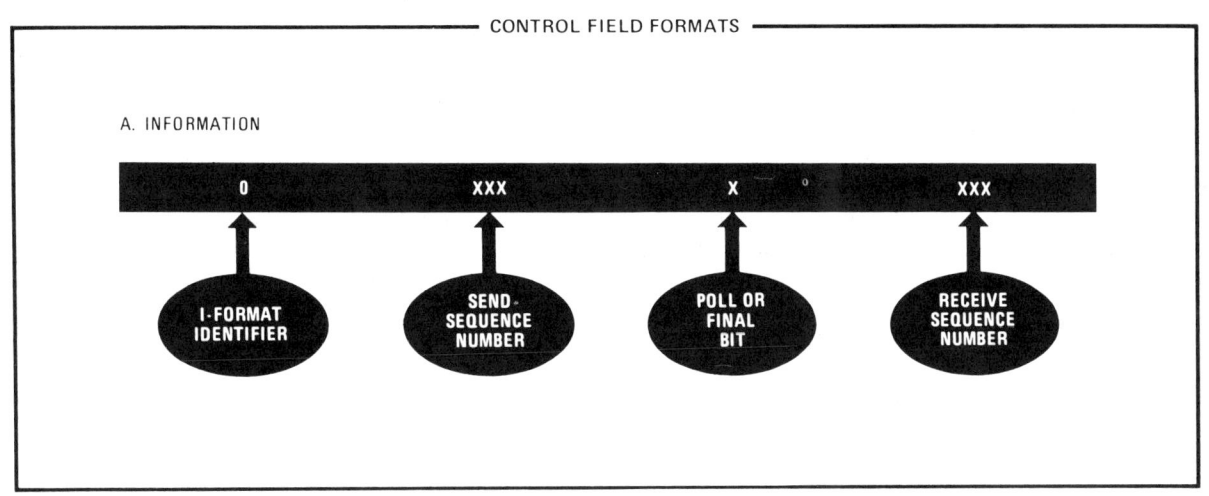

A. INFORMATION

0	XXX	X	XXX
I-FORMAT IDENTIFIER	SEND SEQUENCE NUMBER	POLL OR FINAL BIT	RECEIVE SEQUENCE NUMBER

4A. The control field has three possible formats, as shown in the three diagrams. The format for the information frame keeps track of the numbers of the last frames sent to and received by the station.

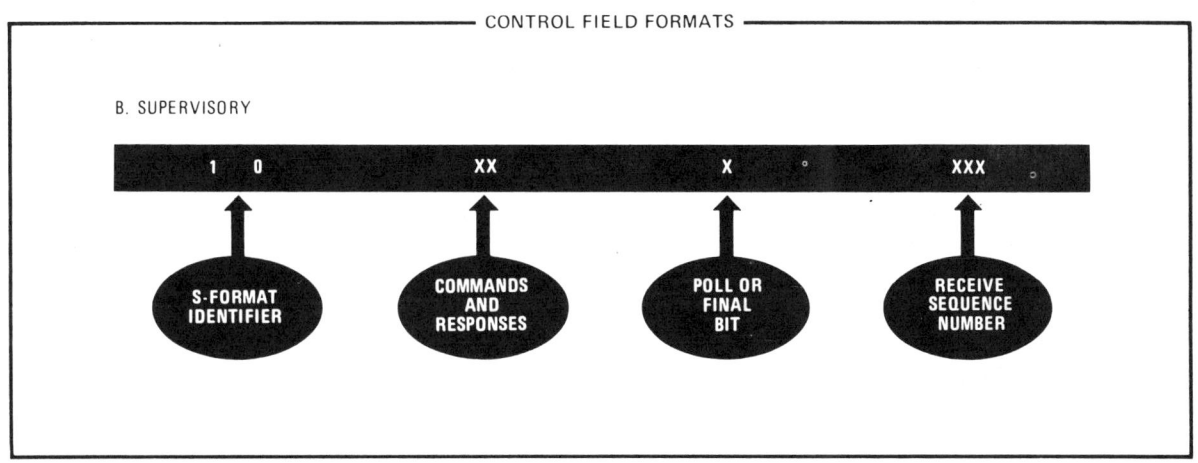

CONTROL FIELD FORMATS

B. SUPERVISORY

1 0	XX	X	XXX
S-FORMAT IDENTIFIER	COMMANDS AND RESPONSES	POLL OR FINAL BIT	RECEIVE SEQUENCE NUMBER

4B. The command and response bit positions in a supervisory frame specify the readiness or inability to receive message frames, or else the complete or selective rejection of one or more received frames.

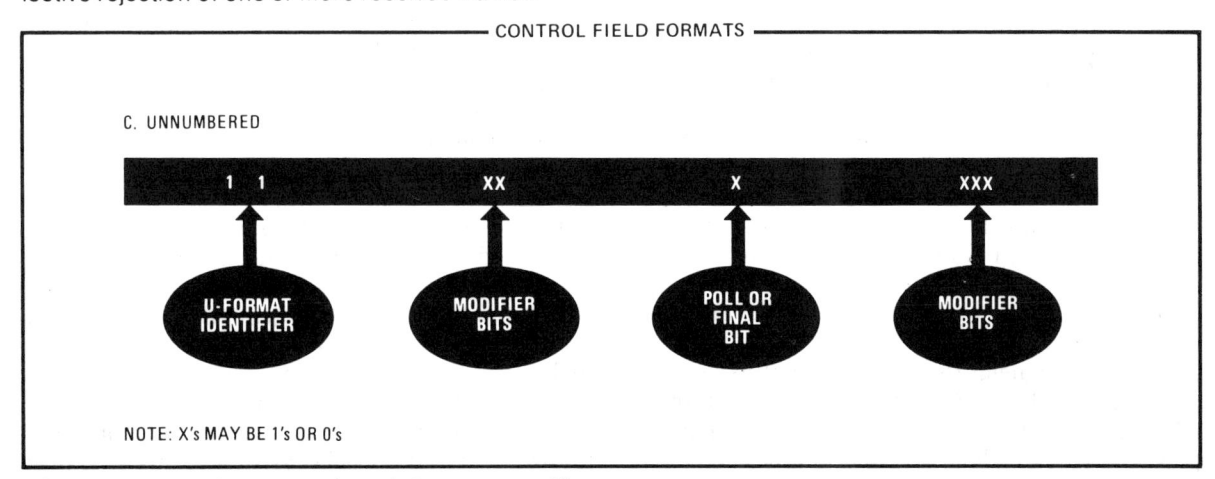

CONTROL FIELD FORMATS

C. UNNUMBERED

1 1	XX	X	XXX
U-FORMAT IDENTIFIER	MODIFIER BITS	POLL OR FINAL BIT	MODIFIER BITS

NOTE: X's MAY BE 1's OR 0's

4C. The format for unnumbered frames specifies any of six unnumbered commands or any of two unnumbered responses, as specified by two groups of modifier bits shown in the diagram.

> ### How much firmware?
>
> The operation of BDLC is controlled by firmware, and the number of features selected for each station determines how much firmware it will need for BDLC.
>
> As an example of a BDLC implementation, a secondary station (TC 3500/DC 140 intelligent terminal) would require 2½ kilobytes of storage to provide: point-to-point or multi-point operation; over switched or nonswitched links; using normal or asynchronous response mode; full- or half-duplex (two-way simultaneous or two-way alternate) transmission; address extension option; six response formats; and frame abort.
>
> For comparison, in a similar terminal Basic would require 1 kilobyte of firmware to provide: point-to-point operation only; over switched or nonswitched lines; full- or half-duplex transmission, and one- to three-digit numbering.

how the message should be processed—as a supervisory frame for protocol control purposes or as a combined control and information frame.

- **Information field.** The information field contains the data being transmitted, which consists of a variable number of bits that need not adhere to any conventional code structure. Although the BDLC procedure does not restrict the length of this field, practical considerations such as the ability of the 16-bit frame check field to detect transmission errors on exceedingly long messages and the cost of buffering tend to limit frame length.
- **Frame check sequence.** The frame check sequence consists of 16 bits used for detection of transmission errors by a process called a cyclic redundancy check. It validates the transmission accuracy of all fields contained between the leading and trailing flags of a frame.

Control field formats

There are three types of frames, each identified by the proper control field. The information frame sends data, a supervisory frame sends commands and responses, and unnumbered frames carry additional link control commands.

The control field of the information frame sent by a primary (Fig. 4a) contains the send-sequence number of the frame being transmitted and also the receive-sequence number expected to be contained in the next information frame from the addressed secondary. The purpose of the receive-sequence number is to inform the other station of the number of the last frame received. If this number is lower than the next send-sequence number at the secondary, then one or more frames are assumed lost, and the transmitting station is thereby informed that the frames that followed must be transmitted once again.

Two control-field formats convey link-control commands and responses. The supervisory format (Fig. 4b) is used when transmitting numbered supervisory frames which are vital to link operation. These frames indicate such situations as readiness to receive (receive ready message), request retransmission of information frames (reject or selective reject), and temporarily interrupt the receiving capability (receive not ready).

One method of polling a secondary station is to send it a ready command with the poll bit set in the control field.

In the event of a transmission error, retransmission is requested by the reject or selective-reject command/response. The selective reject requests only a specified frame to be retransmitted, while the reject command requests retransmission of all frames, starting with frame whose receive-sequence number appears in this frame. In either case all frames numbered below the receive-sequence number are implicitly acknowledged. This method avoids separate acknowledgments.

The unnumbered format (Fig. 4c) is used to transmit supervisory commands or responses for additional link-control functions.

When used in switched networks, the disconnect command causes the addressed secondary to go on-hook, whereas in nonswitched networks it informs the secondary that the primary is suspending operations.

The response-reject command is used by the primary station to report receipt of an invalid response, receipt of a frame with an information field that exceeds the assigned buffer size, or receipt of a response having an invalid sequence count. The status field of the frame indicates the reason for the rejection. The command-reject response provides an equivalent function and is sent by the secondary.

Another group of unnumbered frames includes setting and extension of the normal and asynchronous response modes and acknowledgment of unnumbered frames. The normal-response-mode command sets the addressed secondary to this mode and resets the send and receive sequence numbers to zero to initiate operations in the normal response mode. The extended version of this command stretches the control field by up to 16 bits to accommodate a greater number of unacknowledged frames. The asynchronous extended command performs the same function for operation in the asynchronous mode. An unnumbered acknowledgment message notifies the sending station of receipt of an unnumbered frame. ∎

part 8
data
data
data
data
data
data
data
data
data

network
diagnostics

Network-management centers help keep the system going

Containing both test equipment and operational gear, an NMC, customized to the system's needs, can simplify diagnosis of faults and aid rapid restoration of service

Donald W. Parker
GTE Information Systems Inc.
Silver Spring, Md.

OPERATIONS

Designing and installing a data communications system is only the beginning. Continuing operational requirements must also be kept in mind because time will take its toll, and natural forces tend to degrade or disrupt services and equipment. Then difficulties and delays will arise when trying to pinpoint what is at fault when trouble does occur.

More than likely, the system will be upgraded, eventually, and one piece of equipment will be substituted for another from time to time. For these and other reasons, a data communications system should include a network-management center (NMC) whose main functions are to monitor the network's operation, to permit routine maintenance, and to effect rapid restoral of service when a failure occurs.

Thus, an NMC contains both test and operational equipment. The amount of such equipment in the network-management center depends on many factors: how important the system is to the day-to-day business operation, the number of lines in the network, the number and kinds of terminals used, the geographical dispersion of the network, whether synchronous or asynchronous transmission is used, and even the educational and experience level of the people assigned to keep the network running.

Ideally, a network-management center will be designed at the same time as the data communications system itself. Unfortunately, though, the need for an NMC is often not recognized until network installation has been completed, system operation has become troublesome, and failures start to upset business operations. Even so, an NMC that is retrofitted can prove effective.

As a major benefit, a network-management center facilitates the orderly and convenient packaging of much of the data communications gear at a central office. Figure 1 shows how one unstructured system got out of hand as more lines and terminals were added. The datasets (modems) were simply piled one on top of the other and side by side until the equipment and system operation literally evolved to a precarious position. A person tripping on a wire could send the whole thing crashing down—or at least interfere with the performance of some of the lines.

Figure 2 shows the equivalent data communications equipment, neatly packaged in a conventional 19-inch rack, together with test equipment, to form a network-management center. The NMC in Fig. 2 has two essentially identical bays. Each of five panels at the bottom of

Profile of Network Management Center Test Equipment and Service Restoral Capabilities

	Analog patchfields	Digital patchfields	Circuit-gain measurements	Noise measurements	Envelope-delay measurements	Phase-jitter and hit measurements	Harmonic-distortion measurements	Gain hits and dropout testing	Asynchronous-distortion testing	Bit-error-rate testing	X-Y recorder	Oscilloscope	Monitor and alarm hardware	Back-up equipment switching	Computer-controlled switching	Computer-controlled monitoring	Automatic circuit acquisition	Automatic controlled diagnostics
LEVEL 1 PATCH FACILITY	●	●											X	X				
LEVEL 2 BASIC PATCH AND TEST	●	●	●	X	X	X	X	X	X	X	X	X	X	X				
LEVEL 3 ADVANCED PATCH AND TEST	●	●	●	●	●	●	●	X	X	●	X	X	●	X				
LEVEL 4 COMPUTER-CONTROLLED	●	●	●	●	●	●	●	X	●	●	X	●	X	●	●	⊗		

● MINIMAL X OPTIONAL ⊗ ALTERNATIVE

each bay contains dual 2,400 b/s modems, so that the center handles 20 leased lines. Above the modems are 12 channels of EIA patchfields; then several six-way, four-wire bridging units; and then 24 channels of analog patchfields. A test panel fits into the right bay above the analog patchfields.

The communications equipment is protected against damage, and the NMC makes it easier to diagnose and repair system faults. Such communications gear as datasets, multiplexers, and test equipment can be readily obtained for mounting in 19-inch rack cabinets.

While a network-management center is made up of standard test equipment and operating gear, it is always customized to a particular data communications system. The amount of operational equipment—datasets, for example—is directly related to system size. The variable aspects of network-management centers are the kinds and amount of test and service-restoral equipment they contain, and which can be related, to some extent, to the characteristics of typical data communications systems. The table defines four levels of NMCs.

Simple patchfields

A level 1 NMC would apply to relatively simple networks, for which users do no routine testing of lines or equipment, perhaps because experience shows the systems are quite reliable or because failures likely won't be of great consequence. A level 1 NMC is essentially a patch facility that contains sufficient analog and digital patchfields so that an operator can perform loop-back tests for rapid diagnosis and can call in alternate routing if the line seems to be the cause of the trouble. When trouble is located, the operator can simply plug in another line or substitute a spare dataset for one that may have failed.

Although a level 1 NMC does not contain any test equipment, it may include such service-restoral features as monitor-and-alarm hardware and back-up switching. A monitor-and-alarm circuit alerts the operator, by a warning light or audible alarm, when a piece of equipment has failed or is malfunctioning. For example, certain high-speed datasets check bit error rate (BER) and when the BER exceeds a predetermined value, a signal will appear on the dataset's SIGNAL QUALITY DETECTOR lead (pin 21 on EIA RS-232-C connector). This pin can be wired to the alarm system. Similarly, and depending on system operation, alarms can be initiated on failure of the CARRIER DETECT and DATASET READY signals. If an alarm persists, the operator can restore good service by manually plugging in a spare dataset, using the analog and digital patchfields.

Substituting a new piece of equipment can be made even easier and faster when the network-management center includes back-up switching. Here, electrical relays, operated by one switch, transfer the circuit leads from a deficient equipment to a spare. Inclusion of back-up switching requires more initial design effort, but can be particularly effective—say, for a multiplexer or a front-end processor—when substitution of a spare by an operator would require him to move too many leads. But with back-up switching, all the operator has to do is press one switch. The relays transfer the leads, and the system goes back on line.

As the table shows, a level 2 NMC includes, at a minimum, test equipment to measure circuit gain and steady-state noise, but as indicated, an extensive amount of other test equipment can be included. How much test equipment is added depends on the system itself. For example, if the network operates in an asynchronous-transmission mode (that is, at less than 2,000 bits per second), the NMC might also contain an asynchronous distortion analyzer and a pattern generator to help analyze the distortion. If transmission is synchronous (2,400 b/s or higher), the test-equipment complement would include instruments to measure envelope delay, phase jitter, and bit error rate together with an X-Y recorder and oscilloscope.

Operator needs skills

A level 3 NMC may be needed when a data communications system gets to be fairly extensive, such as one that is used by a medium or large company doing business over a wide geographic area and whose system is characterized by high-speed transmission (although some asynchronous terminals may be present). As the table shows, in addition to the patchfields, this type of NMC would employ instruments to measure noise, envelope delay, phase jitter, harmonic distortion, gain hits, and bit-error rate and would definitely include monitor and alarm hardware. It could also contain the optional facilities noted by Xs in the table.

An important aspect of the level 3 advanced patch-and-test network-management center is that the system designer has included some rather sophisticated test equipment to permit the rapid diagnosis of trouble and quick restoral of service. And the system itself is quite complex. What this means, then, is that the operator who oversees the system via the NMC must possess a comprehensive understanding of data communications theory and equipment, telephone-line characteristics, alternate routings, and the like, to permit him to interpret test results, make the necessary logical deductions, locate faults, and restore service.

The level 4 computer-controlled network-management center contains about the same amount of test and service-restoral equipment as a level 3 center. The main difference is that the level 4 installation contains a minicomputer that automatically performs the monitoring and diagnostic functions, based on readings from the test equipment.

Once programed to meet the system requirements, the minicomputer routinely checks each circuit and piece of equipment, switching in test equipment appropriate to a particular test and automatically analyzing the results. If test runs show faulty operation, the minicomputer can then sound an alarm and even invoke back-up switching of spare equipment. Thus, while a level 4 network-management center would cost more than an equivalent level 3 installation because of the minicomputer and its programing, the operating cost would be less because it can be manned to some extent by a person having lesser skills while still providing a

FIG. 1. PRECARIOUS ■ Piled up modems operate properly, but loose wires and unstructured setup invite trouble.

on incipient or actual degradation of service, and they can perform instant diagnosis to locate a failure node—statistical information of a nature that is more difficult to obtain with a manually operated NMC. (Note that when the NMC is controlled by a computer, the table shows that automatic circuit acquisition can be used in lieu of manual analog and digital patchfields.)

Dial-up restores private lines

Regardless of its level, any communications system in conjunction with its NMC should contain capabilities to restore service rapidly. An obvious, but expensive, way of assuring maximum uptime is to install redundant lines and equipment. It is not unusual, for example, for a company to lease two redundant private (or dedicated) lines for multipoint operation, using one as the primary line and the other as backup. Extra lines can add thousands of dollars a month to operations cost.

However, a viable alternative can be obtained by reverting to dial-up lines when a private line fails. This eliminates the continuing cost of extra private lines, but probably at the penalty of having to operate at reduced transmission speed until the private line can be brought back into service by the telephone company.

Figure 3 shows how multipoint private-line operation can be restored through a six-way, four-wire bridge built into the network-management center and connected to the public, switched telephone system

This service-restoral arrangement requires datasets that are capable of providing the required speed on dial-up lines. Here, two separate phone calls are made from the central station to up to five remote stations. The result is that the bridged up network yields the equivalent of a six-station multipoint full-duplex communications line. Such an arrangement can often suffice for temporary low- to medium-speed service. The network management center in Fig. 2 contains four such bridge circuits. Since they are patchable, they can substitute for any four of 20 high-speed dedicated circuits in the network.

FIG. 2. TOGETHERNESS ■ Bays mount operational and test equipment: bottom racks contain 20 modems equivalent to those in Fig. 1; digital and analog patchfields fit into racks above modems; and right bay includes a meter for measuring circuit gain and a loudspeaker to monitor tones on the communications line.

high-quality of network management. The use of minicomputers in network-management centers is technically feasible. But economically, they can probably only be justified for very large communications systems. When installed, though, they can yield detailed and valuable information, via typed logs and trend curves,

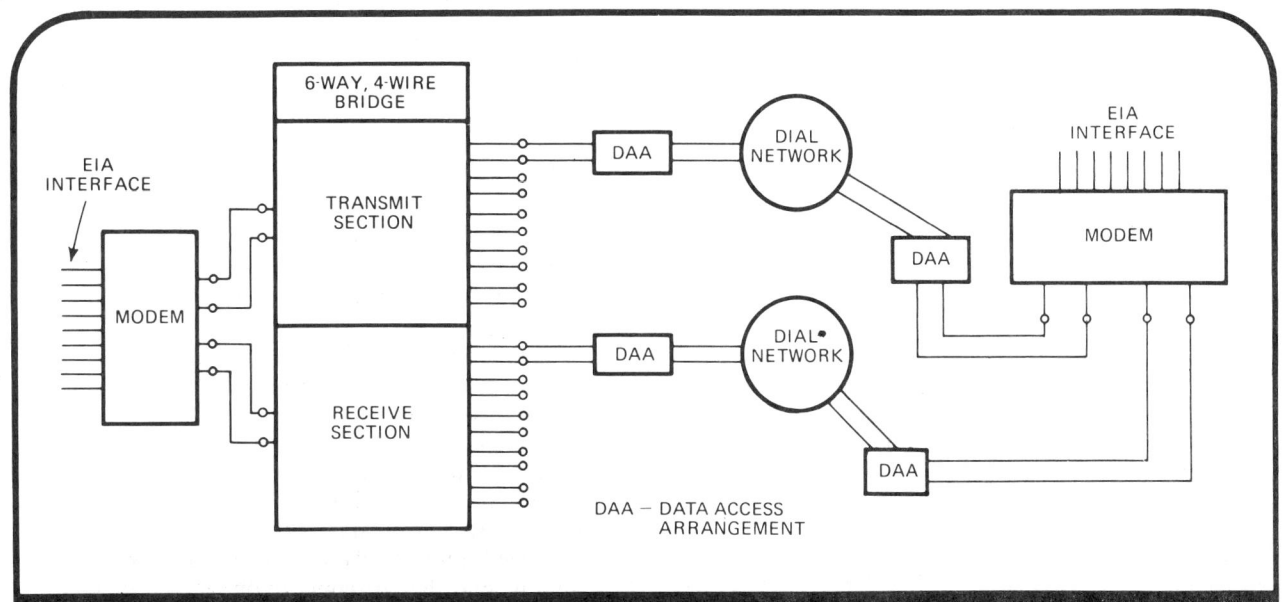

FIG. 3. DIAL-UP MULTIPOINT ■ When a private line goes out of service, a six-way, four-wire bridge, as shown here, can provide full-duplex links by dialing two phone connections between the central station and each of five remote terminals.

Line monitors open a window on data channels

Data and control characters can be captured and displayed, easing hardware and software troubleshooting

Gerald Lapidus
Associate Editor

Diagnostics

When trouble hits a data communications system, finding the fault fast is the paramount problem. That's why an investment in a line monitor—though an outlay of several thousand dollars—may be a wise decision. Such monitors permit a capable operator to isolate faults to a component such as a modem, terminal, or a communications processor. In the hands of an experienced technician, a line monitor can pinpoint problem areas within a defective unit to expedite troubleshooting.

Line monitors trap and display data and control characters to provide a precise picture of line activity, eliminating the difficulty in deciphering pulses viewed on an oscilloscope. In addition, the monitors permit control signals on RS-232-C lines to be checked when there are problems in establishing and maintaining message exchanges.

In a multivendor installation a line monitor can pinpoint a problem so that the proper service organization can be called to make repairs. Down time and servicing costs can be excessive, if the user calls the wrong service man.

Conventional troubleshooting requires the measuring of voltage and current levels, reading phase jitter, and viewing bit streams with oscilloscopes. The proper application of these techniques requires more expertise than is ordinarily available in many communications installations. An equally valuable role of line monitors is frequent on-line testing to check for degradation of service, which often signals incipient failures. Problems found during such routine tests can be corrected before they cause costly, unexpected breakdowns.

All line monitors provide readouts for data and control characters. Information is displayed in one of four ways:
- Light-emitting diodes (LEDs) that indicate the states of individual bits. This type of readout must be converted into the appropriate code.
- Labeled indicator lights.
- Alphanumeric displays.
- Alphanumeric cathode-ray tubes.

Most units permit data or control characters to be trapped and displayed on a selective basis. Trapping may be initiated upon detection of errors, or by decoding specified characters such as a synchronization character. The ability to read the states of the RS-232-C control leads allows the user

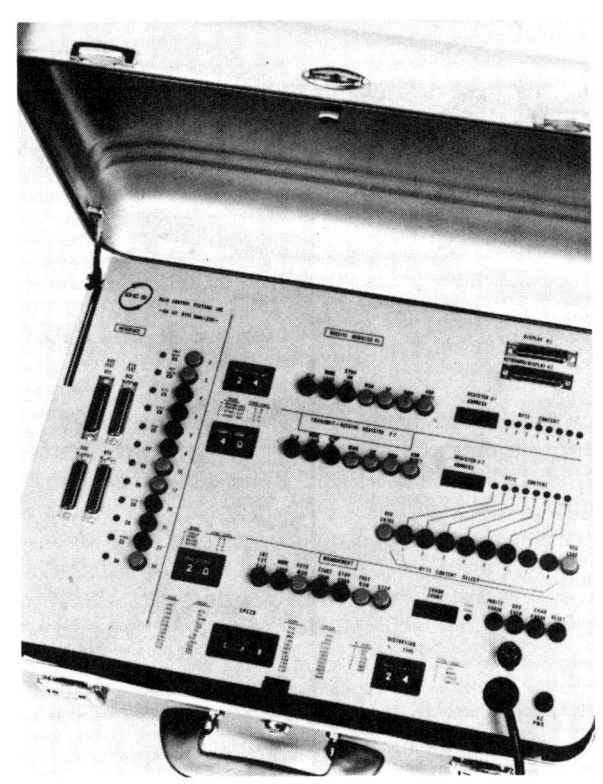

1. Most line monitors are packaged in compact suitcases. Pictured above is Data-Control Systems' Byte Analyzer.

to check the "handshaking" sequence if there are problems in establishing and maintaining message exchanges. Many line monitors also permit these leads to be clamped at desired voltage levels or programed to simulate any aspect of operation.

Some units also provide test messages for polling and answer back, which are used in end-to-end system tests. Parity indications and counts, also common features, are used both in checking out specific problems and in determining line quality. Advanced data link controls such as IBM's SDLC and Digital Equipment Corp.'s DDCMP can be accommodated, so long as the message lengths do not exceed the buffer size of the line monitor.

The compactness of integrated circuits permits line monitors to be built into small enclosures. The makers of units that have binary readouts squeeze them into suitcases small enough to fit under an airplane seat (Fig. 1). Line monitors with cathode-ray tubes are bulkier (Fig. 2), but, nevertheless, they are still fairly small.

Most line monitors are built around serial buffer memories that store the data selected for trapping (Fig. 3). Operation of the buffer is dictated by logic circuits, which route the trapped characters to the memory and direct the readout into the selected mode. All monitors interface to the line with connectors compatible with RS-232-C interface standards. Prices range from about $2,000 for the Digi-Log Systems Inc. Data Line Monitor to $7,500 for Spectron Corp.'s Datascope.

Line monitors are used in two ways: to check out control signals at the RS-232-C interface (establishing and maintaining a proper connection) and to check for the correctness and proper sequencing of control characters and data.

If the message exchange cannot be initiated or maintained, the RS-232-C lines should be checked with the line analyzer at both ends of the data link. If there is difficulty in transmitting, then, for most modems, the line monitor should check the data terminal ready, data set ready, and request to send. Each manufacturer provides for the display of different combinations of RS-232-C lines. The prospective buyer should ensure that the unit he is considering includes those tests deemed necessary for his troubleshooting requirements.

The next level of troubleshooting consists of testing data link control characters and data. Some line monitors trap only control characters, while others also capture data. This should also be considered in selecting a unit.

Some line monitors provide a complete, self-contained testing capability, which includes polling, answer back, and generating test messages. Others can participate in these tests only if other equipment is used to generate signals and responses.

Examples of tests using line monitors are shown in Figures 4 through 8. A discussion of how a typical installation has made use of a line monitor appears in "Learning and testing—a user case history." The characteristics of the line monitors that are commercially available are listed in the table, "Line monitor features," and detailed below.

Intershake glitch catcher

The Intershake (contraction for interface and handshake) DTM-1 Protocol Analyzer has a glitch catcher, which is a latch for monitoring any of 23 pins on the EIA cable for transient conditions, such as carrier dropouts on the received-line-signal de-

2. Cathode-ray-tube monitors trade off compactness for readability. Spectron's Datascope contains many unique features.

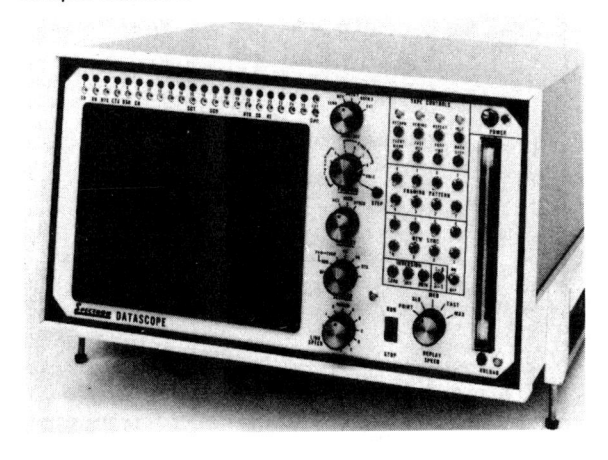

LINE MONITOR FEATURES

MANUFACTURER	MODEL	READOUT METHODS	CODES	TRAPPING CAPACITY	OTHER FEATURES
Atlantic Research Corp. 5390 Cherokee Ave. Alexandria, Va. 22314	Intershake DTM-1	Characters: alphanumeric RS-232-C: binary LEDS (red/green)	Hexadecimal	64 data and control characters	Measures clock frequency Counts glitches Measures RTS/CTS delay Measures LRC and parity-error rates Performs polling and answer back Simulates control-lead conditions
Data-Control Systems Inc. Commerce Dr., Danbury, Conn. 06810	BA-401	Characters: binary or alphanumeric (optional) RS-232-C: binary LEDS	Binary Hexadecimal (optional) ASCII (optional)	256 data and control characters	Indicates loss of synchronization Counts parity errors Tests parallel peripherals Tests error thresholds Performs polling and answer back Simulates control-lead conditions
Data 100 Corp. P. O. Box 1222 Minneapolis, Minn. 55440	Communications analyzer	Control characters: labeled lights RS-232-C: binary lights	EBCDIC IBM SBT CDC 200 UT (EXTERNAL, INTERNAL, or ASCII) UNIVAC DCT 2000	32 control characters	Signal-strength meter Speaker for audible monitoring Counts parity or block errors Generates test messages for local and loopback tests Indicates loss of synchronization
DIGI-LOG Systems Inc. Babylon Rd. Horsham, Pa. 19044	440, 445, 480, 485	Characters: alphanumeric CRT RS-232-C: binary lights	440, 480; ASCII 445, 485; ASCII or EBCDIC	440, 445: 640 data and characters 445, 485: 1,280 data and characters	Requires external video monitor or TV set for character readout 440, 480: asynchronous discipline only 445, 485: synchronous and asynchronous Terminal-only mode reads out only data
Paradyne Corp. 8550 Ulmerton, Rd. Largo, Fla. 33540	810 bisync analyzer	Characters: labeled lights RS-232-C: binary lights	EBCDIC, ASCII (optional)	30 control characters	Indicates status of Auto-DAA signals Signal-strength meter Speaker for audible monitoring
Spectron Corp. Church Rd. and Roland Ave. Moorestown, N.J. 08057	Datascope D-601	Characters: alphanumeric CRT RS-232-C: binary lights	ASCII, hexadecimal, EBCDIC (Special order)	720,000 data and control characters	Negative—image selection Test-point access to EIA lines Polarity inversion Bit-stream reversal Framing-pattern selection Out-of-sync and received data identified
Teletype Corp. 5555 W. Touhy Ave. Skokie, Ill. 60076	TSM-7810S	Characters: alphanumeric RS-232-C: binary LEDS	ASCII	1,024 data and control characters	Audible alarm indicates line activity External clocking for synchronous or isochronous operation Received data identified

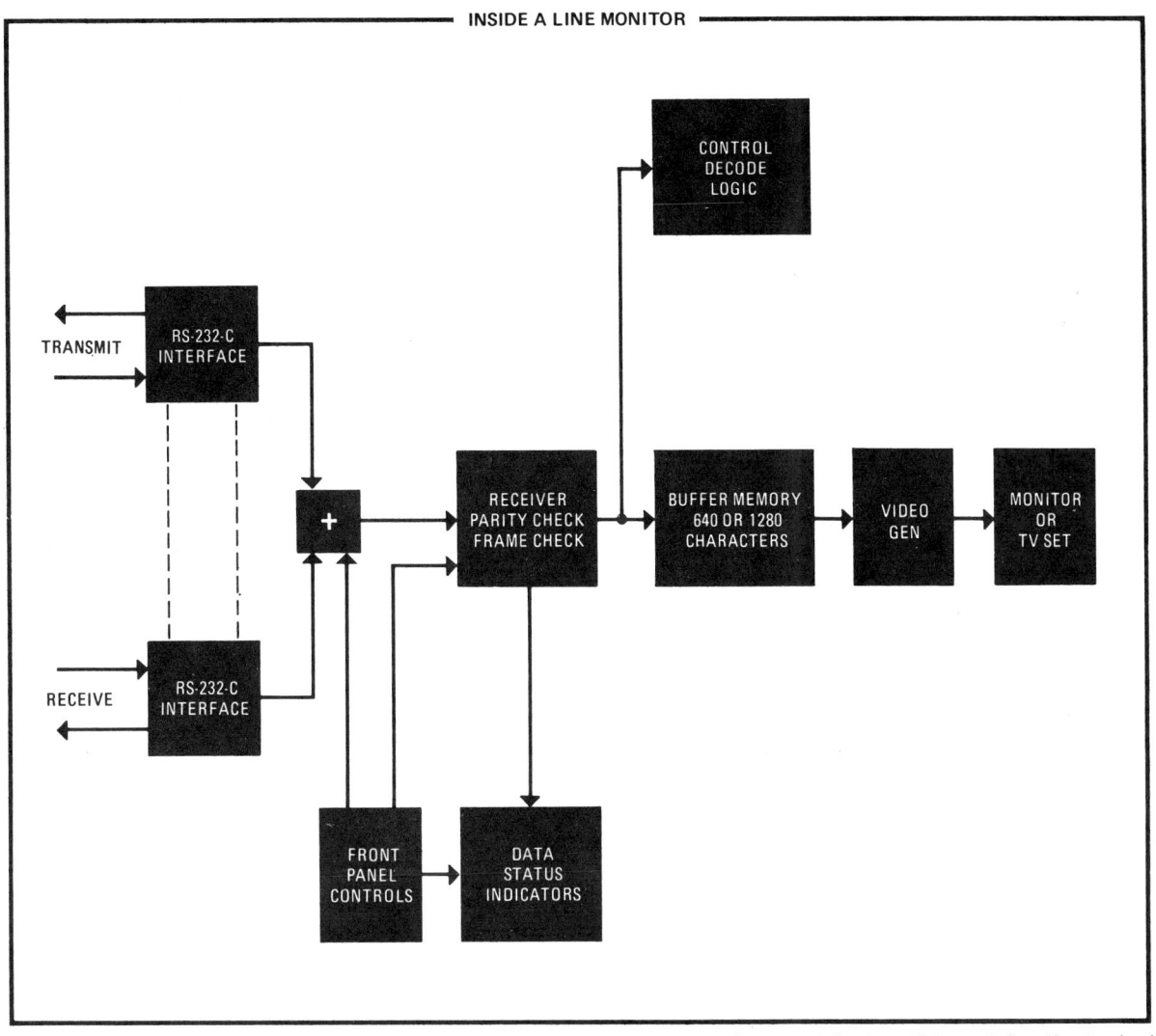

INSIDE A LINE MONITOR

3. Digi-Log Systems' Series 400 contains the typical functional sections found in most line monitors, except that products with indicator lights or alphanumeric displays do not contain the video generators or monitors shown here.

tector (RSLD). By monitoring individual EIA leads while the system is in operation, the user can log the sequence of control lead operation, measure the delay time between the request to send (RTS) and clear to send (CTS), check the frequency of the system clock, and check the correctness of the sync character.

The control leads are read out by a bank of LEDs in a unique manner. Rather than simply switching them on and off for the two binary states, each bit position contains a pair of LEDs, red and green. Red glows for binary 1 and green glows for binary 0. This arrangement, according to the manufacturer, makes for more easily recognizable bit patterns, to improve readability.

The data characters are decoded and displayed in hexadecimal form. Errors are checked by parity and, if desired, by longitudinal redundancy checks (LRC). The LRC character in the data being monitored is compared to an internally generated LRC character, and the presence of errors is indicated by a light. Error counts can be tallied. Data may be selected for trapping by entering the selected character into the analyzer or else by storing a particular position in the character sequence.

Byte analyzer

Data-Control Systems' Byte Analyzer has several novel capabilities. One is that the user can distort a signal being transmitted in order to check the error threshold of the equipment being tested. Pulse-width distortions of 6%, 12%, 25%, 38%, or 44% can be dialed in. Another useful feature is a loss of sync indicator, which is lit when more than eight consecutive character errors are detected, implying that the characters are being improperly framed.

REAL TIME MONITORING

4. Data links can be monitored while in normal operation and error counts logged. In the event of a failure, the last moments of transmission are available for analysis.

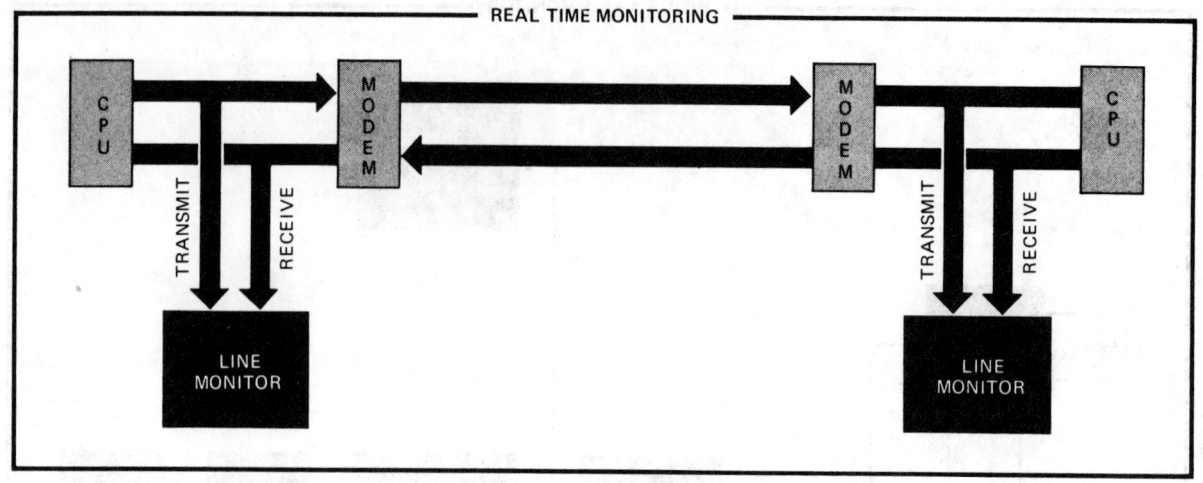

When performing loopback testing—in which data is transmitted to a remote location and returned—the Byte Analyzer also provides a means for calculating system turnaround time. This is done by transmitting a test message in a free-running mode and counting the number of characters that pass between transmission of the first character and its return.

Testing data link controls

The Data 100 Corp. Communications Analyzer is designed specifically for use with IBM's binary synchronous communications (BSC), Control Data's User Terminal, or Univac's DCT 2000 data link controls. A removable legend plate and encoder plug are supplied for the line discipline being tested—the legend plate lists the control codes for the particular data link control and the plug sets up the proper coding.

A complete system check done with the Communications Analyzer consists of three tests: local data set, transmission test, and control code analysis. In addition, there is provision for internal testing to verify that the analyzer is operating properly. The first two tests check parity of all characters, and the control code analysis verifies the legality of the control characters.

The local data set is normally tested first. A selected bit pattern is sent through the data set and turned around. The pattern is then checked with a bit- or block-error count. The bit pattern may be a switch-selectable 8-bit character, a 512-bit random block, or a 2,048-bit random block. Both random blocks are internally generated.

Errors may be intentionally introduced into the data stream by operating a switch that causes one error per block to be inserted. If a single character is being transmitted, its complement can also be selected for transmission. In the transmission test, the message is sent to a remote station (in proper format for the line discipline) and then returned for data-error analysis. The procedure is virtually the same as for the local data set.

The control code analysis checks for the presence of the proper data link control characters. Readout of control characters is in the form of lights alongside the appropriate entries on the legend card. Unlike analyzers with binary indicators, this approach does not require the user to translate the readouts, since they can be read directly.

The buffer of the analyzer contains 32 control characters. Passage of these characters through the analyzer may be monitored on a free-running basis, read out by sequentially stepping through the buffer locations, or be selected.

Doubling as a terminal

The Digi-Log Series 400 Data Line Monitor is a control unit to which the user connects a standard television set or video monitor. In addition to functioning as a diagnostic unit, the monitor also can be used as a read-only CRT terminal, in which only data is read out. Since the monitor displays 16 lines of either 40 or 80 characters per line, 640 or 1,280 characters can be viewed simultaneously.

By viewing a large field of data, the operator can quickly spot missing characters. Also by selecting readout of control characters only, the experienced operator can easily pick out erroneous control characters. In addition to the characters coming down the line, parity and framing errors are indicated by panel lights on the control unit.

When the Series 400 is used as a terminal, control characters are not displayed. Instead, the control characters that apply to the formatting, such as backspace and line feed, control the display in the conventional manner. However, in the data-monitor mode, control characters are read out, but they

> ## Learning and testing—a user case history
>
> Use of a line monitor has taught Donald C. Putnam a lot about his software, and it also allows him to check on the adequacy of repairs. Putnam, technical services manager at the A&P National Data Center in Piscataway, N.J., has a Paradyne 810.
>
> "We had little knowledge of BTAM (IBM's Basic Telecommunications Access Method) when we began applying it, and had no way of looking at what was happening on the line on a step by step basis. We bought the analyzer to debug application programs using BTAM. When we would see something wrong, we would use the trapping capability to capture the erroneous portion of the program before error recovery could mask the error."
>
> As an example of a recent application, Putnam related the following incident on the use of a line monitor in checking repairs:
>
> "An IBM 2701 data adapter unit in St. Louis had failed to generate an ACK (acknowledgment) message. The IBM repairman worked on the system and said that the problem was solved, but the analyzer in Piscataway proved otherwise, as the ACK still failed to appear. After several other repair attempts and checks with the analyzer, the ACK was restored. However, this experience indicates another useful function of line monitor—verifying that the proper repairwork has been done."

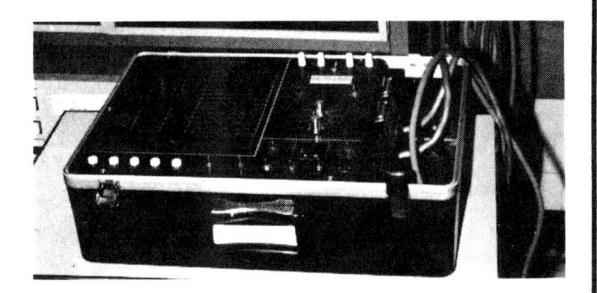

are converted into upper-case alphanumeric characters. SOH, for example, is shown as A, STX is B, and so on.

In addition, all lower-case alphanumeric characters in the data stream are converted to upper case. When the display is used in the diagnostic mode, all incoming characters are read out serially, and each line is completely filled out.

The Series 400 is one of the least expensive line monitors (prices start at $1,945, plus the cost of a display), but it lacks some capabilities found in many other units. There is no provision for responding to selected characters, storing characters for selected readout, generating test messages, or controlling the RS-232-C lines.

Monitoring bisync lines

Paradyne's Model 810 Bisync Analyzer is designed for IBM's BSC systems. It captures and reads out all BSC characters in the data stream and synchronizes to the SYN character.

The readout of control characters is similar to that of the Data 100 unit. One version of the analyzer reads out in EBCDIC. Another version provides for selection of EBCDIC or ASCII.

Especially useful is a group of lights that indicate the status of the lines between the automatic-dial DAA (data-access-arrangement) and the modem in the same way that RS-232-C lines are monitored. These functions are ring indicator in (RI IN), switch hook (SH), data available (DA), off-hook (OH), and coupler cut through (CCT).

A three-position switch allows the operator to stop information from entering the buffer when a selected character in the data stream has been detected. In this way, the operator can analyze the characters before and after the arrival of the character that was selected. The code for selecting a character is entered by a thumbwheel switch, calibrated in hexadecimal code.

Another position of this switch provides for viewing transparent text. In this mode, buffer loading is stopped only when the selected character is preceded by a DLE character.

CRT display

Spectron Corp.'s Datascope combines the appeal of a CRT readout with a number of useful functions to make one of the most expensive ($7,500) line monitors. It replaces a previous CRT monitor made by the company.

Perhaps the most useful capability is an endless-tape cartridge that records all signals as they are being displayed, to provide a continuous record of the traffic. At 2,400 bits per second, the tape can store 40 minutes of traffic, which makes the Datascope especially useful for continuous on-line monitoring, both for documentation and to provide a means of "instant replay" after a failure in the system.

Both transmitted and received data are recorded, along with carrier detect, signal quality, and the request to send signals from the RS-232-C interface. Also recorded is an internally generated event-marker signal, selected by operating a front-panel push button. During replay, these signals are read from the tape and displayed as if they were arriving "live," except that the replay may be slowed down or stopped for close examination of the data at critical moments.

An unusual option permits the unit to be connected to the monitoring point by connectors of

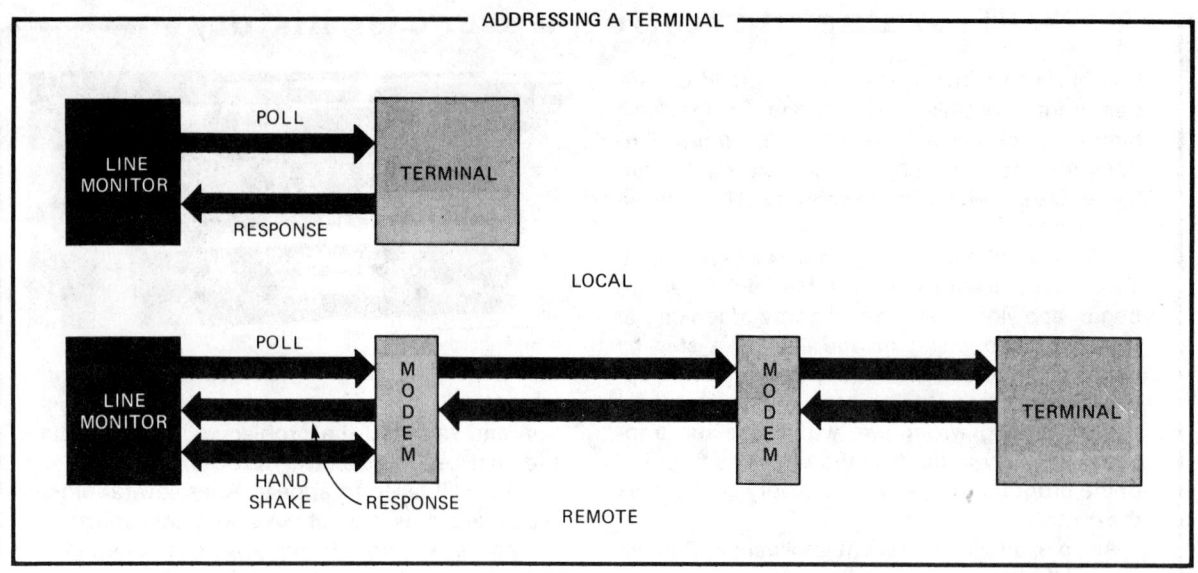

ADDRESSING A TERMINAL

5. A protocol message may be transmitted to a terminal, either locally or remotely to check for the proper response, which is trapped by the line monitor and read out.

mating equipment. Rather than plugging both equipment connectors into the chassis of the monitor, as other line monitors must do, a three-input plug can be inserted between the mating connectors. The third line goes to the Datascope, which may be located at some distance, without disturbing the line characteristics. This is a convenience if the connectors are hard to reach.

The CRT can graphically add supplementary information about the data being monitored. Data that has been received, rather than being transmitted, is identified by an underline; out-of-sync characters are displayed at low intensity; and the character image may be reversed to display black-on-white parity errors, carrier detect, request to send, or the event marker.

The ASCII and hexadecimal display are standard, but the EBCDIC may be specially ordered. For ASCII, upper-case and lower-case characters and punctuation are graphically displayed. For codes shorter than 8 bits, low-order bit positions are filled with zeroes to produce uniform 8-bit characters.

The screen contains a total of 375 data characters, arranged in 15 lines of 25 characters each. The characters are formed on a dot matrix in an 11 × 15 field.

The tape cartridge, which alternates recording among four tracks, records the tape serially, one track at a time, and reverses direction each time the end of the tape is reached. It continues on the next track until all four tracks are full. After completing the fourth track, the tape continues recording on the first track, and old data is replaced with new. Each time a record operation begins or terminates, a tape mark is recorded to identify the beginning of newly recorded information. Data bits from each side of the line and modem-status bits are multiplexed to permit recording on a single track.

Also useful is the capability to reverse characters. Although most systems transmit the data in the proper order, many EBCDIC systems transmit in reverse order. To produce the correct readout, a character-reverse switch is provided.

6. Modems can be tested with either local or remote loopback and parity errors logged. Local testing such as this usually precedes a system test.

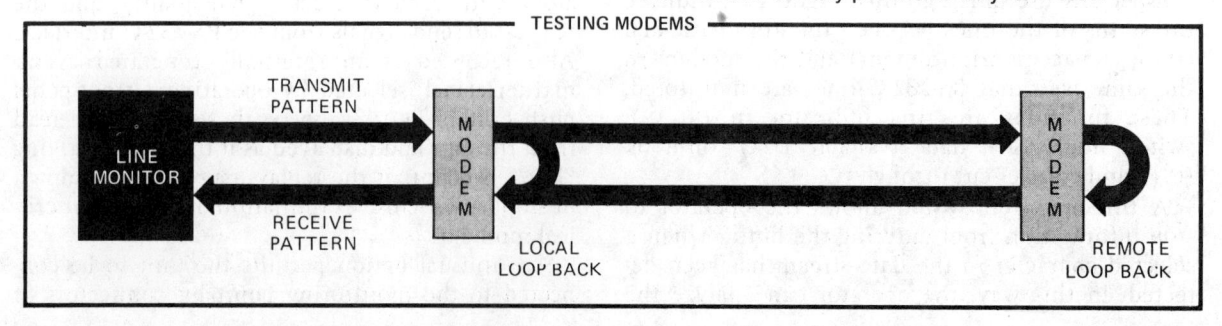

TESTING MODEMS

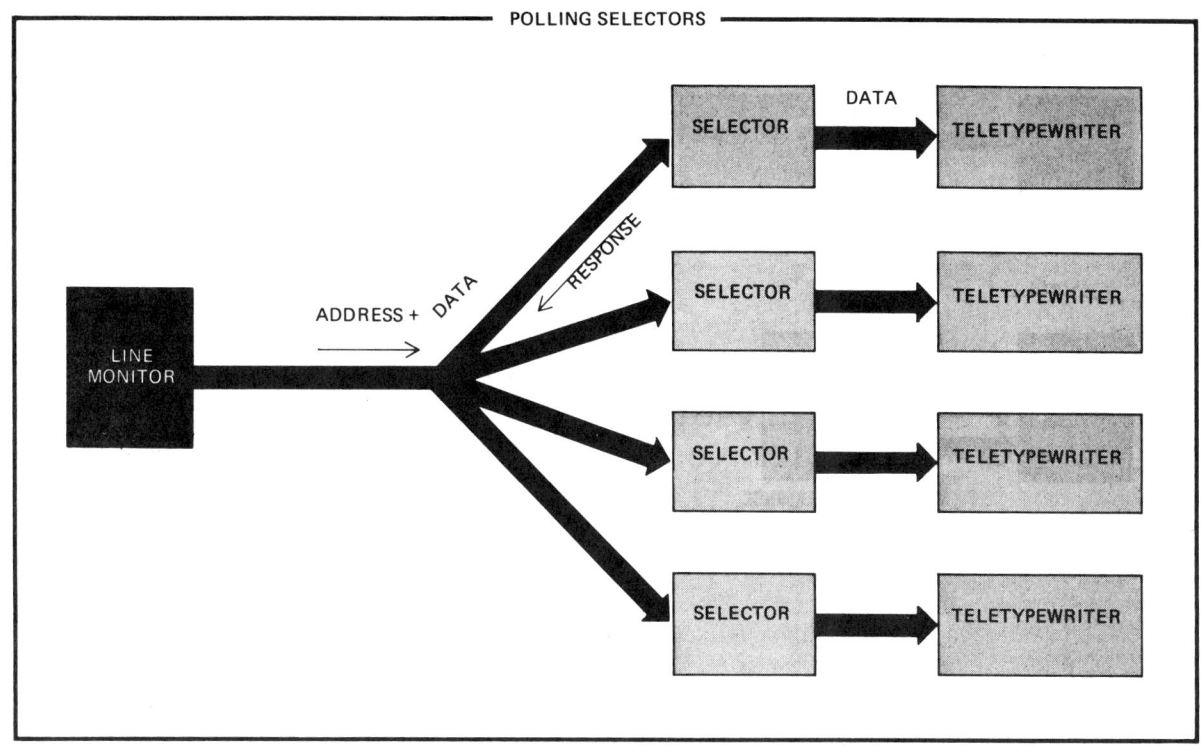

7. Terminals in a cluster may be polled to check for proper addressing, followed by the transmission of test messages.

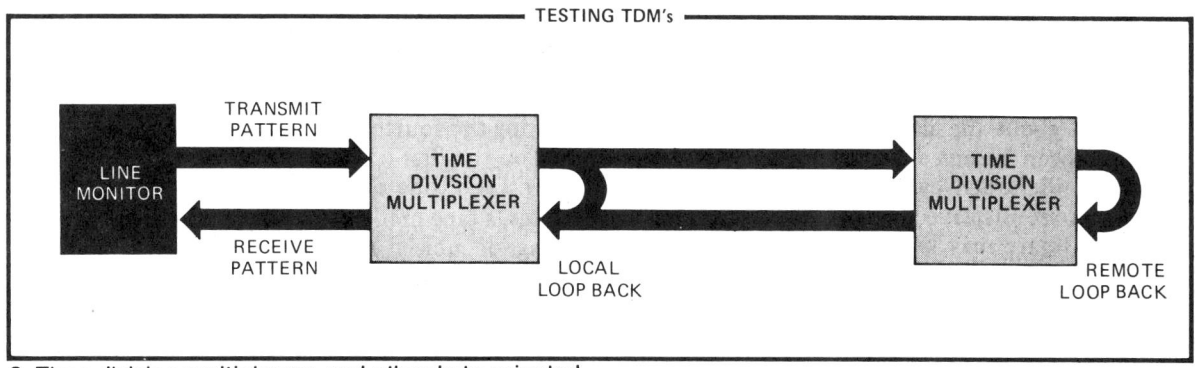

8. Time-division multiplexers and other byte-oriented equipment, such as scramblers and repeaters, can be tested and errors tallied.

Character-framing is provided for 5, 6, 7, and 8-bit synchronous and asynchronous characters, as well as IBM's SDLC.

Two features that are not offered by the Datascope but are available in other sophisticated line monitors are error counters and message generators for polling, answer back, and loopback testing.

Monitor captures 1,024 characters

Teletype Corp.'s TSM-7810S Line Monitor Data Display captures a maximum of 1,024 characters (both data and control) and reads them out serially, 16 at a time. Switches are provided for fast or slow readout, forward or reverse, and for wrapping around the memory. Wraparound permits characters received after the buffer is full to replace data already in the buffer. In addition to being a convenience, the wraparound is also necessary for continuous on-line use.

Indicators show when the buffer is one quarter, one half, three quarters, or completely full. An audible alarm is sounded whenever activity begins on the line. The internal clock provides timing for only 1,200 bits per second. Other speeds can be obtained from timing signals supplied by the modem (up to 9,600 bits per second), and a number of speeds between 75 and 2,400 bits per second are provided by auxiliary plug-in cards. There is no provision for controlling RS-232-C lines, trapping selected characters, or generating test messages. ∎

Centralized diagnostics cuts line downtime to minutes

Leased-line network with 176 sites is designed so faults can be handled by a single operator— and without requiring remote intervention

Gunther Kempin and Paul R. DeVane
Manufacturers Hanover Trust Co.
New York, N.Y.

Application

The inability to identify network problems either easily or quickly is a major weakness of most fast-response data communications systems. Particularly in a bank, time is wasted and customers become impatient whenever a terminal, modem, or line goes down.

Manufacturers Hanover Trust Co., however, handles its passbook savings accounts over an extensive communications network in which faults are diagnosed in a matter of minutes, defective lines replaced by dial backup in only three more minutes, and defective hardware dealt with by its vendor in a couple of hours.

The bank started with a decision to structure two "interrelated" systems simultaneously: the communications network, and remote diagnostics. All the network control and fault diagnosis functions were to be located at a central point, so that they could be kept under the eye of a single skilled operator, and would not need to involve other personnel at outlying bank branches. These concerns were then made fundamental to the design of a communications net linking over 240 terminals at 176 branches throughout the New York metropolitan area.

Other design requirements included fast response time and flexibility for future expansion. The end result is a leased-line, 2-second-response, four-wire communications system, with dial-up lines for automatic backup, and a diagnostic control center at the host computer facility.

The specifications for the network were established by the bank and put out for competitive bidding. The contract was eventually awarded to Intertel Inc., which supplied all modems and bridges, and also designed and built the entire diagnostic center.

The network links 152 remote branch installations, 24 other branch installations, strategically located to serve as bridging points, and the central control installation (Fig. 1).

Each bank site has at least one Olivetti Bank Teller Terminal, paired with a modem hooked up to a line leading to one of the bridging points. Originally, all the lines transmitted at 1,200 bits per second, to cope with a predicted peak traffic load of 6,800 transactions an hour, but six of the lines have since been upgraded to 2,400 b/s because of increased volume. The line discipline used is IBM's

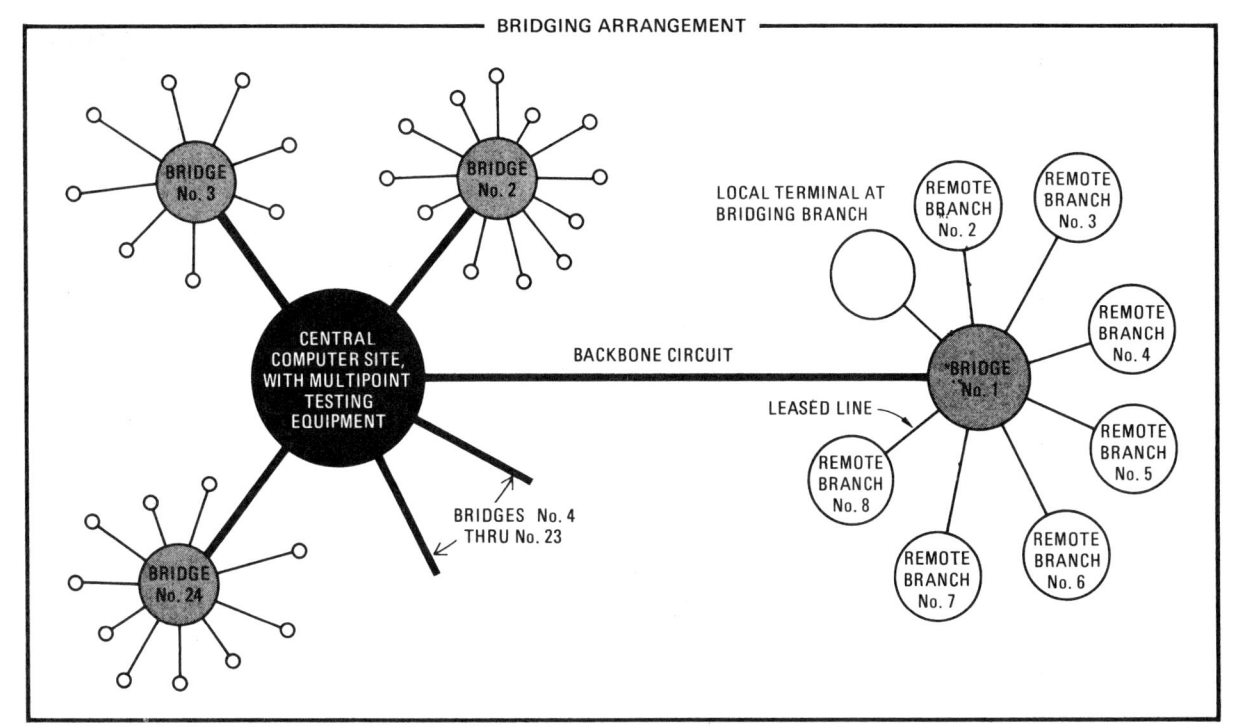

1. To keep costs down, a cluster of terminals feeds into a bridge that is linked to the host computer facility by a single backbone (four-wire) circuit.

BSC (binary synchronous communications). In addition, every banking site has two data access arrangements (DAAs) and an automatic answering and switching capability for dial backup.

The bridging sites tie all the terminals into 24 backbone circuits (leased lines). At these bridging branches, lines from up to eight terminals feed into a 10-way/four-wire bridge through separate ports. Each bridge is gated, so that when the system's constant polling activates one port, the noise from the others is blocked (gated) out and kept from interfering with the signal at the active port. Any bridge can be expanded to handle 17 terminals by having an extra bridging circuit board added and wired to the first bridging board.

Eventually, therefore, the network could service 408 terminals and will be able to accommodate the needs of new branches to be opened over the next several years.

At the control point

From each of these outlying bridges runs a backbone circuit to the bank's Technical Control Center. There the 24 lines eventually interface with an IBM 3705 terminal control unit, emulating an IBM 2703 and itself interfacing with any one of the three IBM 370/165 host computers. A second IBM 3705 is always ready as a spare.

Before the lines reach the control unit, however, each passes through another bridge, which ties it into the dial backup system, and through another modem, which interfaces it not only with the IBM 3705 but also with the multipoint tester (Fig. 2). From this facility, a single operator can quickly observe error conditions, run tests to isolate the problem, and take corrective action to restore a remote site to operation—all without interfering with the normal functioning of the rest of the network.

The multipoint tester allows the operator to select which remote modem is to be tested and to select which test is to be performed. The instrument uses an out-of-band test channel for diagnosing problems on a single line while the rest of the network remains on line, but uses both this test channel and the data channel for other tests that require the backbone circuit to be off line. In both cases, the tester works in conjunction with test abilities and corrective features built into the modems used throughout the system.

At the central site of a multipoint polled network like this one, faults associated with a single remote site normally fall into one of two categories: lack of response from a specific site, or modem streaming, a situation where the terminal's malfunction causes the modem to transmit a constant carrier signal to the central site and prevents identification of the remote site. Examples of the former include: a power failure or malfunction (other than streaming) in the terminal or modem, and an open or highly degraded line.

Test capabilities

All these faults, except a degraded line, can be positively identified by the central multipoint tester

NETWORK CONFIGURATION

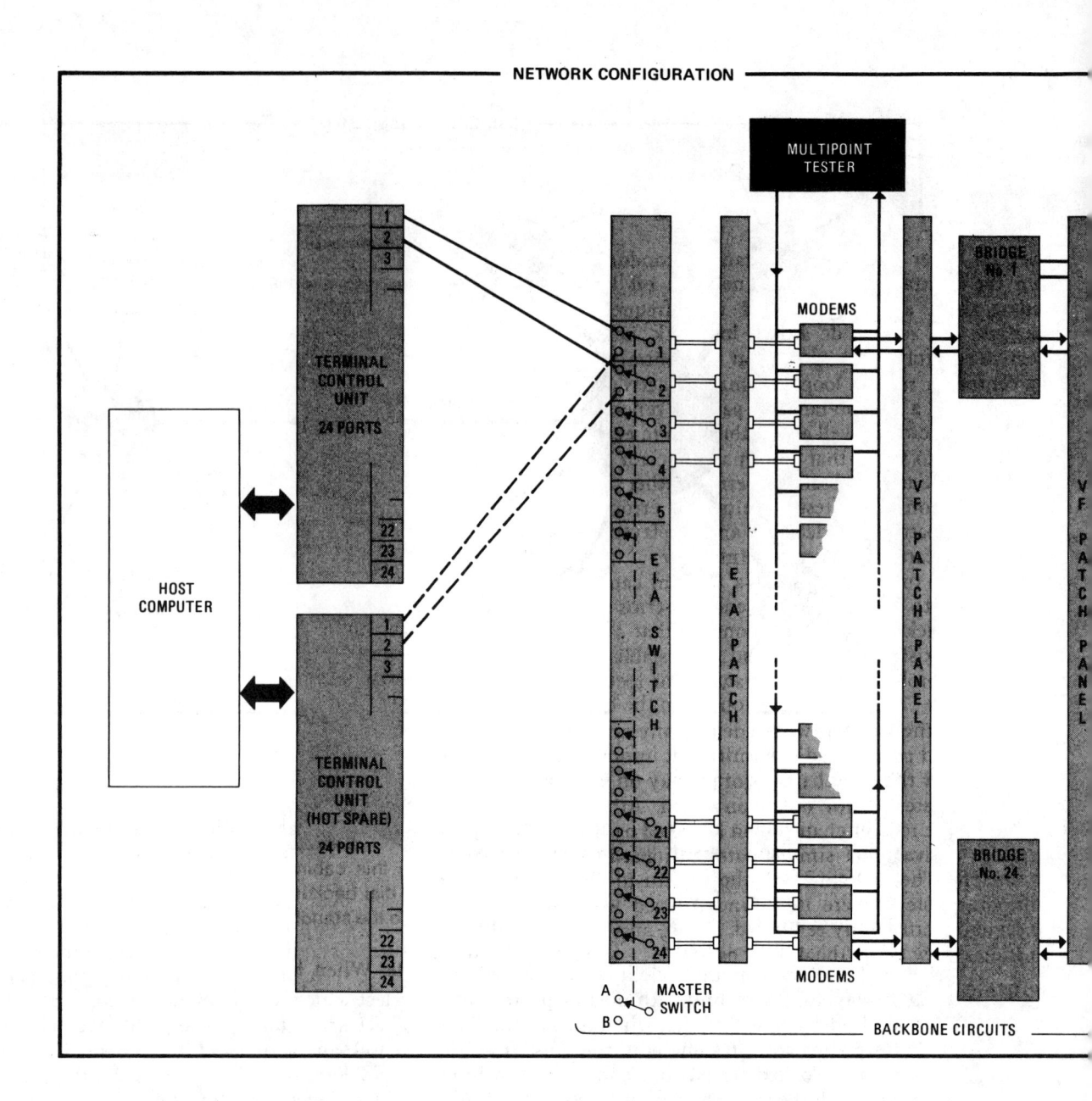

through on-line tests conducted with the aid of certain modem options at both the central and remote sites. The problem can be isolated to the particular site concerned, without the intervention of personnel at the remote site, and while the main data channel continues to poll the remaining remote terminals.

Several other kinds of tests can be conducted from the central site with the network off line. Among them are local dc and voice-frequency loopback, and remote dc and vf loopback. Under selective control by the operator at the central site, any central or remote loopback may be activated. Tests include a pseudo-random pattern generator and error detector, as well as the ability to inject errors in order to verify that a normal "no error" indication is due to the absence of errors rather than the malfunction of the test circuits. The errors in the received pattern are detected and displayed on the error indicator of the multipoint tester.

In addition to loopback tests, a modem can be commanded to run an end-to-end bit-error-rate test. A loopback test can only confirm that a degraded line exists. The end-to-end test establishes whether the problem is on the transmit or receive side of the line. A pattern is sent out from the control center to the modem, which detects any errors in the received pattern and transmits that information back over the test channel for display on the multipoint tester's error detection panel. At the same time, the receive channel can be tested by the operator's activating a similar pattern housed in the modem. The modem sends the pattern to the multipoint tester, where it is synchronized with the same pattern locally generated. Again, any discrepancies show up on the display panel.

Status/control terminals supply the operator with information on the status of the different lines and

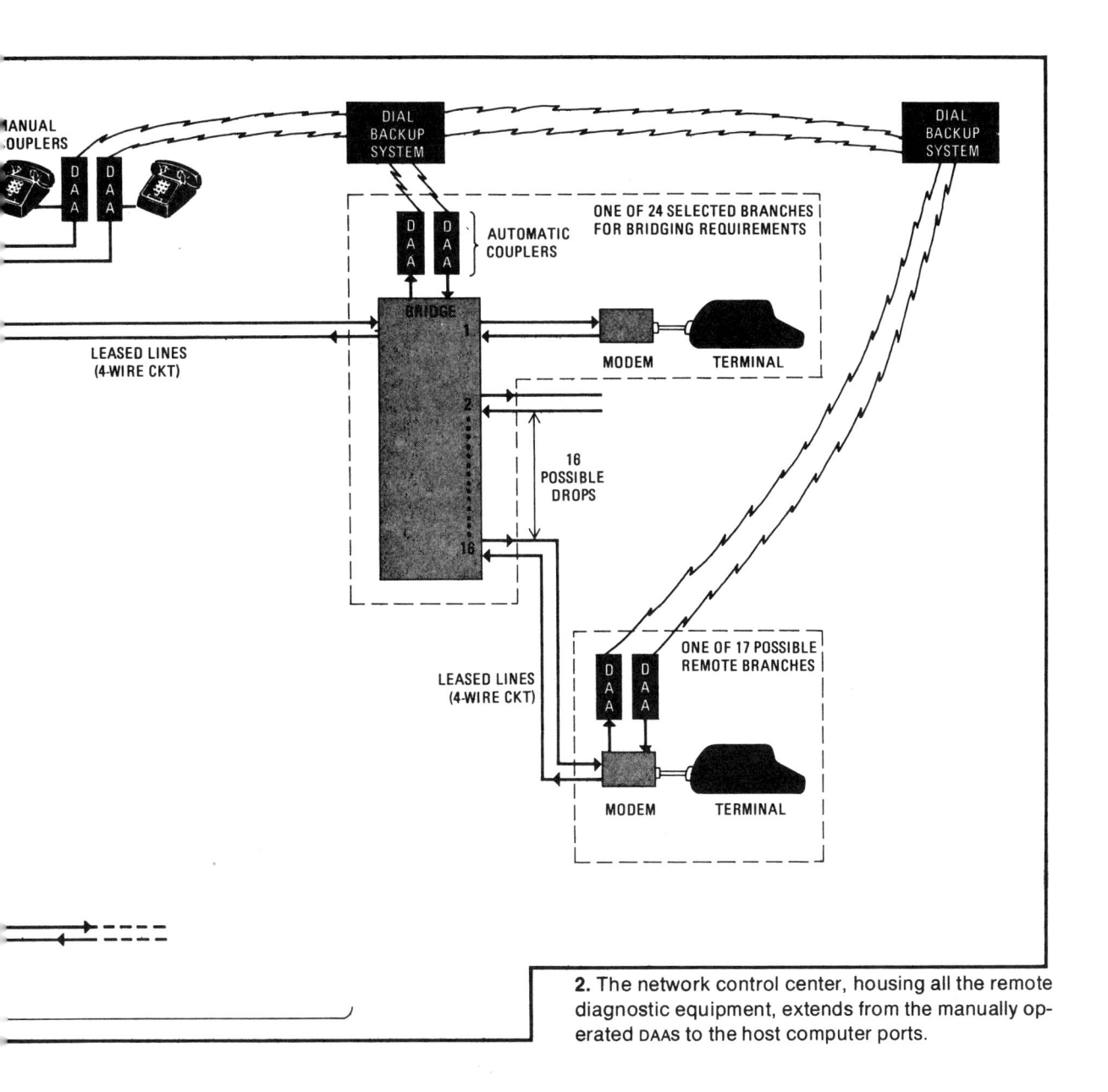

2. The network control center, housing all the remote diagnostic equipment, extends from the manually operated DAAs to the host computer ports.

terminals. He goes into action either when he sees an error message on the control terminal or when he gets a telephone call from a branch office indicating difficulties. In the case of an error message, resulting from a terminal time-out (when there has been no response to a poll), the line control software (BTAM, or Basic Telecommunications Access Method) has already tried to reestablish communication with the terminal seven times.

The error message shows the address of the terminal with a problem. The operator first tries his on-line tests and waits for the results to be flashed on the display.

If the diagnosis indicates that a line is down, he has to switch to dial backup. Since the system is full duplex and has four-wire lines, he places one call to the receive side of the modem at the remote location and another call to the transmit side of the modem. He uses two manual DAAs at the central site in doing this, plus two automatic DAAs at the remote site. When he hears the answer-back tone indicating the modem has answered the two calls, he turns two keys to cut into the line, and the switchover has been completed. Finding the problem takes about two minutes; making the switch takes another three. Thus, a leased-line problem is temporarily overcome within five minutes. The telephone company is then called for repairs.

Switching and patching

Figure 2 indicates the presence of various switching and patching modules that also play a vital role in keeping the network from going down. They include voice-frequency patch modules, an EIA patch module, and an EIA transfer switch.

The vf patches allow the operator to patch, monitor, send, and receive on all 24 vf lines passing from the remote bridging sites through the central bridges and modems. They have 24 full-duplex circuits and three jacks for each circuit—a line jack, equipment jack, and line-monitor patch jack, for use in testing and substituting lines and equipment (see photo above).

The EIA patch module interconnects, monitors, and interchanges equipment that has an RS-232-C type interface. It lets any EIA interface be patched to any modem or controller port.

The prime purpose of both types of patch module is to allow the operator to restore the system rapidly to operation when a line or modem fails. Similarly, the EIA transfer switch allows EIA interface connections to be transferred from the primary IBM 3705 terminal control unit to the spare controller whenever there's a failure in the first unit. If all 24 lines must be transferred, the operator pushes a master button.

Four-wire operation needed

The testing needs of the system were one reason why four-wire operation was preferred to two-wire. A multipoint, two-wire network cannot undergo loopback tests, for example. Instead, end-to-end tests, requiring manual assistance at the remote end must be conducted. This adds to the time and difficulty of resolving problems, conditions a four-wire arrangement largely overcomes.

But the desired response time also was a factor. Simulation studies showed that half-duplex operations required a turnaround time of about 200 milliseconds, which added significantly to response time (defined as the interval elapsing between the last character keyed on a terminal to the first character printed out). In fact, response time came to almost 4 seconds.

However, according to a leading behavioral psychologist who specializes in human-computer interaction, people are unwilling and even unable to perform complex tasks with machines if the machine's responses are delayed more than 2 seconds. For continuity of thinking, or if substantial information must be retained in the human short-term memory while the operator is waiting for a response, a fast reply is essential. So to keep the tellers from becoming frustrated and alienated by the terminals, the four-wire system permitting a 2-second response time was chosen.

Cost savings

Another reason is that two-wire bridging is more expensive and cumbersome. Normally, bridging tends to lower signal intensity, which must then be reestablished by adding devices such as amplifiers on all input and output ports. The addition of such amplifiers on two-wire circuits is far more complex than on four-wire lines.

Line utilization with the four-wire setup is only 15%, opposed to the 25% projected if two-wire had been used. This gives the bank the freedom to add

Patches. From this cabinet, operator can patch-in spare DAAS for dial backup and can switch a line or group of lines to the standby terminal controller.

a greater number of new applications to the communications network as time goes by. Once a single CICS/TCAM (Customer Information Control System/Telecommunications Access Method) monitor was implemented, the network would also be able to accommodate IBM's new synchronous data link control (SDLC), since it has a full-duplex line discipline.

The network, as configured and run today, minimizes the possibility of any locations going down for any lengthy period. Moreover, the multipoint testing capability has the advantage of assigning problems to vendors. Each vendor's products are evaluated every week, based on the printouts from the status terminal. When a particular product starts showing up as faulty too frequently, talks are held with that supplier to discover how to improve matters.

Finally, to keep the testing equipment at peak efficiency, and to handle any maintenance needs that may arise, an Intertel field engineer resides at the control center. ∎

part 9
data data data data data data data data data data

interfaces

Using standard interfaces doesn't always assure trouble free performance

Despite the EIA RS-232-C interface standard, modems and terminals may still not respond correctly to each other's control signals

Karl I. Nordling
Paradyne Corp.
Largo, Fla.

Hardware

During the early planning phase for a data communications network, system designers avoid details in favor of settling major network considerations first—but certain details can't be left to the last. For example, they innocently assume that equipment such as terminals and modems can be plugged together and will work because they conform to interface standards governed by specifications recognized by different equipment designers in different vendor organizations and by different user groups. In the United States the accepted interface standard is the Electronic Industries Association's EIA Standard RS-232-C "Interface Between Data Terminal Equipment and Data Communications Equipment Employing Serial Binary Data Interchange."

The availability of different kinds of communications equipment having interfaces conforming to RS-232 has been a boon to users. RS-232 makes it possible to design and install data networks that will operate well even when equipment is obtained from many different vendors—but it does not necessarily make it easy. What looks simple and straightforward in the planning phase can, in fact, become quite troublesome during the system's engineering, installation, and start-up phases. For one thing, the standard contains certain options and some ambiguities. For another thing, even with the best of intentions, different equipment designers have different perspectives—and, thus, may interpret the standard, including options and ambiguities, in different ways. In short, terminals can be interfaced according to RS-232 and so can modems. Yet some terminal-modem combinations will work together—and others won't, as has been sadly demonstrated so many times in the field over the past few years.

To circumvent future start-up problems, the user system designer must delve deeper into interface operation for each piece of equipment during the equipment selection stage, in particular the step-by-step logical procedure involved in establishing a connection, via modems and data access arrangements, between distant terminals. If the devices are not logically compatible, the result can be a deadlock. Connection will never be established.

Discussed here are some of the most common problems arising at the interface between the terminal and the modem and the interface between

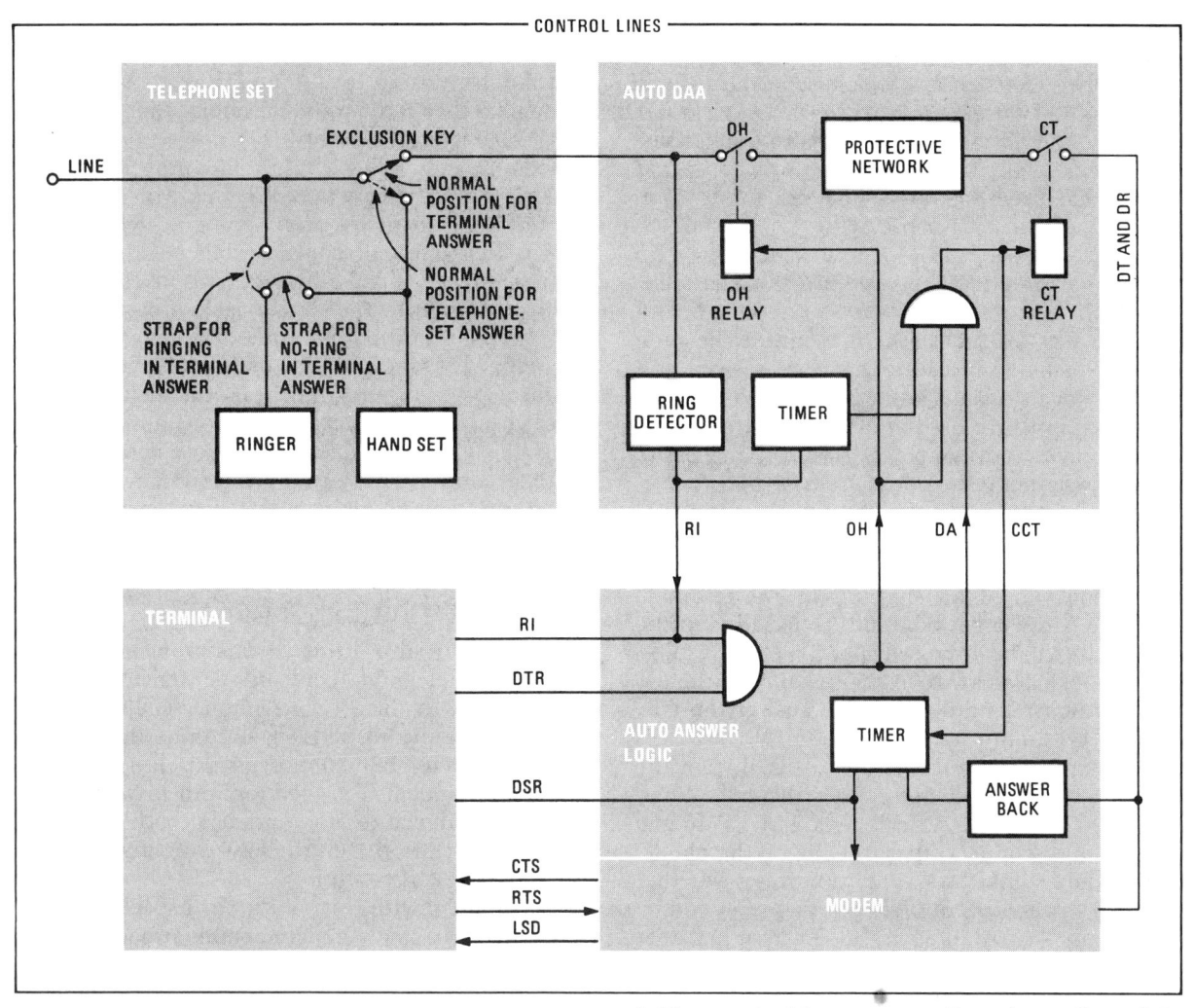

1. The control lines connecting phone line, data access arrangement, auto answer unit, modem, and terminal follow conventions set by EIA RS-232-C standard and the phone company.

the modem and the phone line. (Here the word terminal means any device that sends or receives data, and thus includes such system components as keyboard/display terminals, remote data concentrators, programable front-end processors, and remote job entry terminals. Furthermore, the words modem and data set are used interchangeably.)

The EIA RS-232-C specification is the primary standard governing the interface between the modem and the terminal. The interface control lines are divided into two groups, one concerned with the control sequences for connecting the modem to the line, and the other with the sequences controlling the application (and detection) of signal to the line. On dial-up telephone lines the first group interacts closely with the control lines in the modem-to-phone line (DAA) interface, as shown in Figure 1. Figure 2 contains the logical procedure for connection and disconnection.

The RING INDICATOR (RI) line from the DAA signals the modem that a call is coming in. The modem should respond by raising OH to answer the call and then request to be put on the line by raising the DATA AVAILABLE (DA) line.

The OFF-HOOK (OH) signal from modem to DAA controls the off-hook relay. When the modem detects an incoming call (RI), and if DATA TERMINAL READY is ON, it closes the OH contacts which, in effect, simulates taking the handset off-hook—that is, answers the telephone. However, if DTR is OFF, the modem should not answer the phone.

The DATA AVAILABLE line from the modem is used to signal the DAA to "go to data." It would normally be raised by the modem after raising OH, to put the modem on the line.

The COUPLER CUT THROUGH (CCT) line from the DAA informs the modem that it has been connected to the line and has "gone to data." CCT is turned ON in response to DA from the modem, after an interval of 2 to 5 seconds to allow the automatic message accounting equipment in the cen-

247

tral office to register the call. When the modem receives this CCT signal, it should send the answer-back tone and—after that—turn on DATA SET READY (DSR) to the terminal. (The answer-back tone, 2,025 or 2,225 Hz, informs the call originator that the call has been answered. Modems for auto-answer service must have the ability to generate either tone.)

The RS-232 interface lines are listed in the table. The RING INDICATOR (RI), DATA SET READY, and DATA TERMINAL READY (DTR) lines control the automatic sequencing of call origination, answer, and termination in dial-up links. In leased- or private-line links, RI is not used at all, the state of DTR is used simply to indicate power ON or OFF and DSR to indicate whether test mode is ON or OFF.

The RING INDICATOR originates in the automatic answer DAA and triggers the auto-answer sequence in the modem, provided that DTR is turned on. RI is not normally used for anything by the terminal, since the answer-back circuit is usually in the modem, as will be discussed later.

DATA TERMINAL READY is the signal from the terminal which tells the modem to answer the telephone (ON) and to hang it up (OFF). When DTR is OFF, the modem will ignore the RI signal from the DAA and, hence, will not answer the call. When DTR turns OFF during a call it is a signal to the modem to disconnect: that is, hang up the phone. In leased-line links DTR simply indicates whether terminal power is ON or OFF.

DATA SET READY indicates to the terminal that the modem is in a state in which it is capable of transmitting data. That is, the modem (data set) is connected to the line, it is not in a test mode, its power is ON, and all actions having to do with call establishment and auto answer have been completed. In auto-answer mode, DSR is, in effect, the signal from the modem to the terminal which says, "I've answered the telephone and gone from voice to data." Going back to voice mode by manually operating the exclusion key on the telephone set does take the modem off the line, but does not cause DSR to drop, so the connection remains. Therefore, any time DSR goes OFF during a call, it should be construed as an indication that the connection has been lost and that the terminal should take action to reestablish the call. In leased-line links DSR ON indicates that the modem power is ON and that the modem is not in a test mode.

Auto answer logic

The main things to be concerned with are the auto-answer logic built into the modem and the terminal, and the auto-disconnect problem. RS-232 requires that DTR be ON for the auto-answer sequence to start when the RI signal turns ON. However, certain terminals are designed to not raise DTR unless DSR is ON, so that if the logic in the modem complies with specification, as described above, neither DSR or DTR will ever come ON. One solution to this logical stand-off is to wire the modem so that it answers the phone independent of whether DTR is ON or OFF.

According to RS-232-C, auto disconnect should take place in response to DTR being turned OFF by the terminal. In many cases, however, what happens is that one end hangs up but DTR does not drop at the other end—for such reasons as abnormal termination by the system, lack of proper sign-off from the terminal, or simply an oversight in system design. Also, when someone accidentally dials an auto DAA's number and hears the answer-back tone, the caller will react to this wrong number and hang up, but DTR remains ON. For whatever the cause, as long as DTR stays ON, the modem will not "hang up." The phone is busy for any subsequent calls, so the terminal remains unavailable for either polling or selection.

Partial solution

There is a partial solution to this problem. Some, but not all, telephone company central offices return a signal to the local telephone when the remote end is hung up. This signal can be an ac tone or a momentary interruption of the dc loop current. Some modems designed for auto-answer operation are equipped to sense either or both of these signals and turn off DSR, thus assuring disconnect even without DTR control.

The problem with this solution is its lack of universality. A terminal/modem combination that operates properly at one site when using DSR disconnect (rather than the preferred DTR-controlled disconnect) may fail when duplicated on another site simply because the central office at the new site either does not provide a disconnect signal at all or else the wrong type for a particular modem.

The group of RS-232 control signals that relate to the application and detection of time signal are REQUEST TO SEND (RTS), CLEAR TO SEND (CTS), and LINE SIGNAL DETECT (LSD). The logical interaction of these signals is simple when terminals operate in a two-wire, half-duplex mode and complex when terminals operate a four-wire, full-duplex mode.

REQUEST TO SEND is the signal from the terminal to the modem indicating that the terminal wants to transmit. The modem responds by initiating its transmit mode—which may be a simple process such as starting to send carrier or a complex process such as "training" the modem's automatic equalizer.

CLEAR TO SEND is the signal from the modem to the terminal signifying that the modem has reached a steady-state carrier condition and is ready to transmit. CTS does not imply a positive verification that communications with the other end has been established.

LINE SIGNAL DETECT indicates to the receiving terminal that the receiving modem has detected a line signal (carrier) coming from the other end.

Every terminal designed to run in half-duplex mode raises RTS when it wants to transmit and doesn't start transmitting until CTS turns on. There is no conflict between modem logic and terminal logic. Problems arise in full-duplex operation due to differences of interpretation, data link control (line protocol), and modem technology. These problems occur mainly in high-speed links operating at 4,800 bits per second or faster. To reach the higher net data throughput implicit with high-speed modems requires an operating discipline different from that used for low-speed modems. And the specifications in RS-232 were written with the characteristics of low-speed modems very much in mind. The problem comes about because high-speed modems operating in four-wire, full-duplex service are usually designed to remain in transmission mode at all times. That is, one pair of the wires is continuously transmitting a carrier in one direction and the other pair is continuously transmitting a carrier in the other direction. Therefore, they are not subject to control of the carrier by the RTS line as are low-speed modems.

Terminals for full-duplex, four-wire operation can be divided into at least four different operational categories, the differences depending on how a terminal interacts with RTS, CTS, and LSD:

- The terminal keeps RTS ON continuously and simply requires CTS ON in order to transmit. This is the easy one. The modem will operate properly, as long as it is set up for four-wire operation.

- The terminal turns RTS OFF and ON between blocks and waits for the completion of the RTS-CTS delay before starting to transmit. This type of operation requires that the modem be strapped (wired) to disable RTS control of the transmitter, so that transmission is continuous regardless of the state of that signal. However, RTS still must initiate ON and OFF of CTS after a selectable delay. This RTS-CTS delay could obviously be zero seconds as far as the modem is concerned since all that takes place is the enabling of a gate. However, some terminals are unable to respond if the CTS signal turns ON immediately. Therefore, modems are usually provided with several strap-selectable delays: 0, 8.5, 50, and 150 milliseconds are common. To maximize throughput, the shortest delay that the terminal can handle should be selected.

- The terminal does not turn ON RTS at all, but does require CTS in order to transmit. The fix when using this type of terminal is to rewire the modem to hold CTS ON whenever power is ON and the modem is not in test mode; that is when DSR is ON independent of the ON-OFF state of the RTS signal.

- The terminal turns RTS ON and OFF between blocks, and requires CTS ON and LSD OFF in order to transmit. Known as controlled carrier mode, this type of operation is the hardest to accommodate with high-speed modems. Actual controlled carrier operation on such modems range from inefficient to impossible. Controlled carrier operation is inefficient because turning the transmitter (carrier) from OFF to ON requires time for the modem to train and synchronize, and this delay time reduces net data throughput. Multi-drop polled links using this type of terminal require actual controlled carrier operation, so there is no choice but to tolerate the inefficiency.

Controlled carrier operation is impossible on links that, through time-division multiplexing, divide the modem bit stream into two or more lower speed data channels. This is because by turning OFF the carrier for one channel the entire bit stream is turned OFF, and hence the second channel is interrupted. The result is that neither channel can ever get any data through because the interruption caused by one channel will probably cause errors in every block for the other channel.

Fortunately, there is a solution to the inefficiencies and impossibilities inherent in actual controlled carrier operation: use modems that can simulate controlled carrier operation. Simulated operation can be obtained by sending a self-synchronizing pseudo-random bit pattern on a channel whenever its RTS signal goes OFF. This pattern is then detected by the receiving modem, which causes LSD for that channel to go OFF—but the actual carrier remains ON at all times. The main consideration in a scheme like that is: How many bits should be used in the "carrier OFF" pattern? Too many bits and the delay before LSD goes OFF becomes too great and data throughput decreases; too few bits and the false-alarm probability becomes too great.

For a 30-bit pattern, the probability is 2^{-30}, or about one chance in a billion that any data or control character sequence will match that pattern and cause LSD to drop. In deciding what "false LSD-OFF" probability is low enough, the user must consider the impact when it does occur. If the result in a particular system is that the link comes to a standstill when an unexpected LSD-OFF occurs, then any block that contains the critical pattern will never be transmitted across that link. For a 30-bit pattern on a link transmitting at 9,600 b/s for eight hours a day, the probability is that this will occur about once every three days.

Secondary RTS/CTS

The secondary channel provided on many modems is a derived channel of narrower bandwidth than the primary channel. The secondary channel can serve as a low-speed reverse channel slaved to the primary (forward) channel, or it can function as an independent low-speed channel operating in either direction.

The independent secondary channel is most common in point-to-point dedicated circuits, with both a high-speed synchronous terminal (a remote

RS-232-C Interface Circuit Functions

Pin 1. PROTECTIVE GROUND. Electrical equipment frame and ac power ground.

Pin 2. TRANSMITTED DATA. Data originated by the terminal to be transmitted via the sending modem.

Pin 3. RECEIVED DATA. Data from the receiving modem in response to analog signals transmitted from the sending modem.

Pin 4. REQUEST TO SEND (RTS). Indicates to the sending modem that the terminal is ready to transmit data.

Pin 5. CLEAR TO SEND (CTS). Indicates to the terminal that its modem is ready to transmit data.

Pin 6. DATA SET READY (DSR). Indicates to terminal that its modem is not in a test mode and modem power is ON.

Pin 7. SIGNAL GROUND. Establishes common reference between modem and terminal.

Pin 8. RECEIVED LINE SIGNAL DETECTOR (LSD). Indicates to the terminal that its modem is receiving carrier signals from the sending modem.

Pin 9. Reserved for test.

Pin 10. Reserved for test.

Pin 11. Unassigned.

Pin 12. SECONDARY RECEIVED LINE SIGNAL DETECTOR. Indicates to the terminal that its modem is receiving secondary carrier signals from the sending modem.

Pin 13. SECONDARY CLEAR TO SEND. Indicates to the terminal that its modem is ready to transmit signals via the secondary channel.

Pin 14. SECONDARY TRANSMITTED DATA. Data from the terminal to be transmitted by the sending modem's channel.

Pin 15. TRANSMITTER SIGNAL ELEMENT TIMING. Signal from the modem to the transmitting terminal to provide signal-element timing information.

Pin 16. SECONDARY RECEIVED DATA. Data from the modem's secondary channel in response to analog signals transmitted from the sending modem.

Pin 17. RECEIVER SIGNAL ELEMENT TIMING. Signal to the receiving terminal to provide signal-element timing information.

Pin 18. Unassigned.

Pin 19. SECONDARY REQUEST TO SEND. Indicates to the modem that the sending terminal is ready to transmit data via the secondary channel.

Pin 20. DATA TERMINAL READY (DTR). Indicates to the modem that the associated terminal is ready to receive and transmit data.

Pin 21. SIGNAL QUALITY DETECTOR. Signal from the modem telling whether a defined error rate in the received data has been exceeded.

Pin 22. RING INDICATOR (RI). Signal from the modem indicating that a ringing signal is being received over the line.

Pin 23. DATA SIGNAL RATE SELECTOR. Selects one of two signaling rates in modems having two rates.

Pin 24. TRANSMIT SIGNAL ELEMENT TIMING. Transmit clock provided by the terminal.

Pin 25. Unassigned.

---

job entry terminal, for example) and a low-speed asynchronous terminal (such as a teletypewriter) using the same four-wire link to provide full-duplex transmission for both terminals. In this type of operation the secondary channel actually operates as a low-speed primary channel. Therefore, the secondary pin assignments of the RS-232 connector must be transposed to primary pin assignments. Furthermore, the control lines for each "primary" channel must be truly independent. That is, the secondary CTS must not depend on whether the primary RTS is being held ON or OFF by the primary terminal.

The transposition of the pin assignments can be handled by an adapter cable outside the modem (Fig. 3). The control line independence must be provided for within the modem. If the terminal using the secondary channel requires that DSR be ON, the primary DSR can be used for both channels.

One problem that can occur here is that some units have the NEW SYNC control line on the pin assigned for SECONDARY TRANSMIT DATA. In such a case, the NEW SYNC line must be broken (Fig. 2).

The use of the secondary channel as a reverse channel normally occurs in two-wire links in which the reverse channel carries ACK/NAK control characters in response to the text blocks sent over the primary forward channel. RS-232 states that when a secondary channel is so used, it should be slaved to the primary channel by making the ON/OFF state of the secondary RTS the opposite of the ON/OFF state of the primary RTS. Some modems provide the complementary signals—and some don't. Similarly, some terminals require complementary signals—and others don't since they control secondary RTS in the appropriate fashion. When both modem and terminal control secondary RTS, the combination will work. When neither

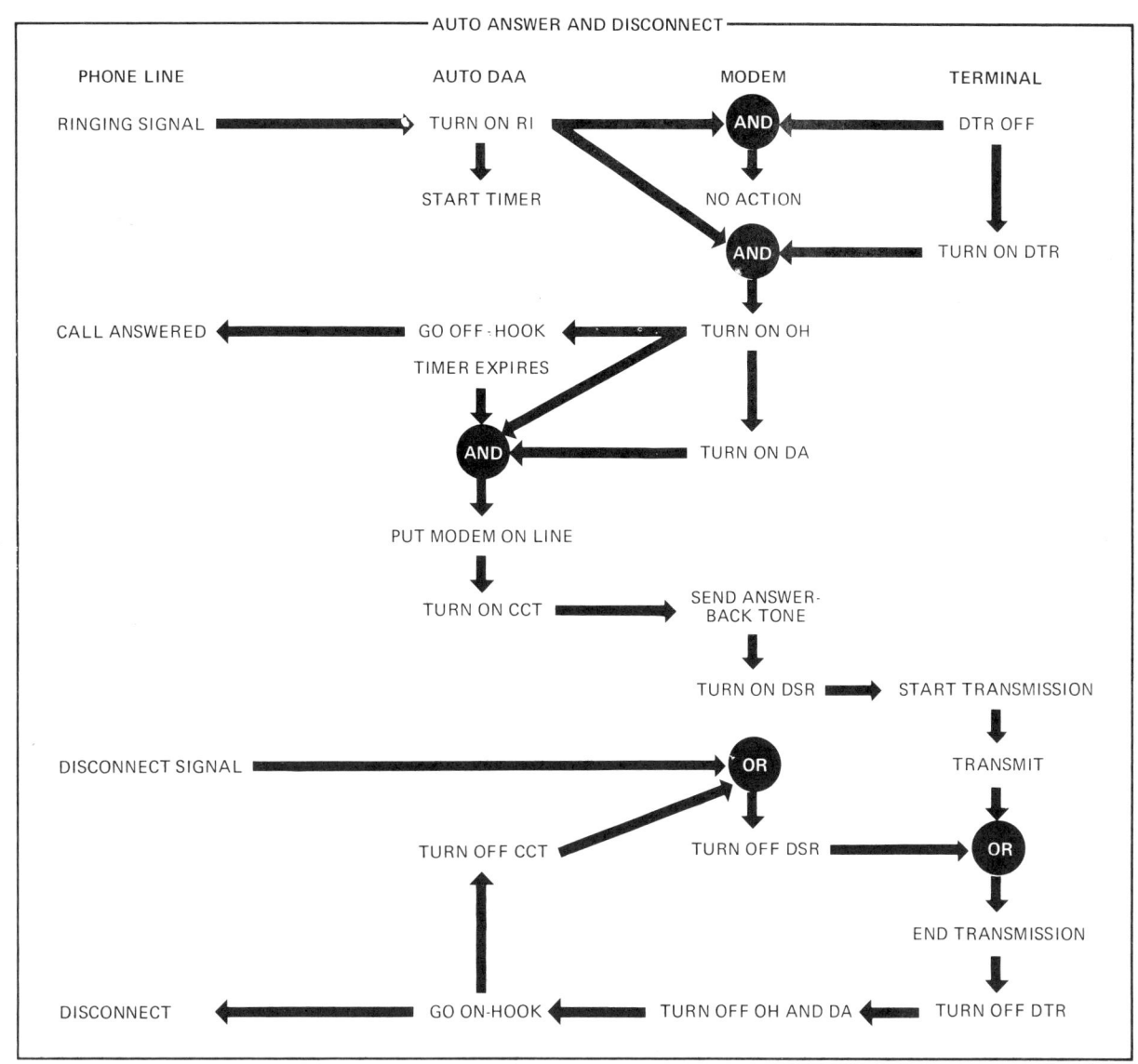

2. Signals on control lines must follow a logical sequence to obtain automatic answer and disconnect, but some terminals and modems may be incompatible and require revisions.

modem or terminal controls secondary RTS the combination will not work.

A potential difficulty can arise when slaving the secondary RTS as prescribed in RS-232: To operate with a reverse channel on a two-wire link, it is necessary that any echo suppressors in the channel be disabled. This is done by sending a 2,025 Hz tone for 400 milliseconds. The total energy in the rest of the band must be at least 12 dB below this 2,025 Hz tone to guarantee that the echo suppressors are disabled. The sending of this disable tone usually takes place automatically when RTS is first turned ON on the primary channel. However, when the secondary channel is slaved to the primary, it will also immediately put energy on the line, in one direction or the other. This energy may exceed the −12 dB limit and thus may prevent the echo suppressor from recognizing the disable tone. The remedy here is to delay secondary RTS ON so that secondary carrier is not transmitted until after the modem has sent the disable tone.

Modem-to-DAA interface

Telephone company tariffs require that when a modem made by an independent company—that is, not the telephone company—is connected to the line then a data access arrangement (DAA) supplied by the telephone company must be interposed between the modem and the line. The telephone company exacts a small monthly charge from the user, the exact cost depending on the type of DAA and the local telephone company tariffs. The data access arrangement has two main functions: It contains a protective network which the telephone company says will circumvent any possible harm that may come to its telephone network or its personnel due to connection of "foreign" devices. And it contains appropriate switching arrangements to permit the phone to be answered and the line to go from voice mode to data mode.

Data access arrangements are available in two types: manual answer or automatic answer of incoming calls. The AT&T designation for the manual DAA is Type 1000A Data Coupler, and its Uniform Service Order Code is CDT. The manual DAA is always used with a telephone set having an exclusion key. The telephone set is used to place and answer calls, and—once the voice connection to the remote site has been made—the exclusion key is used to go from voice to data and back.

The exclusion key on AT&T sets is a white pushbutton in the instrument's cradle. Some non-AT&T sets use a silver button on the front of the cradle to put the telephone on the line (voice), and a red button to put the modem on the line (data).

Adjusting power

The two main problems encountered in the field with manual DAA's is the exclusion key and the power level. The original reason for an exclusion key was to control an extension telephone. For this application, the exclusion key is normally wired at the factory to have "break-before-make" contacts. But this contact arrangement is not satisfactory for voice/data situations, since breaking the dc current loop would cause the telephone to disconnect from the line. Instead, for data communications applications, the telephone set is rewired, usually at installation time, to provide "make-before-break" contacts.

But sometimes the installers forget to make the contact modification. Checking to see whether the DAA will work properly when going from voice to data and back is very simple and can be done before the installer leaves the premises. Simply dial the telephone set's number, wait for it to ring, pick up the handset, make sure a voice connection has been established, and then move the exclusion key to its data position. If the line disconnects, the trouble is most likely that the set's contacts are not properly wired.

As for power, the telephone company specifies that the signal received at the central office must not exceed −12 dBm to avoid creating interference on the line or in the network. Furthermore, the modem transmitter's output (that is, input to the DAA) is restricted to between 0 and −12 dBm by telco specifications. Since different lines will normally have different amounts of power loss between the DAA and the control office each DAA must be individually adjusted as to input power level. Both the DAA and the modem are adjustable, the DAA by telephone company personnel and the modem by the user or by the modem vendor.

The telco installer takes into account local line power loss and straps (wires) the DAA accordingly. The strapping sets a certain maximum power, between 0 and −12 −12 dBm, that can be accepted from the modem. The installer should write that power-limit value on the DAA. The modem output is then adjusted so that its power level matches the DAA's set limit. If the modem power is set above the limit, then the DAA will clip the signal, causing distortion; or if set too low, the noise level may be so great as to raise the bit error rate above acceptable performance.

The important thing here is to make sure that the telco installer has in fact set the DAA power limit correctly and has noted on the DAA the allowable input power limit from the modem. The procedure for setting the power limit applies both to manual DAAs and to automatic-answer DAAs.

Auto answer DAAs

Two automatic-answer DAAs are available: the CBS (or Type 1001A) and the DBT (or Type 1001B) units. The CBS unit uses EIA RS-232-type (voltage) signal levels and is preferred for electronic terminals, multiplexers, concentrators, and computer front-ends. The DBT unit uses contact-closure signalling and is intended for use with simpler and slower electromechanical terminals employing dc current loops. It requires external power, which either the modem or a telephone-company-provided power supply must furnish.

For a user, perhaps the main problem with auto-answer DAAs is confusion because of the many options and alternative modes that he can choose from. Therefore, he must first know what he needs and what he can get. Then the user must make sure he correctly orders what he needs. Finally, the user must make certain that what he receives from the telco is what he ordered, and that those options which are to be accomplished in the field by strapping are in fact wired properly by the installer. The connections between the telephone set and the automatic DAA, the modem, and the terminal are shown in Figure 3.

The standard arrangement for an auto-answer DAA includes a telephone set for dialing. A telephone set is required for call origination unless the system uses an 801-type automatic calling unit. If a telephone set is not required, it must be so specified on the order.

Check the options

Even when the telephone set is included, there are several options related to the ringing arrangement and the exclusion key. In the standard option, the calls are answered automatically by the data terminal with the handset ON HOOK and the link immediately goes into the data mode. The exclusion key is raised to go to voice mode. The standard ringing arrangement for this option is that the telephone

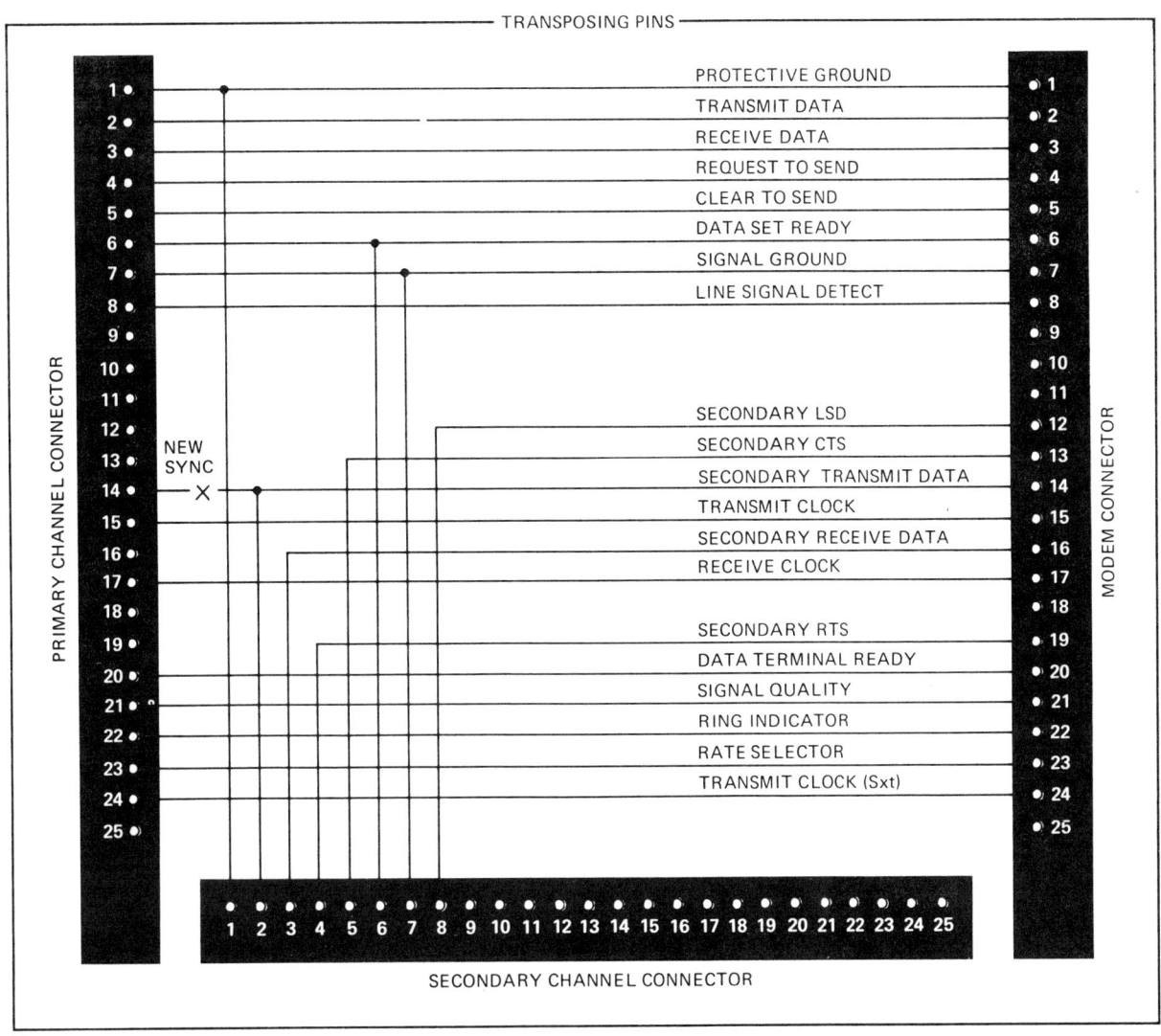

3. Many modems provide a secondary channel that can also serve as a low-speed primary channel operating in either direction; this requires the wiring modifications shown.

does not ring when the call is automatically answered. An optional ringing arrangement can be specified in which the telephone rings when the call is automatically answered. Since the terminal automatically answers the phone in a fraction of a second, the bell hardly rings at all. However, should the terminal not answer because of a malfunction, then the bell will ring in the normal manner and serve as an alarm.

The other option is that the calls are answered on the telephone set. This option is not recommended, however, because it nullifies the auto-answer feature—unless the telephone receiver is always left OFF HOOK and the line transferred to the modem (data mode) by having the exclusion key pulled up. The standard ringing arrangement for this option is that the the telephone rings when the exclusion key is down—that is, in the manual answer mode—and does not ring when the exclusion key is up—that is, in the auto-answer mode. Also, a ringing arrangement can be specified so that the telephone rings when the exclusion key is up (auto-answer mode).

Hopefully, the user will be able to specify what he needs in the way of data access arrangements and will get what he asks for. The important thing is that all strapping made on the DAA in the factory and in the field be done correctly. The proof is whether the telephone (if any), the DAA, the modem, and the terminal follow logical and compatible sequences in call origination, answering, and termination.

The easiest way to prove whether the sequence is correct is simply to dial up the telephone and watch the equipment perform each logical step. To do this is to make sure that terminals, modems, and interface devices have display lights to indicate the ON-OFF states of the lines. ∎

Don't overlook electrical compatibility when mating equipment

Because so many different voltage and current levels are used at the interfaces of data communications equipment, it pays to check the details and avoid the damage

Kuei-Seng Wang and Stephen A. Dalyai
Quindar Electronics Inc.
Springfield, N.J.

ELECTRONICS

It can be shocking enough when trying to start up a data communications link to connect all the equipment, turn on the power, exercise the terminal—and the terminal plays dead. But it can be even more exasperating if a curl of smoke wafts gently out of the equipment, indicating internal damage caused by more serious electrical incompatibility.

If the incompatibility is bad enough to damage the equipment, startup is likely to be delayed for some time. Not only must compatibility requirements be determined, but replacement of the parts can be costly and cause further delay. Ironically, assurance that the terminal is compatible with the line is not difficult; it simply requires an understanding of some essential facts about electrical interfaces between communications equipment, computers, and terminals.

The basic facts, in summary, are these: some types of equipment employ current loops to interchange binary-coded data, and others use voltage interfaces. Obviously, current and voltage interfaces are not compatible with each other. Furthermore, several different values of current are considered standard, and several different voltage levels are used to produce these currents. This situation breeds the prospect for incompatibility. Voltage interfaces are more clearcut because of the adoption of the Electronics Industries Association Standard RS-232-C (see "Where to get RS-232-C," overleaf), which has had a salutary impact because now most modern data communications equipment is designed to its specifications.

However, voltage interfaces that were designed and installed years ago are now being connected to data communications links, and these probably will not be compatible with each other or with RS-232 equipment. Furthermore, current-loop terminals, particularly the popular Teletype senders and receivers and their emulators are used in conjunction with other equipment—such as datasets that operate on voltages. To interface this gear, a translator must convert signals from current to voltage or from voltage to current.

Like so many other aspects of data communications, the present status of electrical interfaces is also the result of evolution. It all started with use of low-speed, electromechanical teletypewriters in common-carrier telegraph systems. And since Teletype machines are the dominant terminals, even today, the current-loop technology associated with these machines is still used in many modern data communications systems. Thus, this

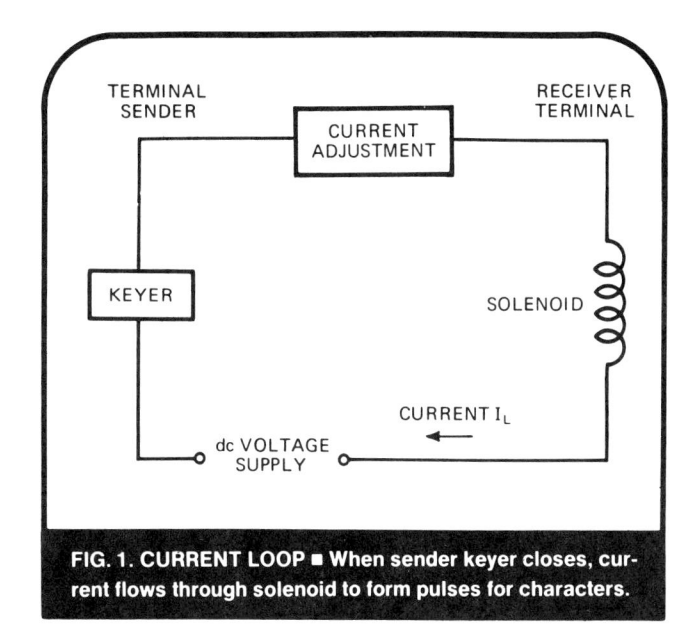

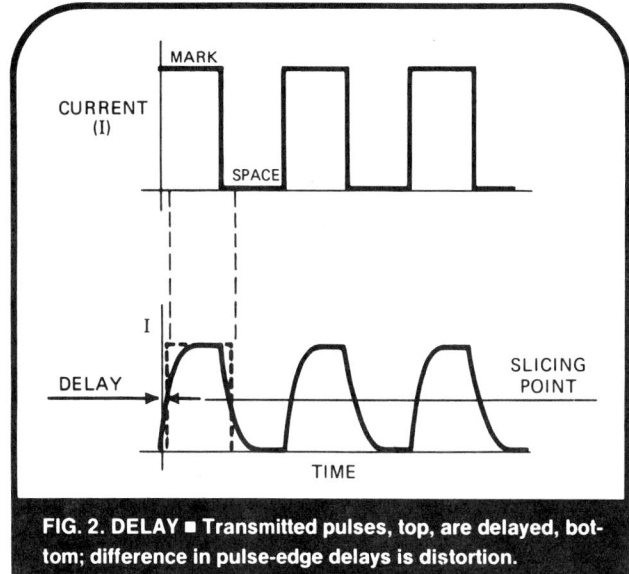

FIG. 1. CURRENT LOOP ■ When sender keyer closes, current flows through solenoid to form pulses for characters.

FIG. 2. DELAY ■ Transmitted pulses, top, are delayed, bottom; difference in pulse-edge delays is distortion.

review of interfaces will begin with the current loop.

Teletypewriter terminals are still electromechanical machines. When receiving, the current through a solenoid causes the machine's mechanism to select and print individual characters formed by specific sequences of openings and closings of the circuit through a switch. When transmitting, the machine's transmitter/distributor converts a character's mechanical representation into an equivalent series of current pulses.

A common current loop, called the neutral telegraph loop, is shown in Fig. 1. The loop consists of a series connection of the sender's keyer (transmitter/distributor), the receiver's solenoid, a direct-current voltage supply, a current adjuster, and the wire in the loop.

The ON current level is determined by the value of the voltage source, divided by the resistances of the wire, the solenoid, and the current adjuster. Since the wire length, or run, would depend on the distance between terminals, and hence determine the resistance, a given circuit would be set at a specified current by the current adjuster. Initially, this adjuster was a rheostat, but now it will more likely be a solid-state automatic current regulator.

The receiver's solenoid follows the ON (closed-circuit) and OFF (open-circuit) current signal produced by the keyer and mechanically translates the solenoid armature movement to select a given character for printing.

A general sequence of ONs and OFFs, also called marks and spaces or 1s and 0s, is shown at the top of Fig. 2. The Baudot code, commonly used with teleprinters that transmit asynchronously, consists of combinations of five 1s and 0s, with each sequence preceded by a start element and followed by a stop element to assure character synchronization between sender and receiver. To confuse further the issue of interfaces, there are numerous versions of the Baudot, or five-level, code.

When machines using two different code versions or two different speeds are electrically compatible, no damage will be done, but messages won't pass between the terminals. However, appropriate registers to accumulate the information bits in a character at the terminal interfaces, together with hardware or software code converters, can solve this problem. Therefore, code and speed compatibility are just as necessary at the interface as current (and voltage) compatibility.

Current differences

Two current standards have evolved over the years: 20-milliampere loops and 62.5-mA loops. (The latter is loosely called a 60-mA loop.) And, the standard voltages that provide these currents are 120, 130, or 260 V dc. As long as the current loops were installed by a common carrier as part of a telegraph service, the user had little interest in electrical parameters.

However, when Teletype machines began to serve as input/output devices for in-house computer systems and later as remote terminals in time-sharing networks and in private—that is, user-owned—data communications networks, then electrical aspects of the current loop became meaningful to users. For one thing, in these applications, the electrical-wire runs were much shorter, usually within the confines of a building, so that the voltage required to produce either 20 mA or 60 mA in the loop was reduced to a lower value ranging from 20 to 60 V dc. Furthermore, additional ramifications were introduced when telephone lines, rather than telegraph lines, became the backbone transmission medium for remote applications.

The telephone line carries ac signals, while the telegraph line carries dc signals, typified by the pulses at the top of Fig. 2. Furthermore, by the use of multiplexing techniques, one telephone line can carry up to 24 75-baud signals of the type generated by Teletype machines and do so at much lower cost than could 24 separate telegraph lines. To take advantage of the telephone-line's cost superiority, though, requires the use of a dataset at each terminal.

Equipment makes signals compatible

A primary function of a dataset is modulation, which is converting voltage pulses representing data into frequencies that can be transmitted over the telephone lines. The reverse operation is demodulation. (Thus, a synonym for dataset is modem, a truncation of the words modulation and demodulation.)

Terminals that produce current pulses therefore need

intermediate equipment to develop the voltage pulses required by the modem circuitry. This interface equipment is shown as the dashed box in Fig. 3. The interface equipment, an electromechanical relay or the solid-state equivalent of a relay, translates the current, I, flowing in the loop connecting it to the terminal into the voltage, V, required by the modulator/demodulator circuit. Interface equipment may be separate or be incorporated within the dataset.

The circuit in Fig. 1 is a simplex loop, actually rarely used in data communications. In practice, terminal operation is either half duplex, which permits the terminal to send or receive, but not both at the same time, or full duplex, which permits the terminal to simultaneously send and receive. The interfacing of a terminal is shown in Fig. 4 for half-duplex transmission and Fig. 5 for full-duplex operation. For the half-duplex mode, two-wire links are needed on the terminal side of the dataset and two wires on the telephone line. Full-duplexing requires three wires on the terminal side and two or four wires on the telephone line, depending on speed and modulation scheme in the dataset. As shown, the interface equipment is assumed to be inside the dataset, along with the voltage source and current adjustment.

As has been mentioned, because of the short distance between the terminal and the dataset, the voltage used to develop the current may be as small as 20 v. Here, then, both the current and voltage ratings must be compatible at both sides of the interface.

What may be overlooked is the specification that tells which side of the interface between the terminal and the intermediate equipment contains the voltage supply. Since the terminal and the interface equipment (or the dataset that provides for a current-loop interface) may be made by different companies, it is not unlikely that the user could wind up with a loop that has no voltage on either side of the interface—and hence no current. It

Where to get RS-232-C

EIA Standard RS-232-C "Interface Between Data Terminal Equipment and Data Communication Equipment Employing Serial Binary Data Interchange" costs $5.10 a copy. Ask also for "Industrial Electronics Bulletin No. 9—Application Notes for EIA Standard RS-232-C" ($2.60), which reviews methods of terminal and dataset operation and introduces graphical descriptions of sequential states of control circuits. Both publications can be obtained from Electronic Industries Association, Engineering Dept., 2001 Eye Street, N.W., Washington, D. C. 20006.

would be even worse, though, if both sides provide the voltage source, for this could lead to an overvoltage in the loop and damage to the components—particularly when the interface equipment uses semiconductor components, which are not nearly as tolerant of voltage or current overloads as are electromechanical relays.

Circuit length limits speed

The speed of teletypewriters in telegraph service is limited to about 75 baud because the resistance, capacitance, and inductance along the circuit creates a lag in the rise and fall of current pulses. In general, the longer the line, the greater the lag. The difference in delay between leading and trailing pulse edges is called distortion. If distortion is great enough, it blurs the distinction between a mark and a space and leads to erroneous decoding of pulse sequences. Pulse rise and fall is fixed by the line's electrical parameters, but the percentage of distortion increases as this pulse rate goes up. Hence, the practical limit of transmission speed works out to be about 75 baud on long lines.

However, when a current-loop terminal interfaces

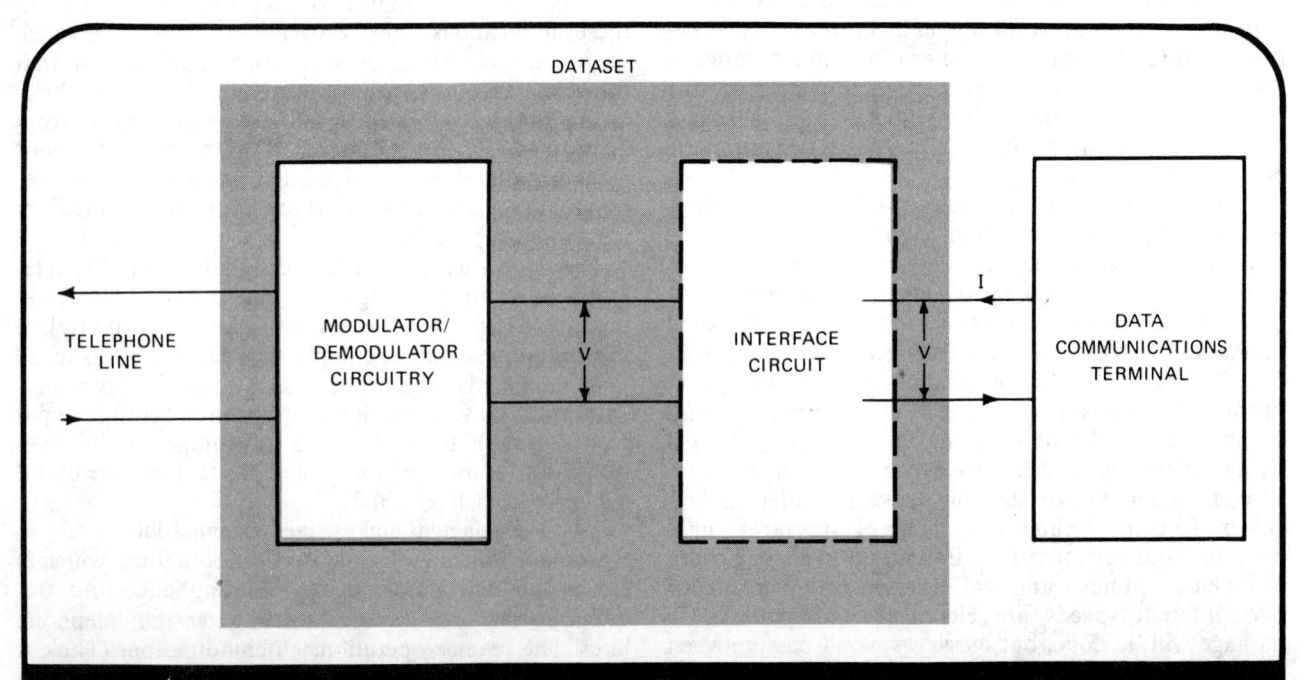

FIG. 3. INTERFACING ■ Terminal can generate current, I, or voltage, V, which—for electrical compatibility—must be translated through the interface circuit to the proper voltage value required by the dataset's modulation circuitry.

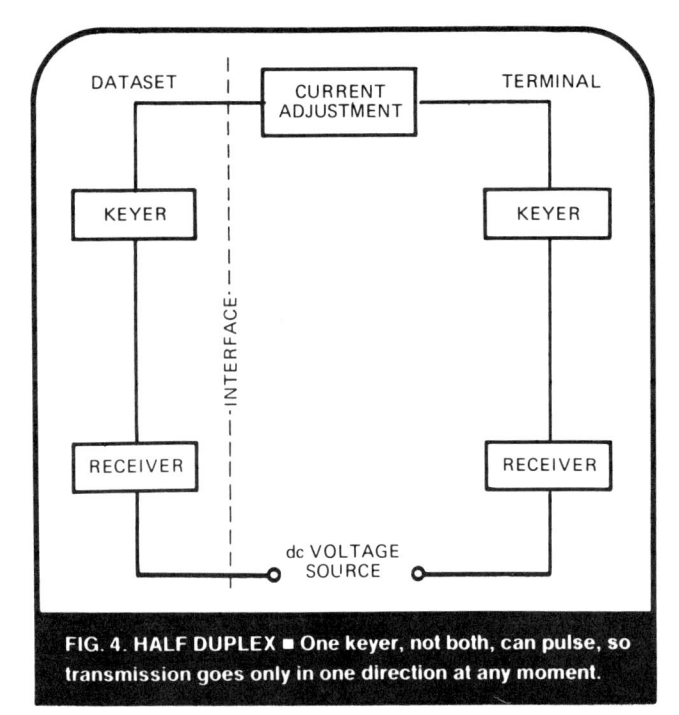

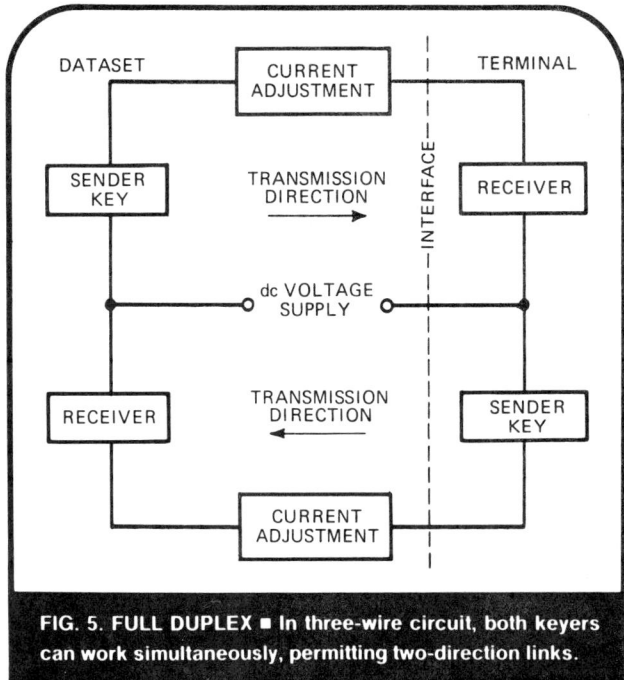

FIG. 4. HALF DUPLEX ■ One keyer, not both, can pulse, so transmission goes only in one direction at any moment.

FIG. 5. FULL DUPLEX ■ In three-wire circuit, both keyers can work simultaneously, permitting two-direction links.

with a dataset—as explained later—the run is short, which reduces the effect of the line's electrical parameters and therefore permits the terminal to operate up to about 110 baud without excessive distortion. Accordingly, increased transmission speed became practical for private networks when terminals were wedded to current-loop technology.

Even so, this speed of 110 baud still was not fast enough for many applications, particularly those in which the terminal was connected to a high-speed remote data processing computer rather than another terminal. Fortunately, faster terminals and faster transmission speeds became available because of advancements in electronics technology: semiconductor circuits operate much faster than relays; integrated circuits provide low-cost memory that can serve as buffers for terminals to accumulate characters slowly and then transmit them at a faster rate; video terminals offer faster writing and larger display areas; and extended code sets—derived from computer design—encourage more efficient, higher-speed transmissions and permit the implementation of more powerful, higher-speed modulation/demodulation schemes.

Newer gear operates on voltage pulses

The newer electronic components and circuits respond to voltage pulses. When the voltage level from the terminal matches the voltage level required by the dataset, the two devices will be compatible and not require the current-to-voltage intermediate equipment shown in Fig. 3 for current-loop terminals.

Voltage compatibility between the dataset and the terminal are assured when equipment is designed according to RS-232-C specifications. This standard defines a mark, or binary 1, as any interface voltage between –3 and –25 v dc, and a space, or binary 0, as any voltage between +3 and +25 v. The standard also specifies the signal and control functions assigned to each of the 25 pins available in an RS-232-C connector. (The control functions, not required in current loops, make sure the datasets at each end of the line are ready to carry the data on the signal leads.)

Before the adoption of RS-232, interface voltages had a variety of levels, to some extent based on the state of electronics technology that existed when the interface was designed. As examples, some interfaces used 0 v to define a binary 1 and +5 v for binary 0. Alternatively, 0 v could represent a binary 0 and +5 v could mean binary 1. A similar confusion existed when the electronic circuits produced signals that were either 0 or –5 v. Furthermore, other types of integrated-circuit devices use +5 v for one binary state and +18 v for the other binary state.

The voltage incompatibility that can arise when trying to interface two of these different voltage and polarity levels can be overcome by introducing intermediate equipment, such as high-speed electromechanical relays and voltage sources or operational amplifiers to bring voltages to proper levels and polarity.

The EIA RS-232-C standard stipulates a maximum data signaling rate of 20,000 bits per second and suggests a limit of 50 feet of cable be run between the dataset and the data terminal to make sure the distributed capacitance between wires does not create excessive distortion of the voltage pulses. However, often the dataset may be located on one floor of a building and the terminal on another floor, thus exceeding a 50-feet run. Here, then, it may be desirable to translate the terminal voltage to either 20 mA or 60 mA, run a neutral loop to the dataset, and then translate the current back to the RS-232 voltage. The Teletype Corp. makes an "EIA-to-Neutral-Loop" converter-modification kit for use when its machines are connected to the RS-232 electrical interface on datasets.

When current loops are run inside a building, the sharp current transitions resulting from current pulses can cause noise in adjacent circuits, so that it is advisable to keep these telegraph loops away from telephone lines carrying either voice or data.

part 10
data data data data data data data data data data

regulations and policy

Regulations and tariffs

Working across state lines, data-communications systems are shaped by intrastate and interstate regulations and tariffs, all undergoing rapid and complex changes

John C. Duffendack,
*ComShare Inc.
Ann Arbor, Mich.*

User demand for improved data-communications systems and reduced costs has had as much impact on communications regulations and tariffs as on communications technology, operations, administration, and maintenance. As a result, three major shifts in regulatory attitude have occurred during the past few years, permitting: more competition for established common carriers, the connection (formerly banned) of customer-owned equipment to telephone lines, and more participation by users in regulatory and tariff proceedings.

Anyone who wants to understand regulations and tariffs completely must wend his way through a labyrinth of seemingly incomprehensible documents. Certainly, such documents require interpretation at both technical and cost levels. Their complexity comes to the fore particularly when a user has some special need, innovative idea, or critical communications requirement, since the user and the common carrier may not agree on interpretation of the tariffs. Still, some order can be made out of the existing situation, and the major aspects of tariffs and regulation of data communications can be summarized.

In 1934, Congress created the Federal Communications Commission (FCC) and defined the FCC's authority for the regulation of the supply of common-carrier interstate communications services. The supply of intrastate communications services, for locations within a state, are regulated by each state's public utility or public services commission. The only exception to intrastate regulation is in Texas, where regulation is performed by each municipality.

Communications common carriers are granted monopolistic powers in a given area. In return, they are required to file, before cognizant regulatory bodies, certificates of necessity and convenience, as well as tariffs for services. The tariff document for communications services records the policies of the communications carrier, which is usually the telephone company, and describes and prices the standard products and services.

As a general rule, the state regulatory commissions adopt the precepts of the Federal Communications Commission. Perhaps more pragmatically, the Bell Telephone operating companies handling intrastate services follow the lead of their owner, the American Telephone and Telegraph Co. (AT&T), which handles interstate traffic.

Two differences between interstate and intrastate regulation of communications are worth noting:
- Changes in an intrastate service, or the introduction of new intrastate services, lag the change in the compa-

rable interstate service by months, if not years.
- Prices (rates) are generally higher for intrastate than for similar interstate services.

A common-carrier service is generally considered intrastate if it can be used for communications wholly within the state. For example, the telephone handset and attached line (local loop) are intrastate services even though calls can be made to locations in other states. Only the rates for interstate calls are subject to Federal regulation. This fact has interesting ramifications when one studies the interconnection of various devices to the switched network, since interconnection is controlled at the state level.

This distinction between intrastate and interstate communications regulation is not merely of legal concern. Cost differences are of major consequence. Consider a specific communications requirement for leased series-3000 voice-grade lines between San Francisco and Los Angeles, and Los Angeles and Phoenix (see figure). The intrastate link at $2,276 per month costs almost four times more than the interstate link at $598 per month—and both links are almost equally long. (In fact, the price of the intrastate line from San Francisco to Los Angeles is about the same as for an interstate line from Los Angeles to New York.) However, if the user requests the carrier to connect the links in Los Angeles, the San Francisco-to-Los Angeles link becomes interstate under the relevant FCC tariff and its cost drops to $589 per month—for an over-all system savings of $1,627 per month. For the telephone company to connect the two links, however, the user must have a legitimate requirement for through service between San Francisco and Phoenix.

Several widely used interstate communications-services tariffs filed with the FCC are listed in the table. Because these tariffs include many common principles, and state tariffs tend to follow Federal tariffs, it will be instructive to study major aspects of one important tariff: AT&T's interstate private line tariff, FCC #260.

FCC tariff #260

USAGE. One of the most significant provisions in this tariff specifies the allowable uses of a leased, point-to-point, voice-grade telephone line, as may be used for dedicated data transmission at speeds up to 9,600 bits per second (b/s). Basically, this provision says that a telephone-service customer is permitted to use this service for transmission of communications to or from various intracompany locations, including those of wholly owned or controlled subsidiaries, where such communications are directly related to corporate business. Further, it allows simultaneous transmission of information relating directly to matters of common-interest parties—in the same general line of business—who are connected to the communications facility. In any case, the lines must be terminated (at least on one end) on the user's property. The important point to note here is that the tariff states that the user may employ his dedicated communications facilities for himself and may not provide a communication service to others. The only exceptions to the above rules are for joint use, discussed later, and for use by the government, members of a commodity or stock exchange, members of an electrical power pool, and for certain uses by a licensed aeronautical-communications company.

RELIABILITY AND RESPONSIBILITY. The tariff limits the liability of the common carrier. It specifies that the common carrier assumes no liability for damages arising out of mistakes, omissions, interruptions, delays, or errors associated with the purchase of communications service, other than an amount equivalent to the proportionate charge for the service period in which the defect occurred. The courts, of course, have held that these tariff provisions do not protect the common carrier in case of its negligence.

The tariffs say nothing about the quality of service the carrier must offer its users. There are no minimum provisions for reliability, error rates, noise levels, and the like. Tariff #260 does specify the bandwidth and delay distortion characteristics for the various levels of conditioning of a voice-grade line. However, this is a far cry from actual specification of those communications-channel characteristics that determine transmission reliability.

Service from a common carrier is provided on a minimum-period basis, generally one month. Generally, credits for service outages are given on the pro rata basis, starting from the time the customer notifies the common carrier of the outage.

Under the tariff, the customer must provide space and power for common-carrier facilities installed on his premises, and permit access to common-carrier employees for the installation and maintenance of equipment.

In addition, of course, the customer is responsible for all payments. By written notice to a customer, the common carrier may immediately discontinue providing communications service for non-payment or for violation of various conditions governing the furnishing of service. Disputes leading to service termination over the past few years have centered around violations of the tariff's interconnection or usage provisions.

JOINT USE. The general intent of the joint-use tariff provision is to allow increased utilization of the country's telecommunications resources, while precluding the sale of communications services by entities other than common carriers. In effect, the primary user is allowed to reduce his communications costs by allowing other users to share his unused facilities without allowing him to profit by the supply of a communications service. The important criteria for joint use are that:

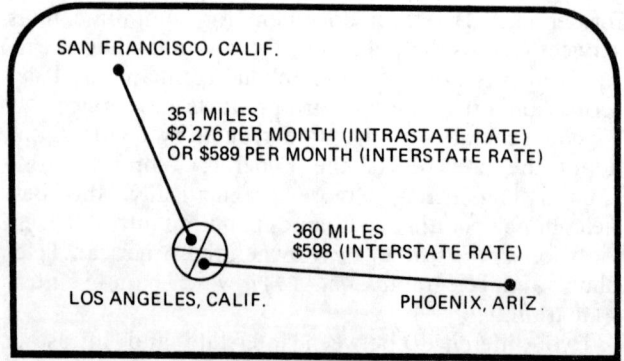

Anomaly. Two separate intrastate links are expensive, but adding a switch in Los Angeles for through traffic cuts overall cost.

SAN FRANCISCO, CALIF.
351 MILES
$2,276 PER MONTH (INTRASTATE RATE)
OR $589 PER MONTH (INTERSTATE RATE)
360 MILES
$598 (INTERSTATE RATE)
LOS ANGELES, CALIF.   PHOENIX, ARIZ.

- The primary user and each joint user have a legitimate need for the facilities (i.e., communications to or from each user and directly relating to his business).
- The joint user has a station or service terminal on his premises and a through connection to all portions of the network he is sharing. The common carrier bills each joint user a pro rata charge for his share of the common-carrier facilities, as specified by the primary user. In general, there is also a flat charge for each joint user on a shared network.

At present, confusion and disagreement exist between the carriers, the regulators, and prominent users as to the applicability of procedures involved in the joint-use arrangements. In looking toward the establishment of a joint-use arrangement, the primary user should obtain approval from both the common carrier and appropriate regulatory bodies. There are some services where joint use is precluded. For example, joint use is specifically not allowed on WATS lines, foreign exchange lines, and certain wideband facilities.

Tariff #260 aside, two major regulatory changes involve network interconnection and the supply of computer/communications services. The 1968 Carterphone ruling by the FCC required the telephone industry to remove its proscription on the connection of customer-owned (or leased) equipment to a telephone service offered by a carrier. This shift in regulatory attitude has resulted in a rapid expansion in the range of options open to the communications user, and has had a significant impact upon the public interest and convenience in allowing more efficient and versatile uses of the telephone network. Now the customer is no longer a total captive of the telephone company or its manufacturing affiliate when it comes to choosing the equipment that will best meet his need for voice or data transmission. However, certain restrictions still govern the use of non-common-carrier, or independent, equipment connected to communications services:

- On the dial-up network, the customer-owned equipment must be connected through a carrier-provided protective device known as a data-access arrangement (DAA), for which there is a charge of approximately $2 to $8 per month. At present there is no requirement for a protective device when connection is made to private-line service.
- Further, various restrictions are imposed upon the signal introduced into the network—whether switched or private—by devices connected to it. These limitations are that the carrier will specify the total signal power permitted within certain frequency bands. Where customer-owned equipment is connected to the switched telephone service, additional criteria must be met to assure proper operation of the common-carrier switching facilities, echo suppressors, and billing equipment.

While at present there is no requirement for a protective device on private lines, tariffs are pending before the FCC for the provision of such a device—in addition to constraints on signal levels. The pending tariffs also require that, on any given circuit, interconnected devices must all be supplied by Bell or all be supplied by independent manufacturers. They cannot be mixed, regardless of their technical compatibility.

In March 1971, the FCC released its Final Decision and Order with respect to regulations that control the interdependent relationship between computer and communications service. The FCC stated that it chooses not to regulate the independent suppliers of computer services and, further, that it will impose certain restrictions on those common carriers who participate in the computer-services business. The decision also defines new categories of service known as hybrid data processing and hybrid communications, depending on whether the intent of the service is primarily data processing or primarily communications, with regulation imposed on the latter. This decision has been affirmed by the FCC after a Motion of Reconsideration and is (as of October 1972) on appeal before a Federal Appellate Court.

Looking ahead

The watchword of the future in data communications is competition. Besides allowing competition from interconnected devices, the FCC has also opened the door to special-services common carriers, such as MCI Communications Inc. and Datran, to construct certain high-traffic routes paralleling those of the telephone companies. With the introduction of competitive carriers has come the disallowance of the original concept of nationwide rate averaging. Further, the FCC has approved open entry for domestic-satellite systems, which will undoubtedly result in the establishment of several competitive long-haul satellite transmission facilities by 1976.

Perhaps the key issue in the future will revolve around clarification and modification of existing tariffs and regulations for joint use. Of particular concern are the questions about the sharing of rates, the sharing of channels, and the brokerage (or reselling) of data-communications services by a company not qualified as a common carrier.

Because of these changes in the regulatory situation, there is now great concern about establishing a fair competitive environment. That is, the regulatory commissions want to be sure that prices for a carrier's non-competitive products do not subsidize the prices for its competitive products, and that carriers do not establish discriminatory pricing between classes of customers.

Thus, carriers and the regulatory commissions are taking action to prepare for this increasingly competitive marketplace. The carriers are becoming more concerned about their costs and cost-related pricing, and the FCC is requiring market information and cost analysis to accompany all proposed tariff changes. This is bound to promote greater operating efficiencies within the carriers and to bring benefits for the users. □

| COMMON INTERSTATE TARIFFS ||
FCC Tariff	Subject
254	Western Union private-line services
255	AT&T administrative rate centers and central offices
259	Wide-area telecommunications service (WATS)
260	AT&T interstate private-line services
263	Long-distance-message telecommunications service

Dilemmas in the nation's telecommunications policy

Bernard Strassburg
Federal Communications Commission
Washington, D.C.

The chief of the FCC's Common Carrier Bureau reveals the major issues in the regulatory environment that have to be settled to provide for fair and equal competition, yet retain the advantages of natural monopoly

ESSAY

Because of the fast growth of remote data systems and the evolution of computer networks, American industry is moving toward greater reliance upon the nation's communication facilities and services. Technology and innovation have impacted the design and application of terminals, peripherals, and even computers connected in networks over both private and switched communications services. This impact has penetrated the regulatory environment of communications and raised a host of fundamental issues that go to the heart of the nation's telecommunications policy.

Indeed, technology may well be challenging the conventional premises of the regulatory process, as well as questioning whether regulation is an adequate tool in resolving the complexities associated with rapid change and innovation. The Federal Communications Commission has been caught up in the interplay between computers and communications, as illustrated by the developments associated with the FCC's Interconnection and Specialized Carrier decisions, as well as its Computer Inquiry.

Until 1968, Bell System practices generally foreclosed customer-owned and -maintained equipment being electrically interconnected to the nation's switched telephone network. This indiscriminate ban was premised on the concept that, with few exceptions, responsibility for the maintenance, installation, innovation, and repair of all equipment connected to the telephone network must remain within the control of the common carrier alone. It followed that customer ownership of any such equipment violated that general policy. Subscribers who persisted in owning their own equipment were exposed to the denial of telephone service, which, in fact, meant denial of hard-wire-coupled data communications transmission over the phone lines.

But in 1968, Bell's interconnection policy was challenged by the FCC's Carterfone decision. That decision overturned the carrier's general ban on customer-owned equipment and held that Bell's blanket restrictions under its tariffs had been unreasonable and unlawful. Bell announced its willingness to cooperate with users and accommodate them by allowing greater access of customer-owned equipment to the telephone exchange—provided that the network would be protected from harm and damage. At the time, the precise method of securing this protection was left unspecified by regulatory authority. AT&T, however, specified as a tariff requirement that such protection must be provided by voice and data couplers (that is, connecting arrangements and network control signaling units) to be provided, installed, and maintained solely by the carrier.

The implications riding on the FCC's Carterfone decision were fundamental and far-reaching. First, the commission broadened the equipment options available to the user of communications service—the need for which alternatives were voiced by the data processing industry in the earlier FCC Computer Inquiry.

Second, the decision provided independent suppliers of terminal and computer equipment with an incentive to develop, manufacture, and sell new hardware—the market clearly spawned by the growing and sophisticated demands of both the telephone user and the data industry. The result was startling. A host of new and improved datasets, multiplexers, communication processors, video terminals, and answering devices became available in the independent market place.

The lesson of the FCC's Carterfone decision has been instructive: before 1968, the market opportunities for certain equipment were foreclosed; after 1968, new opportunities became apparent. Carterfone literally resulted in the creation of a new industry—the interconnect industry, which serves both data and voice markets. And in the best tradition of a competitive market economy, new companies introduced new equipment, expanded capability, and lower prices—changes that ultimately benefited the communications user.

The Bell System's market response and reaction to the FCC's Carterfone decision was at first delayed. But the 1970s now reveal a discernible pattern of response to the interconnect industry. The Bell system reevaluated and revised its prices on existing datasets and introduced a new series of datasets. In the voice field, Bell marketed a new family of private branch exchanges (PBXs), some imported and others manufactured on a crash basis. AT&T realigned its Bell Telephone Laboratories, emphasizing product and marketing development. Recently, Bell introduced a new CRT terminal, through Western Electric's Teletype Corp. subsidiary. Finally, Bell introduced flexible pricing plans and unbundled tariffs dealing with trunk lines, PBX hardware, and station lines.

Amidst this ferment, there are a number of unresolved regulatory questions regarding the data and voice couplers devised by AT&T to protect the operational integrity of its network. From the time of their initiation, these protective arrangements have been variously attacked by the suppliers and users of interconnect equipment as burdensome, unnecessary, and anticompetitive. Thus, the FCC has been exploring various programs of standards and equipment-certification procedures that would permit the user to connect his equipment to the network without the need for carrier-provided couplers, while, at the same time, assuring that the telephone network would not be impaired.

The Bell System initially advocated the establishment of a program of "certification" as the principal, if not sole, precondition to a further liberalization of interconnection. However, of late, certification is no longer central to Bell's position. Bell has now brought to the forefront of its opposition repertoire the demand for a comprehensive examination of the economic consequences of any relaxation in existing interconnection arrangements. As voiced by AT&T chairman John D. deButts at the company's April 1973 stockholder meeting, AT&T ". . . will oppose implementation of any system of certification until the basic questions about its long-term service and economic consequences have been satisfactorily answered . . ."

Further, the National Association of Regulatory Utility Commissioners (Naruc), representing the state regulatory agencies, has supported AT&T's resistance to liberalized interconnection, ostensibly out of a concern for the potentially adverse effects of liberalization on the exchange rates and revenue requirements of the local telephone companies. However, Naruc is participating in the FCC's ongoing Federal-State Joint Board interconnect proceeding, which is expected to treat not only the economic issue, but also whether there should be further expansion in customer options.

In sum, the FCC's policy of promoting a wider range of customer options in the terminal field has called into play an equally wide range of responses from the car-

"should all carriers...have access to the intercity channels of AT&T on uniform terms and conditions?"

riers and state regulators. The various proceedings now before the commission focus on the merits and appropriateness of these responses. The outcome may well be expected to determine whether and to what extent the commission's policy of promoting competitive supply in the interconnect field will produce meaningful and expanded public benefits.

A second policy change by the FCC occurred in 1970: the so-called Specialized Carrier decision. Here, the commission found that the needs of the public, particularly the computer-user industry, were not being satisfied by the Bell System or other communications carriers. The FCC sought to spur the development of specialized carriers to provide the potential for customized communications services and to promote the introduction of a switched digital service on a nationwide basis. The existing carriers opposed such a move on grounds that specialized carriers would serve no unmet need, would duplicate facilities, would waste frequency resources, and threaten rate-averaging principles.

In its Specialized Carrier decision, the commission drew much of its input from the computer industry. Responses to the FCC's Computer Inquiry confirmed the public requirement for new services in the digital and customized-communications market. The validity of this requirement was lent further credence by the studies and recommendations of a Presidential Task Force on Communication Policy.

As in Carterfone, the Specialized Carrier decision impelled a response from the Bell System on a broad front. Bell promptly announced plans for an end-to-end digital-data service. The carrier has recently proposed to the FCC a major restructuring of its private-line voice-grade schedules by which it would establish two separate rating classifications for those services. Bell sought a lower rate classification, to be applied to channels on high-density routes. The higher rate classification would apply to all routes of lesser density. This "Hi-Lo" approach is a major departure from the Bell System's historical practice of basing its uniform rates on nationwide cost-averaging. This tariff shift is in response to the competition that Bell anticipates from the private-line offerings of the specialized carriers.

Feeling threatened by competitive offerings, Bell in recent years has been giving greater and greater emphasis to incremental costs in its rationale for pricing any of its services in competitive markets. At the same time, local Bell System companies have been pressing for increases in the charges for local distribution channels required by the specialized carriers in order to deliver their intercity services to their customers.

Still more recently, Bell has proposed to relax its historic ban against "resale" of its interexchange private-line channels to the limited extent of permitting the lease of such channels by so-called "value-added" data communications carriers. These new entities seek to use the new technology of "packet switching" for the provision of data communications services tailored specifically to the transaction-oriented needs of the data communication user. By permitting "resale" to this limited extent, Bell will promote the greater use of its intercity supply of channels for services which, on the one hand, do not directly compete with any of Bell services, but, on the other hand, more directly compete with the services of the new specialized intercity carriers.

As of Aug. 1, the FCC had not yet decided on the regulatory actions to be taken with respect to the Hi-Lo proposal, nor had the agency determined whether or not the selective relaxation of Bell's resale prohibition is nondiscriminatory and otherwise legal. The resolutions of these proposals are crucial to according a fair and realistic test of competition as a regulatory device to achieve improved, expanded, and more economical communications services. It is clear, however, that the regulatory orchestration of challenge and response has posed a series of public policy questions that now reside at the doorsteps of the Federal Communications Commission. These questions are complex—laden with matters of law, engineering, accounting, and economics.

What are the more critical unresolved issues posed by Bell's responses to the national policy objectives formulated in recent years by the FCC? First, the ultimate question is whether or not Bell's responses will frustrate a fair test of the commission's decisions to introduce competitive sources of supply into both the terminal-equipment market and the specialized-carrier market.

The commission's stated policies have recognized the right of the Bell System and other existing common carriers to participate fully and fairly in the competitive markets. However, if, by the exercise of that right to compete, Bell is successful in blocking or cutting off competitive entry "at the pass," then the FCC's policy objectives to broaden the options of users and to expand the opportunity for innovation will, of course, experience a premature demise.

At the same time, market foreclosure by a variety of strategically timed repricing actions and other selective tariff changes by Bell could also have a devastating effect upon those suppliers of new equipment and services who are in the process of early market penetration. Obviously, this would be a paradoxical cost to pay for the removal of restrictions to market entry.

Second, the commission now finds itself in the position of attempting to locate and define a demarcation line between competitive and noncompetitive markets.

"the FCC has...embarked on a study of corporate structures in the common-carrier industry..."

"the FCC must address itself to the size and dominance of the Bell System"

This definition is, in fact, a major issue. On the one hand, the FCC must define the market that exhibits characteristics of a "natural monopoly," which presumably rules out competitive entry. On the other hand, the commission must define those markets sustaining competition and rivalry.

A third policy question is a corollary to the second. After defining the two markets, the problem then is how to determine the ground rules whereby a regulated carrier can compete fairly and equitably in a market judged to accommodate free and open access. This problem is particularly difficult because of the overriding fact that the Bell System simultaneously occupies and straddles the competitive and monopoly markets.

It is not surprising, then, that the question of a dual-market occupancy raises the issues of cross-subsidization, cost allocation, proper pricing, and the effect of monopoly markets on the growth and viability of competitive markets. The possibility of Bell's monopoly services underwriting and supporting non-telephone services continues as an issue that permeates various rate proceedings and dockets before the FCC and state regulatory agencies.

A fourth question follows from the third. Can regulatory policy protect, by restriction, a new and incipient market, a market that of necessity requires some nurturing until it becomes self-sufficient? The fragility of the new market and industry is abundantly clear: this competitive market currently reflects a very small fraction of Bell's $20 billion annual monopoly revenues. To permit a realistic chance for competitive-market growth and survival, the FCC must address itself to the size and dominance of the Bell System. Should the common carriers be restricted from participating in competitive markets until the infant industry can survive on its own? Is it sufficient to pronounce a policy of "full and fair competition," which promotes "every man for himself," but which, in fact, enables the giant to admonish the midget?

Such a suggestion of market restriction, is of course, hardly unprecedented. The commission in its Domestic Satellite Policy decision of December 1972 ruled that AT&T's satellites could not engage in competitive services for three years. The reasons for such a moratorium were clear and unequivocal in view of the Bell System's established pervasiveness in domestic communications and the potentially adverse effects of that pervasiveness on the opportunity the commission sought to afford others to develop a market for satellite services.

A fifth question also turns on the guideline of "fair and equitable" competition. For example, should all carriers—Bell and non-Bell—have access to the intercity channels of AT&T on uniform terms and conditions? Should the specialized carriers have access to the local-distribution facilities of the Bell System operating companies with the same right and ease of access as does AT&T in reaching the end user?

The issue of defining full and fair competition can also be extended to the nonprofit Bell Telephone Laboratories. Currently operating on an annual budget of some $400 million, which is funded entirely by the general rate-payer, the Bell Telephone Laboratories is devoted to innovation in and advancement of telecommunications technology. As such, it represents the nation's principal, if not the only, research and development facility in the telecommunications field that is supported by nongovernmental funds. Over the years, the fruits of its research and findings have been converted to improvements in the plant, equipment, and services of the Bell System, thereby contributing to the efficiencies and economies of the nationwide telephone network.

But if the Bell System is to participate fully and fairly in the expanding and diversifying communications markets, should not this vital national resource, represented by the laboratories, be accessible to all entities engaged in the development and supply of communications equipment and services? Should the laboratories share the results of research, rather than restricting them to Western Electric, the manufacturing and supply unit of the Bell System? Should the various specialized carriers and other competitors be permitted access to the research and development capabilities of the laboratories on the basis of appropriate contractual arrangements?

The principle of fair and equitable competition inevitably poses such questions. The FCC has recently embarked on a study of the corporate structures in the common-carrier industry and these questions, with many others, will be addressed in that study.

The significant issues that confront the nation's communications industry and markets clearly challenge the effectiveness of the regulatory and other governmental processes to search out timely and meaningful resolutions. The regulatory resources of government at all levels are already overtaxed in dealing with problems of the communications field, and those problems continue to grow in volume and complexity.

The manufacturers and users of data communications systems have a vital stake in regulatory performance. It follows that they cannot afford to take for granted or on faith that governmental actions will be forthcoming and wise. More is required. The importance of the issues requires maximum public support of, and participation in, the regulatory process in the search for the right answers. On those answers ride the potential for future new markets, new services, and new choices.

AT&T's new attitude might have forestalled Justice's suit

Bernard Strassburg
Washington, D.C.

Commentary The antitrust action brought by the U.S. Department of Justice against AT&T will no doubt rate as one of the most significant events of this decade for the communications industry and the public it serves. There is no way of predicting now what that action will ultimately mean—if anything—in terms of restructuring the nation's communications markets. The outcome depends not only on the staying power of the Department of Justice but also on the future attitudes and initiatives of the Bell System.

But, regardless of how we may speculate as to its ultimate outcome, the antitrust suit is an example of how a corporate ideology and parochial values can distort the judgment of management as to where its best interests lie when it is as vulnerable to antitrust challenge as is the Bell System.

That the Bell System has successfully withstood or deterred an antitrust challenge until now, in spite of its dominant position in communications, can be attributed to several factors. First, the Bell System has cultivated and maintained a favorable public image by giving the nation reliable telephone service at rates that have made the service universally accessible. By doing so, it has escaped public and political antipathy to monopoly.

Second, Bell has generally exercised an enlightened restraint in the use of the economic and political muscle inherent in an enterprise of its size and pervasiveness.

Third—and as its most important line of defense—the Bell System has conducted its affairs under the oversight of, and subject to, the policies of the regulatory processes. Although those governmental processes have not always functioned with effectiveness, they have been generally respected by the Bell System and accepted by the public as an adequate safeguard against the excesses of monopoly power.

Then, beginning in 1968, the Federal Communications Commission formulated a series of new policies, which in effect required the Bell System to make room for some competition in supplying selected markets with business and data communications services and equipment. It was the commission's judgment, supported by the evidence of several proceedings, that the dynamics of computer and communications technologies were spawning a diversity of new communication requirements that could not be satisfied by the monolithic "plain old telephone" environment.

These policies unleashed the new interconnect industry which caters to the markets for customer-owned systems and equipment, and it paved the way for the entry of a number of new common carriers specializing in the provision of intercity private-line services and switched data networks.

Initially, AT&T demonstrated a willingness to adapt and compete. Although it questioned the wisdom of the new policies in some respects, AT&T made no appeal to the courts. On the contrary, its prior chairman, Harold I. Romnes, publicly acknowledged that competition could be a public benefit and could spur even Bell to give better service. He conceded that there are valid distinctions to be made between those markets of the Bell System that are natural monopoly—for example, the switched telephone network—and those markets where competition can be productive.

But this initial perspective turned out to be short-lived. The new interconnect industry began to gain momentum in market penetration with PBXs and key systems, datasets, and other terminal devices offering features and price alternatives not previously available to users. The new terrestrial and satellite specialized common carriers began to move into various stages of operational reality, challenging AT&T's preeminence in the field of intercity communications.

Then, in 1972, major changes were made at the

top management and policy-making levels of AT&T. At this time, the Bell System embarked on a course of conduct by which it breached its traditional lines of defense.

What was initially an AT&T attitude of acceptance and accommodation was transformed by AT&T's new leadership into a policy that seemed oriented toward the containment and, arguably, the extinction of competition. Asserting its own right to compete, the Bell System initiated a crash program to develop and market a wide variety of new and diversified lines of customer terminals. It substantially reduced its tariffed prices on data modems and other terminal devices. It launched an intensive advertising and customer-relations program construed by its competitors to undermine consumer confidence in the reliability of the products and maintenance of the independent supplier.

As the new specialized intercity carriers attempted to launch their competitive offerings, they, too, encountered a number of difficulties in dealing with AT&T. At the time the FCC set its policy in this field, it recognized that the new entrants would initially and for some time be dependent upon the local facilities of the Bell System companies in order to terminate their services. The FCC's policy therefore called upon Bell to provide those local-distribution facilities on reasonable terms.

In view of Bell's monopoly control over those essential facilities, the FCC stressed that it would not condone any practice by a Bell company that discriminated in favor of its own affiliate. The new specialized common carriers soon discovered, however, that they were unable to obtain local distribution channels with the same readiness and on similar terms as AT&T's Long Lines department obtained such facilities for its competing intercity private-line services.

These and other counteractions by AT&T to competition had their inevitable consequences. They generated a large number of costly and time-consuming proceedings before the regulatory agencies and the courts. They delayed the efforts of the interconnect industry and the specialized carriers to market their services and products. And potential investors were understandably reluctant to provide needed capital to the new enterprises because of the Bell System's announced opposition to competition. Many aspiring entrants into the interconnect markets gave up the struggle and abandoned their plans.

The Bell System companies found themselves in increasing numbers of civil antitrust damage suits brought by its competitors, who charged Bell with anticompetitive conduct. Finally, AT&T's conduct precipitated a steady stream of complaints into the Antitrust division of the Department of Justice—which has now responded with its massive antitrust action against the Bell System.

In an apparent, but belated, effort to defuse the then impending antitrust action and also after suffering a rebuff in the courts, in early October 1974 AT&T reversed its hard-nosed approach to competition. Since then, it has been in almost continuous negotiation with the specialized carriers, under FCC aegis, to satisfy their many complaints regarding their access to Bell's local-distribution facilities. The negotiations have been going well and seem to demonstrate a renewed willingness by AT&T to adapt to and live within the competitive rules formulated by the FCC.

Some initiatives have also been taken by AT&T in recent months to provide less burdensome interconnect procedures with respect to certain classes of customer equipment and thereby improve the marketing opportunities for such equipment.

One may speculate, therefore, whether Justice would have brought its suit at this time, had AT&T initiated corrective measures much sooner. AT&T's conduct in these areas, among others, is specifically cited in the Government's complaint to support the charge that AT&T is monopolizing telecommunications services and equipment.

In any event, AT&T would appear to be well advised to continue this process of objective self-reexamination. Many of its corporate values and habits are rooted in its monopoly structure. As such, they may no longer be compatible with AT&T's participation in those markets carved out by the FCC as suitable for competition. Certainly, it is in AT&T's best long-term interest to restructure its affairs in keeping with this new environment, thus minimizing the job to be done by the courts. ■

WANTED!

Changing the boundaries between hybrid communications and hybrid data processing can create new services

David J. McKee
Planning Research Corp.
Washington, D.C.

Terence J. McCormick
Office of Telecommunications
U.S. Department of Commerce
Washington, D.C.

Essay

There's untapped wealth in the no-man's land between the regulated hybrid communications and the unregulated hybrid data processing industries. Although the Federal Communications Commission has been reluctant to change its so-called maximum separations policy, the agency has at last started an inquiry into the hybrid services that could be provided by intermediary vendors.

The maximum separations policy is no longer in the public interest. The policy has prevented development of a new industry that could supply a myriad of services making use of many different blends of data processing and data communications. About the only valid point about the present policy is that it prevents cross-subsidization of one kind of service by the other. A new policy should encourage a computer/communications resale industry that allows vendors to offer any combination of remote access data processing (RADP) and message switching services specifically tailored to each customer's application.

Hybrid teleprocessing is the gap between the unregulated data processing and the regulated data communications industries. (Many of the other terms used in this article have been defined in official documents of the FCC, and some are embedded in the U.S. legal code. They are given in the panel "Data definitions").

The potential rewards of hybrid teleprocessing are so great that they have fostered widespread violations of FCC rules—perhaps because violation is so easy and detection is so difficult. Certainly, the most widespread violation is the use of subscribers to timesharing services of their facilities to perform the message switching that, technically, is required to be performed by the common carriers.

The main beneficiary of a shift in the rules would be the small businessman who cannot afford to own a complete teleprocessing network. By a change in the policy that would define workable boundaries between hybrid communications and hybrid data processing, vendors of hybrid teleprocessing services could combine the requirements of a number of small subscribers and lease the required communications facilities in bulk from conventional regulated common carriers. In the public interest, the most attractive solution appears to be continued regulation of common carriers who offer hybrid communications services, but to not regulate data processing vendors who offer hybrid services using circuits leased from the common carriers.

Whatever the FCC's final decision, the computer/communications resale industry should be structured to:
• Respond to user needs.
• Allow easy market entry for new companies.
• Encourage introduction of new services and marketing strategies.
• Permit market forces to encourage realistic competition.
• Speed the transfer of new technological developments to commercial exploitation.
• Require minimum capital investment.

In modifying the maximum separations policy, the FCC should carefully consider the public interest, the users, the Government, the economic viability of the decision, fair competition, monopoly practices, laissez faire entrepreneurship, and meticulously delineate each segment of the industries

a computer/communications resale industry

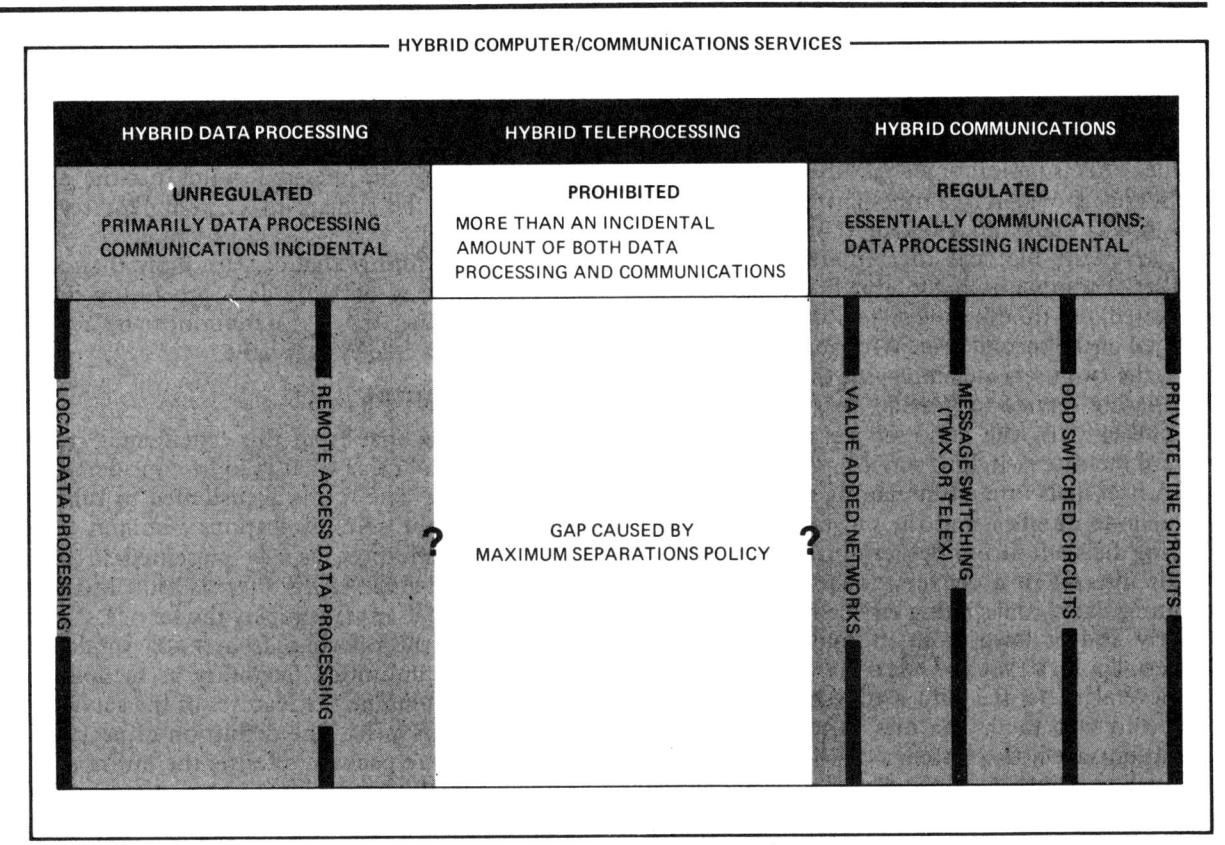

1. The FCC's present maximum separations policy substantially prohibits implementing a variety of hybrid teleprocessing services for public use.

involved as to regulation, nonregulation, or, perhaps even, deregulation.

Figure 1 puts the present maximum separations policy into perspective. Here, hybrid data processing services involve activities in which the use of data communications is incidental. At present, services operating in this manner do not require regulation and, thus, are not under FCC jurisdiction. As the figure shows, remote access data processing falls in the class of hybrid data processing, as does the local data processing, which uses no data communications at all.

Hybrid communications services provide data communications, with data processing essentially an incidental feature. Such offerings are regulated as communications common carriers. Typical hybrid communications offerings include value added networks, message switching (TWX and telex), and switched offerings like direct distance dialing services.

In private data communications networks, the user can do both remote access data processing and message switching without incurring the displeasure of common carriers or running afoul of Federal Communications Commission regulations. The maximum separations policy applies only to public offerings.

At present, a strict interpretation of the maximum separations policy would prohibit the sale of hybrid teleprocessing services. Moreover, the location of the borders between the three types of services have been difficult to define, which leads to a policy interpretation that, at best, results in an uncertain status for hybrid teleprocessing. In turn, this regulatory uncertainty has fostered surreptitious hybrid teleprocessing by some subscribers to available international and domestic timesharing services.

There is a great economic incentive for a subscriber to employ a timesharing company's data communications network for message switching, since in some cases the cost drops to one fourth to one half that charged by a record carrier such as Western Union. The savings are particularly great for international transmissions.

It's simple to achieve surreptitious hybrid teleprocessing:

- A user inputs into a timesharing system, transmits his file of data and saves it in the computer. Then another user dials into the system and reads the file. This is typical of a legal, unregulated database service provided by a timesharing company.

- A user dials into a system, transmits a file of data, saves the file, and then inputs a telephone number of another user. After the first user has disconnected, the timesharing system dials out to the terminal of the second user. With such an arrangement, the two users are employing the unregulated timesharing service to provide, in an illicit manner, substantially the same service as a store-and-forward message switching service.

- A user dials into a timesharing system, inputs a telephone number, and the program the user is running dials out to another terminal. The two terminals interact in a conversational mode. Again, the unregulated timesharing service is being used—wrongly and at lower cost—in competition with Western Union's TWX and telex services.

- Several users dial into a timesharing system, input and save their data files. After all the files have been saved, the system dials out to selected terminals and outputs all the files as a single report. But the regulatory status of such a useful information retrieval service is uncertain, if not actually prohibited, by the maximum separations policy.

Legality in doubt

Uncertainty exists about the legality of public provisioning of message switching by commercial timesharing vendors. The end users are neither organized nor homogeneous enough to force a decision. Commercial timesharing companies cannot afford to offer message switching if doing so would involve interminable regulatory proceedings, and protracted and expensive litigation.

Even if the FCC were to clearly prohibit such message switching, there appears to be no feasible way of policing these activities since—as mentioned—it is difficult, if not impossible, to distinguish message switching from legal teleprocessing activities involving file updates and data processing.

Furthermore, distributed networks that share the teleprocessing load between the public vendor's processors and the customer's processors blur the allocation of functions between their computers, making it even more difficult to draw a distinct line between remote access data processing and message switching. Under these circumstances, to definitely prohibit hybrid teleprocessing altogether would simply drive unregulated message switching underground.

In its 1971 Computer Inquiry (Docket 16979), the FCC ruled that when a message switching service is offered as an integral part of and as an incidental feature of a package offering that is primarily data processing, the service will be unregulated. But when the package offering is oriented essentially to satisfy the communications or message switching requirements of a user—and the data processing feature or function is an integral part of and incidental to the message switching—the entire service will be subject to Government regulation.

For package offerings that contain more than an incidental amount of both data processing and message switching, the FCC determines the regulatory status on a case-by-case basis.

Not so incidental

Several problems arise from this "incidental" ruling. First, it is not clear what is to be considered a package offering. The FCC is accustomed to ruling on offerings of the communications common carriers. In these offerings, usually presented in the form of clearly defined tariffs, there is little flexibility as to how the user may employ the service.

However, in the offerings from RADP vendors, there is almost unlimited flexibility as to how a customer may combine the features of the services to accomplish his tasks. The definition of package is not clear. Is the package offering the entire service offered by the vendor? Is it one particular system? Is it the service as used by one customer? Or is it one task performed by the user? These are some questions not addressed by the FCC decision on its Computer Inquiry. Therefore, the gap between hybrid data processing and hyrid communications remains rather wide and uncertain.

However, a recent decision by the agency has helped define the borders between hybrid communications and hybrid teleprocessing. The decision allowed Packet Communications Inc. (PCI) to construct and operate as a regulated common carrier a value added network (VAN). A VAN leases circuits from an established common carrier, adds value by providing such communications equipment as multiplexers and concentrators, and resells a communications service to end users.

It is emphasized that the VANs are not offering a hybrid teleprocessing service, but rather a hybrid communications service with—at most—incidental data processing functions. Thus, in allowing VANs, the FCC has not yet made a decision concerning hybrid teleprocessing services.

The ruling also allows other business organizations to provide VAN services. Subsequently the ap-

> ## Data definitions
>
> **Data processing** is the use of a computer for the processing of information, as distinguished from circuit or message switching. Processing involves the use of the computer for operations that include the functions of storing, retrieving, sorting, merging, and calculating data according to programed instructions.
>
> **Message switching** is the computer-controlled transmission of messages via communications facilities between two or more points, wherein the contents of the messages remain essentially unaltered.
>
> **Local data processing** service is an offering of data processing wherein communications facilities are not involved in serving the customer.
>
> **Remote access data processing service** is an offering of data processing wherein communications facilities that link a central computer to remote customer terminals provide for the transmission of data between the computer and customer terminals.
>
> **Hybrid service** is an offering of service that combines remote access data processing and message switching to form a single integrated service.
>
> **Hybrid data processing service** is a hybrid service offering wherein the message switching capability is incidental to the data processing function or purpose.
>
> **Hybrid communications service** is a hybrid service offering wherein the data processing capability is incidental to the message switching function or purpose.
>
> **Hybrid teleprocessing** is a hybrid service offering which includes more than an incidental amount of both data processing and message switching.

plications of Telenet Communications Corp. and several other value added carriers were approved.

Economics plays an important role in whether or not hybrid teleprocessing can lead to public offerings by the computer/communications resale industry. Users want to know if new services would be the least expensive way to accomplish the required tasks. Vendors want to determine the commercial potential for these services and the cost of rendering them.

Economic discrimination

Government policy makers would want to study hybrid teleprocessing economics to determine capital requirements and other factors so as to find out if the public interest would best be served by Government regulation of many vendors in open competition, regulation of one or a few monopolistic vendors, or not requiring regulation at all.

Business users are the hardest hit by the lack of hybrid teleprocessing services. Those companies with large remote data processing and message switching requirements, may have enough load to economically justify private hybrid systems. Many companies have gone and will go this route to satisfy their computer/communications requirements. But those companies with small requirements for remote data processing and message switching, cannot justify building their own dedicated computer/communications systems.

The result is economic discrimination against the low volume user. Some options open to them now are not economically viable or are of uncertain legality. Although the businessman can procure data processing services from an RADP vendor and communications services from a common carrier, the cost of separate services could well be prohibitive.

The small user could connect his own computer with a VAN when one becomes available. Such networks are legal and regulated. While this approach may save some communications costs, the user does not enjoy the opportunity to partake of the myriad of specialized programs and other data processing services available from an on-line time-sharing company.

Finally, the subscriber could engage in surreptitious hybrid teleprocessing by clandestine use of the message switching capability inherent in many remote data processing services offered by time-sharing vendors.

Costs are favorable

Whether or not hybrid teleprocessing can be meaningful to the user community depends on the economic health of the vendor. Figure 2 shows unit cost as a function of relative load for a hybrid teleprocessing service constructed around three separate and independent computer/communications systems. One system is dedicated to message switching, another to interactive on-line data processing (similar to timesharing), and the third to remote batch data processing.

In this typical example, the load is composed of equal amounts of the three tasks. Traffic would originate in the 50 largest metropolitan areas in the U.S. A load level of 1 can be supported by one small computer system. More computers (perhaps 100 to 200 large mainframes) and more communications facilities are required when the load level reaches 1,000 units.

To put this into perspective, at a load-level of 1, the monthly rental cost for the data processing equipment and communications facilities would come to about $600,000. At a load level of 1,000, the monthly cost would be $140 million—but this load level far exceeds that needed to satisfy even the most grandiose potential market for hybrid teleprocessing.

Figure 2 also contains the unit cost curve for a

hybrid teleprocessing configuration in which the three services are implemented in one integrated computer/communications system. Such a system is much less costly than that for the three independent, but smaller, systems.

The cost for an integrated system, which starts at $300,000 a month for a load level of 1, is half that for the three independent systems. But at a load level of 1,000, the costs for both configurations are essentially the same. Thus, there is an economic advantage in implementing one multitask system. The reduction in cost for the integrated system is not the results of economics of integration, but from the well-known economies of scale. The unit cost of a large system is less than the unit cost of several small systems, whether their tasks are similar or not.

More detail on costs of integrated hybrid teleprocessing systems are presented in Figure 3. Here, the curve labeled "total system cost" is the same as the curve labeled "multitask hybrid" in Figure 2. This cost is the sum of computer and communications costs, and overhead—but marketing costs and profit are excluded. Again, the computer costs are based on the monthly rental for the number of computers required to handle a given load level. And the communications costs are based on the monthly value of communications processors, other data transmission equipment, and the common carrier tariffs for dial-up and leased circuits.

As might be expected, the unit cost of the total system decreases rapidly at low load levels and gradually flattens at the higher levels. That is, economies of scale exist for multitask hybrid teleprocessing systems out to very high load levels.

The unit cost for data processing computers starts to flatten at load levels requiring about 10

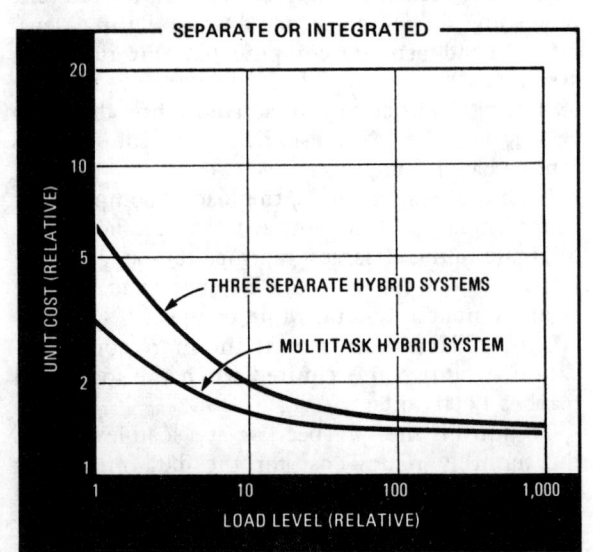

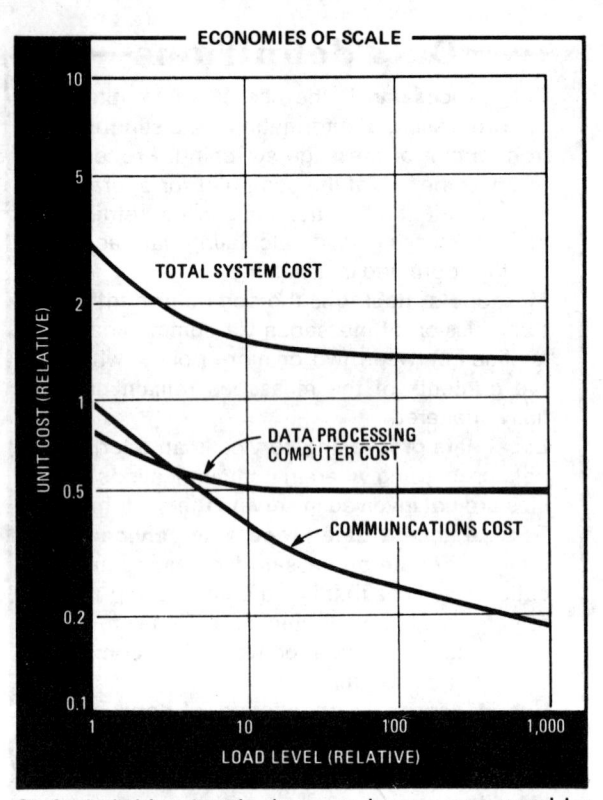

3. As hybrid networks become larger, measured by their load levels, data processing costs remain steady after reaching a certain plateau, but communications costs keep on decreasing.

large computer systems. But unit communications cost continues to decrease, even out to the highest load levels, where it has little impact on the total unit cost. The main reason for the continued reduction in unit communications cost is that a network interconnecting more and more computers makes increasingly efficient use of transmission lines and other communications equipment. Efficiency is improved by load balancing, message queuing during peak loads, and message interleaving. Figure 3 shows that the larger the over-all hybrid teleprocessing system, the less is the unit cost to perform a given task.

The distance between the two curves in Figure 2, particularly at the smaller load levels, indicates that there is sufficient incentive for private enterprise to profitably offer these systems to the public on a multicustomer shared-system basis.

Figure 3 shows that the cost to the hybrid teleprocessing vendor for transacting a customer's load level of 1 would be 30 cost units, say $3, which may be slightly less than the customer's cost for executing the same transaction via a common carrier and a timesharing vendor. However, the cost to the hybrid vendor for transacting 10 load levels—one from each of 10 customers—would be $1.50 per transaction and should mean lower cost to the end user.

With hybrid teleprocessing available, users can

2. The integration of separate hybrid systems into one larger over-all system lowers the unit cost because the one larger system enjoys the benefits of the well known economies of scale.

save money as well as enjoy services previously denied them because of high cost. And vendors can make money providing the services.

Workable boundary

Implementing hybrid teleprocessing will require a change in Government policy. The main requirement of any new policy should be that it clearly delineate a workable boundary between regulated communications common carriage and unregulated remote computer-network services. Some alternative policy options are:
 • Regulating on-line data processing services.
 • Regulating hybrid teleprocessing.
 • Maintaining the status quo.
 • Allowing hybrid teleprocessing when communications services are supplied by a VAN carrier.
 • Allowing hybrid teleprocessing when the vendor does not own the transmission circuits.
 • Deregulating hybrid teleprocessing.
 • Deregulating all public message-switching services.
 • Deregulating all communications services.

The first and last of these alternatives are unacceptable because they are unrealistic and contrary to the public interest. But they do serve as boundaries between which a viable policy can exist.

Three other options mentioned above are discussed next.

DEREGULATE ALL MESSAGE SWITCHING. The total deregulation of message switching as a communications common carrier function is an attractive option. The traditional rationale for regulation of message switching was appropriate when the nation was communications-poor.

But now the nation is becoming rich in communications as the result of the increase in new business markets and the favorable economies of new message switching and networking technologies. Stable free market competition among vendors of message switching services would arise within the computer/communications resale industry.

And if all message switching is deregulated, then hybrid teleprocessing must become deregulated because the data processing component of hybrid teleprocessing is already unregulated.

HYBRID TELEPROCESSING USING VANS. Another reasonable option is to deregulate message switching only in hybrid teleprocessing applications. This can be unconditional.

Or deregulation can apply when the communications network of the hybrid teleprocessing system is provided by a VAN or some similar type of carrier. Under this option, RADP vendors could continue to offer message switching services over common carrier communications networks. But then vendors could not offer message switching on the leased networks to their customers. This means that a data processing vendor could offer message switching only if he "gave up" the communications portion of his service to a VAN or other communications carrier. The vendor would merely be an agent between the communications carrier and the vendor's message switching customers—who also happen to be his data processing customers.

HYBRID TELEPROCESSING OVER LEASED CIRCUITS. This option would constrain hybrid teleprocessing vendors from owning circuits. The vendors would lease them from common carriers, but would not be prevented, as now, from reselling the use of the circuits. The capital-intensive communications carriers would continue to receive regulatory protection for their facilities as an incentive to invest hundreds of millions of dollars in plant. By this same line of reasoning, VANs could also be deregulated, since they would obtain circuits from the regulated common carriers.

Framework for the future

Deregulation of VANs is a necessary first step in establishing new policies. An important characteristic of VANs is that none that have received approval from the FCC have proposed to construct their own transmission facilities. Thus, their capital investment will be small, compared with that of the common carriers. Nor do VANs require radio spectrum allocations. Because ownership of transmission facilities is unnecessary, many industry observers, as well as Packet Communications Inc., have questioned the requirement for regulation of VANs before they can go into business and offer services to the business community.

There is a major objection to a computer/communications resale industry. The objection is that a regulated communications carrier could be drawn into unfair competition with its own customers. The vendors could offer similar services, but they would be unfettered by regulation.

The counter-argument is that resale by the industry as a whole would stimulate offerings that now are largely ignored by the common carriers. And when several companies offer hybrid teleprocessing services, the competition between them would foster the development of specialized, more creative offerings.

Bridging the gap

By aggregating the demands of many small communications users, resale vendors could represent total requirements to the carriers as if they come from a single customer. In this way, vendors could bridge the gap between large customers who can utilize economical bulk communications capacity solely for their own internal use and small customers who cannot justify such capacity by themselves.

Open-entry and resulting competition for hybrid teleprocessing could lead to the duplication of communications services. But duplication of services is less important than is the efficient utiliza-

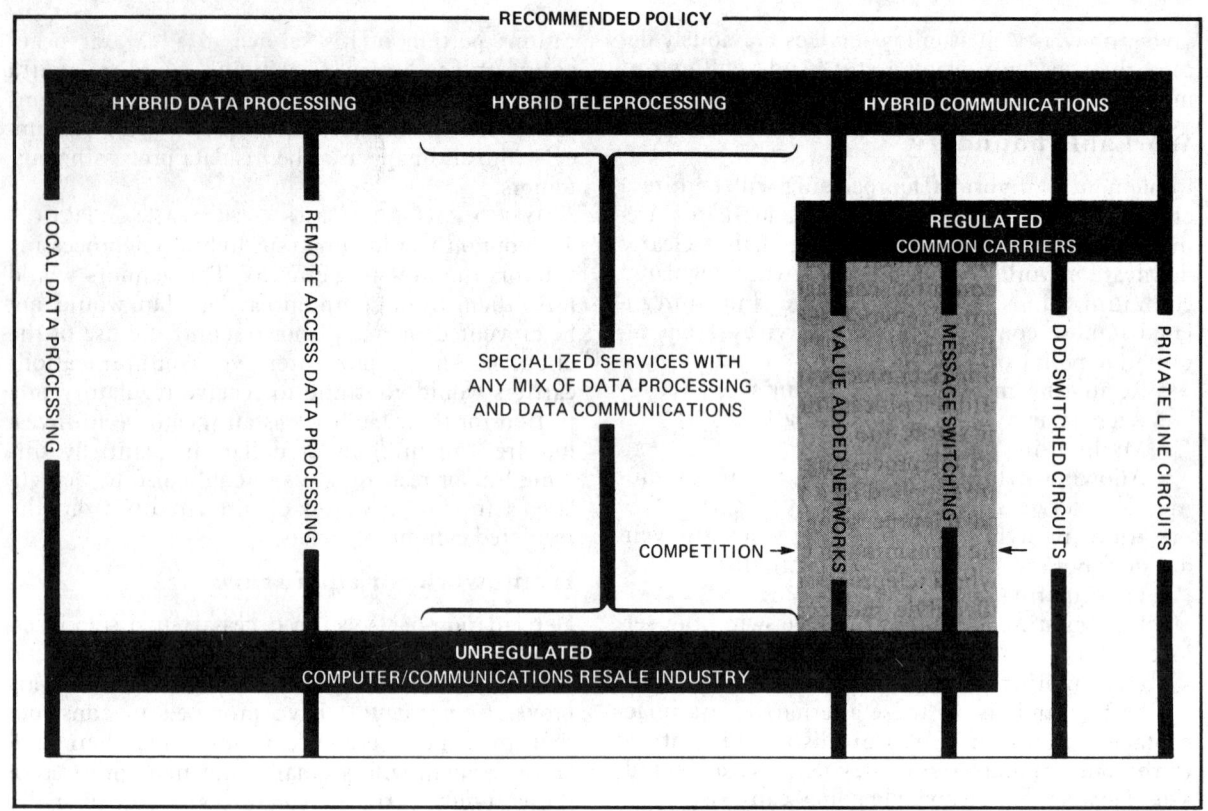

4. Under recommended policy, certain hybrid services—such as value added networks—would become competitive; both regulated and unregulated companies would participate in the market.

tion of the capital-intensive transmission facilities already installed or being constructed in this country. And efficient utilization can be obtained by exploiting the potential economies of scale resulting from larger networks.

Offer all services

The recommended regulatory structure for hybrid services is shown in Figure 4. The regulated conventional and specialized common carriers would be allowed to offer all hybrid communications services. With only one restriction, the entire spectrum of hybrid services (data processing, teleprocessing, and communications) would be open to unregulated resale industry vendors. The restriction would be that these vendors could not own communications circuits, but would have to lease them from the common carriers.

This policy would seem to encourage competition for hybrid communications services between the unregulated resale vendors and the regulated common carriers. However, the nature of hybrid teleprocessing services could foreclose such competition.

On the one hand, the unregulated vendors might not be able to compete with the common carriers in offerings that are basically communications because the regulated common carrier would be able to charge the end users the same that the resale vendor would have to pay.

On the other hand, common carriers make money by increasing their capital investments to expand their rate base, and they probably would not want to compete in the small-investment specialized services markets in which the resale vendors would tend to operate.

Furthermore if an unregulated vendor were able to compete with a regulated common carrier for a given service, it must be questioned why the service was regulated in the first place. The purpose of regulation is to protect the public—not the supplier—where competition is supposedly not possible because of the capital-intensive natural monopoly nature of the service.

The net result of this recommended structure would be to deregulate all hybrid services except those that are of a capital-intensive nature. The timesharing industry is a good example of what the computer/communications resale industry could become. The timesharing industry has responded to user needs, developed innovative service offerings, enjoyed easy market entry, and stimulated stiff competitio ■

The views and interpretations expressed here are those of the authors and should not be interpreted as necessarily representing those of their organizations.

part 11
data
data
data
data
data
data
data
data
data

miscellaneous

Standards for data communications

Equipment and procedures are forced to follow the rules of the road for information transfer, and many organizations are out painting the signs

Glenn C. Hartwig
Data Communications

Specification The standards developed by various governmental agencies, industry groups, and manufacturers not only govern the kinds of equipment the user buys, but also govern most aspects of its operation. These rules, among other things, determine how fast the equipment operates, the type and quality of carrier facilities the user may access, compatibility of the equipment with these facilities as well as equipment from other manufacturers, and the types of codes and protocols that must be followed. However, new standards are not adopted to harass the user who is happy with his communications setup; they are designed to keep up with the state of the art in the field of data transfer.

Although most groups that develop standards insist that complying with documents, rules, and regulations they develop is "voluntary" or merely "recommended practice," the effect on the user is the same as if the standards were handed down by decree. And although members of regulation-making committees may put forth various proposals, argue among themselves as to the best courses of action, and determine the viability of a particular course of action, the user, at the end, is handed a set of rules and specifications he must follow if he wants to transmit information digitally from point A to point B.

Some groups invite public comments and suggestions before presenting a proposed standard to the general committee of the organization for final approval, and some groups offer users a limited opportunity to comment on the proposals. However, others keep the user in the dark until the day he is told that new practices have been instituted. The secretive process may require the user suddenly to send his equipment back to the shop for modification, but seldom do the new practices make in-place equipment obsolete.

But even though a standards-making body may not give advance warning of a change, the user probably won't be put out of business. The pace of adopting new standards is usually slow enough that, by the time they appear, the user has been told by his supplier that innovations are on the way—and the warning is usually accompanied by a sales proposal.

What's more, rarely are changes so drastic that existing equipment is taken off the market or modes of transmission discontinued. The rules of the road that cause the greatest amount of dislocation are seldom called standards, even though they may may carry the same weight. These quasi-standards are the ones announced by equipment manufacturers and common carriers that dictate the rates on various types of lines and what kinds of equipment are needed to operate one manufacturer's equipment with gear supplied by another data communications equipment maker.

Suppliers of data communications equipment compatible with their own mainframes, other terminals, or the central processing units of other manufacturers have publications that describe the equipment, specify operating parameters, and detail operating procedures. Manufacturers' representatives can provide such manuals for most data communications equipment.

However, regardless of whether the standard is developed by a governmental agency, an industry committee, or a manufacturer, the user must wade through stacks of manuals and documents to make sure that his equipment conforms. Or he may have to figure out how to have his equipment modified to keep on operating. To assist the user in his task, DATA COMMUNICATIONS is publishing this list of applicable standards from ANSI, the Electronic Industries Association, the Institute of Electrical and Electronic Engineers, National Communications System, International Standards Organization, and

Developing a standard

Development of data communications standards is an evolutionary process. Proposals for exploiting a technology are generated when it reaches a new plateau. Inventive minds develop ways to raise procedures to a new level of sophistication. As an orderly means of implementing these practices, new standards are developed. And as the procedures go into operation in data communications networks, the new requirements invariably lead to sales of new, higher-technology equipment. However, the head of one of the more important standards-development groups insists that no faction or company dominates standards development.

The process of developing a standard is exemplified by the work of the American National Standards Institute's task group X3S3.4, Link Control Procedures, which developed the Advanced Data Communications Control Procedures. This bit-oriented full-duplex procedure, which will eventually replace character-oriented communications, was first proposed in 1969, but equipment to implement these procedures is only now making its way into the marketplace.

Chairman of task group X3S3.4 is David Carlson, an engineer on the technical staff at Bell Telephone Laboratories, Holmdel Township, N. J. Carlson denies that he or his group actually set the standards, even though the group did a good deal of work in preparing the proposal. The committee had to expand and broaden the scope of the original ideas and extend its potential capabilities before forwarding it.

"We're really people with degrees of technical expertise in an area," Carlson explains. "Because the representation embodied in our membership covers such a broad cross section of the industry—users, manufacturers of mainframes, manufacturers of terminals, providers of telecommunications facilities, and the like—the proposals that come out of our meetings only represent the consensus of a group of technical experts in one field. The standardization process is one that then entails scrutiny of our proposals up a long chain of decision-making."

Carlson also disagrees with the popular view that business competition plays an important role in the attitudes of committee members. "When we meet, I don't think it's possible to say there's a manufacturer's, a carrier's, or a user's group. That would be as much as saying that there exists within a particular segment of the industry a form of group agreement on how something ought to be done. I don't think that generally exists. What you're doing is presenting 18 individual technical experts with a problem of how data-link control might best be structured and relying on their individual expertise to come up with an answer that will serve today's and tomorrow's data communications environment."

Except for a member of a private consulting firm in California, the only "user-oriented" participants in Task Group X3S3.4 include members of the U. S. Office of Telecommunications and "companies that are in the business of providing systems based on the best components available. In that sense, they reflect a user's view."

And rather than being a burden, Carlson considers r than being a burden, Carlson considers a standard is something that everyone is willing to accept and can comfortably live with. "You arrive at that end," Carlson observes, "through all of the various conflicts between all those represented. One company may present its idea better at one meeting, but the length of the standardization process does not permit a favorite proposal to be rammed through before the opposition can marshal stronger arguments."

the common carriers (see associated "Standards-making organizations" panel).

This catalog of data communications standards includes a brief explanation of the scope of each. Prices are supplied when available.

ANSI

These standards are widely used, but on a voluntary basis and sometimes in modified form. Those standards that have been adopted as Federal Information Processing Standards for Government procurement purposes are designated (FIPS) and those adopted for mandatory use by the Department of Defense are designated (DOD). The documents are available from ANSI, 1430 Broadway, New York, N. Y. 10018.

A hardworking group that acts as a ramrod for ANSI is the Computer and Business Equipment Manufacturers Association. Located at 1828 L St. N.W. Washington, D.C., CBEMA appoints ANSI committee members, makes sure work is progressing and, distributes papers on work in progress.

X3.1-1969 SYNCHRONOUS SIGNALING RATES FOR DATA TRANSMISSION (FIPS22). $3.

Provides a group of specific signaling rates for synchronous serial or parallel binary data transmission. These rates apply to data circuits at the interface between data terminal equipment and data communications equipment that operate over nominal 4-kHz voice-bandwidth channels.

X3.4-1968 CODE INFORMATION INTERCHANGE (FIPS 1), $4.50.

Denotes the coded character set to be used for the general interchange of information among information-processing systems, communications systems, and associated equipment.

X3.24-1968 SIGNAL QUALITY AT INTERFACE BETWEEN DATA PROCESSING TERMINAL EQUIPMENT AND SYNCHRONOUS DATA COMMUNICATION EQUIPMENT FOR SERIAL DATA TRANSMISSION, $2.90.

Applies to the exchange of serial binary data signals and timing signals across the interface between data processing terminal equipment and synchronous data communications equipment.

X3.15-1966 BIT SEQUENCING OF THE AMERICAN NATIONAL STANDARD CODE FOR INFORMATION INTERCHANGE IN SERIAL-BY-BIT DATA TRANSMISSION (FIPS 16).

Specifies the bit-sequencing of the ASCII code for serial-by-bit, serial-by-character, data transmission.

X3.16-1966 CHARACTER STRUCTURE AND CHARACTER PARITY SENSE FOR SERIAL-BY-BIT DATA COMMUNICATION IN THE AMERICAN NATIONAL STANDARD CODE FOR INFORMATION INTERCHANGE (FIPS 17), $3.

Defines character structures for both synchronous and asynchronous transmission modes. The relationship of the particular terminals and character structures in data communications is included in the discussion.

X3.25-1968 CHARACTER STRUCTURE AND CHARACTER PARITY SENSE FOR PARALLEL-BY-BIT DATA COMMUNICATION IN THE AMERICAN NATIONAL STANDARD CODE FOR INFORMATION INTERCHANGE (FIPS 18). $3.

Specifies the character structure and sense of character parity for parallel-by-bit, serial-by-character, data communications in the ASCII code, ANSI standard X3.4-1968

X3.28-1971 PROCEDURE FOR THE USE OF THE COMMUNICATION CONTROL CHARACTERS OF THE AMERICAN NATIONAL STANDARD CODE FOR INFORMATION INTERCHANGE IN SPECIFIED DATA COMMUNICATION LINKS, $6.75.

Presents control procedures for specified data communications links that employ the communications-control characters of ASCII code, ANSI standard X3.4-1968.

X3.36-1975 HIGH SPEED SYNCHRONOUS DATA SIGNALING RATES.

Provides a group of specific signaling rates for synchronous, high-speed, serial data transfer. These rates exist on the receive-data and transmit-data circuits of the interfaces between data terminal equipment and data communications equiment that operate over high-speed channels.

X3.44-1974 DETERMINATION OF PERFORMANCE OF DATA COMMUNICATIONS SYSTEMS, $7.75.

Discusses system elements that determine the performance of an information path, as well as evaluation of criteria used to measure performance in a data communications system.

Proposals expected to be adopted as ANSI standards upon completion of balloting in the near future include: X3.57 MESSAGE HEADING FORMATS USING ASCII FOR DATA COMMUNICATIONS, in addition to working papers on NETWORK CONTROL PROCEDURES, PUBLIC DATA NETWORKS, CHANNEL LEVEL INTERFACE, DEVICE LEVEL INTERFACE, AND MINICOMPUTER INTERFACE. Under development at the task-group level is X3S34/589 BIT-ORIENTED DATA LINK CONTROL PROCEDURES, which may result in the formalization, revision, or official recognition, of such industry protocols as IBM's SDLC and Burroughs' BDLC.

EIA

Compliance with EIA standards is voluntary. Some of them duplicate the publications of ANSI. When such duplication exists, it is noted by the corresponding ANSI standard number in parenthesis. Prices are listed when available. EIA standards are available from the association's headquarters at 2001 Eye St. N. W., Washington, D. C. 20006.

RS-232-C, CATALOG 3, INTERFACE BETWEEN DATA TERMINAL EQUIPMENT AND DATA COMMUNICATION EQUIPMENT EMPLOYING SERIAL BINARY DATA INTERCHANGE, $5.10.

Describes interconnection of data terminal equipment and data communications equipment employing serial binary data interchange. The standard defines electrical-signal characteristics, mechanical interface characteristics, and circuits. Included are 13 specific interface configurations intended to meet the needs of 15 defined system applications. A recommended companion document is Industrial Electronics Bulletin No. 9, "Application Notes for EIA Standard RS-232-C."

RS-357, CATALOG 3, INTERFACE BETWEEN FACSIMILE-TERMINAL EQUIPMENT AND VOICE-FREQUENCY DATA COMMUNICATION TERMINAL EQUIPMENT, $3.

Governs analog facsimile systems such as those defined in EIA RS-328 (below) used in common carrier or private voice-band communications channels that require a baseband interface with the data communications terminal equipment. This standard defines a means of exchanging control signals and analog-voltage data signals between a facsimile data terminal and data communications terminal equipment.

RS-366, CATALOG 3, INTERFACE BETWEEN DATA TERMINAL EQUIPMENT AND AUTOMATIC CALLING EQUIPMENT FOR DATA COMMUNICATIONS, $4.50.

Describes the interconnection of data terminal equipment and automatic calling equipment for data communications. It defines electrical-signal characteristics, interface mechanical characteristics, functional descriptions of interchange circuits, and standard interfaces, and it includes recommendations and explanatory notes.

RS-422, ELECTRICAL CHARACTERISTICS OF BALANCED VOLTAGE DIGITAL INTERFACE CIRCUITS, $4.25.

Gives the electrical characteristics of the balanced-voltage digital-interface circuits normally implemented in integrated-circuit technology.

RS-423, ELECTRICAL CHARACTERISTICS OF UNBALANCED VOLTAGE DIGITAL INTERFACE CIRCUITS, $4.25.

Details the electrical characteristics of the unbalanced digital-interface circuits normally implemented in integrated-circuit technology.

RS-269-A, CATALOG 3, SYNCHRONOUS SIGNALING RATES FOR DATA TRANSMISSION (ANSI X3.1-1969), $2.

Provides specific signaling rates for synchronous serial or parallel binary-data transmission over 4-kHz voice-bandwidth channels.

RS-334, CATALOG 3, SIGNAL QUALITY AT INTERFACE BETWEEN DATA PROCESSING TERMINAL EQUIPMENT AND SYNCHRONOUS DATA COMMUNICATION EQUIPMENT FOR SERIAL DATA TRANSMISSION (ANSI X3.24-1968), $2.90.

Applicable to the exchange of serial binary data signals and timing signals across the interface between data-processing terminals and synchronous data communications equipment, as defined in RS-232-C. This standard is valuable when the various pieces of equipment in a system are furnished by different producers.

RS-404, CATALOG 3, STANDARD FOR START-STOP SIGNAL QUALITY BETWEEN DATA TERMINAL EQUIPMENT AND NON-SYNCHRONOUS DATA COMMUNICATIONS EQUIPMENT, $4.

Specifies the quality of serial binary signals exchanged across the interface between start-stop data terminals and a nonsynchronous modems, signal converters, and the like.

RS-363, CATALOG 3, STANDARD FOR SPECIFYING SIGNAL QUALITY FOR TRANSMITTING AND RECEIVING DATA PROCESSING TERMINAL EQUIPMENT USING SERIAL DATA TRANSMISSION AT THE INTERFACE WITH NON-SYNCHRONOUS DATA COMMUNICATIONS EQUIPMENT, $4.30.

Specifies the quality of serial binary signals exchanged across the interface between synchronous or asynchronous data-processing terminals and nonsynchronous data communications equipment, as defined in RS-232-C.

RS-252-A, CATALOG 3, STANDARD MICROWAVE TRANSMISSION SYSTEMS, $6.50.

Describes the baseband characteristics needed to determine compatibility of radio and multiplex equipment. This standard applies to the characteristics of the transmission path between the multiplex baseband send terminals and multiplex baseband receive terminals in both directions of transmission.

RS-328, CATALOG 3, MESSAGE FACSIMILE EQUIPMENT FOR OPERATION ON SWITCHED VOICE FACILITIES USING DATA COMMUNICATION TERMINAL EQUIPMENT, $1.

Provides for interchanges between message-type facsimile equipment operated through modems over switched voice-grade telephone facilities.

RS-368, CATALOG 3, FREQUENCY DIVISION MULTIPLEX EQUIPMENT STANDARD FOR NOMINAL 4-KHZ CHANNEL BANDWIDTHS AND WIDEBAND CHANNELS. $4.40.

Gives criteria for evaluating multiplex-equipment performance. This standard may be said to lay out a communications system incorporating such types of equipment as microwave and cable or enable a user to evaluate and specify multiplex equipment on a terminal or back-to-back basis.

RS-373, CATALOG 3, UNATTENDED OPERATION OF FACSIMILE EQUIPMENT (AS DEFINED IN RS-328 AND RS-357), $1.

Specifies the sequence and duration of control signals at the interface defined in RS-328 and RS-357 to provide for unattended operation of facsimile receivers and transmitters.

RS-411, ELECTRICAL AND MECHANICAL CHARACTERISTICS OF ANTENNAS FOR SATELLITE EARTH STATIONS, $9.20.

Specifies electrical and mechanical characteristics of antennas for satellite stations with high G/T.

IEEE

IEEE standards differ slightly from the others. The IEEE develops standards, recommended practices, and guides only upon recommendations from its members. The publications of the Standards Board come in three degrees of importance: standards, which are documents with mandatory requirements; recommended practices, which present procedures preferred by IEEE; and guides, which suggest good practices but make no clearcut recommendations. These documents are available from IEEE headquarters at 345 E. 47th St., New York, N. Y. 10017

IEEE STD. 312-1970, DEFINITIONS OF TERMS FOR COMMUNICATION SWITCHING, $5.

Is a glossary of terms for mandatory use among IEEE members. This is a collection of all kinds of communications buzz words, with a short definition following each.

IEEE STD. 488-1975 STANDARD DIGITAL INTERFACE FOR PROGRAMABLE INSTRUMENTATION AND RELATED SYSTEM COMPONENTS. Price not yet available.

Deals with systems that use a byte-serial, bit-parallel transfer of digital data among instruments, terminals, and data processors. Of limited use in data communications, the interface system described in the standard is optimized as an interface for communication over a contiguous party-line bus.

NCS

These National Communications System data communications standards are being developed under the Federal Telecommunications Standards Program. When completed, the proposals annotated with an asterisk will be published by the National Bureau of Standards as Federal Information Processing Standards. They will be available from the Office of the Manager, NCS, Washington, D. C. 20305.

*FED STD 1001, SYNCHRONOUS HIGH SPEED SIGNALING RATES BETWEEN DATA TERMINAL EQUIPMENT AND DATA COMMUNICATIONS EQUIPMENT.

Specifies standard signal rates as indicated in the title when data terminal and data communications equipment is operated in the synchronous mode over wideband transmission facilities. The specified rates are 16, 48, 56, 64, 1,344, and 1,544 bits per second.

*FED STD 1031 FUNCTIONAL AND MECHANICAL CHARACTERISTICS OF THE DATA TERMINAL TO DATA COMMUNICATIONS EQUIPMENT INTERFACE.

Specifies the functions to be executed across the interface between data terminal and data communications equipment and designates pin numbers of a new standard connector. Coupled with Fed Std 1020 and 1030, 1031 will gradually phase out the use of EIA Standard RS-232-C by Federal Government departments and agencies.

FED STD 1020 ELECTRICAL CHARACTERISTICS OF BALANCED VOLTAGE DIGITAL INTERFACE CIRCUITS (this standard adopts EIA Standard RS-422).

Specifies the electrical characteristics of digital equipment interfaces using electrically balanced line drivers, transmission lines, and terminators.

*FED STD 1030 ELECTRICAL CHARACTERISTICS OF UNBALANCED VOLTAGE DIGITAL INTERFACE CIRCUITS

(This standard is the same as EIA Standard RS-423). Is identical in scope to Fed Std 1020, except that it deals with unbalanced line drivers, transmission lines, and terminator devices.

*FED STD, unnumbered. BIT-ORIENTED DATA LINK CONTROL PROCEDURES.

Describes bit-oriented machine-to-machine codes and procedures to control information transfer over data transmission links.

*FED STD, unnumbered. MESSAGE FORMATS FOR DATA COMMUNICATION SYSTEMS EMPLOYING BIT-ORIENTED DATA LINK CONTROL PROCEDURES.

Designates standard message formats compatible with bit-oriented data-link-control procedures. This standard and the one listed immediately above are being developed in cooperation with ANSI subcommittee X3S3.

*FED STD 1033 CRITERIA FOR ASSESSING THE PERFORMANCE OF TELECOMMUNICATIONS SYSTEMS SUPPORTING FEDERAL INFORMATION PROCESSING SYSTEMS.

Defines user-oriented performance criteria and measuring techniques for assessing the performance of telecommunications facilities used in direct support of Government data processing systems.

ISO

ANSI, on behalf of the United States, has agreed to use these ISO standards in the U.S. as part of the worldwide data communications standardization process. Prices are listed when available. ISO standards are available through ANSI at 1430 Broadway, New York, N. Y. 10018.

ISO 1177-1973 (E) INFORMATION PROCESSING—CHARACTER STRUCTURE FOR START/STOP AND SYNCHRONOUS TRANSMISSION, $3.15.

Specifies the character structure of the 7-bit coded characters set for serial-by-bit start/stop and synchronous data transfer through modems between the data terminal equipment and data communications equipment.

ISO 1745-1975 (E) BASIC MODE CONTROL PROCEDURES FOR DATA COMMUNICATIONS SYSTEMS.

Tells how to implement the ISO/CCITT 7-bit coded character set for information interchange on data transmission channels. The standard also defines the formats of the transmitted messages and the supervisory sequences in the transmission control procedures. These procedures deal with transmission over one link at a time, but not operation of data links in parallel. This class of control procedures, known as the "basic mode," applies at the interface between data communications equipment and data terminal equipment.

ISO 2110-1972 DATA COMMUNICATIONS-DATA TERMINAL AND DATA COMMUNICATIONS INTERCHANGE CIRCUITS—ASSIGNMENT OF CONNECTOR PINS.

Defines what information will be carried by particular pins in an interconnection plug, much like EIA RS-232-C.

ISO 2111-1972 (E) DATA COMMUNICATION—BASIC MODE CONTROL PROCEDURES—CODE INDEPENDENT INFORMATION TRANSFER.

Defines the means by which a data communications system operating according to the basic mode procedure for 7-bit coded character set defined in ISO/R 1745 (described above) can transfer information messages without code restrictions. Standard extends sections of ISO/R 1745 regarding information transfer and describes other uses for the data link escape (DLE) character.

ISO 2593-1973 (E) CONNECTOR PIN ALLOCATIONS FOR USE WITH HIGH-SPEED DATA TERMINAL EQUIPMENT, $3.15.

Is a guide to data terminal equipment that operates at a data-signaling rate greater than 20,000 bits per second. The standard also correlates the interface-circuit numbers required for use in Government-specified equipment with the pin numbers specified by military standards for connectors used on data communications equipment and the data terminal equipment. The Government/military specification, which requires that data terminal equipment shall be terminated by a 34-pin connector, is available from Navy Publications and Form Center, 5801 Tabor Ave., Philadelphia, Pa. 19120.

ISO 2628-1973 (E) BASIC MODE CONTROL PROCEDURES—COMPLEMENTS

Extends the digital basic-mode control procedures defined in ISO/R 1745 and ISO 2111 to add recovery procedures, abort and interrupt procedures, and multiple-station selection. Systems that conform to ISO/R 1745 do not necessarily have to include the functions described in this international standard. However, systems implementing the functions described in this standard and conforming to ISO/R 1745 and ISO 2111 are required to follow these recommendations.

ISO 2629-1973 (E) BASIC MODE CONTROL PROCEDURES—CONVERSATIONAL MESSAGE TRANSFER.

Defines the means by which a data communications system operating according to the basic-mode control procedures defined in ISO/R 1745 can interchange information messages in a fast conversational manner. Particularly adaptable to inquiry/response systems, this standard extends information transfer, as defined in ISO/R 1745, to allow two stations connected by a data link to reverse their master/slave status, thereby reversing the direction of the information transfer.

ISO is also considering two proposed international standards relating to bit-oriented data-control procedures. The first is tentatively referred to as ISO DIS 3309 HDLC FRAME STRUCTURE. It determines how information carried by bit-oriented data communications facilities is positioned, and the meaning given to the information in each frame. The second is ISO DP 4335 ELEMENTS OF DATA LINK CONTROL PROCEDURES. This draft proposal, which will come up for vote at ISO's October meeting in Washington, D.C., is designed to provide a scope and define applications for synchronous bit-sequence-independent data transmission.

Common carrier standards

These standards are supplied by the Bell System, but MCI and Data Transmission Co. describe theirs as "about the same." Except as noted, copies of all standards are priced at $1.50. They may be ordered from American Telephone and Telegraph Co., Supervisor-Information Distribution Center, 195 Broadway, Room 208, New York, N. Y. 10007.

PUB41001 30-BAUD PRIVATE LINE CHANNELS INTERFACE SPECIFICATION—DECEMBER 1967.

Describes connection requirements of a private-line channel capable of transmitting two-state signals at rates up to 30 b/s for metering, supervisory control, and miscellaneous signaling purposes. End-to-end metallic continuity is not a requirement of this channel and is seldom available. The interface signal is two-state direct current.

PUB41002 45-, 55-, AND 75-BAUD PRIVATE LINE CHANNELS INTERFACE SPECIFICATION—DECEMBER 1967.

Describes connection requirements of three private-line channels capable of transmitting two-state signals up to their respective rated speeds for teletypewriter, data–metering, supervisory control, and miscellaneous signaling purposes. End-to-end metallic continuity is not a requirement of this channel and is seldom available. The interface signal is two-state direct current.

PUB41003 150-BAUD PRIVATE LINE CHANNELS INTERFACE SPECIFICATIONS—FEBRUARY 1968.

Describes connection requirements of a private-line channel capable of transmitting two-state signals at speeds up to 150 b/s for teletypewriter, data-metering, supervisory controls, and miscellaneous signaling purposes. The interface to computers and terminals is the same as EIA RS-232-C.

PUB41004 DATA COMMUNICATIONS USING VOICE-BAND PRIVATE LINE CHANNELS—OCTOBER 1973.

Provides voice-band and channel-connection arrangements for private-line data transmission. Specifies analog impairment limits supported on channels, digital parameters that can be supported by Bell System service, engineering considerations for two-point and multipoint channel usage, and information on maintenance and channel availability are included.

PUB41005 DATA COMMUNICATIONS USING THE SWITCHED TELECOMMUNICATIONS NETWORK—REVISED MAY, 1971.

Details the structure and operation of the DDD network, presents switching and transmission performance data on the network, and discusses topics related to data communications on the switched telecommunications network.

PUB41101 DATA SET 103A INTERFACE SPECIFICATION—FEBRUARY 1967.

Specifications for modem 103A, which provides full-duplex low-speed serial data transmission at rates up to 300 b/s. The 103A, used in conjunction with Data Auxiliary Set 804B1, may be arranged for automatic origination, automatic answering, and alternate voice. This modem is used for Dataphone service and TWX-CE applications.

PUB41012 1A DATA STATION, MULTICHANNEL ARRANGEMENT USED IN PROVISION OF TWO-POINT CHANNELIZING SERVICE—JUNE 1973

Defines a frequency-division multiplexer capable of deriving 75-and/or 150-b/s channels from a 3002 private-line voice-band channel.

PUB41021 DIGITAL DATA SYSTEM—CHANNEL INTERFACE SPECIFICATIONS—PRELIMINARY—MARCH 1973.

Describes the digital data system and the interface required between the channel termination equipment of the DDS, contained in a channel service unit (CSU), and the customer's data terminal. A CSU must be used when the customer's equipment performs coding and decoding, timing recovery, synchronous sampling, formatting, and generation and recognition of control signals.

PUB41102 DATA SET 103A3, 103E, 103G, AND 103H—INTERFACE SPECIFICATION—OCTOBER 1973.

Specifies connectivity requirements for modems 103A3, E, G, and H, which provide full-duplex low-speed serial data transmission at rates up to 300 b/s. The 103E is a basic modem without power supply and controls and is used for multiple set installations. The 103A3 consists of a 103E in a single unit housing and operates with a standard key telephone set. The 103G is similar to the 103E but has an integrated housing and provides a card dialer. The 103H is also similar to the 103E but is designed for mounting in a data terminal.

PUB41103 DATA SET 103F INTERFACE SPECIFICATION—MAY 1964.
Tells how to install modem 103F, which provides full-duplex, low-speed, serial data transmission at rates up to 300 b/s. It is intended for use on private-line channels and does not provide for voice transmission.

PUB41004 DATA SET 113A INTERFACE SPECIFICATION—AUGUST 1973.

Supplies use requirements for modem 113A, an originate-only, full-duplex, low-speed, serial modem for use at rates up to 300 b/s for Dataphone service.

PUB41201 DATA SETS 201A AND B—AUGUST 1969.

Tells how the 201-type modem transmits serial binary data over voice bandwidth facilities using phase-shift-keyed modulation. Operation is synchronous full-duplex or half-duplex and at 2,000 b/s for Dataphone service over the switched network or 2,400 b/s over conditioned private lines.

PUB41105 113-TYPE DATA STATION—INTERFACE SPECIFICATION—OCTOBER 1971.

Supplies use requirements for the 113-type data station, which provides for low-speed serial, frequency-shift-keyed, full-duplex data transmission over the switched telecommunications network.

PUB41202 DATA SETS 202C AND D INTERFACE SPECIFICATIONS—MAY 1964.

Sets requirements for modems 202C and D in a medium-speed binary serial data transmission system. The modems operate up to 1,200 b/s on Dataphone and at 2,400 b/s on conditioned private communications lines.

PUB41203 DATA SET 202E SERIES—REVISED—SEPTEMBER 1971.

Describes the 202E modem series, a modularized family of modem transmitters specified by AT&T as input stations in data-collection systems. They may be used on either the switched telecommunications network or on private lines.

PUB41204 DATA SET 203-TYPE—REVISED—APRIL 1974, $2.25.

Provides information on synchronous transmission and reception of high-speed digital data over the switched telecommunications network or conditioned private-line data channels operating at rates between 1,800 b/s and 10,800 b/s.

PUB41208 DATA SET 202R INTERFACE SPECIFICATION—JUNE 1971, $2.20.

Describes the 202R modem, which is a manual-only, minimum-feature modem that may be used on either the telecommunications network or on private-line channels at line speeds of 1,200 b/s and 1,800 b/s.

PUB41212 DATA SETS 202S AND T INTERFACE SPECIFICATIONS—AUGUST 1974.

Provides connection requirements for obtaining 1,200-b/s asynchronous service on the switched network and 1,800-b/s asynchronous service on private lines. Test capability isolates faults.

PUB41213 DATA SET 209A INTERFACE SPECIFICATION—PRELIMINARY—MAY 1974.

Using this synchronous, binary serial 9,600-b/s modem on a voice-band 4-wire private line.

PUB41301 DATA SET 301B INTERFACE SPECIFICATION—MARCH 1967.

Describes binary serial synchronous data transmission at 40,800 b/s. Modem, which requires a group-bandwidth channel, is used only in private-line applications.

PUB41209 DATA SET 208A INTERFACE SPECIFICATION—NOVEMBER 1973.

Sets requirements for the 208A modem, which provides synchronous, binary, 4,800-b/s service over unconditioned 3002-type four-wire channels. It is suited for multipoint polling applications and point-to-point systems.

PUB41210 DATA SET 201C INTERFACE SPECIFICATION—APRIL 1973.

Tells how modem transmits serial binary data over voice-bandwidth facilities using phase-shift-keyed modulation. Operation is synchronous full-duplex or half-duplex at 2,400 b/s over the switched telecommunications network or private-line channels.

PUB41211 DATA SET 208B INTERFACE SPECIFICATION—AUGUST 1973.

Provides specifications for synchronous, binary 4,800-b/s service with the 208B modem on the switched telecommunications network. Modem is suitable for switched-network systems requiring high throughput.

PUB41405 DATA SETS 402C AND D INTERFACE SPECIFICATION—NOVEMBER 1964.

Tells use and conditions for modems 402C (receiver) and 402D (transmitter) to operate as a medium-speed binary parallel data transmission system for Dataphone service or private lines.

PUB41408 DATA SET 407A INTERFACE SPECIFICATION—NOVEMBER 1973.

Tells how to employ a multifrequency data receiver designed for multiline installations at up to 10 characters per second over the switched telephone network or over unconditioned private lines from Touch Tone telephone sets.

PUB41601 DATA AUXILIARY SET 801A INTERFACE SPECIFICATIONS—MARCH 1964.

Specifies connection requirements for a dc dial-pulse automatic-calling unit which permits a business machine to place calls over the switched telecommunications network.

PUB41602 DATA AUXILIARY SET 801C INTERFACE SPECIFICATION—SEPTEMBER 1965.

Provides technical details of a Touch Tone automatic-calling unit, which enables a business machine to place calls over the switched telecommunications network.

PUB41450 DIGITAL DATA SYSTEM—DATA SERVICE UNIT INTERFACE SPECIFICATIONS—PRELIMINARY—MARCH 1973.

Tells how the data-service unit for data services through the digital data system provides plug-for-plug interchangeability with existing EIA type D or E interfaces at synchronous speeds of 2,400, 4,800, and 9,600 b/s.

PUB41704 86B-TYPES DATA SELECTIVE CALLING SYSTEM STATIONS—NOVEMBER 1968, $1.65.

Provides the same information as PUB41703, except that no provision is made for use where the subscriber's computer controls any line function.

PUB41705 DATA LINE CONCENTRATOR SYSTEM (DLCS) ARRANGEMENT—MAY 1971.

Describes system arrangements that provide for the connection of a number of stations to a smaller number of computer-communications ports.

PUB41706 MODEL 37 TELETYPEWRITER STATIONS FOR DATAPHONE SERVICE—SEPTEMBER 1968, $1.55.

Describes the operating and line characteristics of the M37 teletypewriter station operating on Dataphone service.

PUB41707 DATASPEED TYPE-4 SYSTEM—SEPTEMBER 1969, $1.90.

Provides installation requirement for a paper-tape transmission system operating at either 1,050 or 1,200 words per minute.

PUB41702 85A1 AND 2 DATA SELECTIVE CALLING SERVICE STATIONS—OCTOBER 1971, $2.25.

Defines requirements for 85A data selective calling service stations used to provide eight-level, half-duplex, private-line selective calling systems. The 85A1 operates at 100 words per minute with model 33 or 35 equipment, and the 85A2 operates at 150 words per minute with model 37 equipment. Traffic flow is controlled by a subscriber's computer that sequentially polls the station transmitter. ∎

Standards-making organizations

The mission of Subcommittee X3S3 of Sectional Committee X3 on Computers and Information Processing of the American National Standards Institute is to define the characteristics of digital data-generating and -receiving systems that function with communications systems. It is responsible for developing and recommending standards for data communications. ANSI standards deal with the quality and characteristics of data during transmission.

ANSI, located at 1430 Broadway, New York, N. Y. 10018, describes itself as a clearinghouse of voluntary national and international standards. It is also the U.S. representative to the International Standards Organization and will often either approve international standards for use in the United States or tailor ISO proposals to the needs of American industry. Domestic ANSI standards are often adopted as Federal Information Processing Standards (FIPS) for Government procurement purposes, and some are adopted for mandatory use by the Department of Defense.

Interface standards. The Electronic Industries Association is the United States' trade association of electronic-equipment manufacturers. EIA, which is a member of ANSI, has three working groups concerned with data transfer.

The data communications and data terminal group, Committee TR-30, stresses interface compatibility. Included is the interface between terminals and modems, signal quality at the interface, and data-signaling speeds.

Committee TR-29, the facsimile engineering committee, is concerned with development of interface standards between communication and facsimile-terminal equipment.

The purview of Committee TR-37 is the development and upgrading of standards dealing with electrical interfaces between common-carrier-provided connectors, such as AT&T's Data Access Arrangement (DAA) and user equipment.

Users of data communications may have a say in the workings of EIA committees by requesting the Association, at 2001 Eye Street, N.W., Washington, D.C. 20006, to send them EIA subcommittee documents. These documents are circulated to anyone interested after they have been wrangled through the subcommittee and are ready for presentation to the general committee. Comments are then received, reviewed and resolved before the documents are sent to the general committee for a vote. Comments on subcommittee reports can, and have, caused documents to be reworked. The user, in effect, does have a voice in the way a final standard appears.

IEEE standards. The Institute of Electrical and Electronic Engineers' Computer Society at 345 E. 47th St., N.Y., N.Y. 10017, operates a standards committee and a committee on computer communications, both of which are concerned with data communications standards. IEEE membership is required for participation on standards-making committees, but institute standards are available for purchase by non-members. The IEEE's subcommittee on computer communications is developing standards for interfacing between man and machines. A working group of this subcommittee is now drafting a standard on the man-machine interface for display-type data terminals.

Government standards. The Federal Government's Telecommunications Standards Program is working on data communications standards, which, when completed, are published by the National Bureau of Standards and are known as Federal Information Processing Standards (FIPS). These standards are adopted by the Department of Defense's National Communications System for all branches of the Government to use in the procurement of data communications systems. Most of the Government's telecommunications standards, which are developed by ANSI and EIA, pertain to data communications. Most Federal standards are compatible with industry standards or are being developed in cooperation with industry.

International standards. The primary international body concerned with worldwide data communications standards is technical committee 97, subcommittee 6 of the International Organization for Standardization. Membership in ISO is limited to recognized national standards-making bodies. There are now 62 full members of ISO and 19 correspondent members—developing countries that don't yet have formal standards-making bodies of their own.

The job of ISO TC97/SC6 includes constant appraisal of data communications through all types of telecommunications media, as well as the ability of its members to provide themselves with data communications to link data processing systems.

Intergovernmental standards. Also working in international data communications standards-making is the Consultative Committee on International Telegraph and Telephone. The CCITT weighs all technical, operational, and tariff matters at the governmental level for all types of information transfer. Voting power is vested in Class A members only. The Department of State votes for the U. S.

Microcomputers unlock the next generation

Microcomputers are now starting to have a big impact on line interface units and terminals, with communications processors coming later

Charles J. Riviere and Patrick J. Nichols
Telcom Inc.
McLean, Va.

Technology

Because of its excellent cost effectiveness, the microcomputer promises to play increasingly important roles in a whole range of data communications equipment. For users, microcomputers can and will provide powerful data communications functions at low cost with high reliability. And the design of microcomputer-based equipment permits modifications and upgrades to be made in the field with ease, so that data networks can benefit quickly from operational improvements brought about by such developments as advanced data link controls.

Already, the microcomputer has appeared in such data communications gear as terminals and line interface units characterized by relatively slow speeds. Some terminal controllers also now contain microcomputers.

Furthermore, when the microprocessor—the central device in a microcomputer— becomes even faster and cheaper, microcomputers will challenge the minicomputer in executing many of the higher throughput functions now taking place in such communications processors as programable front ends and remote data concentrators.

Manufactured by at least 10 domestic semiconductor companies, the microprocessor has already gone through several stages of improvement, particularly in extended word length, faster instruction execution times, and reduced cost, size, and power requirements. But despite the enthusiasm that can be generated in behalf of applying microcomputers to data communications— from the terminal to the communications processor— the present roles for microcomputers must be placed in perspective. They aren't always cost effective. Pragmatically, the micro-

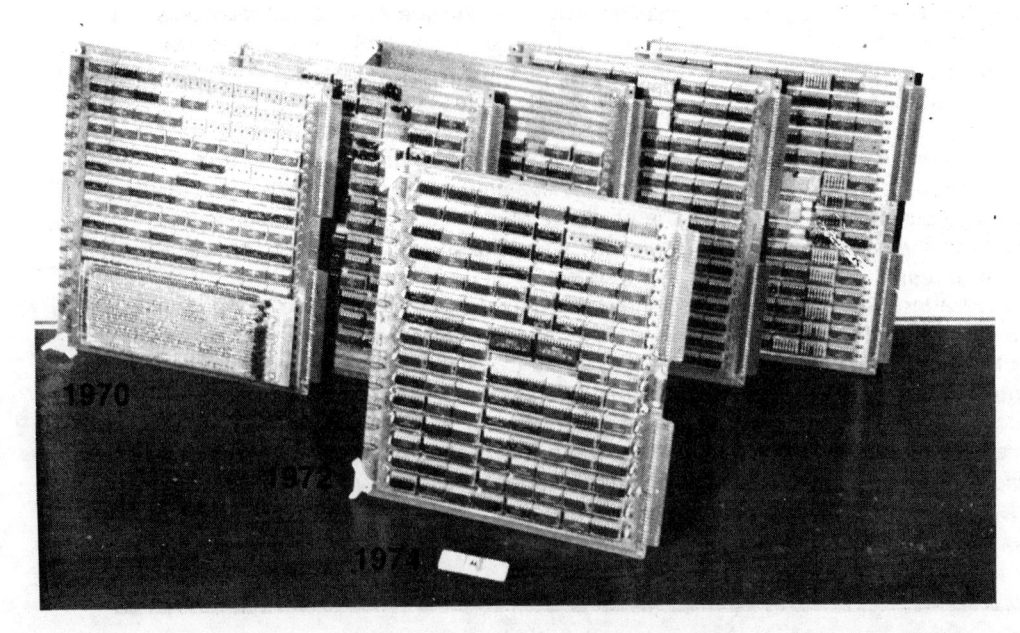

THE MICROCOMPUTER

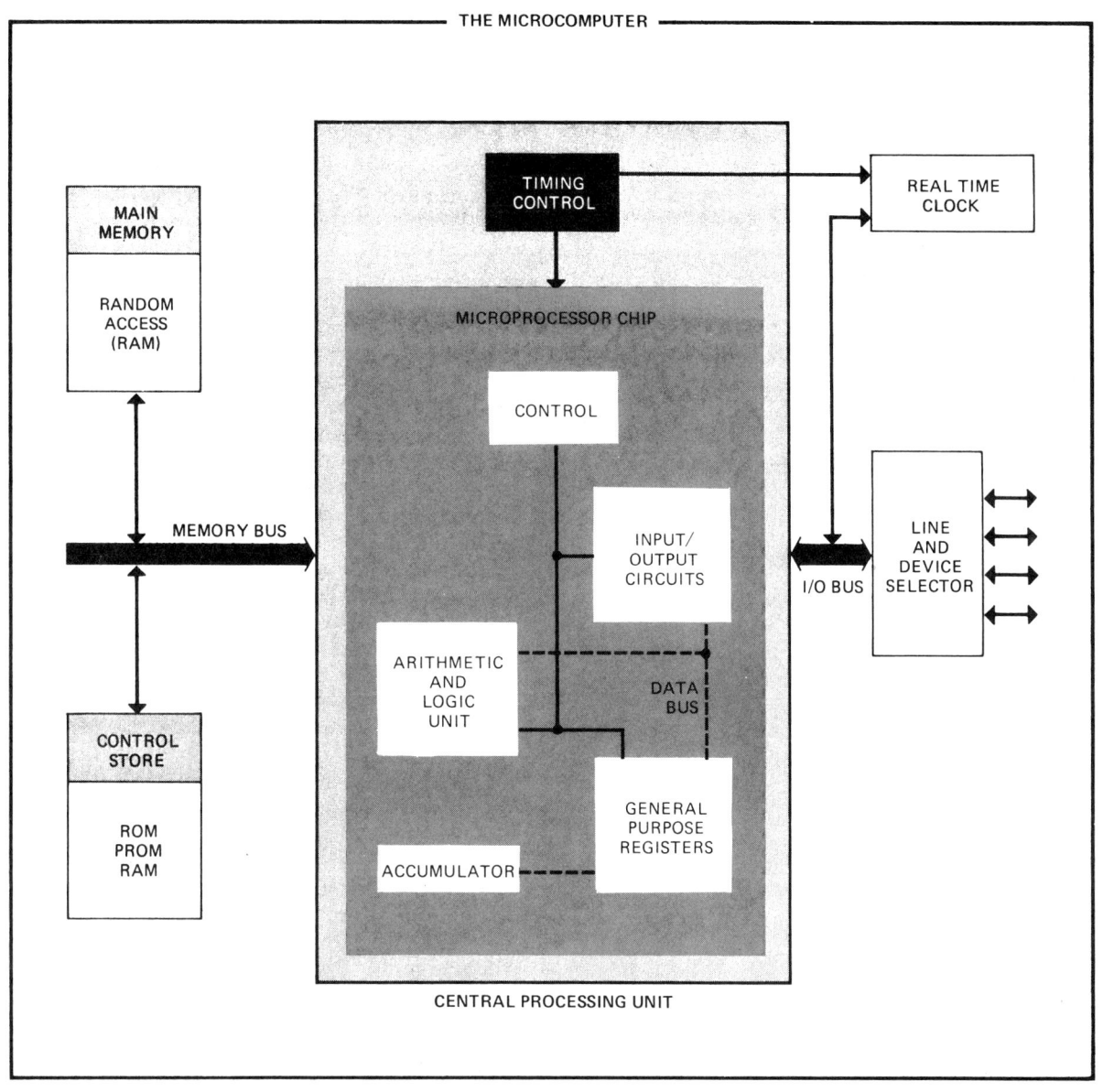

1. With architecture quite similar to that of larger computers, the microcomputer works only with programs stored in firmware memories.

computer is simply one of three ways to implement the numerous functions, such as code conversion and error control, that must be carried out on a data communications link. The two other alternatives are the minicomputer and the hardwired device.

All these approaches will be examined here for cost effectiveness in carrying out data communications functions. The present and future roles of microcomputers will then become apparent.

Chronologically, hardwired devices appeared first. Their main characteristic is that, once built, they lack programability, restricting versatility for both the equipment maker and the end user. The implementation of hardwired devices ranges from electromechanical relays, to discrete semiconductor devices (diodes and transistors), to large scale integrated (LSI) circuits. The lead photo shows three generations of equipment, with the tiny microprocessor (front) performing the same work as the two progenitors. In large quantities, LSI hardwired devices can be cost-effective and reliable. However, other hardwired devices require many discrete components and thus can be unreliable, inflexible, and expensive—by today's standards.

Next came the stored program minicomputer. After a long series of product and software improvements and price cuts, the minicomputer today sells for less than $10,000. For that price, the minicomputer will contain a central processing unit, input/output channels, power supply, oper-

287

ator's console, and a fast core memory able to store 16,000 8-bit characters (bytes). The minicomputer is and will remain a powerful piece of equipment in the data communications environment, but the uses to which it is put will change as microcomputers make inroads.

Firmware added

Responding to software commands stored in its main memory, the minicomputer exhibits excellent programability, so that it readily implements any of a number of functions required in processing incoming and outgoing data traffic.

Later developments added firmware to the minicomputer to speed up execution of certain programs and reduce software overhead. One firmware approach, resulting in a microprogramed minicomputer, uses a read only memory (ROM) as the control memory. The ROM contains a specific instruction set "burned in" at the factory. In short, the ROM constitutes the instruction set. The ROM is not subsequently alterable by the designer or end user. But different ROM's can be inserted into a minicomputer for different applications.

The other firmware approach, called writable control store, results in a user or designer microprogramable minicomputer. Here, a portion of the minicomputer's control store is a user-alterable random access memory (RAM).

The microcomputer, programed only by firmware, is designed around a microprocessor chip which contains an arithmetic and logic unit, and limited input/output circuitry. The chip also has some internal general purpose registers allowing certain basic instructions to be performed without the chip having to waste time while accessing external memory. These registers also store intermediate results for those instructions that do require access to external memory.

As shown by Figure 1, the microprocessor chip and a timing control unit constitute the equivalent of a central processing unit (CPU). The microcomputer contains a control store for microinstructions, a main memory for microprogram and data storage, a real-time clock, and a line selector (multiplexer) that directs the streams of input and output data.

The control and main memories, usually large scale integrated (LSI) circuits, will each be on one or more chips, depending on memory size. All packages making up the microcomputer are mounted on one small printed circuit card. Figure 2 shows a microcomputer serving as a data concentrator (multiplexer) that can merge about 20 10-characters-per-second lines onto one 4,800 bits-per-second line.

Conventional architecture

The arithmetic and logic unit (ALU) is an electronic device that requires a particular set of electrical signals at its input to execute a particular arithmetic or logic instruction. All instructions in a given set are contained in the control memory. The main memory, in turn, contains the program for a particular application function that tells the control memory which instructions are to be exe-

2. Multiplexer. National Semiconductor's IMP-8C uses microprocessor to handle 20 10-character-second lines.

cuted and in what order. During the execution interval, the ALU may be told by the control memory to store and retrieve data from the main memory.

Although microcomputer architecture is rather conventional, everything is smaller—word length, memory size, and instruction repertoire, as well as physical size. Microcomputers use microinstructions, which are less versatile than the conventional instruction in larger computers. When invoked, a microinstruction in the control memory determines which electrical gates in the ALU to turn on and off, and in what sequence. An example of a microinstruction is: ADD B. In the Intel 8080 microprocessor this can be executed within the general purpose registers on the ALU chip. Figure 3 shows how this is done. The fastest clock cycle for the 8080 is 0.5 microseconds, and the ADD B instruction takes four steps, so that the total execution time is 2 microseconds for this register-to-register ADD microinstruction (Table I).

Certain higher level instructions, not part of the basic microinstruction repertoire, can be made by performing several microinstructions in sequence. For example, the MULTIPLY instruction is executed using ADD and SHIFT microinstructions in a repeated fashion. The sequence for accomplishing such a higher level microinstruction is called a microcode which is also stored in the control memory. Neither microinstructions nor microcodes are available to the user. They can be either standard or customized, but once burned into the programable read only memory (PROM) they cannot be altered.

The sequence in which microinstructions and microcodes are invoked for a particular application function, such as code conversion, and the sequence in which application functions occur for a given set of tasks, such as required in a line interface unit, are together called a microprogram. For flexibility, the microprogram is stored in the main (random access memory) store. But for fixed, repetitive functions, the microprogram can be in a custom-designed control store for fast execution.

MOS vs. bipolar

Note in Table I that, on average, the execution time for the first seven instructions is about two to four times slower for the microcomputer than for the minicomputers. The reason is that the microprocessor used here is based on metal oxide semiconductor (MOS) electronics technology while the microcomputers use devices based on intrinsically faster bipolar technology.

Note also in Table I that the MULTIPLY and DIVIDE instruction execution times are far superior in the minicomputers, which employ sophisticated programing, whereas the microcomputer uses microinstructions and microcodes that need many time-consuming iterations and memory references.

Generally, the speed of a microcomputer is limited by the speed of the microprocessor. A single-chip microprocessor can be a 4-, 8-, or 12-bit device, with a basic cycle time ranging from 2 to 20 microseconds. The Intel 8080 package typifies the fastest 2 microsecond 8-bit microprocessor available now.

The 8-bit microprocessor appears adequate for

3. General purpose registers inside the micro-processor help execute basic microinstructions. For example, internal registers perform an ADD B in four clock intervals for a total instruction execution time of two microseconds.

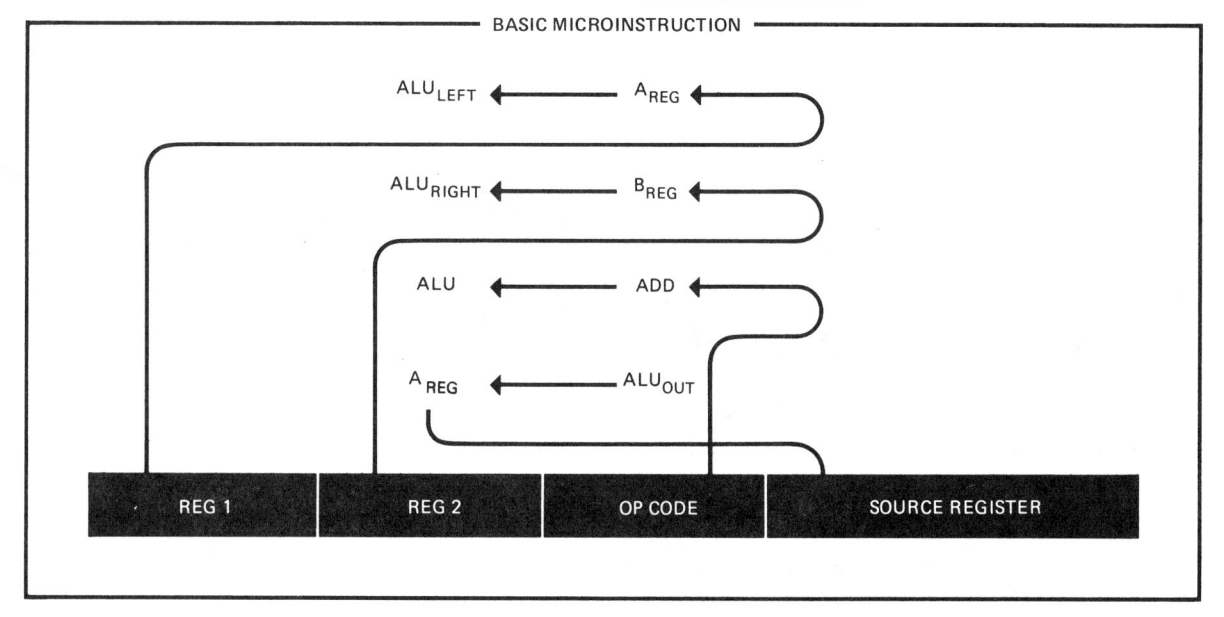

TABLE I

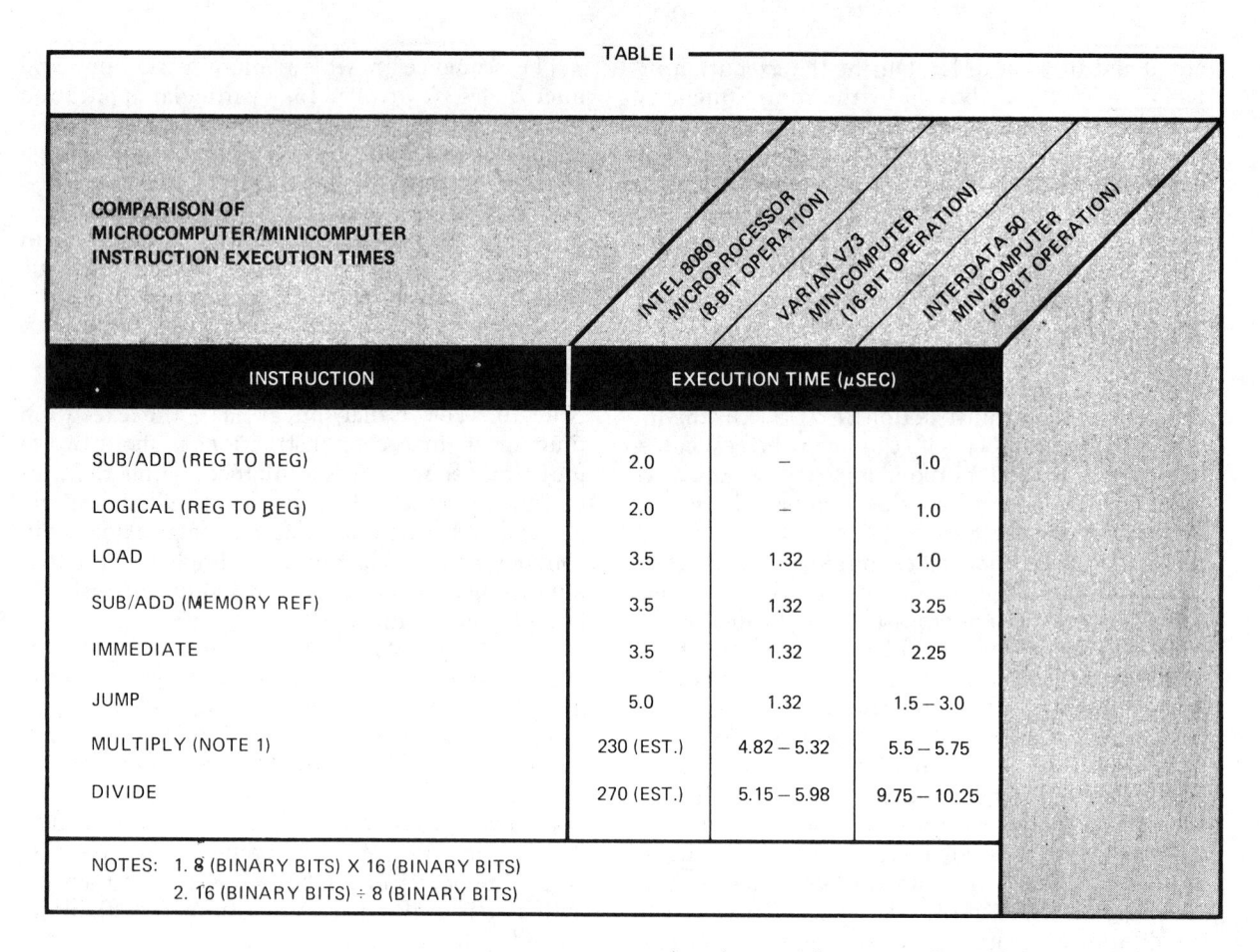

COMPARISON OF MICROCOMPUTER/MINICOMPUTER INSTRUCTION EXECUTION TIMES

INSTRUCTION	INTEL 8080 MICROPROCESSOR (8-BIT OPERATION)	VARIAN V73 MINICOMPUTER (16-BIT OPERATION)	INTERDATA 50 MINICOMPUTER (16-BIT OPERATION)
	EXECUTION TIME (μSEC)		
SUB/ADD (REG TO REG)	2.0	—	1.0
LOGICAL (REG TO REG)	2.0	—	1.0
LOAD	3.5	1.32	1.0
SUB/ADD (MEMORY REF)	3.5	1.32	3.25
IMMEDIATE	3.5	1.32	2.25
JUMP	5.0	1.32	1.5 – 3.0
MULTIPLY (NOTE 1)	230 (EST.)	4.82 – 5.32	5.5 – 5.75
DIVIDE	270 (EST.)	5.15 – 5.98	9.75 – 10.25

NOTES: 1. 8 (BINARY BITS) X 16 (BINARY BITS)
2. 16 (BINARY BITS) ÷ 8 (BINARY BITS)

most data communications applications.

To give an idea of cost, the Intel 8080 micro-processor package sells for about $300 each. Adding sufficient ROM, RAM, and other devices to convert the microprocessor into a microcomputer would cost $150 to $500 more, depending on the number of application functions desired.

Consider, for example, a line interface unit (LIU) that can be used either with a remote terminal or a communications processor. What may be involved is the implementatation of binary synchronous communications data link control, including a cy-

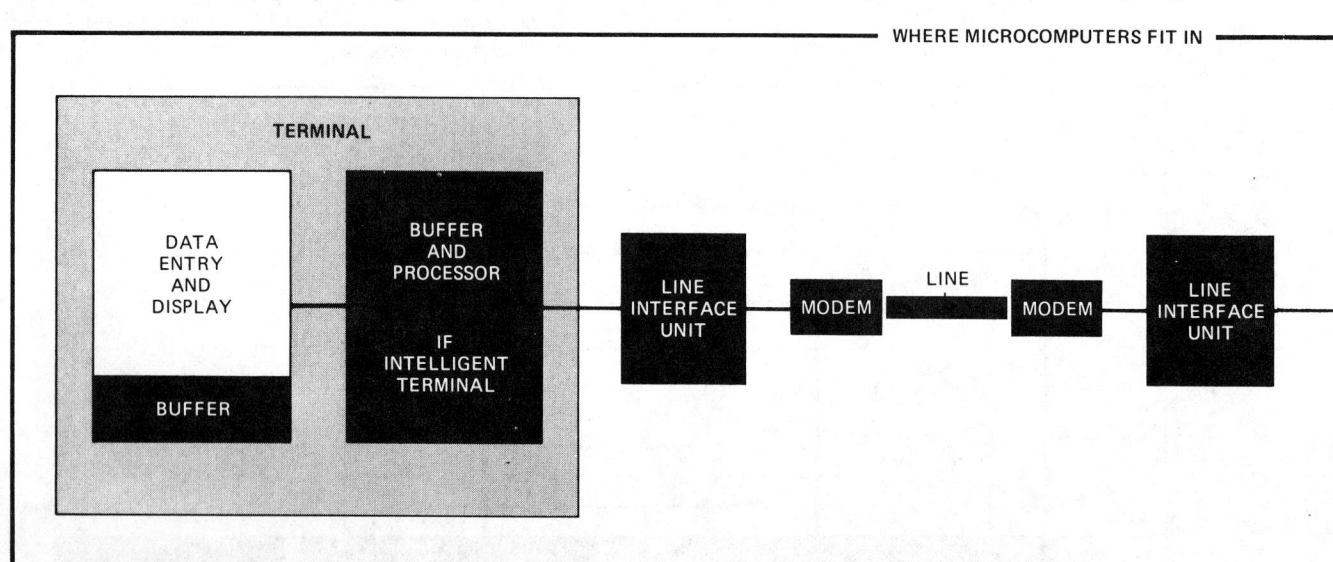

WHERE MICROCOMPUTERS FIT IN

4. Solid boxes show where, in a data communications channel, microcomputers make their early appearance, with shaded boxes showing later uses.

TABLE II

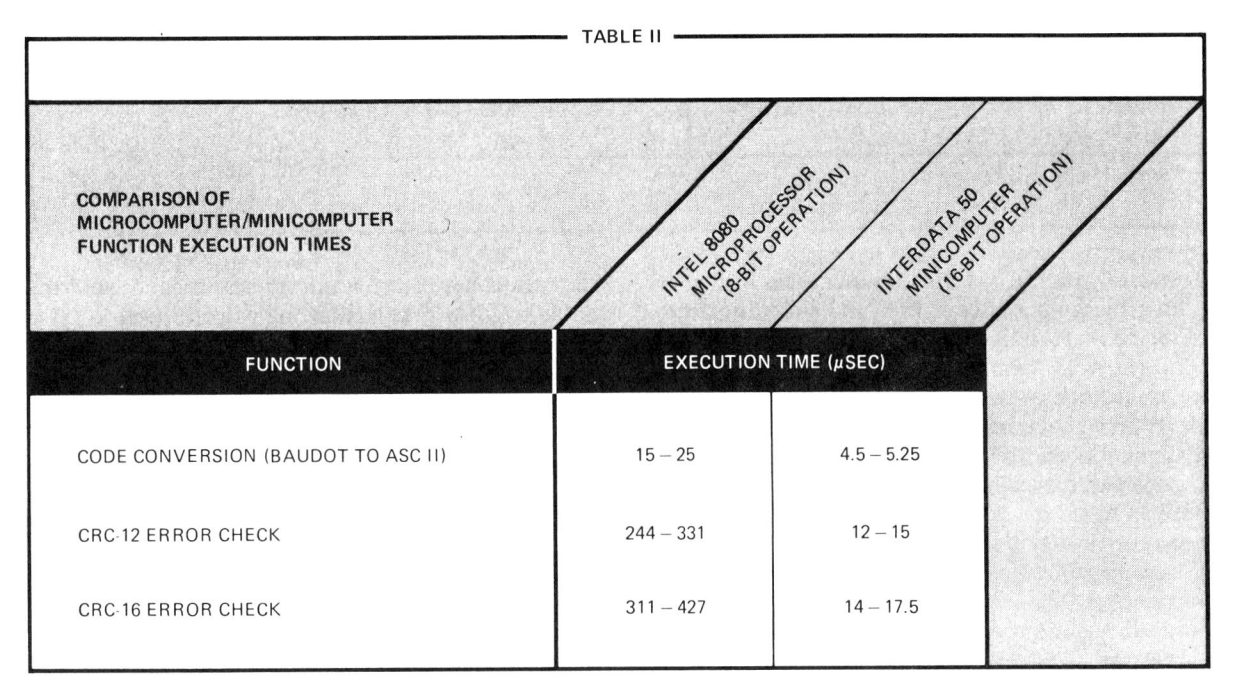

COMPARISON OF MICROCOMPUTER/MINICOMPUTER FUNCTION EXECUTION TIMES

FUNCTION	INTEL 8080 MICROPROCESSOR (8-BIT OPERATION) EXECUTION TIME (μSEC)	INTERDATA 50 MINICOMPUTER (16-BIT OPERATION)
CODE CONVERSION (BAUDOT TO ASC II)	15 – 25	4.5 – 5.25
CRC-12 ERROR CHECK	244 – 331	12 – 15
CRC-16 ERROR CHECK	311 – 427	14 – 17.5

clic redundancy check for error control, and code conversion. The cyclic redundancy check (CRC) is the most time consuming task. As shown in Table II, a microcomputer built around the Intel 8080 microprocessor takes about 370 microseconds on average to perform CRC-16 algorithm.

Suppose the line operates at 9,600 bits per second. That is, 1,200 8-bit characters appear on the line each second. Therefore, the total line taken each second is 1,200 x 0.370 milliseconds, or 444 milliseconds each second. That is, the CRC operation occupies only 44.4% of the available time.

Saturated or not

Suppose, also, that each incoming character is in Baudot code and must be converted into ASCII code with parity. As Table II shows, such a code conversion takes between 15 and 25 microseconds. Using an average of 20 microseconds, converting 1,200 8-bit characters each second takes 24 milliseconds, or about 2.4% of the available time.

In short, a line interface unit based on a microcomputer, for the situation described, can have plenty of time to perform all its functions and have processing time left over. That is, at 9,600 b/s the LIU is not throughput limited. In fact, such an LIU could handle 19,200 b/s transmission and still be marginally below saturation. With a slight change in the firmware, such a line interface unit could also handle up to 16 1,200 b/s lines.

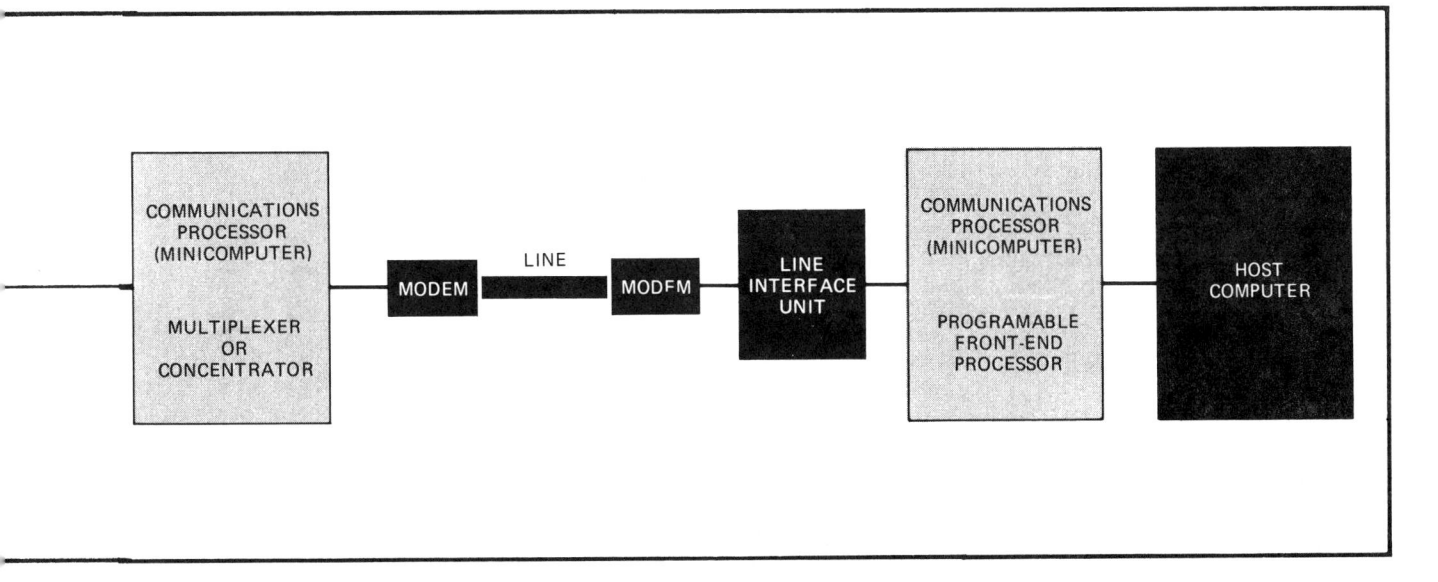

Table II reveals that, despite the versatility of the microcomputer-based line interface unit, the minicomputer can do the job faster, and probably at less prorated cost. Why, then, would a microcomputer be considered at all? The answer depends, for one thing, on whether the line interface functions are to take place at a terminal or at a minicomputer-based communications processor.

If throughput is relatively small, and enough unused memory is available, it may be more effective to have such interface operations as data link control, error control, and code conversion performed by the minicomputer. There is little economic penalty for handling the additional load.

But if the minicomputer is nearing saturation, and its processor must handle more throughput because of a network expansion, then the microcomputer, in the form of a self-standing line interface unit (LIU) as a low-cost preprocessor for the minicomputer, tends to be the logical choice. So, spending about $500 for a multifunctional LIU, as described, can be more economical than upgrading the minicomputer with a faster processor, more memory, or both.

The merit of using a microprocessor-based LIU at the terminal is more obvious. Here only certain prescribed functions—such as error control, line protocol, and perhaps code conversion— are all that's required. In a case like this, a minicomputer's capability is so great that using one would be a case of overkill, and a waste of money.

As already mentioned, the three alternatives to implementing data communications functions—hardwired devices, minicomputers, and microcomputers—will each yield a different cost effectiveness. Table III provides comparative insight. It lists 12 common data communications functions. Each alternative is rated for cost effectiveness, assuming for the moment that just the indicated function is provided by the equipment. (The ratings given are subjective, based on experience as well as some calculations.)

Note that certain functions can't be done at all by hardwired devices, as indicated by a zero in the table's matrix. On the other hand, the table also shows that implementing CRC error control is very cost effective (three stars) for the hardwired device, not so cost effective (one star) if that's all the microcomputer has to do, and a definite case of "overkill" (open square) if that's all the expensive minicomputer has to do.

Error control

Consider in more detail the task of calculating a CRC control character. As mentioned, the defined microcomputer can handle a 19,200 b/s line, but in doing so it would use up almost all available processing time. If, instead, the line's transmission rate is 56,000 b/s, the microcomputer would run out of time (saturate) and could not service the redundancy check for each data block. (At 370 microseconds per CRC calculation, the microcomputer would saturate at 21,600 b/s.)

Hardwired devices provide an out for such a situation. Instead of computing each cyclical redundancy check, an alternative is to add to the microcomputer a special shift register, designed specifically for CRC. Motorola's MC 8503 universal polynomial generator is typical. It is a nonprogramable integrated circuit package. Here, the hardwired redundancy check is started using only one instruction from the microcomputer's control memory to the CRC generator. But the arithmetic and logic unit (ALU) itself is free to go on to its other tasks after issuing its START instruction. If the total ALU START interval for a redundancy check is, say, five microseconds, then the error checking of the 7,000 8-bit data characters arriving each second at the line interface unit will take less than 5% of the LIU's throughput capability.

Adding the polynomial generator to the microcomputer does raise the microcomputer's cost, but it frees about 40% of the LIU throughput capability for other chores.

The concept of displacing one type of equipment, say memory, with another type of equipment, say a microcomputer, must take into account such factors as cost, and effect on throughput. For example, an 8-bit word of semiconductor memory costs about, say, 30 cents. The 8-bit microprocessor costs about $300. Therefore, a microprocessor package is the cost equivalent of 1,000 words of memory. But in fact, the programable microcomputer can perform certain tasks that would need a memory capacity many more times the cost of the microprocessor.

For example, a given task could be done with a lookup table. This would require perhaps 1,000 words of main memory store dedicated to the task. Instead, the output solutions to given inputs can be obtained by implementing an equivalent algorithm using the microcomputer. Doing so would use up slightly more time but reduce memory requirements to 20 words, for a savings of about $290.

Table III also shows whether a function is located at a terminal, a communications processor, or both. The table shows that the microprocessor is very cost effective (three stars) or cost effective (two stars) for such functions as data compression, buffering, buffer management, format control, and text editing. Thus, the microcomputer proves very beneficial as a processor in advanced programable terminals—as proven by recent market entries.

Such functions as code conversion, error control, and handshaking taking place at the terminal's line interface can also be accomplished in an overall cost effective manner with microcomputers.

Another look at the data in Table III reveals

TABLE III

EFFECTIVENESS OF IMPLEMENTATION OF DATA COMMUNICATIONS FUNCTIONS

FUNCTION	MINICOMPUTER	HARDWIRED DEVICES	MICROCOMPUTER	T	CP
AUTO SPEED DETECTION	☐	★★	★★★	—	✓
DATA COMPRESSION	☐	○	★★	✓	✓
AUTO POLL	★★	★	★★★	—	✓
BUFFER (CRT DISPLAY)	☐	★★	★★★	✓	✓
BUFFER MANAGEMENT	★★	○	★★	✓	✓
SCHEDULING TASKS	★★★	○	★	—	✓
DATA LINK CONTROL	☐	★★	★	✓	✓
HANDSHAKING	☐	★★	★★★	✓	✓
ERROR CONTROL (CRC)	☐	★★★	★	✓	✓
CODE CONVERSION	☐	★★★	★★	✓	✓
FORMAT CONTROL	☐	○	★★★	✓	✓
TEXT EDITING	★★	○	★★	✓	—

★★★ VERY COST EFFECTIVE   ○ CAN'T BE DONE
★★ COST EFFECTIVE   T TERMINAL
★ NOT COST EFFECTIVE   CP COMMUNICATIONS PROCESSOR
☐ OVERKILL

---

again that the minicomputer is an expensive device to serve simply as a line interface for a terminal. But when the minicomputer is intended to implement all the functions of a communications processor, it is very cost effective.

However, should a communications processor near saturation, a next step would be to remove certain of the low speed functions—CRC calculation, for example—from the minicomputer and embed them in a line interface unit (a microcomputer) serving as a preprocessor. If so, the communications processor's LIU would be substantially similar to the terminal's LIU.

Figure 4 sums up the prospects for microcomputers in data communications. As shown by the color boxes, microcomputers will, in the near term, be popular as line interface units for both terminals and communications processors. And they will serve as processors in some terminals.

The semiconductor industry has a remarkable record of improving its products rapidly while reducing unit costs. Within five years the cost of a microprocessor will likely drop to $20 to $40. Furthermore, new microcomputer architectures may be developed that could exploit the microprocessor even more effectively.

Microcomputers will thus not only prove valuable as preprocessors (LIU's), but powerful microcomputer-based communications processors may themselves become available—and for about $500. This future situation is indicated by the shaded boxes in Figure 4. Thus microcomputers will penetrate into more areas of data communications and lead to such benefits as more proficient terminals with little, added cost and improved link protocol. And more extended use of low-cost microcomputer-based concentrators and multiplexers will make better economic use of lines. ∎

What's needed today in a high-level data communications language

Software costs will be burdensome unless a faster way of programing communications computers is developed soon

Gilbert Held
Honeywell Informations
Systems Inc.
McLean, Va.

Software

The time is ripe for development of a computer language structured especially for data communications. Such a language would be on the same order as Fortran, Cobol, and other well-known procedure-oriented languages, or POLs, and would provide the same programing efficiencies. As things stand now, computer programing for data communications must rely on an assembly language or an extended version of languages' tailored specifically for other applications. Indeed, communications programing is comparatively primitive and expensive, a situation that belies the sophisticated design and advanced technology of modern communications systems.

A communications POL (or CPOL as it will be called here) would, first of all, enable communications processors to be programed at a higher level. It would also permit batch computers to be freed of the inefficiencies inherent in using non-communications higher-level languages to support communications applications. A CPOL that ranks with Fortran and Cobol also would allow a programer to select a language most efficiently suited for a given application: Fortran (formula translation) for scientific work, Cobol (common business oriented language) for business, and CPOL for communications—all on the same computer.

The value of POLs is well established, as witness the popularity of Fortran and Cobol. In general, procedure-oriented languages are easy to learn and use, hence they reduce training and program-writing time. They also simplify program checkout, maintenance, and documentation, and speed program conversion from one computer to another. Overall, POLs save expensive programing manpower, the cost of which keeps going up.

However, POLs are inefficient in object code. They also require more memory, they extend program run time, and sometimes they cannot express all the operations required in specialized applications. But the tendency for POLs to require larger memories and to take longer than assembly languages to execute is mitigated by the sharp cost reduction in computer memory in recent years and the significant increase in computer speed.

The argument is made here, then, that the advantages of a CPOL outweigh the disadvantages, and that industry should start structuring such a higher-level language. What follows may provide some insight into the scope of such a project and offers some specific suggestions. The subject is addressed with an eye to the basic relationship between different levels of programing, to the essential characteristics required of a CPOL, to a review of the major functions and equipment in a communications-computer environment, and to a sample of suggested CPOL programing statements. A typical task programed in CPOL is included.

In early digital computer applications, programing was accomplished by logically and electrically inserting binary digits into the computer's main memory. Here, a particular string of, say, 8 or 16 bits constitutes a language instruction such as "store the result." The computer hardware then interprets each coded instruction for program execution. Such numeric coding, called machine-language programing, makes excellent use of computer resources. But the effort itself is so tedious and time-consuming that few computers are now still programed this way.

Assembly-language programing uses easy-to-remember symbolic notations, or mnemonics, instead of strings of 1s and 0s, to call out an operation or instructions. The mnemonic ADD (for addition), for example, when inserted by the programer is interpreted by the computer so as to generate the corresponding machine-language instructions. That is, there is a one-to-one correspondence in that each assembly language mnemonic symbol

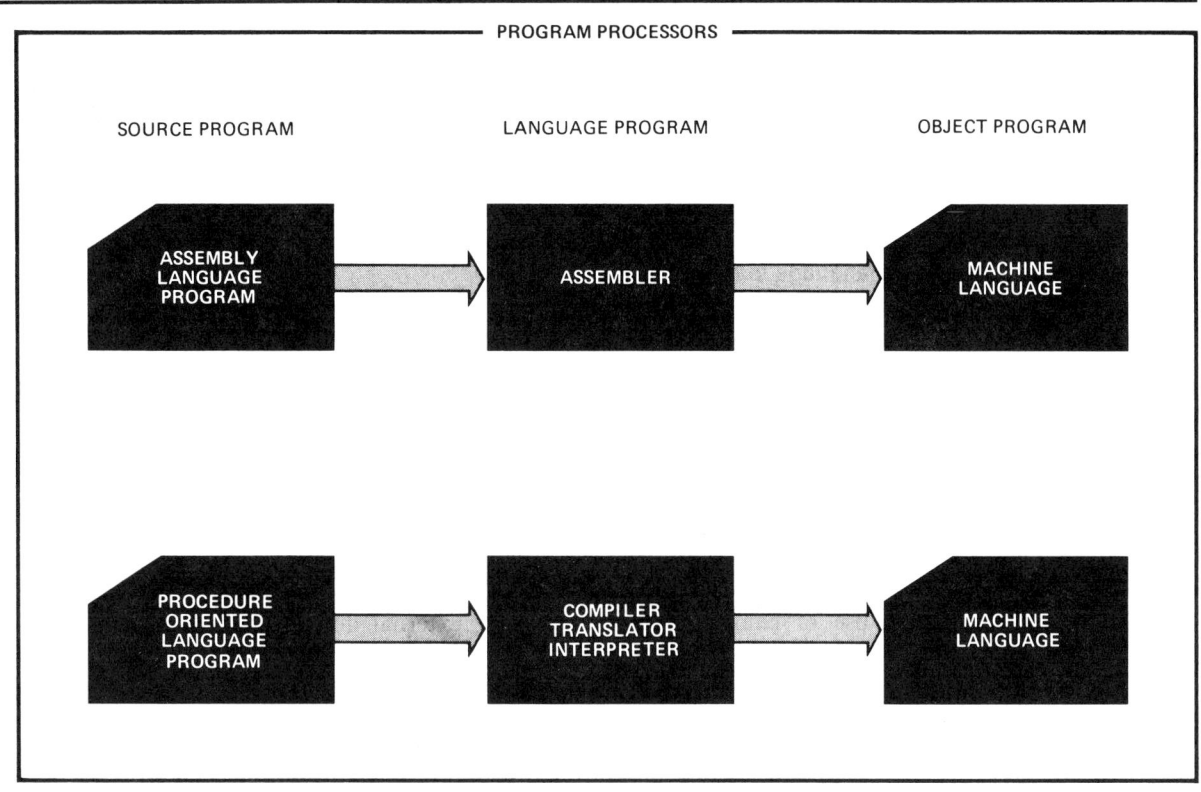

1. Whether a source program is in assembly or program-oriented language, a computer needs an object program in terms of machine language.

produces an equivalent machine-language instruction. The result of the assembly translation is called an object program. The object program is then ready to execute the source (application) program (Fig. 1, top).

The chief advantage of the assembly language is that the programer only needs to remember and use simple symbols instead of a long string of 1s and 0s. The net result is that assembly-language programing is easier, faster, and more accurate than machine-language programing.

Applications programed in assembly language appropriate to a particular computer are, however, usually valid only for that computer or generic family of computers. So, if circumstances (network expansion, for example) justify a changeover to another vendor's computer, the application must be reprogramed for the new computer. In other words, machine-language and assembly-language programs are both "machine-dependent."

With a procedure-oriented language, the programing of a particular application becomes independent of the computer—provided that the computer vendor can supply a compiler, translator, or interpreter that converts the POL program into a machine-language object program suited to his computer (Fig. 1, bottom).

With a POL, programers can produce with one statement a segment of a program that in assembly language would require many more individual statements. Figure 2 (next page) shows the relationship between machine-language, assembly-language, and POL programing for a simple application involving the summing of two numbers. What the programer has to insert for each of these methods is shown in the tinted boxes. Note that in both the machine-language and assembly-language methods, the programer writes five "lines" of instructions and two lines of data. The POL requires the same two lines of data but just one statement. (The com-

LANGUAGE COMPARISON		MACHINE LANGUAGE
MEMORY LOCATION	FUNCTIONAL DESCRIPTION OF INSTRUCTION	MACHINE LANGUAGE EQUIVALENT (BINARY)
770	PICK UP DATA IN LOC 774	0000100111111110
771	ADD DATA IN LOC 775	0001100111111101
772	STORE RESULT	0001000111111110
773	STOP EXECUTION	0000000000000000
774	DATA (DECIMAL 1)	0000000000000001
775	DATA (DECIMAL 100)	0000000001100100
776	RESULTS GO HERE	0000000001100101

2. Programing the summing of two numbers is much easier in a program-oriented language than in either machine or assembly languages.

puter's translator program automatically converts this statement to the equivalent machine-language instructions.)

In more complex programs, five POL statements could perform the equivalent of, say, 50 or more assembly-language instructions. This is why POL programing beats assembly-language programing, from the viewpoint of programer speed.

Early procedure-oriented languages were certainly less efficient in the use of computer resources and slower in program execution because the POL program may invoke more individual machine instructions than would be required by the equivalent assembly-language program. But POLs have substantially improved in efficiency, speed, and scope. Such popular POLs as Cobol, Fortran, Basic (beginners all symbolic instruction code), RPG (report program generator), and PL/1 (programing language one) have been developed to meet specific business and scientific requirements. But none is designed for programing data communications applications.

Language characteristics

To be effective and practical, a POL for communications must have two essential characteristics. It should first of all be specialized, containing primarily those statements related to the execution of communications programs. That is, it should concentrate on the statements involving bit manipulation, shift, and control, and avoid superfluous statements (such as those for the solution of a complex equation) that would reduce efficiency.

Second, a CPOL program must be able to be run on any eligible computer. For this, each computer maker addressing the communications market would have to develop a translator that would convert a CPOL-programed application into an object program for his particular computer (Fig. 3, next spread). This is, of course, exactly what computer makers have done with respect to Fortran, Cobol, and other procedure-oriented languages.

In fact, CPOL would handle a computer/communications system's external components in a manner similar to the way Fortran or Cobol handles computer peripherals. Thus, when using CPOL, a programer would specify, by unique statements, a block of data to be transferred as well as the corresponding external device, such as a synchronous single-line controller. This is in much the same manner that a Fortran programer specifies a variable list and the peripheral unit on which the list data is to be written. Like Fortran translators, then, each CPOL translator would have its own set of input/output (I/O) drivers (subroutines) appropriate to the operations of a corresponding set of terminals, peripheral devices, and lines.

Besides the characteristics of the computer itself, CPOL statements must also accommodate a variety of computer interfaces, controllers, and line interfaces (Fig. 4, next spread).

The computer interface is usually determined by the type of communications subsystem and the necessary input/output transfer rate. Interfacing may occur at the computer's I/O bus via such devices as direct memory control (DMC) and direct memory access (DMA).

Data transferred on the I/O bus is bit-serial and under control of the program. In the DMC mode, data transfers are effected independently of program control, and data blocks are transferred on a word basis (bit parallel) to and from any portion of main memory. The DMA mode employs a word-oriented, direct-to-memory, medium- or high-speed channel. This allows a computer to transfer data from several devices and lines concurrently on a timeshared basis.

Controllers can range from a single modified

ASSEMBLY LANGUAGE			PROCEDURE ORIENTED LANGUAGE
SYMBOLIC LOCATION	MNEMONIC OPERATION	OPERAND	
	LDA	DATA1	
	ADD	DATA2	
	STA	RESULT	
	STOP		
DATA1 (=1)	DEC	1	DATA (=1)
DATA2 (=100)	DEC	1 0 0	DATA (=100)
RESULT	BSS	1 0 1	RESULT = DATA1 + DATA2 1 0 1

shift register, designed to send and receive data from one communications line, to those designed to handle many synchronous communications lines. Character assembly and error checking can be undertaken entirely in the hardware, with fully matured characters then transferred to the computer. As an alternative, the controller can use a minimum of hardware and rely on software to assemble the characters inside the computer.

Lastly, the line interface unit provides the electrical and logical compatibility for connecting communications-lines terminating equipment, such as modems, to the computer's other equipment.

In actuality, then, a CPOL will require a repertoire of many statements in order to be compatible with all reasonable hardware options.

Major communications functions

Functions performed in a data communications environment can be categorized under three headings: line control, message blocking, and message interpretation. Line control includes the initiation and control of transmissions, error checking and detection, and the detection, deletion, and insertion of such control characters as end of transmission (EOT), synchronization (SYNC), idle (IDLE), and start of header (SOH).

Message blocking, which is the assembly of character strings into complete messages (blocks), includes the allocation and release of buffer space, the assigning, reserving, and releasing of communication input/output devices to related buffers, and the filling of the appropriate buffers with the desired character string.

Message interpretation includes those tasks associated with a course of action required by the content of a message, such as code conversion, message routing, and automatic dialing.

Each functional area requires both unique and general program statements for implementation. The broad range of tasks therefore need a broad range of statements, specific to data communications. Typical CPOL statements for several function areas are explained next.

Implementing the language

Data communications programs written in CPOL would consist of a number of tasks, with each task having one or more program statements. Further, since two or more tasks may have a subtask in common, a method of interrelating tasks and subtasks becomes necessary. (The concept of task/subtask is similar to the main program/subroutine relationship of other procedure oriented languages.) Thus, two CPOL program statements are:

TASK IDENTY (arg1, arg2, . . . argn)
SUBTASK IDENTY (arg1, arg2, . . . argn)

Here, IDENTY is the name of the task or subtask and arg1, arg2 . . . argn are arguments that can be passed between tasks and subtasks. A specific TASK IDENTY is TASK READ.

A CALL statement links tasks and subtasks. Likewise, the conclusion of a subtask requires a RETURN statement. Formats for these statements are:

CALL IDENTY (arg1, arg2, . . . argn)
RETURN

where the CALL statement passes control to the subtask IDENTY, and the RETURN statement terminates the subtask and returns control back to the originating task.

In the real-time data-communications environment, programs or portions thereof must be executed according to their relative urgency. Therefore it becomes necessary to have a method of associating a section of coding with the appropriate interrupt, as well as scheme for noninterrupt code. An ENABLE statement permits a task to be interruptible. And, at certain times it may be required

that an interruptible task be changed to a noninterruptible task through the use of an INHIBIT statement. The format for these statements are:
ENABLE IDENTY
INHIBIT IDENTY
where IDENTY again is the designation for a task.

Interrupts of ongoing tasks to execute another task of higher priority can be caused by non-time-related hardware or software events, or at any predetermined time interval. The statement:
CONNECT INTERRUPT e IDENTY
means that an interrupt event occurring on interrupt line e will stop the ongoing task and initiate the task designated by IDENTY. That is, this kind of CONNECT statement links the interrupt line to the new task.

Two types of interrupts can initiate tasks at predetermined times, one for elapsed time and the other for real time. The elapsed-time interrupt:
CONNECT TIME IDENTY (e_1, e_2, 3_3)
links the computer's real-time clock to the task denoted by IDENTY (such as TRANSMIT). Here the argument (or parameter) e_1 stands for basic time increments, and e_1 can have four different values, each denoting a particular time interval:
$e_1 = 1$ 1/60th-second intervals
$e_1 = 2$ 1-second intervals
$e_1 = 3$ 1-minute intervals
$e_1 = 4$ 1-hour intervals

The expression e_2 specifies the number of basic intervals (defined by e_1) before the first execution of the task denoted by IDENTY. The number of times the task IDENTY is to be executed is specified by e_3.

As an example, to execute the task TRANSMIT at a time equal to 5 minutes after the program starts, and every 5 minutes thereafter for 1 hour, the following statement would be specified:
CONNECT TIME TRANSMIT (3, 5, 12)
The real-time CONNECT statement is:
CONNECT CLOCK IDENTY (e_1, e_2, e_3, e_4, e_5, e_6, e_7)
This statement would interrupt any ongoing task and cause the new task designated by IDENTY to be executed at time e_1 hours, e_2 minutes, and e_3 seconds. The expression e_4 specifies the number of times the task is to be executed, and e_5, e_6, and e_7 specify the time in hours, minutes, and seconds between each execution. So executing the task POLL at 18:30 and every 30 minutes thereafter until 20:00 hours would use the following statement:
CONNECT CLOCK POLL (18, 30, 0, 4, 0, 30, 0)

Interrupt and connect functions must be also disabled. The statement DISCONNECT IDENTY re-

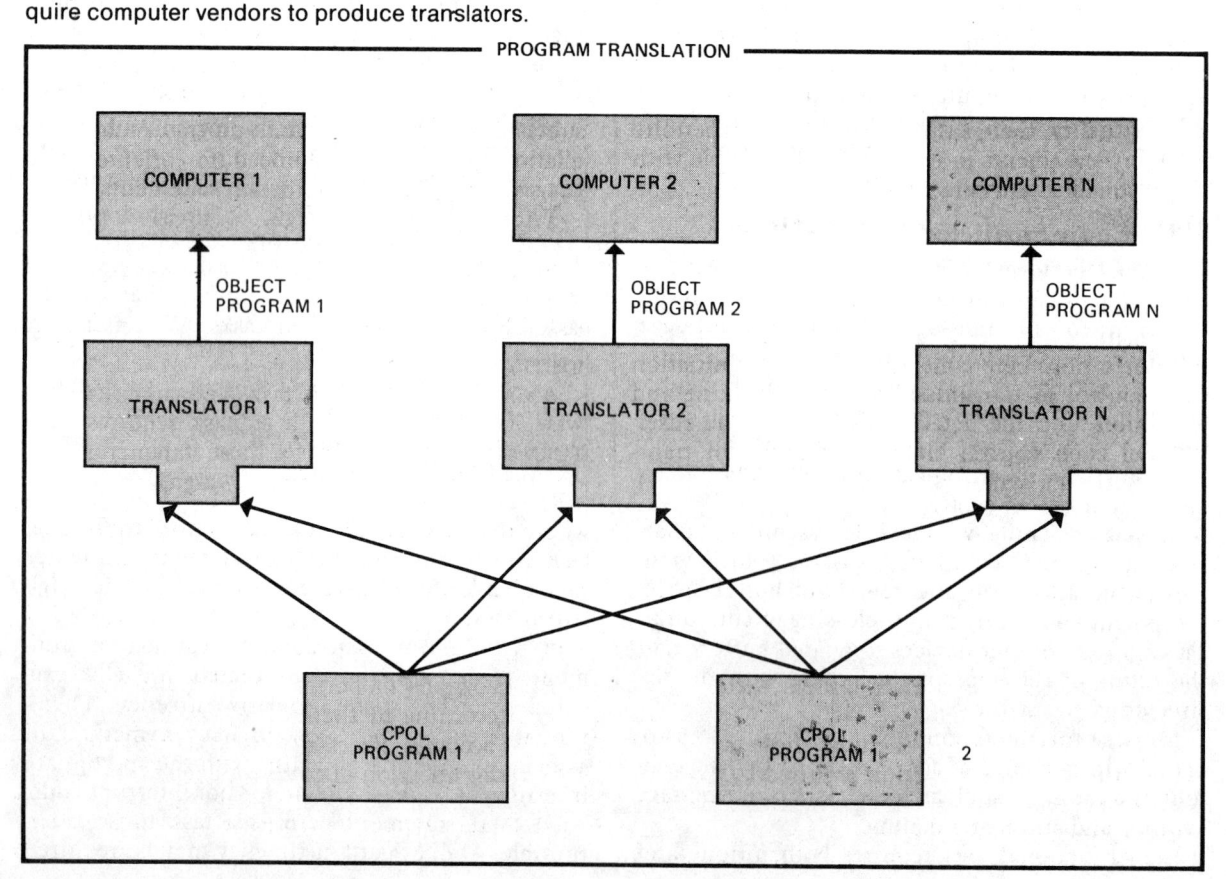

3. The development of an industry-accepted communications procedure-oriented language will require computer vendors to produce translators.

PROGRAM TRANSLATION

COMPUTER 1 — OBJECT PROGRAM 1 — TRANSLATOR 1 — CPOL PROGRAM 1

COMPUTER 2 — OBJECT PROGRAM 2 — TRANSLATOR 2 — CPOL PROGRAM 1 2

COMPUTER N — OBJECT PROGRAM N — TRANSLATOR N

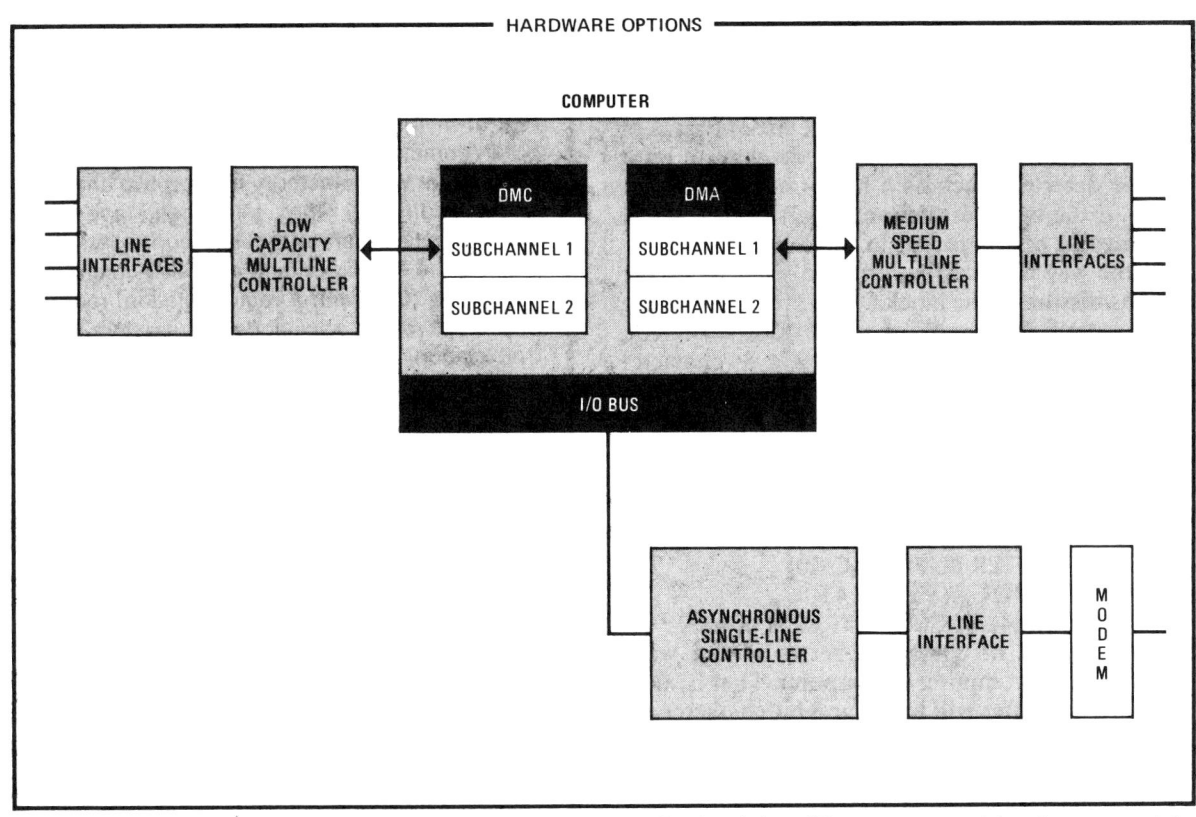

4. Each of the different types of hardware used in communications links would require individual statements in CPOL to call them into play under control of the computer's operating system.

moves the linkage between an interrupt line and the task IDENTY. This statement can also remove the linkage between any CONNECT statement and the designated tasks and, thus, it can disconnect the clock. The statement DISABLE IDENTY prohibits a specified task from being interrupted until an ENABLE IDENTY statement is activated.

Buffer management

Buffers derived from the main memory of a communications processor or a host computer provide a temporary memory space, so that, for instance, characters can be accumulated and stored during sending and receiving intervals, or data can be manipulated by appropriate programs.

A method of defining the size of a buffer and allocating it somewhere in the main memory becomes necessary. Since communications networks operate dynamically, some buffer space will be required throughout a program execution, and other buffers will be needed only during certain periods of program operation. The two statements to satisfy these requirements are:
RESERVE BUFFER IDENTY, W
RELEASE BUFFER IDENTY
Thus, the first programing statement reserves W words of main memory to be the buffer labeled IDENTY until the second statement releases the main memory assigned to that buffer.

The inputs and outputs, including peripherals and communications lines, can be associated with buffers through two more CPOL statements:
FILL BUFFER IDENTY (device, p), END = SN
XMIT BUFFER IDENTY (device, p), END = SN
Using the first statement, the buffer labeled IDENTY will be filled by the input from a device (including a line). The input can be specified by the device address, or through a mnemonic corresponding to an ASSIGN statement; the latter according to the protocol specified by p as designated by a statement number (SN). Control will be transferred to another task designated by another statement number when the buffer is filled (END).

The second statement sends the contents of the buffer labeled IDENTY to the specified device or line according to a protocol specified by p. On completion of transmission, control will jump to a task specified by the SN.

Message handling

The protocol statement contains details of exactly how an incoming or outgoing data stream should be handled with respect to such factors as message blocking, synchronization, error detection and control, and the stripping of control characters from the stream. The general protocol statement is:

SN PROTOCOL (list), ERROR = SN

Here, the first SN is the statement number referred to as the p-value in an XMIT, FILL, or other statement. The word "list" refers to the list of details of the particular protocol, including the way an error can be detected, such as a block-check character, (BCC). If the BCC indicates an error, then control of the ongoing task is passed to another SN which details the way the error is corrected—such as by the retransmission of the block.

By way of an example of a PROTOCOL statement, assume that a computer contains a 64-character buffer, called (BUF1), and the contents of that buffer are to be transmitted, via a single synchronous line controller (SSLC), preceded by four synchronizing characters (SYNC) and followed by two BCCs and an end-of-transmission character (EOT). The two statements to execute the data transfer are:

    XMIT BUFFER BUF1 (SSLC, 10)
  10 PROTOCOL (SYNC(4);
        DATA(64);BCC(2));EOT,ERROR=SN

The number of characters per computer word will vary from computer to computer. That is, an 8-bit-word computer will have one 8-bit character per word, and a 32-bit computer will have four such characters. Therefore, the input and output operations will be on a character-by-character basis, with the CPOL translator for the particular computer performing the necessary characters-per-word assembly and disassembly operations.

Once characters get into the main memory as computer words, the words may have to be manipulated, that is moved from one buffer space to another, to, for example, assemble and disassembled messages. Changing buffer space would be accomplished by such CPOL statements as:

  COPY BUFFER IDENTY$_1$, LOC$_1$, N$_1$; IDENTY$_2$, LOC$_2$
  MERGE BUFFER IDENTY$_1$ ... IDENTY$_{N-1}$; IDENTY$_N$
  PIECE BUFFER IDENTY$_N$; IDENTY$_1$, ... IDENTY$_{N-1}$

The first statement will cause N$_1$ computer words, starting at location LOC$_1$ relative to the beginning of the buffer labeled IDENTY$_1$, to be copied into the buffer labeled IDENTY$_2$ starting at location LOC$_2$ relative to the beginning of that buffer. Thus, transferring the first 40 words of an 80-word buffer labeled START to the middle 40 words of a 120-word buffer labeled END, could be initiated by:

  COPY BUFFER START, 1,40;END, 41

The MERGE-BUFFER statement assembles the contents of the buffers labeled IDENTY$_1$ through IDENTY$_{N-1}$ into the buffer labeled IDENTY$_N$. This is done on word-by-word basis, starting with the first character contained in the buffer labeled IDENTY$_1$. Conversely, the PIECE BUFFER statement will disassemble the contents of the buffer labeled IDENTY$_N$ into separate pieces (that is, words) and place one word in each of the buffers labeled IDENTY$_1$

CPOL PROGRAMMING EXAMPLE

Fig. 5. In this programing example, the computer/communications configuration includes a computer with a memory holding two characters per computer word, a magnetic tape unit, and a single synchronous line controller. The task for the computer is to establish communications at 18:30 with a remote station, receive one block of 64 ASCII-coded characters from that station, convert this block to BCD-coded characters, and finally transfer the block of data to the magnetic tape unit. First, the task is initialized—that is, all the parameters are defined. Then the task, to be called LOGGER, is defined by appropriate statements from the communications procedure-oriented language:

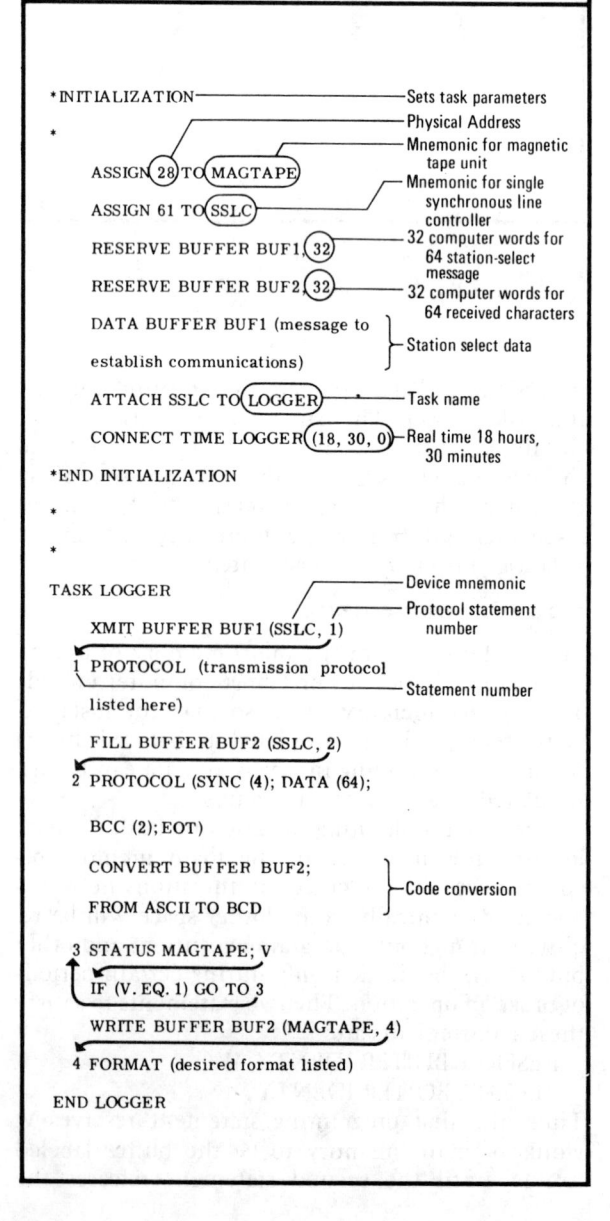

```
*INITIALIZATION                          — Sets task parameters
                                         — Physical Address
*                                        — Mnemonic for magnetic
     ASSIGN 28 TO MAGTAPE                  tape unit
     ASSIGN 61 TO SSLC                   — Mnemonic for single
                                           synchronous line
                                           controller
     RESERVE BUFFER BUF1, 32             — 32 computer words for
                                           64 station-select
                                           message
     RESERVE BUFFER BUF2, 32             — 32 computer words for
                                           64 received characters
     DATA BUFFER BUF1 (message to
     establish communications)           ⎱ Station select data
     ATTACH SSLC TO LOGGER               — Task name
     CONNECT TIME LOGGER (18, 30, 0)     — Real time 18 hours,
                                           30 minutes
*END INITIALIZATION
*
*
TASK LOGGER                              — Device mnemonic
                                         — Protocol statement
     XMIT BUFFER BUF1 (SSLC, 1)            number
   1 PROTOCOL (transmission protocol
     listed here)                        — Statement number

     FILL BUFFER BUF2 (SSLC, 2)
   2 PROTOCOL (SYNC (4); DATA (64);
     BCC (2); EOT)

     CONVERT BUFFER BUF2;                ⎱ Code conversion
     FROM ASCII TO BCD

   3 STATUS MAGTAPE; V
     IF (V. EQ. 1) GO TO 3

     WRITE BUFFER BUF2 (MAGTAPE, 4)
   4 FORMAT (desired format listed)
END LOGGER
```

through IDENTY$_{N-1}$.

Reading and writing data to and from input/output devices can use these statements:

READ BUFFER IDENTY (device, F)
WRITE BUFFER IDENTY (device, F)

The READ statement will cause a block of data from the specified device to be transferred to the buffer labeled IDENTY under the format specified by F. The format would, for example, locate each incoming character in alternate cells in the buffer. Conversely, the WRITE statement would cause the contents of the buffer IDENTY to be transferred to the specified device under the format designated by F.

The code conversion statement:

CONVERT BUFFER IDENTY; FROM XXX TO YYY

would change the contents of the buffer IDENTY from the code designated by XXX to the code designated by YYY. As an example, to convert the contents of a buffer named LOG from ASCII code to BCD code would simply require the statement:

CONVERT BUFFER LOG; FROM ASCII TO BCD

(Here ASCII stands for American Standard Code for Information Interchange, and BCD stands for binary coded decimal.)

Assigning devices

As program execution continues, input/output devices must be assigned from moment to moment so data can be read into or written from an appropriate device. Which devices and when they are invoked must be determined beforehand. Device assignment is part of the initialization for a task.

A clear way to assign devices is by the statement:

ASSIGN e to M

Here the programer assigns the physical address, e, to the device M. For example:

ASSIGN 28 TO MAGTAPE

relates physical address 28 to a magnetic-tape unit whose mnemonic is MAGTAPE.

Once the device is assigned, transfer of data between the device and an appropriate buffer is accomplished by the FILL and XMIT programing statements described earlier. For example:

FILL BUFFER BUF2 (SSLC, 2)

means that data from a single synchronous line controller (whose mnemonic is SSLC) is transmitted to the buffer (whose mnemonic is BUF2), using a protocol defined by statement number 2 of the ongoing task.

Once the device has been assigned to the computer through a physical address, the device can be attached to the task through the statement:

ATTACH device TO task; p

so that the device is available for the exclusive use of that task. Several application tasks may be running concurrently, however, with one or more tasks from each program requiring the use of the same device. So there can be uncertainty about whether the device will, in fact, be available for attachment when the ATTACH statement is executed.

What happens next depends on the value of priority, p, in the ATTACH statement. If p is set at 0, for example, then the ATTACH statement will be honored only if the device is currently unattached. But if the priority is set at some level, say 10, then the statement having this priority will be honored, provided no other task with a higher priority seeks to use that device. All lower-priority statements seeking the same device will therefore have to wait in queue.

After a task has finished with a device, the device can be released by the statement:

DETACH device FROM task

A typical statement is:

DETACH MAGTAPE FROM LOGGER

Figure 5 contains a program for a task, along with commentary in color to clarify the steps.

CPOL's place in hierarchy

As envisioned, then, CPOL would take its place as a free-standing, standardized procedure-oriented language with the same rank in the hierarchy of programing languages as Fortran and Cobol (Fig. 6). All such languages are under the control of a computer's operating system (OS), which is responsible for orchestrating all the individual programs, devices, and lines for the application programs being run concurrently on the computer.

Furthermore, if the OS is modified to allow linkages between CPOL and the other POLs, then portions of a major program being run on the computer can mesh with each other and can be programed with equal efficiency in the most appropriate program-oriented language. ∎

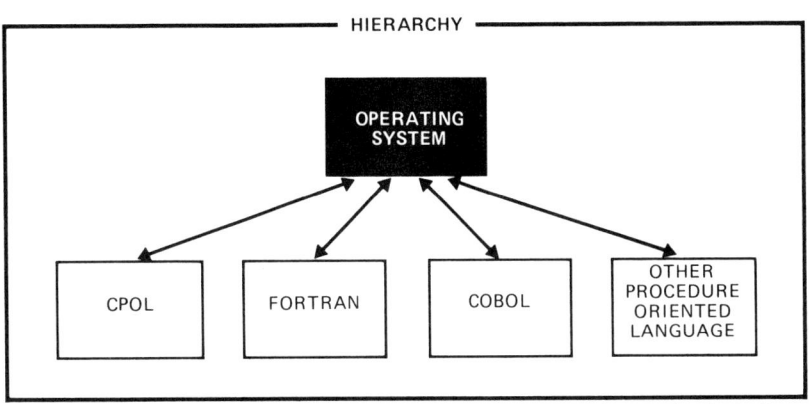

6. The communications procedure-oriented language would rank with Fortran, Cobol, and the like, and the operating system could link them all.

KEY SUBJECT LOCATOR

Access method	67, 70
ACK (acknowledge accurate block)	12, 51, 171
Acoustic data coupler	47, 48
Address, channel	67
ARQ (automatic request for repeat)	169
Asynchronous	6, 9, 10
Availability	8, 152-156
Binary synchronous communications	194-200
Bit	8
Bits per second	9, 47-53
Block length, optimum	13, 14
Block (of characters)	8, 13
Buffer	8, 83-88, 94, 108, 109
Burroughs data link control	220-226
Carriers, common	259-261
Carriers, specialized	11, 261
CCITT (International Consultative Committee for Telegraph and Telephone)	61-63
Centralized network	14
Channel, derived	60-62, 65, 96-101
Channel, reverse	51
Channel, secondary	52
Channel, voice-grade	10, 49
Channel, wideband	49
Character	14
Code, code conversion	6, 81, 104, 136
Communications processors	15, 27, 102-109
Concentrators	83-88, 104
Conditioning, line	60-65, 188-192
Contention	11, 165
Cyclic redundancy check (CRC)	170, 171
Data access arrangement	50, 51, 250-252, 261
Data base management	131-137
Data compression	36, 40
Data link controls	194-200, 201-210, 211-219, 220-226

Dataset (see Modem)	
Decentralized network	14
Digital data communications message protocol (DDMCP)	211-219
Distributed network	21, 113
Dropout, line	82
Duplex (see Half duplex and Full Duplex)	
Echo suppressor	9, 10, 51
EIA (Electronic Industries Association)	8, 246-253
Equalization	52
Error burst	171
Error control	6, 12, 13, 106, 169-173, 180-187
Error detection codes (see Cyclic redundancy check, Forward error correction, Longitudinal redundancy check, Vertical redundancy check)	
Error rate	9
Facsimile	38-40
Formatting	35, 38-40
Forward error correction	169, 171
Four-wire circuit	16, 52, 62
Frequency-division multiplexing (FDM)	61-63
Front-end processors	14, 66-71, 72-79, 104
Full duplex	9, 10, 13
Half duplex	10, 181, 187
Handshaking	51
Hybrid communications, computing	268-274
Interfaces (electrical)	246-257
Joint-use tariff	259-261
Limited distance modems	54-58
Line drivers	54-58
Line monitors	232-239
Lines, dial up (or switched)	9, 11

Lines, leased (or private or dedicated)	8, 9, 60-65
Lines, voice-grade	8, 11, 60-65
Link, logical	141
Longitudinal redundancy check (LRC)	170
Loop network	114
Memory	80, 81, 105, 135, 172
Message switcher	89-95, 104
Microcomputers	15, 16, 286-293
Microwave transmission	174-178
Minicomputers	3, 8, 35, 66-71, 107
Modems (see also Limited distance modems)	8, 10, 29, 49-53, 180-187
Multidropping	53
Multiplexing (see also Frequency-division multiplexing and Time-division multiplexing)	11, 28, 60-65
Multipoint circuit	10, 52
Multiporting	51
NAK (acknowledge block in error)	12, 51, 171
Network design	7-14, 19, 111-117, 118-124, 164-167
Network diagnostics	228-231, 232-239, 240-244
Network economics	125-130
Network optimization	8, 13, 118-124
Network simulation	72-79
Operating system	9
Outage, line	83
Packet switching	115, 138-145
Parity bit	82, 170, 171
Parity check	12, 36, 170, 171
Point-to-point circuit	10, 61, 166
Polling	11, 91, 106
Private lines	10
Programs (see Software)	
Protocol (see also Data link control)	71, 81, 82, 172

Redundancy (equipment)	157-163
Regulations	15, 25, 259-261, 262-265, 268-274
Reliability	8, 157-163
Remote data concentrators	80-88
RS-232-C interface	246-253, 254-257
Satellites	15, 23
Serial transmission	9
Standards	276-285
Software	15, 22, 66-72, 102-109, 294-301
Store and forward	81, 82, 94
Synchronous	9
Synchronous data link control (SDLC)	201-210
Tariffs	118-124, 259-261
Teleprinter	96-101
Terminals	6, 8, 9, 15, 26, 32-37
Terminals, CRT display	33, 34, 37, 41-45
Terminals, facsimile	38-40
Terminals, interactive	33
Terminals, keyboard/printer	33
Terminals, nonprogramable	34, 37
Terminals, programable	32, 104
Terminals, remote job entry (batch)	33, 37, 172
Throughput	8, 9, 12, 13, 51, 172, 180-187
Time-division multiplexing (TDM)	62-64
Traffic volume	5
Turnaround time	12, 13, 51, 171, 172, 181
Two-wire circuit	9, 10, 51, 62
Value-added networks	116, 146-151
Vertical-redundancy check (VRC)	170
WATS (Wide area telecommunications service)	14, 38-40